Markov Processes for Stochastic Modeling

STOCHASTIC MODELING SERIES

Series Editors

Laurence Baxter *State University of New York at Stony Brook*, USA
Marco Scarsini *Universita D'Annuzio*, Italy
Moshe Shaked *University of Arizona*, USA
Shaler Stidham, Jr *University of North Carolina*, USA

G. Samorodnitsky and M.S. Taqqu
Stable Non-Gaussian Processes: Stochastic Models with Infinite Variance

K. Sigman
Stationary Marked Point Processes: An Intuitive Approach

P. Guttorp
Stochastic Modeling of Scientific Data

M. Neuts
Algorithmic Probability

A. Shwartz and A. Weiss
Large Deviations for Performance Analysis: Queues, Communications and Computing

M. Kijima
Markov Processes for Stochastic Modeling

Markov Processes for Stochastic Modeling

MASAAKI KIJIMA
Associate Professor
Graduate School of Systems Management
University of Tsukuba
Tokyo, Japan

Springer-Science+Business Media, B.V.

First edition 1997

Originally published by Chapman & Hall in 1997.
Softcover reprint of the hardcover 1st edition 1997

ISBN 978-0-412-60660-1 ISBN 978-1-4899-3132-0 (eBook)
DOI 10.1007/978-1-4899-3132-0

A catalogue record for this book is available from the British Library

∞ Printed on permanent acid-free text paper, manufactured in accordance with ANSI/NISO Z39.48-1992 and ANSI/NISO Z39.48-1984 (Permanence of Paper).

Contents

Preface ix

1 Introduction 1
- 1.1 Stochastic processes 1
- 1.2 The Markov property 2
- 1.3 Some examples 6
- 1.4 Transition probabilities 13
- 1.5 The strong Markov property 19
- 1.6 Exercises 21

2 Discrete-time Markov chains 25
- 2.1 First passage times 25
- 2.2 Classification of states 30
- 2.3 Recurrent Markov chains 38
- 2.4 Finite Markov chains 51
- 2.5 Time-reversible Markov chains 58
- 2.6 The rate of convergence to stationarity 64
- 2.7 Absorbing Markov chains and their applications 75
- 2.8 Lossy Markov chains 84
- 2.9 Exercises 95

3 Monotone Markov chains 101
- 3.1 Preliminaries 102
- 3.2 Distribution classes of interest 112
- 3.3 Stochastic ordering relations 121
- 3.4 Monotone Markov chains 129
- 3.5 Unimodality of transition probabilities 135
- 3.6 First-passage-time distributions 142
- 3.7 Bounds for quasi-stationary distributions 147

3.8 Renewal processes in discrete time 150
3.9 Comparability of Markov chains 158
3.10 Exercises 163

4 Continuous-time Markov chains 167
4.1 Transition probability functions 167
4.2 Finite Markov chains in continuous time 173
4.3 Denumerable Markov chains in continuous time 183
4.4 Uniformization 195
4.5 More on finite Markov chains 200
4.6 Absorbing Markov chains in continuous time 208
4.7 Calculation of transition probability functions 216
4.8 Stochastic monotonicity 224
4.9 Semi-Markov processes 229
4.10 Exercises 235

5 Birth–death processes 243
5.1 Boundary classification 244
5.2 Birth–death polynomials 248
5.3 Finite birth–death processes 253
5.4 The Karlin–McGregor representation theorem 261
5.5 Asymptotics of birth–death polynomials 266
5.6 Quasi-stationary distributions 271
5.7 The decay parameter 281
5.8 The M/M/1 queue 287
5.9 Exercises 290

A Review of matrix theory 295
A.1 Nonnegative matrices 295
A.2 ML-matrices 298
A.3 Infinite matrices 300

B Generating functions and Laplace transforms 303
B.1 Generating functions 303
B.2 Laplace transforms 307

C Total positivity 313
C.1 TP_r functions 313
C.2 The variation-diminishing property 316

References 319

Symbols 329

Author index **331**

Subject index **334**

Preface

This book presents an algebraic development of the theory of countable state space Markov chains with discrete- and continuous-time parameters. A Markov chain is a stochastic process characterized by the Markov property that the distribution of future depends only on the current state, not on the whole history. Despite its simple form of dependency, the Markov property has enabled us to develop a rich system of concepts and theorems and to derive many results that are useful in applications. In fact, the areas that can be modeled, with varying degrees of success, by Markov chains are vast and are still expanding.

The aim of this book is a discussion of the time-dependent behavior, called the *transient behavior*, of Markov chains. From the practical point of view, when modeling a stochastic system by a Markov chain, there are many instances in which time-limiting results such as stationary distributions have no meaning. Or, even when the stationary distribution is of some importance, it is often dangerous to use the stationary result alone without knowing the transient behavior of the Markov chain. Not many books have paid much attention to this topic, despite its obvious importance.

From the theoretical point of view, however, the transient behavior of Markov chains is much more difficult than its stationary counterpart, and is often inaccessible except in some special cases. To overcome this difficulty, some measures have been considered that evaluate, e.g., the speed of convergence to stationarity and the distance from stationarity. First-passage-time distributions and quasi-stationary distributions, the conditional distributions restricted to a subset of the state space, also provide some insight into the study of transient behaviors. This book contains several modern ideas such as monotone processes, reversibility, rate of convergence and quasi-stationary distributions, together with some numerical methods for studying the 'transient behavior' of Markov chains. It is hoped that this book will provide a useful reference for applied probabilists and those mathematically literate practitioners in other fields who wish to apply Markov chains in their research.

This book is organized as follows. In Chapter 1, we begin with an introduction of some basic concepts of Markov chains. Chapter 2 concerns discrete-time Markov chains defined on a countable state space. Reversibility, rate of convergence and quasi-stationary distributions along with classical results about first passage times, classification of states and recurrent and absorbing Markov chains are discussed in some detail. Chapter 3 is devoted to monotone Markov chains. The properties of monotonicity are important both theoretically and practically because they lead to a variety of structural insights. In particular, they are a basic tool for deriving many useful inequalities. Stochastic orderings and the theory of total positivity play prominent roles in this chapter. In Chapter 4, we discuss continuous-time Markov chains; classical results on transition probability functions are first studied and then the parallelism between discrete-time and continuous-time Markov chains is discussed via uniformization. Finally, in Chapter 5, we consider birth–death processes in detail; we discuss boundary classification at infinity, the connection between orthogonal polynomials and birth–death processes, the Karlin–McGregor representation of transition probability functions, the decay parameter and quasi-stationary distributions of birth–death processes. Markovian queues are also considered as specific examples of birth–death processes

It is assumed that the reader is familiar with the elements of probability theory, real analysis and linear algebra as found, for example, in parts of Feller (1957), Bartle (1976) and Noble and Daniel (1977) respectively. Furthermore, an algebraic treatment of nonnegative matrices, notions of generating functions and Laplace transforms, and the theory of total positivity are essential in developing materials contained in this book. We provide a concise summary of these theories in the appendices for the reader's convenience.

In conclusion, I would like to thank, first, my wife Mayumi and my family. I would also like to thank Laurence Baxter, Naoki Makimoto, Phil Pollett, Eugene Seneta, Erik van Doorn, and Ward Whitt for their generous support and helpful comments. Without their encouragement and friendship over the years, I could not have completed this book. Technical contributions by Kimiaki Aonuma and Sheila Shepherd are also greatly appreciated. Lastly, I dedicate this book to my teachers, Julian Keilson, Hidenori Morimura and Ushio Sumita for their strict but generous instructions since I first met them.

1

Introduction

Stochastic processes are sequences of random variables generated by probabilistic laws. The word 'stochastic' comes from the Greek and means 'random' or 'chance'. Markov processes are a class of stochastic processes that are distinguished by the Markov property and have many applications in, for example, operations research, biology, engineering, and economics. In this chapter, we introduce some basic concepts of Markov processes.

1.1 Stochastic processes

A *stochastic process*, or, simply, a process, is a family of random variables $\{X_t\}$, where t denotes a parameter running over a suitable index set $\mathcal{T}$. We shall often write the process as $\{X_t,\ t \in \mathcal{T}\}$ when emphasis on the index set is required. The parameter t usually represents time, but different situations may arise. For example, t may be a distance from the origin in a plane, in which case X_t may represent the number of points randomly scattering in the plane whose distances from the origin are less than t. However, for simplicity of presentation, we refer to the parameter t as time and call $\{X_t\}$ a *discrete-time* process, or a process in discrete time, if the index set is $\mathcal{T} = \mathcal{Z}_+ \equiv \{0, 1, 2, \cdots\}$, and a *continuous-time* process, or a process in continuous time, if the index set is $\mathcal{T} = R_+ \equiv [0, \infty)$. Other situations will not appear in this book.

Stochastic processes are distinguished by their *state space* $\mathcal{N}$, the range of possible values for the random variables X_t, by their index set $\mathcal{T}$, and by the dependence structure between the $X_t,\ t \in \mathcal{T}$. Throughout this book, we assume that the state space $\mathcal{N}$ is either finite or denumerably infinite. The joint distribution of $(X_{t_1}, \cdots, X_{t_n})$ is known through the dependence structure for all n and all $t_1 < \cdots < t_n$, where $t_i \in \mathcal{T},\ i = 1, \cdots, n$. Markov processes, the main theme of this book, are characterized by a special form of dependence structure that makes them a useful class of stochastic processes in stochastic modeling.

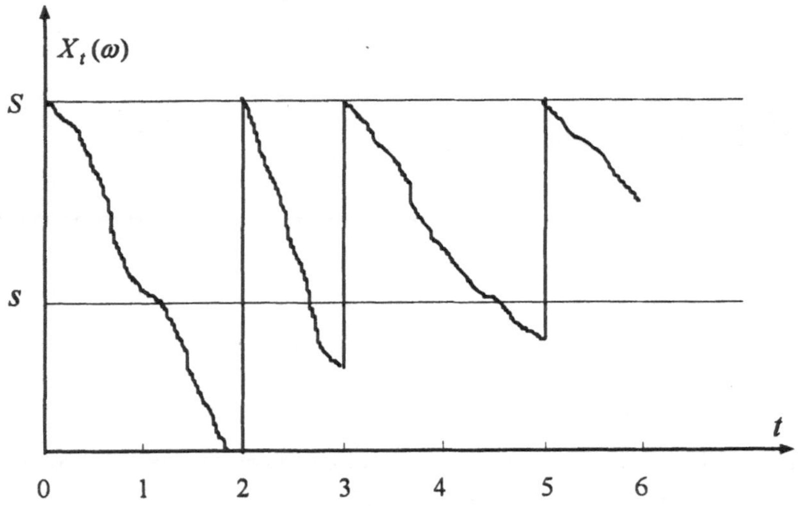

Figure 1.1 *An example of a sample path.*

Let $(\Omega, \mathcal{F}, P)$ be a (canonical) probability space on which a stochastic process $\{X_t\}$ is defined. For each $\omega \in \Omega$, a function $X_t(\omega)$ with respect to t, denoted by $\{X_t(\omega),\ t \in \mathcal{T}\}$, is called a *sample path* or *realization* of the process $\{X_t\}$. Figure 1.1 depicts a sample path arising from an inventory model, where X_t denotes an inventory level of a given product at time t (see Example 1.1 below). The realization of the process $\{X_t\}$ up to time t is $\{X_s(\omega),\ s \leq t\}$. If the present time is t, any time such that $s < t$ is called a *past* while any time such that $s > t$ is a *future*. This past–present–future unidirectionality is implicit in our case of a one-dimensional index set $\mathcal{T}$. At this point, it is worth noting that future distributions of any stochastic process can be determined in principle through a given dependence structure, provided that its realization up to the present is known. As will be formally defined shortly, a Markov process is a stochastic process whose future behavior can be determined independently of the past.

1.2 The Markov property

A sequence of random variables $\{X_n\}$ is called *independent* if, for each n and (Borel) subsets $A_1, \cdots, A_n$ of the real line $R \equiv (-\infty, \infty)$,

$$P\left[\bigcap_{i=1}^{n} \{X_i \in A_i\}\right] = \prod_{i=1}^{n} P[X_i \in A_i].$$

Independence leads to such classical limit theorems as the strong law of large numbers, the central limit theorem and the law of the iterated logarithm. However, as one can imagine, the independence assumption is very restrictive in many practical situations. For example, nobody would agree with the hypothesis that the number of births for the next year in a country will be independent of the current population.

In the context of stochastic processes, the opposite of independence is to assume that the distribution of a future depends on the complete history. To make the point precise, consider a discrete-time stochastic process $\{X_t\}$ with state space $\mathcal{N} = \mathcal{Z} \equiv \{\cdots, -2, -1, 0, 1, 2, \cdots\}$. Then, with full generality, we may assume that for every n the distribution of X_{n+1} depends on the whole history $\{X_0 = i_0, \cdots, X_n = i_n\}$. However, analytical tractability is completely lost.

In contrast, the Markov property asserts that the distribution of X_{n+1} depends only on the current state $X_n = i_n$, not on the whole history. Formally, the process $\{X_n\}$ is called a *Markov process* if, for each n and every $i_0, \cdots, i_n$ and $j \in \mathcal{N}$,

$$P[X_{n+1} = j | X_0 = i_0, \cdots, X_n = i_n] = P[X_{n+1} = j | X_n = i_n]. \tag{1.1}$$

Here $P[Y = y | X = x]$ denotes the conditional probability defined by

$$P[Y = y | X = x] = \frac{P[Y = y,\ X = x]}{P[X = x]},$$

provided $P[X = x] > 0$. The Markov property was proposed by A.A. Markov (1856–1922) as part of his work on generalizing the classical limit theorems of probability. Note that the independence assumption is

$$P[X_{n+1} = j | X_0 = i_0, \cdots, X_n = i_n] = P[X_{n+1} = j]. \tag{1.2}$$

The difference between the Markov property (1.1) and the independence assumption (1.2) does not perhaps seem significant at first sight. However, as we shall see, the Markov property enables us to develop a rich system of concepts and theorems and to derive many results that are useful in applications.

Given the history $\{X_0 = i_0, \cdots, X_n = i_n\}$, the Markov property (1.1) suggests that the current state $X_n = i_n$ is enough to determine all distributions of the future. To see this, the chain rule for conditional probabilities yields

$$\begin{aligned}
& P[X_{n+1} = i_{n+1}, \cdots, X_{n+m} = i_{n+m} | X_0 = i_0, \cdots, X_n = i_n] \\
= \; & P[X_{n+1} = i_{n+1} | X_0 = i_0, \cdots, X_n = i_n] \\
& \times P[X_{n+2} = i_{n+2} | X_0 = i_0, \cdots, X_n = i_n, X_{n+1} = i_{n+1}] \\
& \quad \vdots \\
& \times P[X_{n+m} = i_{n+m} | X_0 = i_0, \cdots, X_n = i_n, \cdots, X_{n+m-1} = i_{n+m-1}]
\end{aligned}$$

for all $m = 1, 2, \cdots$. But, from the Markov property (1.1), the right-hand

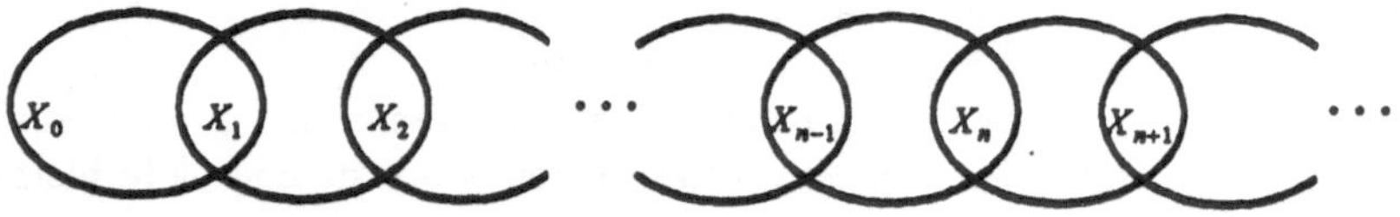

Figure 1.2 *Chain dependence in a Markov process.*

side of the above equation becomes

$$\begin{aligned}&P[X_{n+1} = i_{n+1}|X_n = i_n]\\ &\quad \times P[X_{n+2} = i_{n+2}|X_{n+1} = i_{n+1}]\\ &\quad \vdots\\ &\quad \times P[X_{n+m} = i_{n+m}|X_{n+m-1} = i_{n+m-1}],\end{aligned}$$

whence

$$\begin{aligned}&P[X_{n+1} = i_{n+1}, \cdots, X_{n+m} = i_{n+m}|X_0 = i_0, \cdots, X_n = i_n]\\ = \; &P[X_{n+1} = i_{n+1}, \cdots, X_{n+m} = i_{n+m}|X_n = i_n] \qquad (1.3)\end{aligned}$$

for all $m = 1, 2, \cdots$, as claimed. Thus, once the current state is known, prediction of future distributions cannot be improved by adding any knowledge of the past. Note, however, that this does not imply that the past lacks information about the future behavior (although this is in fact true for the independent case). The past *does* affect the future through the present state. Figure 1.2 depicts the situation where each oval reveals that two random variables are related through (1.1). One can observe that the random variables X_n are connected by a *chain*.

Another consequence of the Markov property (1.1) is

$$\begin{aligned}&P[X_0 = i_0, \cdots, X_{n-1} = i_{n-1},\\ &\qquad X_{n+1} = i_{n+1}, \cdots, X_{n+m} = i_{n+m}|X_n = i_n]\\ = \; &P[X_0 = i_0, \cdots, X_{n-1} = i_{n-1}|X_n = i_n] \qquad (1.4)\\ &\qquad \times P[X_{n+1} = i_{n+1}, \cdots, X_{n+m} = i_{n+m}|X_n = i_n].\end{aligned}$$

That is, the past $\{X_0, \cdots, X_{n-1}\}$ and the future $\{X_{n+1}, \cdots, X_{n+m}\}$ are conditionally independent given the present $X_n = i_n$. Conversely, (1.4) characterizes the Markov property (1.1) and, hence, the conditional independence property (1.4) is an equivalent notion to that of the Markov property. The reader is asked to prove this fact in Exercise 1.1.

Markov processes often appear as a form of recursive relations between random variables. We shall show this through the following example.

Example 1.1 Consider an inventory in which a product is stocked and assume that replenishment of the stock takes place at the end of periods labeled by $\mathcal{T} = \{0, 1, 2, \cdots\}$. The total demand for the product during period n is a random variable D_n and, when demand exceeds the inventory on hand, sales are lost. The inventory is examined at the end of each period and the (s, S)-*policy* is used for ordering. The policy is to order up to S units of product whenever the inventory level is below s, $s \leq S$; no order is placed if the inventory level is s or greater. A picture of the dynamics of the inventory level was shown in Figure 1.1. Let X_n represent the inventory level at the end of period n. According to the ordering policy, the inventory levels at two consecutive periods are connected by

$$X_{n+1} = \begin{cases} \{X_n - D_{n+1}\}_+ , & s \leq X_n \leq S, \\ \{S - D_{n+1}\}_+ , & 0 \leq X_n < s, \end{cases}$$

where $\{x\}_+ = \max\{0, x\}$. Now suppose that for each n demand D_{n+1} is independent of the past $(X_0, \cdots, X_n)$. Then, since

$$\begin{aligned} & P[X_{n+1} = j | X_0 = i_0, \cdots, X_n = i_n] \\ = & \begin{cases} P[\{X_n - D_{n+1}\}_+ = j | X_0 = i_0, \cdots, X_n = i_n], & s \leq X_n \leq S, \\ P[\{S - D_{n+1}\}_+ = j | X_0 = i_0, \cdots, X_n = i_n], & 0 \leq X_n < s, \end{cases} \\ = & \begin{cases} P[\{X_n - D_{n+1}\}_+ = j | X_n = i_n], & s \leq X_n \leq S, \\ P[\{S - D_{n+1}\}_+ = j], & 0 \leq X_n < s, \end{cases} \end{aligned}$$

the process $\{X_n\}$ constitutes a Markov process in discrete time with state space $\mathcal{N} = \{0, 1, \cdots, S\}$.

We next turn to the continuous-time case. In what follows we shall denote a continuous-time process by $\{X(t)\}$ and a discrete-time process by $\{X_n\}$. A continuous-time process $\{X(t)\}$ with state space $\mathcal{N} = \mathcal{Z}$ is called a Markov process if, for each $t \in \mathcal{T} = R_+$, $s > 0$, and for each state $j \in \mathcal{N}$,

$$\begin{aligned} & P[X(t+s) = j | X(u) = x(u),\ 0 \leq u \leq t] \\ = \ & P[X(t+s) = j | X(t) = x(t)], \end{aligned} \tag{1.5}$$

where $\{x(u),\ u \leq t\}$ denotes the history of $\{X(t)\}$ up to time t. Results similar to (1.3) and (1.4) hold for the continuous-time case, too. That is, distributions of the future of the process can be determined once the present is known, and the past and the future are conditionally independent given the present.

Terminology. In this book, we refer to a Markov process as a *Markov chain* for both discrete- and continuous-time cases, when the state space is either finite or denumerably infinite, which is our setting here. Some standard textbooks such as Meyn and Tweedie (1993), Nummelin (1984) and Revuz (1984) refer to a Markov process as a Markov chain when the time parameter is discrete even if the state space is general. Iosifescu (1980) uses the term 'Markov chain' for the discrete-time case and the term 'Markov process' for the continuous-time case.

1.3 Some examples

Many random phenomena in, e.g., operations research, biology, engineering and economics can be modeled by Markov processes. In this section, we provide several such examples to motivate study of the use of Markov processes in stochastic modeling.

Example 1.2 Let $Y_1, Y_2, \cdots$ be independent, identically distributed (abbreviated to IID) random variables and define

$$X_n = \sum_{i=1}^{n} Y_i, \quad n = 1, 2, \cdots; \quad X_0 \equiv 0.$$

The discrete-time process $\{X_n\}$ is a *partial-sum process* associated with the IID random variables Y_n. Since

$$X_{n+1} = X_n + Y_{n+1}, \quad n = 0, 1, \cdots,$$

the independence assumption on Y_n yields

$$P[X_{n+1} \le y | X_0 = x_0, \cdots, X_n = x] = P[X_n + Y_{n+1} \le y | X_n = x]$$

for all $x, y \in R$, which shows that the process $\{X_n\}$ is Markovian.

Now suppose that the distribution function of Y is G. It follows that

$$P[X_{n+1} \le y | X_n = x] = P[Y_{n+1} \le y - x | X_n = x] = G(y - x).$$

Hence, denoting the distribution function of X_n by F_n, we have

$$F_{n+1}(y) = \int_{-\infty}^{\infty} G(y - x) dF_n(x), \quad n = 0, 1, \cdots. \tag{1.6}$$

Since $F_0(x) = U(x)$, the unit step function, i.e., $U(x) = 1$ for $x \ge 0$ and $U(x) = 0$ for $x < 0$, the probabilistic law of X_n can be obtained by the recursive relation (1.6).

Example 1.3 Let $U_1, U_2, \cdots$ be IID positive random variables representing the lifetimes of successively replaced systems. The partial-sum process $\{T_n\}$, where $T_n = \sum_{i=1}^{n} U_i$ with $T_0 \equiv 0$, represents the time epochs of failure. For each $t \in R_+$, define

$$X(t) = t - T_n, \quad T_n \le t < T_{n+1}.$$

The random variable $X(t)$ is the age of the system in operation at time t. The continuous-time process $\{X(t)\}$ is called an *age process* associated

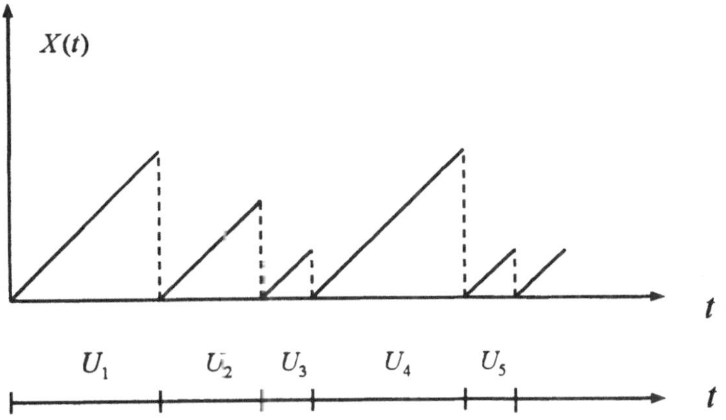

Figure 1.3 *A sample path of the age process.*

with the partial-sum process $\{T_n\}$. A typical sample path of the age process is depicted in Figure 1.3. Now, suppose that $X(t) = u,\ 0 \le u \le t$, and that $T_n \le t < T_{n+1}$ for some $n \in \mathcal{Z}_+$. Let V_1 be the remaining lifetime of the system in operation at time t. The survival probability of V_1 conditional on $X(t) = u$ is given by

$$P[V_1 > x | X(t) = u] = \frac{P[U_{n+1} > u + x]}{P[U_{n+1} > u]}, \quad x \ge 0.$$

Define $V_k = U_{n+k},\ k = 2, 3, \cdots$ and consider the partial-sum process $\{S_n\}$, where $S_n = \sum_{i=1}^{n} V_i$ with $S_0 \equiv 0$. The distribution of V_1 may differ from that of $V_k,\ k = 2, 3, \cdots$. For each $s \in R_+$, let

$$Y(s) = s - S_n, \quad S_n \le s < S_{n+1}.$$

The stochastic behavior of the continuous-time process $\{Y(s)\}$ is completely determined by that of $\{S_n\}$, which depends only on the current value $X(t) = u$. Note that

$$X(t+s) = \begin{cases} u + s, & V_1 > s, \\ Y(s), & V_1 \le s. \end{cases}$$

This means that the current value $X(t) = u$ suffices to determine the distribution of the future value $X(t+s)$ for all $s > 0$. Therefore, the age process $\{X(t)\}$ is Markovian in continuous time. The reader interested in age processes and their related topics should refer to, e.g., Çinlar (1975) or Ross (1983).

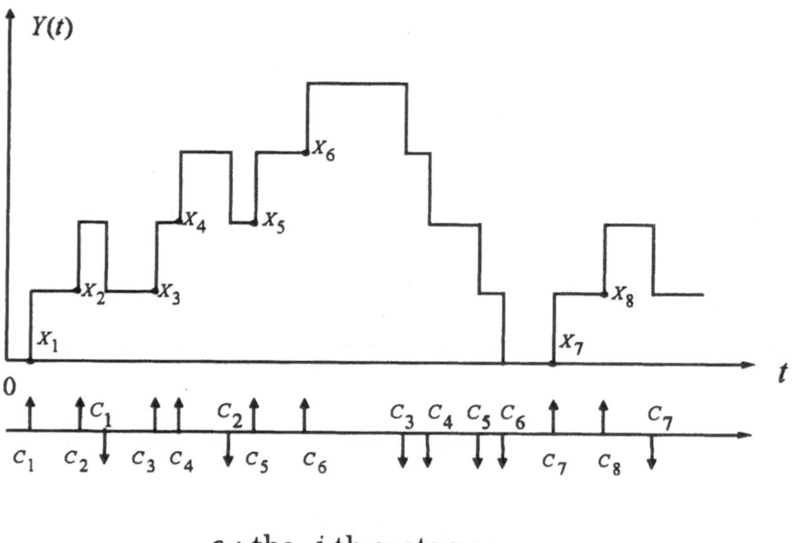

Figure 1.4 *A sample path of the queue-size process.*

Example 1.4 Consider a GI/M/1 queue in which the arrival times T_n constitute a partial-sum process, the service times are independent and exponentially distributed, and there is a single server with no restriction on the queue size. The arrival process and the service times are mutually independent. The continuous-time queue-size process $\{Y(t),\ t \geq 0\}$ is not Markovian, but can be studied through a Markov chain in discrete time. Let X_n be the value of Y just prior to the nth arrival, i.e., $X_n = Y(T_n-)$. The relation between X_n and $Y(t)$ is pictured in Figure 1.4. Let D_n be the number of *potential* departures during the period $[T_{n-1}, T_n)$. We then have

$$X_{n+1} = \{X_n + 1 - D_{n+1}\}_+, \quad n = 0, 1, \cdots. \tag{1.7}$$

Now, because of the memoryless property (see (4.20) in Section 4.2 for details) of exponential service times, D_{n+1} depends only on the value $U_{n+1} \equiv T_{n+1} - T_n$, which is independent of the past $(X_0, \cdots, X_n)$, by assumption. In fact, $D_1,\ D_2, \cdots$ are IID random variables, since so are the U_ns. It follows that the process $\{X_n\}$ is Markovian in discrete time. It should be noted that n is not a time index. The process $\{X_n\}$ is called an *embedded process.*

Specifically, let F be the distribution function of the interarrival time U and let μ be the service rate. Note that

$$a_k \equiv P[D = k] = \int_0^\infty P[D = k | U = t] dF(t), \quad k = 0, 1, \cdots.$$

Since the service times are exponentially distributed, we have

$$P[D=k|U=t]=\frac{(\mu t)^k}{k!}\,e^{-\mu t},$$

whence

$$a_k=\int_0^\infty \frac{(\mu t)^k}{k!}\,e^{-\mu t}dF(t), \quad k=0,1,\cdots. \tag{1.8}$$

It follows from (1.7) that

$$P[X_{n+1}=j|X_n=i]=\begin{cases} a_{i+1-j}, & 1\le j\le i+1,\\ \sum_{k=i+1}^{\infty} a_k, & j=0,\\ 0, & \text{otherwise,}\end{cases} \tag{1.9}$$

for each $i,\ j\in \mathcal{Z}_+$.

Example 1.5 In this example, we consider an M/G/1 queue, i.e., the interarrival times are independent and exponentially distributed, the service times are IID and the arrival process and the service times are mutually independent. There is a single server and no restriction on the queue size is imposed. Again, the continuous-time queue-size process $\{Y(t),\ t\ge 0\}$ is not Markovian, but can be studied through a Markov chain in discrete time. Let X_n denote the number of customers just after the nth service completion and let A_n be the number of arrivals during the nth service time S_n. It is easily seen that

$$X_{n+1}=\begin{cases} A_{n+1}, & X_n=0,\\ X_n+A_{n+1}-1, & X_n\ge 1,\end{cases} \qquad n=0,1,\cdots.$$

The process $\{X_n\}$ is Markovian in discrete time due to the memoryless property of exponential interarrival times. More specifically, let G be the distribution function of the service time S and let λ be the arrival rate. Then, as in Example 1.4, we have

$$b_k\equiv P[A=k]=\int_0^\infty \frac{(\lambda t)^k}{k!}\,e^{-\lambda t}dG(t), \quad k=0,1,\cdots, \tag{1.10}$$

and

$$P[X_{n+1}=j|X_n=i]=\begin{cases} b_j, & i=0,\ j\ge 0,\\ b_{j-i+1}, & i\ge 1,\ j\ge i-1,\\ 0, & \text{otherwise,}\end{cases} \tag{1.11}$$

for each $i,\ j\in \mathcal{Z}_+$. For detailed discussions of queueing theory, the reader is referred to, e.g., Asmussen (1987), Kleinrock (1975) or Wolff (1989).

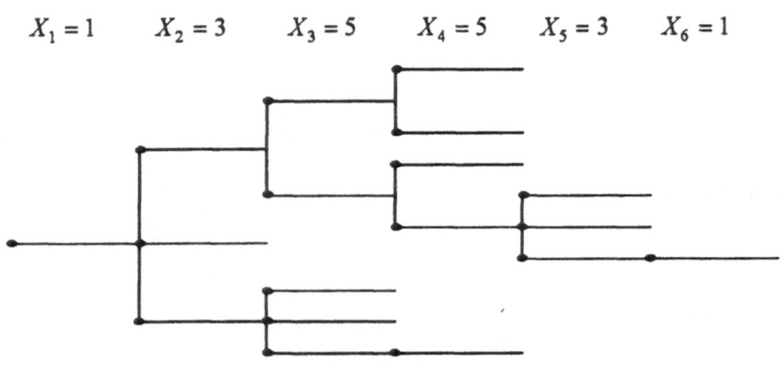

Figure 1.5 *A sample path of the branching process.*

Example 1.6 Suppose that an organism at the end of its lifetime produces a random number of offspring. We assume that all offspring act independently and the numbers of offspring are identically distributed. The lifespans of all organisms are assumed to be the same. A possible realization is depicted in Figure 1.5. Let X_n be the population size at the nth generation. $X_n = 0$ implies $X_{n+m} = 0$ for all $m \geq 0$ so that the organism is extinct. If $X_n \geq 1$, the ith individual, $i = 1, \cdots, X_n$, produces the random number ξ_i of offspring. It follows that

$$X_{n+1} = \sum_{i=1}^{X_n} \xi_i, \quad n = 0, 1, \cdots,$$

where the empty sum $\sum_{i=1}^{0}$ is interpreted as zero. The process $\{X_n\}$ is called a *branching process*. By assumption, the ξ_i are IID random variables. Hence, the distribution of X_{n+1} depends only on the current value X_n and the process $\{X_n\}$ is a Markov chain in discrete time. Exercise 1.5 asks the reader to derive the moment generating function of X_n. For detailed discussions of branching processes, see Athreya and Ney (1972).

Example 1.7 Consider a simple learning experiment that involves a sequence of trials under similar conditions. It is natural to regard the subject's behavior or response on trial n as a random variable, R_n say. *Stochastic learning models* assume that the distribution of R_n is determined by another random variable X_n which represents the subject's state of learning at trial n. In typical models, the responses will exhibit complicated statistical interdependence, but the state sequence $\{X_n\}$ will be simple. Suppose that on each trial, a human subject predicts whether or not a lamp on a panel placed in front of the subject will flash. Flashes on different trials are independent. Let X_n be the probability of predicting a flash on trial n. A *linear* learning model assumes that if the subject predicts a flash and it occurs, X_n increases to

$$X_{n+1} = (1-\theta_1)X_n + \theta_1.$$

If the subject predicts a flash and it does not occur, X_n decreases to

$$X_{n+1} = (1-\theta_2)X_n.$$

An unpredicted flash increases X_n to

$$X_{n-1} = (1-\theta_3)X_n + \theta_3,$$

while the prediction that a flash does not occur reduces X_n to

$$X_{n+1} = (1-\theta_4)X_n.$$

The parameters $\theta_i \in [0,1)$, $i = 1, \cdots, 4$, control the rates of learning. It is obvious that the process $\{X_n\}$ is Markovian with the state space $[0,1]$, and such a model is termed a *Markovian learning model.* For a complete discussion of this topic, see Norman (1974).

Example 1.8 Suppose that initially we have a white and b black balls in an urn. We draw a ball at random and, before drawing the next ball, we replace the one drawn, adding also s balls of the same color. Let X_n denote the number of white balls obtained in the first n drawings. Thus $n - X_n$ black balls have been drawn to date, and the number of white balls in the urn is $a + sX_n$ while the number of black balls is $b + s(n - X_n)$. Since drawing is random, the probability of drawing a white ball at the next step depends only on the number of black and white balls in the urn. To be more specific, we have

$$P[X_{n+1} = X_n + 1 | X_n = i] = \frac{a+si}{a+b+sn}$$

and

$$P[X_{n+1} = X_n | X_n = i] = \frac{b+s(n-i)}{a+b+sn}.$$

Hence, $\{X_n\}$ is a Markov chain in discrete time, and this model is called the Pólya urn model.

Example 1.9 Suppose that a stochastic process $\{X_n\}$ does not satisfy the Markov property (1.1) but does satisfy

$$\begin{aligned} & P[X_{n+1} = j | X_0 = i_0, \cdots, X_{n-1} = i_{n-1}, X_n = i_n] \\ = \;& P[X_{n+1} = j | X_{n-1} = i_{n-1}, X_n = i_n]. \qquad (1.12) \end{aligned}$$

This property is called the *second order Markov property* and $\{X_n\}$ is a second order Markov process. A higher order Markov process is defined similarly. A second order Markov process can be converted to an ordinary (first order) Markov process by defining the two-dimensional process $Y_n = (X_n, X_{n+1})$. To see this, one has from (1.12) that

$$\begin{aligned} & P[X_n = i_n,\ X_{n+1} = i_{n+1} | X_0 = i_0, \cdots, X_n = i_n] \\ = \ & P[X_n = i_n,\ X_{n+1} = i_{n+1} | X_{n-1} = i_{n-1}, X_n = i_n], \end{aligned}$$

whence

$$\begin{aligned} & P[Y_n = (i_n, i_{n+1}) | Y_0 = (i_0, i_1), \cdots, Y_{n-1} = (i_{n-1}, i_n)] \\ = \ & P[Y_n = (i_n, i_{n+1}) | Y_{n-1} = (i_{n-1}, i_n)]. \end{aligned}$$

Thus, $\{Y_n\}$ satisfies the Markov property (1.1) as claimed (see Exercise 1.9 for a related problem). Higher order Markov processes can similarly be converted to ordinary Markov processes.

A pth order Markov process often appears in the form of

$$X_n = \phi_1 X_{n-1} + \phi_2 X_{n-2} + \cdots + \phi_p X_{n-p} + \varepsilon_n,$$

where ϕ_i, $i = 1, \cdots, p$, are constant, and ε_n are IID (normal) random variables with mean 0 representing a 'noise'. The discrete-time process $\{X_n\}$ is called an *autoregressive* (AR) process of order p in time series analysis. Note that X_n is regressed on previous values $X_{n-1}, \cdots, X_{n-p}$ of itself; hence the model is autoregressive. For detailed discussions of time series analysis, the reader is referred to, e.g., Box and Jenkins (1976).

Example 1.10 Consider a Markov process $\{X_n\}$ on a finite state space $\mathcal{N}$. Let f be a mapping from $\mathcal{N}$ to another set $\mathcal{A}$. The mapping may not be an injection, that is, it may be the case that $f(i) = f(j)$ for $i \neq j$. Suppose that $\{X_n\}$ is not directly observable but that the related process $\{f(X_n)\}$ can be observed. The mapped process $\{f(X_n)\}$ is often called a ***hidden Markov process*** and has considerable practical importance in the context of information theory. In this regard, the set $\mathcal{A}$ is called an ***output alphabet.*** The hidden Markov process is in general *not* Markovian (see Exercise 1.10). If for every initial distribution the hidden Markov process $\{f(X_n)\}$ is indeed Markovian, then the underlying Markov process $\{X_n\}$ is called ***lumpable*** with respect to the mapping f. Lumpability conditions have been studied in detail under various situations. The interested reader may consult Kemeny and Snell (1960) and Iosifescu (1980).

Of related interest is the ***identifiability problem***. In practice, there may exist different underlying Markovian information sources that generate the same output process. Hence, it is of great importance to show when this happens for seemingly different input processes. In a recent paper by Ito, Amari and Kobayashi (1992), the identifiability problem was completely resolved.

1.4 Transition probabilities

In this section, we study some elementary properties of the conditional probability

$$P[X_{n+1} = j | X_n = i],$$

i.e. the right-hand side of the Markov property (1.1). Here and hereafter, it is assumed that the state space $\mathcal{N}$ under consideration is either finite or denumerably infinite.

Let $\{X_n\}$ be a Markov chain and define

$$p_{ij}(n, n+1) = P[X_{n+1} = j | X_n = i], \quad n = 0, 1, \cdots. \tag{1.13}$$

The conditional probability $p_{ij}(n, n+1)$ is called the (one-step) *transition probability* from state i to state j at time n. The matrix

$$\mathbf{P}(n, n+1) = (p_{ij}(n, n+1))$$

is the (one-step) *transition matrix* at time n. According to the Markov property (1.1), the transition probability $p_{ij}(n, n+1)$ is in fact equal to

$$p_{ij}(n, n+1) = P[X_{n+1} = j | X_0, \cdots, X_{n-1}, X_n = i].$$

The m-step transition probabilities at time n are defined by

$$p_{ij}(n, n+m) = P[X_{n+m} = j | X_n = i], \quad n = 0, 1, \cdots, \tag{1.14}$$

and the corresponding m-step transition matrix at time n is

$$\mathbf{P}(n, n+m) = (p_{ij}(n, n+m)).$$

In particular, we write

$$p_{ij}(m) = p_{ij}(0, m); \quad \mathbf{P}(m) = \mathbf{P}(0, m),$$

i.e. the m-step transition probabilities and matrix, respectively, at time 0. It should be noted that transition matrices may not be square, as the next example illustrates.

Example 1.11 In the Pólya urn model described in Example 1.8, let X_n denote the number of white balls obtained in the first n drawings. Note that the possible values for X_n are $\{0, 1, \cdots, n\}$. From the result given in Example 1.8, when $X_n = i$, the transition probabilities are

$$p_{ij}(n, n+1) = \begin{cases} \dfrac{a + si}{a + b + sn}, & j = i + 1, \\[2ex] \dfrac{b + s(n-i)}{a + b + sn}, & j = i, \\[2ex] 0, & \text{otherwise}. \end{cases}$$

Hence, the one-step transition matrix $\mathbf{P}(n, n+1)$ at time n is $n \times (n+1)$ with the components given above.

The one-step transition probabilities determine the m-step transition probabilities. To see this, note that

$$\begin{aligned} p_{ij}(n,n+2) &= P[X_{n+2}=j|X_n=i] \\ &= \sum_k P[X_{n+1}=k,\ X_{n+2}=j|X_n=i] \\ &= \sum_k P[X_{n+1}=k|X_n=i]\,P[X_{n+2}=j|X_n=i,\ X_{n+1}=k], \end{aligned}$$

where the sum is taken over all states possible for X_{n+1}. Here the third equality follows from the chain rule for conditional probabilities. But, by the Markov property, we have

$$p_{ij}(n,n+2)=\sum_k p_{ik}(n,n+1)\,p_{kj}(n+1,n+2), \quad n=0,1,\cdots.$$

The above equations can be written in matrix form as

$$\mathbf{P}(n,n+2)=\mathbf{P}(n,n+1)\,\mathbf{P}(n+1,n+2),$$

regardless of whether or not the transition matrices are square.* Moreover, by an induction argument, it follows that

$$\mathbf{P}(n,n+m)=\mathbf{P}(n,n+1)\mathbf{P}(n+1,n+2)\cdots\mathbf{P}(n+m-1,n+m) \quad (1.15)$$

for all $m=1,2,\cdots$. Note that $\mathbf{P}(n,n)=\mathbf{I}$, the identity matrix, for all n. Also note that, since the matrix product is *not commutative*, i.e., $\mathbf{AB}\neq\mathbf{BA}$ for two (even square) matrices in general, the order of transition matrices in the right-hand side of (1.15) is crucial.

It is easily seen from (1.15) that

$$\mathbf{P}(m,n)=\mathbf{P}(m,\ell)\,\mathbf{P}(\ell,n), \quad m\le\ell\le n, \quad (1.16)$$

or, equivalently,

$$p_{ij}(m,n)=\sum_k p_{ik}(m,\ell)\,p_{kj}(\ell,n), \quad m\le\ell\le n. \quad (1.17)$$

Equation (1.17) is known as the *Chapman–Kolmogorov equation.*

Definition 1.1 A (not necessarily square) matrix $\mathbf{A}=(a_{ij})$ is said to be *stochastic* if

$$a_{ij}\ge 0 \quad \text{and} \quad \sum_j a_{ij}=1 \quad \text{for all } i,\ j.$$

In this case, we write $\mathbf{A}\in\mathcal{S}$, i.e., $\mathcal{S}$ denotes the class of stochastic matrices. If $\sum_j a_{ij}\le 1$ then $\mathbf{A}$ is said to be *substochastic.* A substochastic matrix is called *strictly substochastic* if $\sum_j a_{ij}<1$ for at least one state i.

* Throughout this book, the size of a matrix or vector is always such that all matrix operations involved are well defined.

The above definition states that a stochastic (substochastic, respectively) matrix is a nonnegative matrix whose row sums are unity (less than or equal to unity). In matrix notation, $\mathbf{A}$ is stochastic (substochastic) if and only if

$$\mathbf{A} \geq \mathbf{O} \quad \text{and} \quad \mathbf{A1} = \mathbf{1} \;\; (\mathbf{A1} \leq \mathbf{1}), \tag{1.18}$$

where $\mathbf{O}$ denotes the *zero matrix*, i.e. the matrix whose components are all zero, and $\mathbf{1}$ the column vector whose components are all unity. Note that (1.18) reveals that the Perron–Frobenius eigenvalue of any finite, square stochastic matrix is unity and the corresponding right eigenvector is $\mathbf{1}$ (see Appendix A.1).

The next lemma shows some closure properties of the class $\mathcal{S}$. It is clear that the same results hold for the class of substochastic matrices.

Lemma 1.1 *Suppose* $\mathbf{A}$, $\mathbf{B} \in \mathcal{S}$. *Then* $\mathbf{AB} \in \mathcal{S}$ *and* $\lambda\mathbf{A} + (1-\lambda)\mathbf{B} \in \mathcal{S}$ *for all* $\lambda \in [0, 1]$.

Proof. The nonnegativity of $\mathbf{AB}$ and $\lambda\mathbf{A} + (1-\lambda)\mathbf{B}$ follows at once. We have

$$(\mathbf{AB})\mathbf{1} = \mathbf{A}(\mathbf{B1}) = \mathbf{A1} = \mathbf{1},$$

where the associativity holds since the matrices involved are nonnegative (see Theorem A.8 in Appendix A.3). Also,

$$(\lambda\mathbf{A} + (1-\lambda)\mathbf{B})\mathbf{1} = \lambda\mathbf{A1} + (1-\lambda)\mathbf{B1} = \mathbf{1}.$$

Hence, the conditions in (1.18) hold for each case. □

The next result may appear to be trivial but is important in subsequent developments. The proof is left to the reader as an exercise (see Exercise 1.11).

Theorem 1.1 *For each* n, *the* m*-step transition matrix* $\mathbf{P}(n, n+m)$ *is stochastic for all* $m \in \mathcal{Z}_+$.

Let $\alpha_i \equiv P[X_0 = i]$ and let $\boldsymbol{\alpha} = (\alpha_i)$. The column vector $\boldsymbol{\alpha}$ is called the *initial distribution* of $\{X_n\}$ Note that $\boldsymbol{\alpha}$ is a *probability vector* in the sense that $\alpha_i \geq 0$ and $\sum_i \alpha_i = 1$ or, in matrix notation,

$$\boldsymbol{\alpha} \geq \mathbf{0} \quad \text{and} \quad \boldsymbol{\alpha}^\top \mathbf{1} = 1, \tag{1.19}$$

where $\mathbf{0}$ denotes the *zero vector*, i.e. the column vector whose components are all zero, and $\top$ the transpose. We shall denote by $\mathcal{P}$ the class of probability vectors. Defining the unconditional probabilities by

$$\pi_i(n) = P[X_n = i], \quad n = 0, 1, \cdots,$$

the column vector $\boldsymbol{\pi}(n) = (\pi_i(n))$ is called the *state distribution* of $\{X_n\}$ at time n. Of course, $\boldsymbol{\pi}(0) = \boldsymbol{\alpha}$, the initial distribution. From (1.16), we have

$$\boldsymbol{\pi}^\top(n+1) = \boldsymbol{\pi}^\top(n)\,\mathbf{P}(n, n+1), \quad n = 0, 1, \cdots. \tag{1.20}$$

It follows that the state distribution can be computed by

$$\boldsymbol{\pi}^\top(n) = \boldsymbol{\alpha}^\top \mathbf{P}(n), \quad n = 0, 1, \cdots. \tag{1.21}$$

Note that, since $\mathbf{P}(n)$ is stochastic, one has $\boldsymbol{\pi}(n) \geq \mathbf{0}$ and $\boldsymbol{\pi}^\top(n)\mathbf{1} = 1$ from (1.21), whence $\boldsymbol{\pi}(n) \in \mathcal{P}$ for all $n = 0, 1, \cdots$.

If one-step transition probabilities are independent of time n, the Markov chain $\{X_n\}$ is said to be ***temporally homogeneous***, or ***homogeneous*** (***non-homogeneous***, otherwise) for short. We shall assume that $\{X_n\}$ is homogeneous so that we can define

$$p_{ij} \equiv P[X_{n+1} = j | X_n = i], \quad n = 0, 1, \cdots; \tag{1.22}$$

cf. (1.13). For a homogeneous Markov chain, the transition matrix does not depend on time and we denote it by $\mathbf{P} = (p_{ij})$. That is,

$$\mathbf{P} = \mathbf{P}(n, n+1), \quad n = 0, 1, \cdots.$$

Note that the transition matrix $\mathbf{P}$ is necessarily ***square***, and the possible states for every X_n are the same. In what follows, we shall deal with homogeneous Markov chains only. Detailed discussions of nonhomogeneous Markov chains may be found in Seneta (1981) for the discrete-time case and in Iosifescu (1980) for the continuous-time case.

Throughout this book, we shall use the following notation. For a Markov chain $\{X_n\}$ with state space $\mathcal{N}$, the conditional probability of event A given $X_0 = i$ is denoted by

$$P_i[A] = P[A | X_0 = i], \quad i \in \mathcal{N},$$

and the corresponding expectation operator by E_i. Then, due to homogeneity, the transition probability from state i to state j is written as

$$p_{ij} = P_i[X_1 = j] = P[X_{n+1} = j | X_n = i].$$

When the initial distribution of $\{X_n\}$ is $\boldsymbol{\alpha} \in \mathcal{P}$, the conditional probability of event A is denoted by $P_\alpha[A]$ and the corresponding expectation operator by E_α. The state probability is then given by

$$\pi_i(n) = P_\alpha[X_n = i].$$

Note that we write $P_i[A]$ rather than $P_{\delta_i}[A]$ when $\boldsymbol{\alpha} = \boldsymbol{\delta}_i$, i.e. the ith unit vector which is the probability vector having a point mass on $i \in \mathcal{N}$.

The m-step transition matrix in (1.15) becomes

$$\mathbf{P}(n, n+m) = \mathbf{P}^m, \quad m = 0, 1, \cdots,$$

i.e. the mth power of $\mathbf{P}$, which is independent of n. The m-step transition probability at time n is therefore independent of n and is the (i,j)th component of $\mathbf{P}^m$. We shall write

$$p_{ij}(m) \equiv P_i[X_m = j] = P[X_{n+m} = j | X_n = i]; \tag{1.23}$$

cf. (1.14). The Chapman–Kolmogorov equation (1.17) can be expressed as

$$p_{ij}(m+n) = \sum_k p_{ik}(m)\, p_{kj}(n), \tag{1.24}$$

or, in matrix form,

$$\mathbf{P}^{m+n} = \mathbf{P}^m \mathbf{P}^n, \tag{1.25}$$

a trivial multiplicative identity of matrices; cf. (1.16).

Transition matrices play a central role in the theory of discrete-time Markov chains. Let $\boldsymbol{\alpha} = (\alpha_i)$ be the initial distribution of a Markov chain $\{X_n\}$ with transition matrix $\mathbf{P} = (p_{ij})$. We will show that the initial distribution $\boldsymbol{\alpha}$ and the transition matrix $\mathbf{P}$ together determine joint distributions of the form

$$P[X_{n_1} = i_1, X_{n_2} = i_2, \cdots, X_{n_m} = i_m] \tag{1.26}$$

for all $m = 1, 2, \cdots$ and all $n_1 < n_2 < \cdots < n_m$. For this purpose, it suffices to evaluate

$$P[X_0 = i_0, X_1 = i_1, \cdots, X_n = i_n], \quad n = 1, 2, \cdots,$$

from which the marginals (1.26) can be obtained. To this end, the chain rule of conditional probabilities in conjunction with repeated application of the Markov property and homogeneity yields

$$P[X_0 = i_0, X_1 = i_1, \cdots, X_n = i_n] = \alpha_{i_0}\, p_{i_0,i_1} \cdots p_{i_{n-1},i_n}. \tag{1.27}$$

Hence the joint distribution (1.26) can be determined by $\boldsymbol{\alpha} = (\alpha_i)$ and $\mathbf{P} = (p_{ij})$ only. Note the natural interpretation of the right-hand side of (1.27) as the probability of starting from state i_0, moving from i_0 to i_1, then to $i_2, \cdots$, and finally from i_{n-1} to i_n.

Although more care is needed in stating the next theorem, we have the following intuitively plausible result. A rigorous proof may be found in Chung (page 7, 1967).

Theorem 1.2 *Let* $\mathbf{P}$ *be a square, stochastic matrix and let* $\boldsymbol{\alpha}$ *be a probability vector whose indices run over* $\mathcal{N}$. *Then, there exists a probability space* $(\Omega, \mathcal{F}, P)$ *and a homogeneous Markov chain* $\{X_n\}$ *in discrete time defined on it with the state space* $\mathcal{N}$, *the initial distribution* $\boldsymbol{\alpha}$ *and the transition matrix* $\mathbf{P}$.

Recall from Theorem 1.1 that $\mathbf{P}^m$ is stochastic for all $m = 1, 2, \cdots$. According to Theorem 1.2 above, one might guess that any stochastic matrix is the m-step transition matrix of a Markov chain. However, this claim is not valid in general. The reader should consult Exercise 1.12 for a counterexample to this claim.

Before closing this section, we determine the transition matrices of some examples given in Section 1.3 (see also Exercises 1.2–1.4).

Example 1.12 In the GI/M/1 queue given in Example 1.4, let X_n be the queue size just before the nth arrival. The process $\{X_n\}$ is a Markov chain with state space $\mathcal{N} = \mathcal{Z}_+$. The transition probabilities of $\{X_n\}$ are given by (1.9). Hence, defining $\overline{A}_k = \sum_{i=k}^{\infty} a_i$, where the a_k are given by (1.8), the transition matrix is given by

$$\mathbf{P} = \begin{pmatrix} \overline{A}_1 & a_0 & 0 & 0 & \cdots \\ \overline{A}_2 & a_1 & a_0 & 0 & \cdots \\ \overline{A}_3 & a_2 & a_1 & a_0 & \cdots \\ \vdots & \vdots & \vdots & \ddots & \ddots \end{pmatrix}. \tag{1.28}$$

Note the special zero structure of $\mathbf{P} = (p_{ij})$, viz. $p_{ij} = 0$ for all $j > i + 1$. Such a matrix is often called a *lower Hessenberg matrix.*

Example 1.13 Consider now the M/G/1 queue described in Example 1.5. In this case, the number of customers just after the nth departure forms a discrete-time Markov chain on $\mathcal{Z}_+$. (Recall that n does not represent a time.) The transition matrix is given by

$$\mathbf{P} = \begin{pmatrix} b_0 & b_1 & b_2 & b_3 & \cdots \\ b_0 & b_1 & b_2 & b_3 & \cdots \\ 0 & b_0 & b_1 & b_2 & \cdots \\ 0 & 0 & b_0 & b_1 & \cdots \\ \vdots & \vdots & \vdots & \ddots & \ddots \end{pmatrix}, \tag{1.29}$$

where the b_k are given by (1.10). It should be noted that the first and second rows are identical. A matrix with $p_{ij} = 0$ for all $i > j + 1$ is called an *upper* Hessenberg matrix.

Another example of a Markov chain that we shall often encounter in this book is the following.

Example 1.14 A one-dimensional *random walk* is a homogeneous Markov chain on the countable state space $\mathcal{N} = \mathcal{Z}$ in which a *particle*, if it is in state i, can in a single transition either stay at i or move to one of the adjacent states $i - 1$ or $i + 1$. Let $q_i > 0$ be the downward transition probability, $r_i \geq 0$ the probability of no transition (self-transition), and $p_i > 0$ the upward transition probability. That is,

$$\begin{aligned} q_i &= P[X_{n+1} = i - 1 | X_n = i], \\ r_i &= P[X_{n+1} = i | X_n = i], \\ p_i &= P[X_{n+1} = i + 1 | X_n = i], \end{aligned}$$

and $p_i + r_i + q_i = 1$ for all $i \in \mathcal{N}$. The transition matrix is given by

$$\mathbf{P} = \begin{pmatrix} \ddots & \ddots & \ddots & \vdots & \vdots & \vdots & \\ \cdots & q_{-1} & r_{-1} & p_{-1} & 0 & 0 & \cdots \\ \cdots & 0 & q_0 & r_0 & p_0 & 0 & \cdots \\ \cdots & 0 & 0 & q_1 & r_1 & p_1 & \cdots \\ & \vdots & \vdots & \vdots & \ddots & \ddots & \ddots \end{pmatrix}.$$

Note the special zero structure. Random walks are useful because they frequently serve as good discrete approximations to physical processes describing the motion of diffusing particles.

1.5 The strong Markov property

The Markov property may hold at some random times as well. Working in the discrete-time setting, consider a stochastic process $\{X_n\}$ with state space $\mathcal{N} = \mathcal{Z}$. A random variable T is called a *stopping time* of $\{X_n\}$ if, for each n, the occurrence of the event $\{T \le n\}$ is determined by $X_0, \cdots, X_n$, i.e., there exists a function f such that

$$\mathrm{I}_{\{T \le n\}} = f(X_0, \cdots, X_n),$$

where I_A denotes the indicator function of event A, meaning $\mathrm{I}_A(\omega) = 1$ if $\omega \in A$ and $\mathrm{I}_A(\omega) = 0$ otherwise. A typical example of stopping times is the first passage time of $\{X_n\}$, viz.

$$T = \inf\{n \ge 1 : X_n \in S\}$$

for some $S \subset \mathcal{N}$. The validity of the Markov property at stopping times is called the *strong Markov property*. The process satisfying the strong Markov property is characterized by

$$P[X_{T+1} = j | X_0, \cdots, X_{T-1}, X_T = i] = P[X_{T+1} = j | X_T = i] \qquad (1.30)$$

for every stopping time T. Here the random variable X_T is defined as $X_{T(\omega)}(\omega)$ for each $\omega \in \Omega$. It can be shown from (1.30) that the past and the future of the process are conditionally independent given the present, where the stopping time T plays the role of 'present' (see Exercise 1.14).

The next result proves that any discrete-time Markov chain has the strong Markov property.

Theorem 1.3 *For a homogeneous Markov chain $\{X_n\}$ in discrete time with transition matrix $\mathbf{P} = (p_{ij})$, let T be a stopping time. If T is finite with probability one, then*

$$P[X_{T+1} = j | X_0, \cdots, X_{T-1}, X_T = i] = P[X_{T+1} = j | X_T = i],$$

whenever the left-hand side is defined.

Proof. Since the events $\{T = k\}$ are disjoint and $P[T < \infty] = 1$, it follows that

$$\begin{aligned} & P[X_0, \cdots, X_{T-1}, X_T = i, X_{T+1} = j] \\ = & \sum_{k \geq 0} P[X_0, \cdots, X_{T-1}, X_T = i, X_{T+1} = j, T = k] \\ = & \sum_{k \geq 0} P[X_0, \cdots, X_{k-1}, X_k = i, X_{k+1} = j, T = k] \\ = & \sum_{k \geq 0} P[X_0, \cdots, X_{k-1}, X_k = i, T = k] \\ & \qquad \times P[X_{k+1} = j | X_0, \cdots, X_{k-1}, X_k = i, T = k]. \end{aligned}$$

But, from the Markov property and homogeneity, we have

$$P[X_{k+1} = j | X_0, \cdots, X_{k-1}, X_k = i, T = k] = p_{ij}.$$

Hence,

$$\begin{aligned} & P[X_0, \cdots, X_{T-1}, X_T = i, X_{T+1} = j] \\ = & \; p_{ij} \sum_{k \geq 0} P[X_0, \cdots, X_{k-1}, X_k = i, T = k] \\ = & \; p_{ij} P[X_0, \cdots, X_{T-1}, X_T = i], \end{aligned}$$

so that

$$P[X_{T+1} = j | X_0, \cdots X_{T-1}, X_T = i] = p_{ij}.$$

Similarly, we have

$$\begin{aligned} & P[X_T = i, X_{T+1} = j] \\ = & \sum_{k \geq 0} P[X_k = i, T = k] P[X_{k+1} = j | X_k = i, T = k], \end{aligned}$$

whence

$$P[X_T = i, X_{T+1} = j] = p_{ij} P[X_T = i]$$

or, equivalently,

$$P[X_{T+1} = j | X_T = i] = p_{ij},$$

proving the theorem. □

Turning to the continuous-time case, the situation becomes more complicated. Consider a continuous-time stochastic process $\{X(t)\}$ with a countable state space. A random variable T is a stopping time for $\{X(t)\}$ if, for each t, the event $\{T \leq t\}$ is determined by the history up to time t. Numerous sufficient conditions are known for the strong Markov property, but we do not pursue them here. The interested reader should consult, e.g., Freedman (Section 7.4, 1971). If $\{X(t)\}$ is sufficiently smooth, then the Markov property holds for all stopping times. Any finite homogeneous Markov chain $\{X(t)\}$ in continuous time has the strong Markov property.

1.6 Exercises

Exercise 1.1 Show that the property (1.4) is equivalent to the Markov property (1.1).

Exercise 1.2 In Example 1.1, suppose that $D_1, D_2, \cdots$ are IID nonnegative, integer-valued random variables. Suppose, further, that S and s are positive integers. Show that the process $\{X_n\}$ is a homogeneous Markov chain. Furthermore, assuming that D has a common distribution

$$P[D = k] = \frac{\lambda^k}{k!} e^{-\lambda}, \quad k = 0, 1, \cdots,$$

i.e. the *Poisson distribution* with mean λ, obtain the one-step transition matrix when $s = 2$ and $S = 4$.

Exercise 1.3 In Exercise 1.2, instead of following an (s, S)-policy, we use a (q, Q)-policy. That is, if the stock level at the end of each period is less than q units, we order Q additional units. Otherwise, no ordering will take place. Show that the process $\{X_n\}$ is a homogeneous Markov chain and obtain the one-step transition matrix when $q = 2$ and $Q = 4$ (Hillier and Lieberman, 1990).

Exercise 1.4 In Example 1.3, suppose U has a discrete distribution

$$P[U = k] = r_k > 0, \quad k = 1, \cdots, N,$$

where $N < \infty$. Let X_n denote the age of the system in operation at time $n = 0, 1, \cdots$. Prove that the process $\{X_n\}$ is a homogeneous Markov chain, and determine the one-step transition matrix of $\{X_n\}$.

Exercise 1.5 In Example 1.6, let $g(s) = E[e^{s\xi}]$ be the *moment generating function* of ξ, and let $\phi_n(s)$ be the moment generating function of X_n. Assuming that the moment generating functions exist in a neighborhood of the origin, show that $\phi_{n+1}(s) = \phi_n(\log g(s))$.

Exercise 1.6 A subject can make one of two responses A_1 and A_2. Associated with the responses are a set of N stimuli S_i, $i = 1, \cdots, N$. Each stimulus is conditioned to one of the responses. A single stimulus is sampled with probability $1/N$ and the subject responds according to the stimulus sampled. Reinforcement occurs at each trial with probability π, $0 < \pi < 1$, independently of the history of the process. When reinforcement occurs, the stimulus sampled does not alter its conditioning state. In the opposite case, the stimulus becomes conditioned to the other response. Consider the Markov chain whose state variable is the number of stimuli conditioned to response A_1. Determine the transition matrix of this Markov chain (Karlin and Taylor, 1975).

Exercise 1.7 In the Pólya urn model (Example 1.8), let Y_n be, in turn, the ratio of white balls to black balls in the first n drawings. Show that the process $\{Y_n\}$ is Markovian. What is the expected value $E[Y_n]$?

Exercise 1.8 Consider two urns A and B containing a total of N balls. An experiment is performed in which a ball is equally likely to be selected among the N balls. Then an urn is selected at random (A is chosen with probability p and B is selected with probability $q = 1 - p$) and the ball drawn is placed in this urn. The state at each trial is represented by the number of balls in urn A. Determine the transition matrix for this Markov chain.

Exercise 1.9 Consider a discrete-time process $\{X_n\}$ which takes the values 0, 1, or 2. Suppose

$$P[X_{n+1} = j | X_0 = i_0, \cdots, X_n = i] = \begin{cases} p_{ij}^1, & n \text{ is even}, \\ p_{ij}^2, & n \text{ is odd}, \end{cases}$$

where $\mathbf{P}^k = (p_{ij}^k)$, $k = 1, 2$, are stochastic. Is $\{X_n\}$ Markovian? If not, then show how, by enlarging the state space, we may transform it into a Markov chain (Ross, 1989).

Exercise 1.10 In Example 1.10, let $\mathcal{N} = \{0, 1, 2\}$, $\mathcal{A} = \{a, b\}$, where $a \neq b$, and $f(0) = f(1) = a$, $f(2) = b$. Let $\{X_n\}$ be a Markov chain on $\mathcal{N}$ with transition matrix $\mathbf{P} = (p_{ij})$ and define $Y_n = f(X_n)$. Compute

$$P[Y_2 = a | Y_0 = a, \, Y_1 = a] \quad \text{and} \quad P[Y_2 = a | Y_1 = a]$$

to conclude that $\{Y_n\}$ is not Markovian unless $p_{02} = p_{12}$.

Exercise 1.11 For a nonhomogeneous Markov chain $\{X_n\}$, prove that the transition matrix $\mathbf{P}(n, m)$ is stochastic for all $n, \; m \in \mathcal{Z}_+$.

Exercise 1.12 Show that the stochastic matrix

$$\mathbf{A} = \begin{pmatrix} 1-a & a \\ b & 1-b \end{pmatrix}, \quad 0 < a, \; b < 1,$$

is the two-step transition matrix of a Markov chain if and only if $a + b \leq 1$. (Hint: Obtain $\mathbf{P}$ such that $\mathbf{P}^2 = \mathbf{A}$ and check whether $\mathbf{P}$ is stochastic.)

Exercise 1.13 Let $\boldsymbol{\alpha}^\top = (1, 0, 0)$ be the initial distribution and let

$$\mathbf{P} = \begin{pmatrix} 0.2 & 0.6 & 0.2 \\ 0.3 & 0.2 & 0.5 \\ 0.7 & 0.1 & 0.2 \end{pmatrix}$$

be the transition matrix of a Markov chain. Determine the state distributions $\boldsymbol{\pi}(n)$ for $n = 1, 2, 3$.

Exercise 1.14 Under the same conditions as given in Theorem 1.3, show that

$$\begin{aligned} & P[X_0, \cdots, X_{T-1} = j, \, X_{T+1} = k, \cdots, X_{T+m} | X_T = i] \\ = \; & P[X_0, \cdots, X_{T-1} = j | X_T = i] \; P[X_{T+1} = k, \cdots, X_{T+m} | X_T = i] \end{aligned}$$

for all $m = 1, 2, \cdots$, whenever the left-hand side is defined.

Exercise 1.15 Consider a consumer's brand choice. At purchase, t, if the consumer buys the brand under consideration then we denote $X_t = 1$ and $X_t = 0$ otherwise. We assume that the probability v_t of buying the brand at time t follows the *linear learning model*

$$v_t = \begin{cases} \alpha - \beta + \lambda v_{t-1}, & \text{if } X_{t-1} = 1, \\ \alpha + \lambda v_{t-1}, & \text{if } X_{t-1} = 0, \end{cases}$$

where the parameters are positive and $\alpha + \beta + \lambda \leq 1$. Determine the limiting probability $\lim_{t \to \infty} v_t$ of buying the brand (Massy, Montgomery and Morrison, 1970).

Exercise 1.16 (Wald's identity) Suppose that $X_1, X_2, \cdots$ are IID random variables having a finite mean and that N is a stopping time for $\{X_n\}$ such that $E[N] < \infty$. Prove that

$$E\left[\sum_{i=1}^{N} X_i\right] = E[N]E[X].$$

(Hint: Let $I_n = 1$ if $N \geq n$ and $I_n = 0$ otherwise. Then $\sum_{i=1}^{N} X_i = \sum_{n=1}^{\infty} X_n I_n$. By assumption, I_n is independent of X_n.)

2

Discrete-time Markov chains

This chapter concerns discrete-time Markov chains defined on a finite or denumerably infinite state space $\mathcal{N}$. The Markov chains under consideration are assumed to be homogeneous. We assume without loss of generality that the state space consists of nonnegative integers $\mathcal{N} = \{0, 1, \cdots, N\}$, where $N < \infty$ or $N = \infty$.

2.1 First passage times

For a Markov chain $\{X_n\}$ with state space $\mathcal{N}$ and transition matrix $\mathbf{P} = (p_{ij})$, let T_j be the first time that the chain $\{X_n\}$ visits the state $j \in \mathcal{N}$, i.e., $T_j = \inf\{n \geq 1 : X_n = j\}$, with the understanding that the infimum of the empty set is ∞. The random variable T_j is called the *first passage time* to state j and is a stopping time dependent on the realization of $\{X_n\}$ (see Section 1.5). That is, given $\omega \in \Omega$, $T_j(\omega) = n$ if and only if $X_k(\omega) \neq j$ for $k = 1, \cdots, n-1$ and $X_n(\omega) = j$. Define

$$f_{ij}(n) = P_i[T_j = n], \quad n = 1, 2, \cdots.$$

The distribution $\{f_{ij}(n)\}$ with respect to n is called the *first-passage-time distribution* of $\{X_n\}$ from state i to state j, which may be *defective* in the sense that $P_i[T_j < \infty] < 1$. If $i = j$, the distribution $\{f_{jj}(n)\}$ is called a *return-time distribution* for state j; $f_{ij}(n)$ can be calculated recursively via

$$f_{ij}(n) = \begin{cases} p_{ij}, & n = 1, \\ \sum_{k \neq j} p_{ik} f_{kj}(n-1), & n \geq 2. \end{cases} \tag{2.1}$$

To see this, we have, for $n \geq 2$,

$$\begin{aligned} f_{ij}(n) &= \sum_{k \neq j} P_i[X_1 = k, X_2 \neq j, \cdots, X_{n-1} \neq j, X_n = j] \\ &= \sum_{k \neq j} P_i[X_1 = k]\, P_i[X_2 \neq j, \cdots, X_{n-1} \neq j, X_n = j | X_1 = k]. \end{aligned}$$

But, by the Markov property and homogeneity, it follows that

$$f_{ij}(n) = \sum_{k \neq j} p_{ik} P_k[X_1 \neq j, \cdots, X_{n-2} \neq j, X_{n-1} = j],$$

proving (2.1) for $n \geq 2$. The case where $n = 1$ is trivial. This type of argument specific to homogeneous Markov chains is called ***first step analysis***. That is, the method proceeds by enumerating the possibilities that can arise at the end of the first transition, and then invoking the law of total probability coupled with the Markov property and homogeneity to establish a characterizing relationship among the unknown variables or functions. We shall frequently make use of this method.

The transition probabilities and the first-passage-time distributions are connected through the following identities. We denote *Kronecker's delta* by δ_{ij}, meaning that $\delta_{ij} = 1$ for $i = j$ and $\delta_{ij} = 0$ for $i \neq j$.

Theorem 2.1 *For a Markov chain* $\{X_n\}$ *with transition matrix* $\mathbf{P} = (p_{ij})$,

$$p_{ij}(n) = \begin{cases} \delta_{ij}, & n = 0, \\ \displaystyle\sum_{\nu=1}^{n} f_{ij}(\nu)\, p_{jj}(n-\nu), & n \geq 1. \end{cases}$$

Proof. Suppose $n \geq 1$. Since $P_i[T_j \leq n | X_n = j] = 1$, we have

$$\begin{aligned} P_i[X_n = j] &= \sum_{\nu=1}^{n} P_i[T_j = \nu,\, X_n = j] \\ &= \sum_{\nu=1}^{n} P_i[T_j = \nu]\, P_i[X_n = j | T_j = \nu]. \end{aligned}$$

But, by the strong Markov property (Theorem 1.3), we see that

$$P_i[X_n = j | T_j = \nu] = P_i[X_n = j | X_\nu = j],$$

which, according to homogeneity, is equal to $p_{jj}(n - \nu)$. □

Let $f_{ij}(0) = 0$ for all $i,\ j \in \mathcal{N}$, and define

$$f_{ij}^*(z) = \sum_{n=0}^{\infty} f_{ij}(n) z^n$$

for all real z for which the series converges. The function $f_{ij}^*(z)$ with respect to z is called the *generating function* of $f_{ij}(n)$. Some elementary properties of generating functions are summarized in Appendix B.1. The generating function of $p_{ij}(n)$ is given by

$$p_{ij}^*(z) = \sum_{n=0}^{\infty} p_{ij}(n) z^n$$

for all real z for which the series converges. Since $0 \le p_{ij}(n),\ f_{ij}(n) \le 1$, the above two generating functions exist for $|z| < 1$. When $i \neq j$, Theorem 2.1 yields

$$\begin{aligned} p^*_{ij}(z) &= \sum_{n=0}^{\infty} z^n \sum_{\nu=0}^{n} f_{ij}(\nu)\, p_{jj}(n-\nu) \\ &= \sum_{\nu=0}^{\infty} z^{\nu} f_{ij}(\nu) \sum_{n=\nu}^{\infty} z^{n-\nu} p_{jj}(n-\nu), \end{aligned}$$

where the second equality follows by the interchange of the summations, which is allowed by Theorem B.1. Thus,

$$p^*_{ij}(z) = f^*_{ij}(z)\, p^*_{jj}(z), \quad |z| < 1, \quad i \neq j. \tag{2.2}$$

Similarly (see Exercise 2.2), we have

$$p^*_{jj}(z) - 1 = f^*_{jj}(z)\, p^*_{jj}(z), \quad |z| < 1, \tag{2.3}$$

or

$$p^*_{jj}(z) = \frac{1}{1 - f^*_{jj}(z)}, \quad |z| < 1, \tag{2.4}$$

which is valid since $|f^*_{jj}(z)| < 1$ if $|z| < 1$.

Lemma 2.1 *For real numbers a_n, $n = 0, 1, \cdots$, let $\phi(z) = \sum_{n=0}^{\infty} a_n z^n$ for $|z| < 1$. If $a_n \ge 0$ and $\lim_{z\to 1-} \phi(z) = a \le \infty$, then*

$$\sum_{n=0}^{\infty} a_n = a,$$

where $\lim_{z\to 1-}$ means that z approaches 1 from values less than 1.

Proof. Since $\sum_{n=0}^{\infty} a_n z^n \le \sum_{n=0}^{\infty} a_n$ for $0 < z < 1$, the case where $a = \infty$ is trivial. Suppose $a < \infty$. Then, $\sum_{n=0}^{\infty} a_n z^n < a < \infty$ for $0 < z < 1$. Hence, $\sum_{n=0}^{N} a_n \le a$ for all N. Since $\sum_{n=0}^{N} a_n$ is bounded, monotonically increasing in N, it has a finite limit, b say. But from Abel's theorem (see Theorem B.2), we conclude that $b = a$, completing the proof. □

The next result is a special case of a deeper result, called *Fubini's theorem*. Suppose that for a pair (i, j) of nonnegative integers, one has an element x_{ij} in R. The (m, n)th *partial sum* is defined by

$$s_{mn} = \sum_{j=0}^{n} \sum_{i=0}^{m} x_{ij}.$$

The *double series* $\sum_{i,j} x_{ij}$ is said to converge to $x \in R$ if for any $\varepsilon > 0$ there exists an integer $N(\varepsilon)$ such that $|s_{mn} - x| < \varepsilon$ for all m, $n \ge N(\varepsilon)$. The proof can be found in any standard textbook on analysis, e.g., Bartle (page 311, 1976).

Lemma 2.2 (i) *If* $x_{ij} \geq 0$ *for all* $i,\ j \in \mathcal{Z}_+$, *then*

$$\sum_{i,j} x_{ij} = \sum_{i=0}^{\infty}\sum_{j=0}^{\infty} x_{ij} = \sum_{j=0}^{\infty}\sum_{i=0}^{\infty} x_{ij},$$

which can diverge.

(ii) *If one of the iterated series*

$$\sum_{i=0}^{\infty}\sum_{j=0}^{\infty} |x_{ij}|, \quad \sum_{j=0}^{\infty}\sum_{i=0}^{\infty} |x_{ij}|$$

is convergent, then the other is also convergent and

$$\sum_{i,j} x_{ij} = \sum_{i=0}^{\infty}\sum_{j=0}^{\infty} x_{ij} = \sum_{j=0}^{\infty}\sum_{i=0}^{\infty} x_{ij}.$$

Let N_j be the number of visits to state j for the Markov chain $\{X_n\}$, so that

$$N_j = \sum_{n=0}^{\infty} \mathrm{I}_{\{X_n = j\}},$$

where I_A denotes the indicator function of event A. For a given $\omega \in \Omega$, if $N_j(\omega)$ is finite, then the Markov chain $\{X_n\}$ eventually leaves state j never to return. If $N_j(\omega) = \infty$, $\{X_n\}$ visits state j repeatedly. Since

$$p_{ij}(n) = P_i[X_n = j] = E_i[\mathrm{I}_{\{X_n = j\}}],$$

it follows that

$$\sum_{n=0}^{\infty} p_{ij}(n) = \sum_{n=0}^{\infty} E_i[\mathrm{I}_{\{X_n = j\}}] = E_i[N_j], \tag{2.5}$$

where the second equality follows by Fubini's theorem. We then have

$$p_{ij}^* \equiv p_{ij}^*(1) = E_i[N_j], \quad i,\ j \in \mathcal{N},$$

the expected number of visits to state j starting from state i. Also,

$$f_{ij}^* \equiv f_{ij}^*(1) = \sum_{n=1}^{\infty} f_{ij}(n) = P_i[T_j < \infty], \quad i,\ j \in \mathcal{N}.$$

It follows from (2.2), Abel's theorem and Lemma 2.1 that $p_{ij}^* = f_{ij}^*\, p_{jj}^*$ or

$$E_i[N_j] = P_i[T_j < \infty]\, E_j[N_j], \quad i,\ j \in \mathcal{N}, \tag{2.6}$$

while, from (2.4), we have $p_{jj}^* = (1 - f_{jj}^*)^{-1}$ or

$$E_j[N_j] = \frac{1}{1 - P_j[T_j < \infty]}, \quad j \in \mathcal{N}, \tag{2.7}$$

where the right-hand side of (2.7) is interpreted as ∞ if $f_{jj}^* = 1$. Therefore,

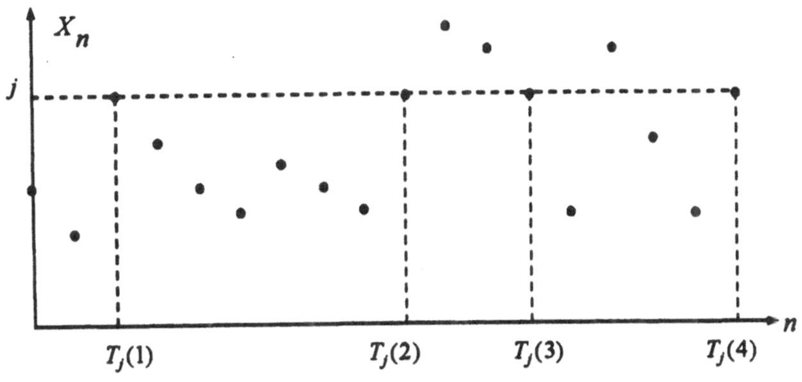

Figure 2.1 *The successive indices of visits.*

as a result of (2.7), we observe that $f_{jj}^* = 1$ if and only if the expected number of returns to state j is infinite.

Let $\mathbf{N}$ be a square matrix with components $E_i[N_j]$, $i, j \in \mathcal{N}$. Then, from (2.5), we have

$$\mathbf{N} = \mathbf{I} + \mathbf{P} + \mathbf{P}^2 + \cdots,$$

whence

$$\mathbf{NP} = \mathbf{PN} = \mathbf{N} - \mathbf{I},$$

or

$$\mathbf{N}(\mathbf{I} - \mathbf{P}) = (\mathbf{I} - \mathbf{P})\mathbf{N} = \mathbf{I}. \tag{2.8}$$

It should be noted that the matrix $(\mathbf{I} - \mathbf{P})$ may not be invertible.

Let $T_j(1), T_j(2), \cdots$ be the successive indices $n \geq 1$ for which $X_n = j$ as long as there are such n. In the case where $N_j(\omega) = m < \infty$, we define $T_j(k+1) - T_j(k) = \infty$ for all $k \geq m$. If $N_j(\omega) = \infty$, there are infinitely many such n. Figure 2.1 depicts the successive indices of visits to state j. Note that $T_j = T_j(1)$ in the preceding notation. It is clear that, as for the first passage time T_j, every $T_j(k)$ is a stopping time. Hence, by the strong Markov property, the future after $T_j(k)$ is independent of the past before $T_j(k)$ given $X_{T_j(k)}$. On the other hand, we have $X_{T_j(k)} = j$ on event $\{T_j(k) < \infty\}$, by definition. Hence, every time $\{X_n\}$ visits state j, the past loses all its influence on the future. In particular, we have the next result. It can be proved by mimicking the proof of Theorem 1.3 (see Exercise 2.3).

Theorem 2.2 *Consider the event $\{T_j(k) < \infty\}$ for some $k \geq 1$. Then*

$$P_i[T_j(k+1) - T_j(k) = m | T_j(1), \cdots, T_j(k) = n] = f_{jj}(m)$$

for $m = 1, 2, \cdots$.

An immediate consequence is the following.

Corollary 2.1 *Let* N_j *be the number of visits to state* j. *Then*

$$P_j[N_j = m] = (1-\rho)\rho^{m-1}, \quad m = 1, 2, \cdots,$$

where $\rho = f^*_{jj}$.

Proof. By definition, $N_j = m$ if and only if $T_j(k) < \infty$ for $k \le m$ and $T_j(m+1) = \infty$. According to Theorem 2.2, the events

$$\{T_j(k+1) - T_j(k) < \infty\}, \ k < m, \ \text{and} \ \{T_j(m+1) - T_j(m) = \infty\}$$

are independent and their probabilities are given by f^*_{jj} and $1 - f^*_{jj}$ respectively. □

From Corollary 2.1 above, we conclude that $P_j[N_j < \infty] = 1$ if and only if $f^*_{jj} \equiv P_j[T_j < \infty] < 1$ and that $P_j[N_j = \infty] = 1$ if and only if $f^*_{jj} = 1$. If $N_j = \infty$ with probability one, then $E_j[N_j] = \infty$. From (2.7) or, of course, from Corollary 2.1, we know that $E_j[N_j] = \infty$ if and only if $f^*_{jj} = 1$. Therefore, in this case, $E_j[N_j] = \infty$ if and only if the actual number of returns to state j is infinite with probability one.

2.2 Classification of states

The classification of states is a starting point of the theory of Markov chains. In this section, we provide results that are essential for later developments in this chapter.

Our first classification is fundamental and concerns the crucial issues of whether or not return to a state is certain and, when return is certain, whether the mean time to return is finite or infinite. Throughout this section, we use the same notation as in Section 2.1, unless otherwise specified.

Definition 2.1 (i) State j is called *recurrent* if $f^*_{jj} = 1$. If $f^*_{jj} < 1$ then j is called *transient*.

(ii) A recurrent state j is said to be *positive* if $E_j[T_j] < \infty$. Otherwise, it is called *null*.

State j is recurrent if and only if, starting from j, the probability of returning to j is unity. Since $f^*_{jj} = 1$ if and only if $p^*_{jj} = \infty$, by (2.7), the expected number of returns to j is infinite. Thus, state j is recurrent if and only if $E_j[N_j] = \infty$, which is also equivalent to $P_j[N_j = \infty] = 1$. On the other hand, if state j is transient, then $f^*_{jj} < 1$ and there is a positive probability $1 - f^*_{jj}$ of never returning to j (see Corollary 2.1). In this case, the total number of returns to state j is finite with probability one, and (2.7) implies that the expected number of returns to j is also finite. Thus, state j is transient if and only if $E_j[N_j] < \infty$, which is equivalent to $P_j[N_j < \infty] = 1$.

To clarify the classification given in Definition 2.1, we consider the random walk given in Example 1.14.

Example 2.1 Let $\{X_n\}$ be the random walk considered in Example 1.14. Suppose that $p_i = p > 0$, $q_i = q > 0$ and $r_i = 0$ for all i. That is, at each transition the particle moves upward with probability p and downward with probability $q = 1 - p$. It is readily seen that

$$p_{00}(2n) = \frac{(2n)!}{n!\,n!}(pq)^n$$

and $p_{00}(2n+1) = 0$ for all $n = 0, 1, \cdots$. Using Stirling's formula (see Exercise 2.5), i.e.

$$n! \approx n^n e^{-n}\sqrt{2\pi n},$$

we have for sufficiently large n

$$p_{00}(2n) \approx \frac{(4pq)^n}{\sqrt{\pi n}}.$$

Note that $4pq \le 1$ and an equality holds if and only if $p = q = 1/2$. Hence $p^*_{00} = \sum_{n=0}^{\infty} p_{00}(n) = \infty$ if and only if $p = 1/2$. This argument applies to every state. Therefore, from the above discussions, we conclude that every state of the random walk is recurrent if and only if $p = q = 1/2$.

Now suppose $p = q = 1/2$. Let $\mu_i = E_i[T_0]$, where T_0 denotes the first passage time to state 0. The mean return time to state 0 satisfies the recursive relationship

$$\mu_0 = \frac{1}{2}(1 + \mu_1) + \frac{1}{2}(1 + \mu_{-1}) = 1 + \mu_1,$$

where the first equality follows from first step analysis and the second from symmetry. On the other hand, first step analysis also yields

$$\mu_1 = \frac{1}{2} + \frac{1}{2}(1 + \mu_2).$$

But, by the spatial homogeneity, we have $\mu_2 = 2\mu_1$. Hence $\mu_1 = 1 + \mu_1$, which is impossible unless $\mu_1 = \infty$. Therefore, if $p = q = 1/2$, then the mean return time $E_0[T_0]$ is infinity and state 0 is null recurrent.

If state j is transient, then we have $E_i[N_j] < \infty$ for all $i \in \mathcal{N}$ since, by (2.6),

$$E_i[N_j] = f^*_{ij} E_j[N_j] \le E_j[N_j] < \infty.$$

By (2.5), $E_i[N_j]$ is the sum of $p_{ij}(n)$ with respect to $n \ge 0$. Therefore $E_i[N_j]$ can be finite only if $p_{ij}(n) \to 0$ as $n \to \infty$. If j is recurrent, then $E_j[N_j] = \infty$ and we are unable to decide whether or not the transition probability $p_{jj}(n)$ tends to zero. As we shall see later, it turns out that if j is null recurrent, then the fact that the expected time between successive returns is infinite implies $p_{jj}(n) \to 0$ as $n \to \infty$.

We thus have the following.

Theorem 2.3 *If state j is transient or null recurrent, then*

$$\lim_{n\to\infty} p_{ij}(n) = 0$$

for every $i \in \mathcal{N}$.

In the random walk case given in Example 2.1, we saw that $p_{jj}(2n) > 0$ and $p_{jj}(2n+1) = 0$ for $n = 0, 1, \cdots$, which shows that there is a periodic structure. On the other hand, if a self-transition is allowed for a single state i, then the random walk with initial state j can reach state i and remain there for any length of time before returning to state j, so that there is no periodicity. Thus, recurrent states are classified further as follows.

Definition 2.2 A recurrent state j is said to be *periodic with period d*, if $d \geq 2$ is the greatest common divisor of all integers $n \geq 1$ for which $p_{jj}(n) > 0$.* If there is no such $d \geq 2$, then j is called *aperiodic*.

State j is said to be *accessible* from state i, in which case we write $i \to j$, if there is some integer $n \geq 0$ such that $p_{ij}(n) > 0$. Thus, for $i \neq j$, $i \to j$ if and only if $p^*_{ij} = \sum_{n=0}^{\infty} p_{ij}(n) > 0$ and hence if and only if $f^*_{ij} > 0$. Accessibility does not concern the actual value of the probability, but only whether or not it is zero. Note that in order for $i \to j$, there must exist a sequence $\{i_1, i_2, \cdots, i_n\}$ of states such that $p_{i,i_1} p_{i_1,i_2} \cdots p_{i_n,j} > 0$.

Definition 2.3 (i) A set of states is called *closed* if no state outside it is accessible from any state inside it.

(ii) A state forming a closed set by itself is called *absorbing*.

(iii) A closed set is said to be *irreducible* if no proper subset of it is closed.

(iv) A Markov chain is called *irreducible* if its only closed set is the set of all states.

Some immediate consequences of Definition 2.3 are the following. State j is absorbing if and only if $p_{jj} = 1$. A Markov chain is irreducible if and only if all states are accessible from one another. If C is a closed set, then deleting from the transition matrix $\mathbf{P}$ those rows and columns corresponding to states not in C leaves another stochastic matrix $\widetilde{\mathbf{P}}$. If C consists of a single closed set, that is, C is irreducible, then the chain restricted to C is an irreducible Markov chain with the transition matrix $\widetilde{\mathbf{P}}$.

If two states i and j are accessible from each other, i.e., both $i \to j$ and $j \to i$ hold, then the states i and j are said to *communicate* and we write $i \leftrightarrow j$. The concept of communication is an equivalence relation:

(reflexivity) $i \leftrightarrow i$;

(symmetry) If $i \leftrightarrow j$ then $j \leftrightarrow i$;

(transitivity) If $i \leftrightarrow j$ and $j \leftrightarrow k$ then $i \leftrightarrow k$.

* If $p_{jj}(n) = 0$ for all $n \geq 1$, we define $d = 0$.

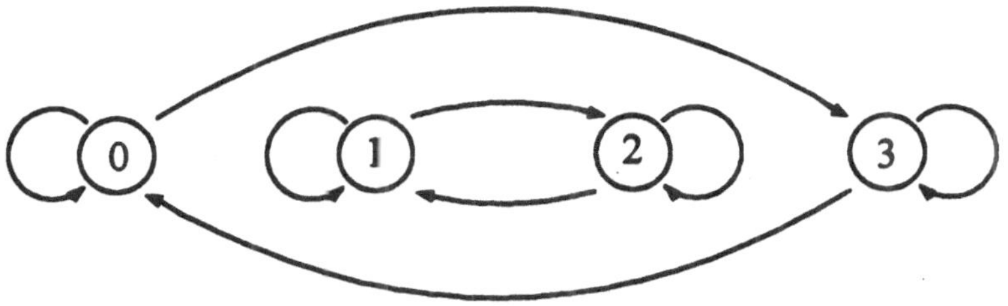

Figure 2.2 *A transition diagram.*

Since $p_{ii}(0) = 1 > 0$, reflexivity holds. Symmetry follows at once from the definition of communication. The proof of transitivity hinges on the next result, whose proof is left to the reader (see Exercise 2.7).

Lemma 2.3 *If $i \to j$ and $j \to k$ then $i \to k$.*

We partition the state space into equivalence classes based on communication. The states in an equivalence class are those which communicate with each other. It is possible, starting from one class, to enter some other class with positive probability. However, it is not possible to return to the initial class; for otherwise, the states in the two classes communicate so that they form a single class. Hence irreducibility of a Markov chain can be phrased as it has a single equivalence class. The next two examples illustrate the concept.

Example 2.2 Consider a Markov chain with state space $\{0, 1, 2, 3\}$ and transition matrix

$$\mathbf{P} = \begin{pmatrix} 1/2 & 0 & 0 & 1/2 \\ 0 & 1/2 & 1/2 & 0 \\ 0 & 3/4 & 1/4 & 0 \\ 1/3 & 0 & 0 & 2/3 \end{pmatrix}.$$

Since the actual values of the transition probabilities are irrelevant to accessibility, the *transition diagram* corresponding to $\mathbf{P} = (p_{ij})$ may be more useful for identifying equivalence classes. The transition diagram depicted in Figure 2.2 is constructed in such a way that a directed arrow from state i to state j indicates that $p_{ij} > 0$. It is, then, obvious that the Markov chain has two classes $\{0, 3\}$ and $\{1, 2\}$, and it is reducible.

Example 2.3 Suppose an individual (player A) with fortune k plays a game against an adversary (player B), and has probability p_k of winning one unit, probability q_k of losing one unit, and probability r_k of a draw in each contest. Suppose that they start a series of contests with initial fortunes a and b, respectively, and let $N = a + b$. Let X_n represent player A's fortune after n contests, so that $N - X_n$ describes player B's fortune. Then the process $\{X_n\}$ is a random walk on the state space $\{0, 1, \cdots, N\}$. Note that once the process reaches either state 0 or state N, it remains there forever because those states represent *ruin* for one of the players. Player A is ruined when the process reaches state 0, while player B is ruined when the process is in state N. We are particularly interested in which player is ruined first. This problem is often termed the *gambler's ruin problem.*

The transition matrix takes the form

$$\mathbf{P} = \begin{pmatrix} 1 & 0 & 0 & 0 & \cdots & 0 \\ q_1 & r_1 & p_1 & 0 & \cdots & 0 \\ 0 & q_2 & r_2 & p_2 & \cdots & 0 \\ \vdots & & \ddots & \ddots & \ddots & \vdots \\ 0 & \cdots & 0 & q_{N-1} & r_{N-1} & p_{N-1} \\ 0 & \cdots & 0 & 0 & 0 & 1 \end{pmatrix}.$$

The two states 0 and N act as *absorbing* states. It is easy to see that the random walk has three classes $\{0\}$, $\{1, \cdots, N-1\}$, and $\{N\}$. It is possible, starting from the second class, to enter the other classes, but it is not possible to return to the second class from the other classes.

Before proceeding to our main result of this section, we state a useful lemma.

Lemma 2.4 *If state j is recurrent and $j \to i$, then state i is recurrent.*

Proof. Under the condition, we have $i \to j$, since, otherwise, j cannot be recurrent. Hence, there is an integer ℓ such that $p_{ij}(\ell) > 0$. Let m be an integer such that $p_{ji}(m) > 0$ and let $a \equiv p_{ij}(\ell)p_{ji}(m) > 0$. The Chapman–Kolmogorov equation is

$$p_{ij}(m+n) = \sum_k p_{ik}(m)p_{kj}(n), \quad m,\ n = 0, 1, \cdots. \tag{2.9}$$

Repeated application of (2.9) yields

$$p_{ii}(\ell + n + m) \geq p_{ij}(\ell)p_{jj}(n)p_{ji}(m) = a\,p_{jj}(n). \tag{2.10}$$

It follows that

$$p_{ii}^* \geq \sum_{n=0}^{\infty} p_{ii}(\ell + n + m) \geq a \sum_{n=0}^{\infty} p_{jj}(n) = a\,p_{jj}^*.$$

Since j is recurrent, we have $p_{jj}^* = \infty$. Also, $a > 0$. Hence $p_{ii}^* = \infty$, so that state i is recurrent. □

Here is the main result of this section.

Theorem 2.4 *Suppose $i \leftrightarrow j$.*

(i) *If state i is positive recurrent then state j is positive recurrent.*

(ii) *If state i is null recurrent then state j is null recurrent.*

(iii) *If state i is transient then state j is transient.*

(iv) *If state i is aperiodic then state j is aperiodic. If state i is periodic with period $d \geq 2$ then state j is periodic with period d.*

Proof. If i is recurrent, then, by Lemma 2.4, j is recurrent. The proof of Lemma 2.4 in conjunction with the assumption $i \leftrightarrow j$ also implies that if i is transient then j is transient. Now, suppose i is null recurrent so that, from Theorem 2.3, $\lim_{n\to\infty} p_{ii}(n) = 0$. Let ℓ, m and a be defined as in the proof of Lemma 2.4. Then, from (2.10), we also have $\lim_{n\to\infty} p_{jj}(n) = 0$. Since j is recurrent, this can happen only if it is null recurrent, proving parts (i)–(iii).

To prove part (iv), suppose i is periodic with period d. Then, since

$$p_{ii}(\ell + m) \geq p_{ij}(\ell)\, p_{ji}(m) = a > 0,$$

and since i has period d, $\ell + m$ must be a multiple of d. If n is not a multiple of d then neither is $\ell + n + m$. Hence, from (2.10),

$$0 = p_{ii}(\ell + m + n) \geq a\, p_{jj}(n) \geq 0,$$

so that $p_{jj}(n) = 0$ whenever n is not a multiple of d. It follows that j must be periodic with period $d' \geq d$. But, if j is periodic with period d', then, reversing the roles of j and i, we conclude that i is periodic with period $d \geq d'$. Thus $d = d'$ and the proof is complete. □

Theorem 2.4 reveals that positive recurrence, null recurrence, transience, and periodicity are all *class properties*, i.e., all the states belonging to the same class have the same property. This property is often referred to as the *solidarity property*. We therefore have the following.

Corollary 2.2 *For an irreducible Markov chain, either all states are positive recurrent, or all are null recurrent, or all are transient. Either all recurrent states are aperiodic or all are periodic with the same period.*

Definition 2.4 An irreducible Markov chain is called *positive recurrent* (*null recurrent* or *transient*, respectively) if all the states are positive recurrent (null recurrent or transient). A recurrent Markov chain is called *aperiodic* or *periodic with period d* if all the states are aperiodic or periodic with period d respectively.

For a finite Markov chain, Theorem 2.4 yields the following important result.

Corollary 2.3 *For a finite Markov chain $\{X_n\}$, no state is null recurrent, and not all states are transient.*

Proof. If some state were null recurrent, then all the states in the same class, C say, would be null recurrent, by Theorem 2.4, so that, by Theorem 2.3, $\lim_{n\to\infty} p_{ij}(n) = 0$ for all $i, j \in C$. Now, for each $i \in C$, we have $\sum_{j\in C} p_{ij}(n) = 1$ for all $n = 0, 1, \cdots$, since any recurrent class is closed (why?). Hence

$$1 = \lim_{n\to\infty} \sum_{j\in C} p_{ij}(n) = \sum_{j\in C} \lim_{n\to\infty} p_{ij}(n) = 0,$$

where interchanging the summation and the limit is allowed because the state space is finite. Hence, there can be no null recurrent states. A similar proof holds for the transient case. □

According to Corollary 2.3, there must exist at least one positive recurrent class for any finite Markov chain. If a finite Markov chain has r positive recurrent classes, its transition matrix $\mathbf{P}$ has the canonical form

$$\mathbf{P} = \begin{pmatrix} \mathbf{P}_1 & \mathbf{O} & \cdots & \mathbf{O} & \mathbf{O} \\ \mathbf{O} & \mathbf{P}_2 & \cdots & \mathbf{O} & \mathbf{O} \\ \vdots & \vdots & \ddots & \vdots & \vdots \\ \mathbf{O} & \mathbf{O} & \cdots & \mathbf{P}_r & \mathbf{O} \\ \mathbf{T}_1 & \mathbf{T}_2 & \cdots & \mathbf{T}_r & \mathbf{T} \end{pmatrix}, \tag{2.11}$$

where each $\mathbf{P}_i$, $i = 1, \cdots, r$, denotes a stochastic matrix corresponding to the recurrent class i, $\mathbf{T}_i$ is a substochastic matrix of transition probabilities from the class of transient states to the recurrent class i, and $\mathbf{T}$ is a strictly substochastic matrix of transition probabilities within the transient class. For an irreducible finite Markov chain, we have the following result.

Corollary 2.4 *For a finite, irreducible Markov chain, there is only one positive recurrent class and no other classes.*

At this point, we provide a result that is useful in determining whether a given Markov chain is recurrent or transient. A more powerful result can be found in Meyn and Tweedie (Chapter 8, 1993).

Theorem 2.5 *Let $\{X_n\}$ be an irreducible Markov chain with state space $\mathcal{Z}_+$ and transition matrix $\mathbf{P} = (p_{ij})$. Then $\{X_n\}$ is transient if and only if the system of equations*

$$\sum_{j=0}^{\infty} p_{ij} y_j = y_i, \quad i = 1, 2, \cdots,$$

has a bounded, nonconstant solution.

Proof. Suppose $\{X_n\}$ is transient. Then $f_{i0}^* < 1$ for some $i \neq 0$ since, otherwise, state 0 would be recurrent. From (2.1), we have

$$f_{i0}^* - p_{i0} = \sum_{n=2}^{\infty} \sum_{j=1}^{\infty} p_{ij} f_{j0}(n-1) = \sum_{j=1}^{\infty} p_{ij} f_{j0}^*,$$

where the second equality follows by Lemma 2.2(i). Now let $y_0 = 1$ and $y_i = f^*_{i0}$ for $i = 1, 2, \cdots$. This $\{y_i\}$ is the desired bounded, nonconstant solution. The proof of the converse is left to the reader (Exercise 2.26). □

Note that if $\{y_i\}$ is a solution of the system in Theorem 2.5, then $\{y_i - c\}$ is also a solution of the system for any constant c, since

$$\sum_{j=0}^{\infty} p_{ij}(y_j - c) = \sum_{j=0}^{\infty} p_{ij} y_i - c \sum_{j=0}^{\infty} p_{ij} = y_i - c.$$

Hence, defining $y_0 = 0$, Theorem 2.5 can be restated as follows. A given Markov chain is transient if and only if the system of equations

$$\sum_{j=1}^{\infty} p_{ij} y_j = y_i, \quad i = 1, 2, \cdots, \tag{2.12}$$

has a bounded, nonconstant solution.

Example 2.4 Liu (1994) considered a Markov chain with state space $\mathcal{Z}_+$ and transition matrix

$$\mathbf{P} = \begin{pmatrix} q_0 & p_0 & 0 & 0 & \cdots \\ q_1 & q_1 & p_1 & 0 & \cdots \\ q_2 & q_2 & q_2 & p_2 & \cdots \\ \vdots & \vdots & \vdots & \vdots & \ddots \end{pmatrix},$$

where $p_i q_i > 0$ and $p_i + (i+1)q_i = 1$ for $i = 0, 1, \cdots$. To determine whether the Markov chain is transient or recurrent, we can use Theorem 2.5. Consider the system of equations

$$y_i = q_i \sum_{j=1}^{i} y_j + p_i\, y_{i+1}, \quad i = 1, 2, \cdots.$$

It is not difficult to show that

$$y_{n+1} = \left(1 + \sum_{j=1}^{n} \frac{q_j}{p_1 \cdots p_j}\right) y_1, \quad n = 1, 2, \cdots.$$

Hence, if $y_1 \neq 0$ and

$$\sum_{j=1}^{\infty} \frac{q_j}{p_1 \cdots p_j} < \infty,$$

then the solution $\{y_i\}$ is bounded and nonconstant.

Finally, we consider periodic Markov chains. The next theorem provides the canonical form for a periodic transition matrix. The proof may be found in, e.g., Çinlar (1975) or Seneta (1981).

Theorem 2.6 *Let $\{X_n\}$ be a recurrent Markov chain with transition matrix* $\mathbf{P}$*. If $\{X_n\}$ has period d, then, after rearranging states appropriately,* $\mathbf{P}$ *can be written as*

$$\mathbf{P} = \begin{pmatrix} \mathbf{O} & \mathbf{P}_1 & \mathbf{O} & \cdots & \mathbf{O} \\ \mathbf{O} & \mathbf{O} & \mathbf{P}_2 & \cdots & \mathbf{O} \\ \vdots & \vdots & \vdots & \ddots & \vdots \\ \mathbf{O} & \mathbf{O} & \mathbf{O} & \cdots & \mathbf{P}_{d-1} \\ \mathbf{P}_d & \mathbf{O} & \mathbf{O} & \cdots & \mathbf{O} \end{pmatrix},$$

where each $\mathbf{P}_i$ denotes a stochastic matrix.

The next example illustrates the result of Theorem 2.6.

Example 2.5 Consider a Markov chain with state space $\{0, 1, 2, 3\}$ and transition matrix

$$\mathbf{P} = \begin{pmatrix} 0 & 1 & 0 & 0 \\ q & 0 & p & 0 \\ 0 & q & 0 & p \\ 0 & 0 & 1 & 0 \end{pmatrix}, \quad p + q = 1, \quad p,\ q > 0.$$

Clearly, the Markov chain has period 2. After rearranging the states as $\{0, 2, 1, 3\}$, the transition matrix can be written as

$$\widetilde{\mathbf{P}} = \begin{pmatrix} 0 & 0 & 1 & 0 \\ 0 & 0 & q & p \\ q & p & 0 & 0 \\ 0 & 1 & 0 & 0 \end{pmatrix}.$$

Note that

$$\widetilde{\mathbf{P}}^2 = \begin{pmatrix} q & p & 0 & 0 \\ q^2 & p + pq & 0 & 0 \\ 0 & 0 & q + pq & p^2 \\ 0 & 0 & q & p \end{pmatrix},$$

so $\widetilde{\mathbf{P}}^2$ has two classes and is reducible.

2.3 Recurrent Markov chains

In this section, we consider recurrent Markov chains on the state space $\mathcal{N} = \mathcal{Z}_+$. Finite Markov chains are treated in the next section. Recall that recurrence is a class property. Since any set of recurrent states forms a closed set, it is sufficient to consider one irreducible recurrent set of states. We shall assume throughout this section that the Markov chain under consideration is irreducible and recurrent. In the following, we let $\{X_n\}$ be such a Markov chain with state space $\mathcal{Z}_+$ and transition matrix $\mathbf{P} = (p_{ij})$.

Recall from Theorem 2.3 that $\lim_{n\to\infty} p_{ij}(n) = 0$ if state j is transient or null recurrent. Our first result concerns the limit of $p_{ij}(n)$ as $n \to \infty$ for recurrent states. Two important lemmas are needed.

Lemma 2.5 *Let $\{a_n\}$ and $\{b_n\}$ be nonnegative sequences such that*

$$\sum_{n=0}^{\infty} a_n = 1, \quad \sum_{n=0}^{\infty} b_n < \infty.$$

Suppose that the greatest common divisor of all integers $n \geq 1$ for which $a_n > 0$ is 1. If the renewal equation

$$u_n - \sum_{k=0}^{n} a_{n-k} u_k = b_n, \quad n = 0, 1, \cdots,$$

is satisfied by a bounded sequence $\{u_n\}$ of real numbers, then the limit $\lim_{n\to\infty} u_n$ exists and

$$\lim_{n\to\infty} u_n = \frac{\sum_{n=0}^{\infty} b_n}{\sum_{n=0}^{\infty} n a_n},$$

where the right-hand side is interpreted as 0 if $\sum_{n=0}^{\infty} n a_n = \infty$.

Proof. We assume that the limit $\lim_{n\to\infty} u_n$ exists. A proof of the existence is beyond the scope of this book (see, e.g., Karlin and Taylor, Chapter 3, 1975). Let $u(z) = \sum_{n=0}^{\infty} u_n z^n$ for $|z| < 1$, i.e. the power series of $\{u_n\}$. The power series $a(z)$ and $b(z)$ for $\{a_n\}$ and $\{b_n\}$, respectively, are defined similarly. If $\{u_n\}$ satisfies the renewal equation, Theorem B.4 yields

$$u(z) = \frac{b(z)}{1 - a(z)}, \quad |z| < 1.$$

Under the assumptions, Theorem B.5 implies that

$$\lim_{n\to\infty} u_n = \lim_{z\to 1-} (1 - z) u(z) = \lim_{z\to 1-} \frac{(1 - z) b(z)}{1 - a(z)}.$$

The desired result follows from L'Hospital's rule and Lemma 2.1. □

The next result is referred to as the *dominated convergence theorem*. See, e.g., Williams (1991) for a more general version of the dominated convergence.

Lemma 2.6 *For real numbers $a_i(n)$ defined on $\mathcal{Z}_+ \times \mathcal{Z}_+$, suppose that the limit $a_i = \lim_{n\to\infty} a_i(n)$ exists for each i, and that $\sum_i a_i b_i$ and $\sum_i a_i(n) b_i$, $n \in \mathcal{Z}_+$, exist, where b_i is nonnegative. If there exists a sequence $\{M_i\}$ such that $|a_i(n)| \leq M_i$ for all n and that $\sum_i M_i b_i < \infty$, then*

$$\lim_{n\to\infty} \sum_{i=0}^{\infty} a_i(n) b_i = \sum_{i=0}^{\infty} a_i b_i.$$

Proof. For any $\varepsilon > 0$, let K be such that $\sum_{i=K}^{\infty} M_i b_i < \varepsilon$, from which one obtains

$$\left|\sum_{i=K}^{\infty} a_i(n) b_i\right| < \varepsilon, \quad n \in \mathcal{Z}_+.$$

Since $a_i = \lim_{n\to\infty} a_i(n)$, we have

$$\lim_{n\to\infty} \sum_{i=0}^{K} a_i(n) b_i = \sum_{i=0}^{K} a_i b_i.$$

It follows that

$$\left|\sum_{i=0}^{\infty} a_i(n) b_i - \sum_{i=0}^{\infty} a_i b_i\right| \le \left|\sum_{i=0}^{K} a_i(n) b_i - \sum_{i=0}^{K} a_i b_i\right| + 2\varepsilon,$$

which is less than 3ε for sufficiently large n. □

The following consequence of the dominated convergence theorem is often useful and is called the *monotone convergence theorem.*

Corollary 2.5 *For $a_i(n)$ defined on $\mathcal{Z}_+ \times \mathcal{Z}_+$, suppose that $a_i(n)$ is monotone in n for each i and that $\sum_i a_i(n) b_i$, $n \in \mathcal{Z}_+$, exist, where b_i is non-negative. Define $a_i = \lim_{n\to\infty} a_i(n)$. If $\sum_i a_i b_i$ exists then*

$$\lim_{n\to\infty} \sum_{i=0}^{\infty} a_i(n) b_i = \sum_{i=0}^{\infty} a_i b_i.$$

Proof. Suppose that $a_i(n)$ is monotonically increasing, i.e., $a_i(0) \le a_i(1) \le \cdots \le a_i$. Then, assuming $a_i(0) \ge 0$, we can take $M_i = a_i$ in Lemma 2.6 and the result follows. The other case follows similarly. □

Let T_j be the first passage time to state j and denote by $\mu_j \equiv E_j[T_j]$, the *mean return time* to state j. The next theorem is often referred to as the *basic limit theorem.*

Theorem 2.7 *Let $\{X_n\}$ be aperiodic. Then*

$$\lim_{n\to\infty} p_{ij}(n) = \frac{1}{\mu_j}$$

for all i, $j \in \mathcal{N}$, where the right-hand side is interpreted as 0 if $\mu_j = \infty$.

Proof. Suppose $i = j$. Then, by Theorem 2.1, we identify $u_n = p_{jj}(n)$, $a_n = f_{jj}(n)$ and $b_n = \delta_{n0}$. Hence, applying Lemma 2.5, we have

$$\lim_{n\to\infty} p_{jj}(n) = \frac{1}{\sum_{n=0}^{\infty} n\, f_{jj}(n)} = \frac{1}{E_j[T_j]}.$$

Next suppose $i \ne j$. Then, by Theorem 2.1 again,

$$p_{ij}(n) = \sum_{k=1}^{\infty} f_{ij}(k)\, p_{jj}(n-k),$$

where $p_{jj}(n) = 0$ for all $n < 0$. Since $p_{jj}(n-k)$ is bounded and convergent as $n \to \infty$, the dominated convergence theorem allows one to interchange the limit and the summation to yield

$$\lim_{n\to\infty} p_{ij}(n) = \lim_{n\to\infty} \sum_{k=1}^{\infty} f_{ij}(k) p_{jj}(n-k) = \sum_{k=1}^{\infty} \frac{f_{ij}(k)}{E_j[T_j]}.$$

The result follows since $\sum_{k=1}^{\infty} f_{ij}(k) = 1$ for recurrent states. □

An interpretation of the basic limit theorem is that the long-run frequency of the Markov chain being in state j is the inverse of the mean return time to state j. If the Markov chain $\{X_n\}$ is positive recurrent and aperiodic, then $\mu_j = E_j[T_j] < \infty$, so that the limiting probability $\lim_{n\to\infty} p_{ij}(n)$ is positive, by Theorem 2.7. If $\{X_n\}$ is null recurrent and aperiodic, Theorem 2.7 implies $\lim_{n\to\infty} p_{ij}(n) = 0$. In the case where $\{X_n\}$ is recurrent and periodic with period $d \geq 2$, we can still show that

$$\lim_{n\to\infty} p_{jj}(nd) = \frac{d}{\mu_j}$$

(see Exercise 2.11). Since $p_{ij}(nd+m) = 0$ for $m = 1, \cdots, d-1$, if $\{X_n\}$ is null recurrent then the limit of $p_{ij}(n)$ as $n \to \infty$ exists and is equal to 0 whether or not the chain is aperiodic. This proves Theorem 2.3 for the null recurrent case.

Recall that if state j is transient then $P_j[T_j = \infty] > 0$ so that $\mu_j = E_j[T_j] = \infty$. Therefore, the result given in Theorem 2.7 is consistent for all cases. That is, if $\{X_n\}$ is irreducible and aperiodic, then

$$\lim_{n\to\infty} p_{ij}(n) = \frac{1}{\mu_j}, \quad i,\ j \in \mathcal{N},$$

see Theorem 2.3.

Suppose that a Markov chain $\{X_n\}$ has an initial distribution $\boldsymbol{\alpha} = (\alpha_i)$. Then, from (1.21), the state distribution $\boldsymbol{\pi}(n) = (\pi_i(n))$ of $\{X_n\}$ is given by

$$\pi_j(n) \equiv P_\alpha[X_n = j] = \sum_{i\in\mathcal{N}} \alpha_i p_{ij}(n), \quad n = 0, 1, \cdots.$$

The next result is an immediate consequence of the dominated convergence theorem.

Corollary 2.6 *For a Markov chain $\{X_n\}$, suppose that $\lim_{n\to\infty} p_{ij}(n) = \pi_j \geq 0$ for all $i \in \mathcal{N}$. Then, for any initial distribution $\boldsymbol{\alpha}$, we have*

$$\lim_{n\to\infty} \pi_j(n) = \pi_j.$$

For real numbers a_n, $n = 0, 1, \cdots$, if the limit of $\sum_{k=0}^{n-1} a_k/n$ as $n \to \infty$ exists, it is called the *Cesàro limit* of the sequence $\{a_n\}$. It is well known (and, in fact, readily verified) that if the sequence $\{a_n\}$ has a limit, a say, then

$$\lim_{n\to\infty} \frac{1}{n+1} \sum_{k=0}^{n} a_k = a.$$

Hence, if the transition probability $p_{ij}(n)$ converges to π_j as $n \to \infty$, then

$$\lim_{n\to\infty} \frac{1}{n+1} \sum_{k=0}^{n} p_{ij}(k) = \pi_j. \tag{2.13}$$

Note that the Cesàro limit may exist without the ordinary limit of the sequence. That is, (2.13) does not imply the convergence of the transition probability $p_{ij}(n)$, as the periodic case shows.

Definition 2.5 A nonnegative, nonzero vector $\boldsymbol{\nu} = (\nu_i)$ is said to be *subinvariant* with respect to $\mathbf{P} = (p_{ij})$ if

$$\nu_i \geq \sum_{j=0}^{\infty} \nu_j p_{ji}, \quad i \in \mathcal{N}. \tag{2.14}$$

If the inequality is replaced by an equality, $\boldsymbol{\nu}$ is said to be *invariant* with respect to $\mathbf{P}$.

Lemma 2.7 *Let $\boldsymbol{\pi} = (\pi_i)$ be nonnegative and nonzero.*

(i) *If $\boldsymbol{\pi}$ is subinvariant with respect to $\mathbf{P}$, then $\boldsymbol{\pi}$ is strictly positive componentwise.*

(ii) *If, in addition, $\sum_{i=0}^{\infty} \pi_i < \infty$, then $\boldsymbol{\pi}$ is invariant with respect to $\mathbf{P}$.*

Proof. (i) Suppose $\pi_j > 0$. For any i, there is some n such that $p_{ji}(n) > 0$ since $j \to i$. Iterating (2.14), we have

$$\pi_i \geq \sum_{j=0}^{\infty} \pi_j p_{ji}(n),$$

whence $\pi_i \geq \pi_j p_{ji}(n) > 0$.

(ii) Suppose that a strict inequality holds in (2.14) for some i. Then, summing both sides over i and noting that $\sum_{j=0}^{\infty} p_{ij} = 1$, we obtain

$$\sum_{i=0}^{\infty} \pi_i > \sum_{i=0}^{\infty} \sum_{j=0}^{\infty} \pi_j p_{ji} = \sum_{j=0}^{\infty} \pi_j \sum_{i=0}^{\infty} p_{ji} = \sum_{j=0}^{\infty} \pi_j,$$

which is a contradiction, where the interchange of the summations is permissible by Fubini's theorem. □

Here is the main result of this section.

Theorem 2.8 *Suppose that the Markov chain $\{X_n\}$ with transition matrix $\mathbf{P} = (p_{ij})$ is irreducible and aperiodic. Then $\{X_n\}$ is positive recurrent if* *and only if there exists a probability vector* $\boldsymbol{\pi} = (\pi_i)$ *which is invariant with respect to* $\mathbf{P}$*, i.e.,*

$$\pi_i = \sum_{j=0}^{\infty} \pi_j p_{ji}, \quad i \in \mathcal{N}; \quad \sum_{i=0}^{\infty} \pi_i = 1. \tag{2.15}$$

If there is such a $\boldsymbol{\pi}$*, then it is strictly positive componentwise, there are no other solutions, and*

$$\pi_j = \lim_{n \to \infty} p_{ij}(n)$$

for all $i, j \in \mathcal{N}$*.*

Proof. By the Chapman–Kolmogorov equation (2.9), we have

$$p_{ij}(n+m) = \sum_{k=0}^{\infty} p_{ik}(n)\, p_{kj}(m) \geq \sum_{k=0}^{N} p_{ik}(n)\, p_{kj}(m)$$

for all $n, m = 0, 1, \cdots$, where N is an arbitrary positive integer. Suppose that all the states are positive recurrent. Then, from Theorem 2.7, the limit of $p_{ij}(n)$ as $n \to \infty$, π_j, say, exists and is positive. It follows that

$$\pi_j = \lim_{n \to \infty} p_{ij}(n+m) \geq \sum_{k=0}^{N} \lim_{n \to \infty} p_{ik}(n)\, p_{kj}(m) = \sum_{k=0}^{N} \pi_k p_{kj}(m) \tag{2.16}$$

and, as $m \to \infty$,

$$\pi_j \geq \sum_{k=0}^{N} \pi_k \lim_{m \to \infty} p_{kj}(m) = \pi_j \sum_{k=0}^{N} \pi_k.$$

Since $\pi_j > 0$ and N is arbitrary, we must have $\sum_{k=0}^{\infty} \pi_k \leq 1$. Taking $m = 1$ in (2.16) and letting $N \to \infty$, one sees that $\boldsymbol{\pi} = (\pi_i)$ is subinvariant with respect to $\mathbf{P}$ and hence it is positive componentwise and is invariant with respect to $\mathbf{P}$, by Lemma 2.7. Iterating (2.15) yields

$$\pi_i = \sum_{j=0}^{\infty} \pi_j p_{ji}(n), \quad n = 1, 2, \cdots. \tag{2.17}$$

Since $p_{ji}(n)$ is bounded, $\lim_{n \to \infty} p_{ji}(n)$ exists and $\sum_{j=0}^{\infty} \pi_j \leq 1$, the dominated convergence theorem shows in fact that $\sum_{j=0}^{\infty} \pi_j = 1$. It remains to verify that there is only one solution. Let $\boldsymbol{\nu} = (\nu_i)$ be a solution to (2.15). Then, as for (2.17), we have

$$\nu_i = \sum_{j=0}^{\infty} \nu_j \lim_{n \to \infty} p_{ji}(n) = \pi_i \sum_{j=0}^{\infty} \nu_j = \pi_i,$$

whence the solution is unique.

Conversely, suppose that there is a solution $\boldsymbol{\pi} = (\pi_i)$ to (2.15). Suppose also that the states are not positive recurrent. Then $\lim_{n\to\infty} p_{ij}(n) = 0$ for all $i,\ j \in \mathcal{N}$. But (2.17) still holds for all i. It follows that

$$\pi_i = \sum_{j=0}^{\infty} \pi_j \lim_{n\to\infty} p_{ji}(n) = 0, \quad i \in \mathcal{N},$$

which contradicts $\sum_{j=0}^{\infty} \pi_j = 1$. □

From Theorems 2.7 and 2.8, we know that if the Markov chain $\{X_n\}$ is irreducible, positive recurrent and aperiodic, then the limit of $p_{ij}(n)$ as $n \to \infty$ exists which is positive and independent of the initial state i. This property is often termed 'ergodicity'.

Definition 2.6 A Markov chain is called *ergodic* if it is irreducible, positive recurrent and aperiodic.

Let $\boldsymbol{\pi} = (\pi_i)$ be the probability vector given in Theorem 2.8. Invariance with respect to $\mathbf{P}$ can be written in matrix form as

$$\boldsymbol{\pi}^\top = \boldsymbol{\pi}^\top \mathbf{P}. \tag{2.18}$$

Now suppose that the initial distribution of $\{X_n\}$ is equal to $\boldsymbol{\pi}$. Then, from (1.21) and (2.18), we have the state distribution $\boldsymbol{\pi}(n) = \boldsymbol{\pi}$ for all $n = 0, 1, \cdots$. Hence the Markov chain $\{X_n\}$ is *stationary* in the sense that if X_0 has distribution $\boldsymbol{\pi}$ then X_n has the same distribution $\boldsymbol{\pi}$ for all n. A solution to (2.15), if it exists, is called a *stationary* (or *invariant*) *distribution*, and the system of equations (2.15) is called the *stationary equation*. Note that if $\boldsymbol{\pi}$ is a solution to (2.18) then any constant multiple of $\boldsymbol{\pi}$ is also a solution to (2.18). From Theorem 2.8, however, adding the equation $\boldsymbol{\pi}^\top \mathbf{1} = 1$ makes the solution unique. Therefore, in solving the stationary equation (2.15), it would be best to solve (2.18) first and then normalize the solution to satisfy the constraint $\boldsymbol{\pi}^\top \mathbf{1} = 1$.

Example 2.6 Let $\{X_n\}$ be the irreducible random walk considered in Example 1.14, but now $q_0 = 0$ so that the state space is restricted to nonnegative integers $\mathcal{Z}_+$. The transition matrix takes the form

$$\mathbf{P} = \begin{pmatrix} r_0 & p_0 & 0 & 0 & \cdots \\ q_1 & r_1 & p_1 & 0 & \cdots \\ 0 & q_2 & r_2 & p_2 & \cdots \\ 0 & 0 & q_3 & r_3 & \cdots \\ \vdots & \vdots & \vdots & \vdots & \ddots \end{pmatrix}.$$

Equation (2.18) is written componentwise as

$$\pi_i = \pi_{i-1} p_{i-1} + \pi_i r_i + \pi_{i+1} q_{i+1}, \quad i = 0, 1, \cdots, \tag{2.19}$$

where $\pi_{-1} = p_{-1} = 0$. Let $\pi_0 = 1$ and define

$$\pi_i = \pi_{i-1} \frac{p_{i-1}}{q_i} = \frac{p_{i-1} \cdots p_0}{q_i \cdots q_1} > 0, \quad i = 1, 2, \cdots. \tag{2.20}$$

Then, by using the facts that $r_i = 1 - p_i - q_i$ and $q_0 = 0$, it is easy to verify that the π_i satisfy (2.19), whence $\boldsymbol{\pi} = (\pi_i)$ is invariant with respect to $\mathbf{P}$. If $\sum_{i=0}^{\infty} \pi_i < \infty$ then the normalization yields a solution to (2.15). Hence the random walk $\{X_n\}$ is ergodic if and only if $\sum_{i=0}^{\infty} \pi_i < \infty$, provided $r_0 \neq 0$. The assumption $r_0 \neq 0$ makes the random walk aperiodic. Note that the self-transition probabilities r_i play no role in the stationary distribution.

Example 2.7 In this example we consider Liu's Markov chain given in Example 2.4. Equation (2.18) implies

$$\pi_{i+1} = p_i \pi_i + \sum_{k=i+1}^{\infty} q_k \pi_k, \quad i = 0, 1, \cdots; \quad \pi_0 = \sum_{k=0}^{\infty} q_k \pi_k. \tag{2.21}$$

Let

$$\pi_i = (i+1) p_0 \cdots p_{i-1} \pi_0, \quad i = 1, 2, \cdots.$$

It is a simple matter to prove that the π_i satisfy (2.21). Therefore, assuming that the Markov chain is irreducible and aperiodic, it is ergodic if and only if

$$\sum_{i=1}^{\infty} (i+1) p_0 \cdots p_{i-1} < \infty.$$

Combining this result with the results given in Example 2.4, the classification of states of this Markov chain is complete.

Example 2.8 Let the transition matrix $\mathbf{P}$ be given by (1.28), i.e., of a GI/M/1 queue. Suppose that $a_0 > 0$ and $a_0 + a_1 < 1$. Then the Markov chain with the transition matrix $\mathbf{P}$ is irreducible and aperiodic. Similarly to Example 2.7, equation (2.18) implies

$$\pi_{i+1} = \sum_{k=0}^{\infty} a_k \pi_{i+k}, \quad i = 0, 1, \cdots; \quad \pi_0 = \sum_{k=0}^{\infty} \overline{A}_{k+1} \pi_k. \tag{2.22}$$

In order to find the stationary distribution, we shall try $\pi_i = \sigma^i$ for some $0 < \sigma < 1$, $i = 0, 1, \cdots$, because we can then reduce (2.22) to an equation involving $g(z) = \sum_{n=0}^{\infty} a_n z^n$, the generating function of $\{a_n\}$. The first part of the equations of (2.22) is satisfied if there is such a σ that

$$\sigma = g(\sigma), \quad 0 < \sigma < 1. \tag{2.23}$$

If such a σ exists, then

$$\begin{aligned}
\sum_{k=0}^{\infty} \left(\sum_{m=k+1}^{\infty} a_m \right) \sigma^k &= \sum_{m=1}^{\infty} a_m \sum_{k=0}^{m-1} \sigma^k \\
&= \sum_{m=1}^{\infty} \frac{1-\sigma^m}{1-\sigma} a_m \\
&= \frac{1 - a_0 - (g(\sigma) - a_0)}{1-\sigma} = 1,
\end{aligned}$$

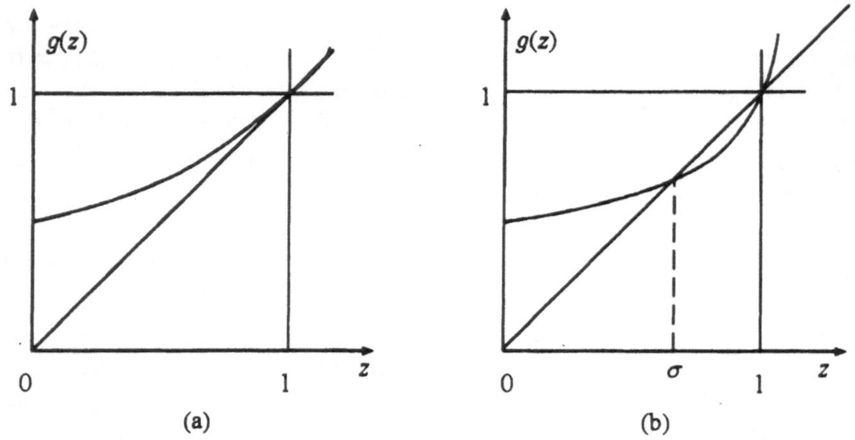

Figure 2.3 *The generating function* $g(z)$.

whence all the equations in (2.22) are satisfied.

Now consider the generating function $g(z) = \sum_{n=0}^{\infty} a_n z^n$. Since $g(0) = a_0 > 0$, $g(1) = 1$ and $g(z)$ is convex in z, if

$$g'(1) = \sum_{n=1}^{\infty} n a_n > 1,$$

then there exists a unique σ satisfying (2.23). See Figure 2.3(b). In this case, we have $\pi_i = \sigma^i > 0$ and $\sum_{i=0}^{\infty} \pi_i < \infty$. Therefore, the Markov chain is ergodic if and only if $\sum_{n=1}^{\infty} n a_n > 1$, in which case the limiting probabilities are given by

$$\lim_{n \to \infty} p_{ij}(n) = (1-\sigma)\sigma^j, \quad j = 0, 1, \cdots.$$

Exercise 2.9 asks the reader to derive the stationary distribution of an M/G/1 queue.

Let a Markov chain $\{X_n\}$ be ergodic with state space $\mathcal{N}$ and transition matrix $\mathbf{P} = (p_{ij})$. Let $\boldsymbol{\pi} = (\pi_i)$ be the stationary distribution satisfying (2.15). Define

$$p_{ij}^R = \frac{\pi_j p_{ji}}{\pi_i}, \quad i,\ j \in \mathcal{N}, \tag{2.24}$$

which is well defined since $\pi_i > 0$. Denoting the diagonal matrix with diagonal elements π_i by $\boldsymbol{\pi}_{\mathrm{D}}$, the above equation can be written succinctly in matrix form as

$$\mathbf{P}_{\mathrm{R}} = \boldsymbol{\pi}_{\mathrm{D}}^{-1} \mathbf{P}^{\top} \boldsymbol{\pi}_{\mathrm{D}}, \tag{2.25}$$

where $\mathbf{P}_\mathrm{R} = (p_{ij}^R)$. It is obvious that $\mathbf{P}_\mathrm{R} \geq \mathbf{O}$. Moreover,

$$\mathbf{P}_\mathrm{R}\mathbf{1} = \boldsymbol{\pi}_\mathrm{D}^{-1}\mathbf{P}^\top \boldsymbol{\pi} = \boldsymbol{\pi}_\mathrm{D}^{-1}(\boldsymbol{\pi}^\top \mathbf{P})^\top = \boldsymbol{\pi}_\mathrm{D}^{-1}\boldsymbol{\pi} = \mathbf{1},$$

where the associativity of matrix products is guaranteed because the matrices involved are nonnegative. Hence the matrix $\mathbf{P}_\mathrm{R}$ is stochastic.

Definition 2.7 For an ergodic Markov chain $\{X_n\}$ with transition matrix $\mathbf{P}$, the matrix $\mathbf{P}_\mathrm{R} = (p_{ij}^R)$ is called the *dual* of $\mathbf{P}$. A Markov chain with the dual transition matrix $\mathbf{P}_\mathrm{R}$ is called the *time reversal* of $\{X_n\}$.

The term 'dual' is used because $(\mathbf{P}_\mathrm{R})_\mathrm{R} = \mathbf{P}$ (see Lemma 2.8 below). The term 'time reversal' becomes clear in Section 2.5. From (2.24), we have

$$p_{ij}^R(n) = \frac{\pi_j p_{ji}(n)}{\pi_i}, \quad n = 1, 2, \cdots,$$

where $\mathbf{P}_\mathrm{R}^n = (p_{ij}^R(n))$. Hence, irreducibility and aperiodicity of the time reversal are inherited from the original Markov chain.

Lemma 2.8 *If a Markov chain $\{X_n\}$ is ergodic then so is its time reversal. The stationary distributions of the two Markov chains are the same.*

Proof. Let $\boldsymbol{\pi} = (\pi_i)$ be the unique solution to (2.15). It is enough to show that $\boldsymbol{\pi}$ is also invariant with respect to the dual $\mathbf{P}_\mathrm{R}$. Now we have

$$\sum_{j=0}^{\infty} \pi_j p_{ji}^R = \sum_{j=0}^{\infty} \pi_j \frac{\pi_i p_{ij}}{\pi_j} = \sum_{j=0}^{\infty} \pi_i p_{ij} = \pi_i,$$

since $\sum_{j=0}^{\infty} p_{ij} = 1$. □

Intuitively, since π_i represents the long-run frequency of the ergodic Markov chain being in state j, the original Markov chain and its time reversal *must* have the same stationary distribution.

For a real-valued function $f(i)$ on $\mathcal{N}$, we write $\mathbf{f} = (f_i)$, where $f_i = f(i)$.

Theorem 2.9 *Suppose that a Markov chain $\{X_n\}$ with transition matrix $\mathbf{P}$ is ergodic. Then, for any bounded function $\mathbf{f} = (f_i)$ on $\mathcal{N}$, we have*

$$\lim_{n\to\infty} E_i[f(X_n)] = \sum_{k=0}^{\infty} \pi_k f_k = \boldsymbol{\pi}^\top \mathbf{f},$$

independently of the initial state i, where $\boldsymbol{\pi} = (\pi_i)$ is the stationary distribution of $\{X_n\}$.

Proof. First note that

$$E_i[f(X_n)] = \sum_{j=0}^{\infty} p_{ij}(n) f_j, \quad n = 1, 2, \cdots.$$

Since $\mathbf{f}$ is bounded, we can assume without loss of generality that $f_j \geq 0$.

Now let $\mathbf{P}_{\mathrm{R}}$ be the dual of $\mathbf{P}$. Then, one has

$$E_i[f(X_n)] = \sum_{j=0}^{\infty} \frac{\pi_j f_j p_{ji}^R(n)}{\pi_i}, \quad n = 1, 2, \cdots.$$

We know that $\lim_{n\to\infty} p_{ji}^R(n) = \pi_i$ from Lemma 2.8 and Theorem 2.8. It follows from the dominated convergence theorem that

$$\lim_{n\to\infty} \sum_{j=0}^{\infty} \frac{\pi_j f_j p_{ji}^R(n)}{\pi_i} = \sum_{j=0}^{\infty} \frac{\pi_j f_j \pi_i}{\pi_i} = \boldsymbol{\pi}^{\top}\mathbf{f},$$

and the theorem follows. □

A slightly more general result than (2.13) is that if $E_i[f(X_n)]$ converges to $\boldsymbol{\pi}^{\top}\mathbf{f}$ as $n \to \infty$, then

$$\lim_{n\to\infty} \frac{1}{n+1} E_i\left[\sum_{k=0}^{n} f(X_k)\right] = \boldsymbol{\pi}^{\top}\mathbf{f}. \tag{2.26}$$

That is, if $f(j)$ is the reward received whenever the Markov chain $\{X_n\}$ is in state j, then the expected average reward in the long-run converges to the constant $\boldsymbol{\pi}^{\top}\mathbf{f}$. Also, it should be expected that, for any bounded function $\mathbf{f}$ on $\mathcal{N}$, we have

$$\lim_{n\to\infty} \frac{1}{n+1} \sum_{k=0}^{n} f(X_k) = \boldsymbol{\pi}^{\top}\mathbf{f} \tag{2.27}$$

almost surely. This result is known as the *strong law of large numbers* for ergodic Markov chains. We omit a proof of this result. Equation (2.27) means that, as for (2.26), if $f(j)$ is the reward received whenever the chain $\{X_n\}$ is in state j, then the actual average reward, as well as the expected average reward, in the long-run converges to the constant $\boldsymbol{\pi}^{\top}\mathbf{f}$. The reader is referred to Meyn and Tweedie (1993) for more general results. In particular, the boundedness on $\mathbf{f}$ is a strong assumption if $\mathbf{f}$ is to be interpreted as a cost function. Such a restriction can be removed. For the central limit theorem in ergodic Markov chains, see, e.g., Kurtz (1981) and Lacey and Philipp (1990).

Taking $\mathbf{f}$ as $f(j) = 1$ and zero otherwise in (2.27), we have the following result.

Corollary 2.7 *Suppose that a Markov chain $\{X_n\}$ is ergodic with stationary distribution $\boldsymbol{\pi} = (\pi_i)$. Then*

$$\lim_{n\to\infty} \frac{1}{n+1} \sum_{k=0}^{n} \mathrm{I}_{\{X_k = j\}} = \pi_j$$

almost surely.

Corollary 2.7 states that the average number of visits to state j during the first n steps is approximately equal to π_j for large n. This is of practical value in estimating the stationary distribution $\boldsymbol{\pi} = (\pi_j)$ from observed data.

So far, we have assumed that the chain $\{X_n\}$ is aperiodic. Suppose now that $\{X_n\}$ is positive recurrent and periodic with period d. Let $\mathbf{P}$ be its transition matrix. Then, as Example 2.5 suggests (and, in fact, can be readily seen from Theorem 2.6) the dth power of $\mathbf{P}$ is of the form

$$\mathbf{P}^d = \begin{pmatrix} \mathbf{Q}_1 & \mathbf{O} & \cdots & \mathbf{O} \\ \mathbf{O} & \mathbf{Q}_2 & \cdots & \mathbf{O} \\ \vdots & \vdots & \ddots & \vdots \\ \mathbf{O} & \mathbf{O} & \cdots & \mathbf{Q}_d \end{pmatrix},$$

where each $\mathbf{Q}_i$ is a stochastic matrix. Thus, the Markov chain corresponding to $\mathbf{P}^d$ has d closed sets, $B_1, \cdots, B_d$, say, each of which is ergodic. It follows from Theorem 2.8 that the limit of $\mathbf{Q}_i^n$ as $n \to \infty$ exists separately. On the other hand, if $i \in B_k$ then

$$P_i[X_m \in B_\ell] = 1, \quad \ell = k + m \pmod{d}.$$

Hence, $p_{ij}(n)$ does not converge while the limit of $p_{ij}(nd + m)$ as $n \to \infty$ exists, depending on the initial state i. That is,

$$\lim_{n\to\infty} p_{ij}(nd+m) = \begin{cases} \pi_j, & i \in B_k,\ j \in B_\ell,\ \ell = k + m \pmod{d}, \\ 0, & \text{otherwise}, \end{cases} \tag{2.28}$$

where the limiting probabilities π_j are the unique solution of

$$\pi_i = \sum_j \pi_j p_{ji}, \quad \sum_i \pi_i = d. \tag{2.29}$$

The proof of this result is left to the reader (see Exercise 2.11).

The next result generalizes (2.13).

Theorem 2.10 *Suppose that $\{X_n\}$ is positive recurrent and periodic with period d. Then*

$$\lim_{n\to\infty} \frac{1}{n+1} \sum_{k=0}^{n} p_{ij}(k) = \frac{\pi_j}{d}, \quad i,\ j \in \mathcal{N},$$

where the π_j are the solution of (2.29).

Proof. For any positive integer K, we have

$$\sum_{k=1}^{K} p_{ij}(k) = \sum_{n=0}^{[K/d]-1} \sum_{m=1}^{d} p_{ij}(nd + m) + \sum_{m=1}^{K-d[K/d]} p_{ij}(d[K/d] + m),$$

where $[x]$ denotes the largest integer less than or equal to x. It follows from

(2.28) that

$$\begin{aligned}
\lim_{K\to\infty} \frac{1}{K+1}\sum_{k=0}^{K} p_{ij}(k)
&= \frac{1}{d}\lim_{K\to\infty}\frac{1}{(K+1)/d}\sum_{n=0}^{[K/d]-1}\sum_{m=1}^{d} p_{ij}(nd+m) \\
&= \frac{1}{d}\sum_{m=1}^{d}\lim_{K\to\infty}\frac{1}{(K+1)/d}\sum_{n=0}^{[K/d]-1} p_{ij}(nd+m) \\
&= \frac{\pi_j}{d},
\end{aligned}$$

as claimed. □

We note that results similar to the aperiodic case also hold for the periodic case. For example, (2.26) can be transformed to

$$\lim_{n\to\infty}\frac{1}{n+1}E_i\left[\sum_{k=0}^{n} f(X_k)\right] = \frac{\boldsymbol{\pi}^{\mathsf{T}}\mathbf{f}}{d}$$

for any function $\mathbf{f}$, where $\boldsymbol{\pi}$ is the solution of (2.29), provided that $\boldsymbol{\pi}^{\mathsf{T}}\mathbf{f}$ converges absolutely. In fact, the results hold under less restrictive conditions; i.e., when the Markov chain $\{X_n\}$ is only recurrent. We state these results without proof.

Theorem 2.11 *Let $\{X_n\}$ be a recurrent Markov chain with state space $\mathcal{N} = \mathcal{Z}_+$ and transition matrix $\mathbf{P} = (p_{ij})$. Then:*

(i) *There exists a strictly positive vector $\boldsymbol{\pi}$ which is invariant with respect to $\mathbf{P}$, i.e., (2.18) has a strictly positive solution. Any other solution is a constant multiple of $\boldsymbol{\pi}$;*

(ii) *For any functions $\mathbf{f}$ and $\mathbf{g}$ on $\mathcal{N}$ for which the sums $\boldsymbol{\pi}^{\mathsf{T}}\mathbf{f}$ and $\boldsymbol{\pi}^{\mathsf{T}}\mathbf{g}$ converge absolutely and $\boldsymbol{\pi}^{\mathsf{T}}\mathbf{g} \neq 0$, we have*

$$\lim_{n\to\infty}\frac{\sum_{m=0}^{n} E_i[f(X_m)]}{\sum_{m=0}^{n} E_j[g(X_m)]} = \frac{\boldsymbol{\pi}^{\mathsf{T}}\mathbf{f}}{\boldsymbol{\pi}^{\mathsf{T}}\mathbf{g}},$$

independently of $i,\ j \in \mathcal{N}$, and

$$\lim_{n\to\infty}\frac{\sum_{m=0}^{n} f(X_m)}{\sum_{m=0}^{n} g(X_m)} = \frac{\boldsymbol{\pi}^{\mathsf{T}}\mathbf{f}}{\boldsymbol{\pi}^{\mathsf{T}}\mathbf{g}}$$

almost surely.

Theorem 2.11(i) states that any recurrent Markov chain $\{X_n\}$ has a positive invariant vector $\boldsymbol{\pi}$ which is unique up to constant multiples. Recall that if $\{X_n\}$ is positive recurrent then $\boldsymbol{\pi}^{\mathsf{T}}\mathbf{1}$ is finite and $\boldsymbol{\pi}$ is a constant multiple of the stationary distribution. Conversely, positive recurrence is characterized by the existence of such a finite, positive invariant vector. The random walk with $p = q = 1/2$ considered in Example 2.1 is recurrent and $\boldsymbol{\pi} = \mathbf{1}$ is a positive invariant vector. However, $\boldsymbol{\pi}^\top \mathbf{1}$ is not convergent and so the *symmetric* random walk is not positive recurrent. It is worth noting that the existence of an invariant vector alone does not imply that $\{X_n\}$ is recurrent.

2.4 Finite Markov chains

In this section, we assume that the state space is finite and given by $\mathcal{N} = \{0, 1, \cdots, N\}$. Suppose that a finite Markov chain $\{X_n\}$ has one recurrent class. Then, from (2.11), the transition matrix takes the form

$$\mathbf{P} = \begin{pmatrix} \mathbf{Q} & \mathbf{O} \\ \mathbf{R} & \mathbf{T} \end{pmatrix}, \tag{2.30}$$

where the submatrix $\mathbf{Q}$ corresponds to the recurrent class and $\mathbf{T}$ the set of transient states. Note that both $\mathbf{Q}$ and $\mathbf{T}$ are square but $\mathbf{R}$ may not be. Also, $\mathbf{Q}$ is stochastic while $\mathbf{T}$ and $\mathbf{R}$ are strictly substochastic and nonzero. By induction, it is readily seen that

$$\mathbf{P}^n = \begin{pmatrix} \mathbf{Q}^n & \mathbf{O} \\ \mathbf{R}_n & \mathbf{T}^n \end{pmatrix}, \quad n = 1, 2, \cdots, \tag{2.31}$$

where $\mathbf{R}_1 = \mathbf{R}$ and

$$\mathbf{R}_{n+1} = \mathbf{R}_n \mathbf{Q} + \mathbf{T}^n \mathbf{R}, \quad n = 1, 2, \cdots.$$

The Markov chain will eventually leave the set of transient states and approach equilibrium within the recurrent class. Hence, as regards the limiting probabilities, we need only consider the recurrent class. In fact, if the recurrent class is aperiodic, then, letting $\boldsymbol{\pi}$ be the stationary distribution of $\mathbf{Q}$, the invariant vector with respect to $\mathbf{P}$ in (2.30) is given by $(\boldsymbol{\pi}^\top, \mathbf{0}^\top)$. See Exercise 2.13 for the limiting distribution.

In what follows, we assume that the finite Markov chain $\{X_n\}$ is irreducible. When the chain is aperiodic, we have the following important result. The proof is taken from Iosifescu (1980).

Lemma 2.9 *Let $\{X_n\}$ be an irreducible and aperiodic Markov chain with transition matrix $\mathbf{P}$. Then there exists some integer k such that $\mathbf{P}^k$ has no zero components.*

Proof. Since

$$p_{ii}(n + n') \geq p_{ii}(n)\, p_{ii}(n'),$$

the set $\{n : p_{ii}(n) > 0\}$ is closed under addition. Hence, aperiodicity implies that there exists an integer n_i such that $p_{ii}(n) > 0$ for all $n \geq n_i$. Put $M = \max_i n_i$. Also, irreducibility implies that there are integers n_{ij} such that $p_{ij}(n_{ij}) > 0$ for all i and j. So, putting $N = \max_{i,j} n_{ij}$, N as well as

M is finite since the state space is finite. Now, let $n \geq M + N$ and suppose that there exists a pair of states i and j such that $p_{ij}(n) = 0$. Since

$$p_{ij}(n) \geq p_{ij}(n_{ij})\, p_{jj}(n - n_{ij}),$$

and since $M + N \geq n_j + n_{ij}$ so that $n - n_{ij} \geq n_j$, the right-hand side in the above inequality is positive, which is a contradiction. This proves the lemma. $\square$

A finite stochastic matrix with the property stated in Lemma 2.9 is often called *regular* (or *primitive*, see Definition A.1). From the proof of Lemma 2.9, if $\mathbf{P}^k$ has no zero components then neither does $\mathbf{P}^n$ for all $n \geq k$. In fact, let $\delta \equiv \min_{i,j} p_{ij}(k) > 0$ so that $\mathbf{P}^k \geq \delta \mathbf{E}$, where $\mathbf{E}$ denotes the matrix whose components are all unity. Then, since $\mathbf{P}$ is stochastic, we have $\mathbf{PE} = \mathbf{E}$, and so

$$\mathbf{P}^{k+1} = \mathbf{P}\,\mathbf{P}^k \geq \delta\,\mathbf{PE} = \delta\,\mathbf{E} > \mathbf{O}.$$

Recall that an irreducible Markov chain is ergodic if it is positive recurrent and aperiodic. From Corollary 2.4, any finite, irreducible Markov chain is positive recurrent. Thus, a finite Markov chain is ergodic if it is irreducible and aperiodic, that is, if its transition matrix is regular.

Let $\{X_n\}$ be ergodic with regular transition matrix $\mathbf{P}$. Then, from the Perron–Frobenius (PF) theorem (Theorem A.1), we know that 1 is the simple PF eigenvalue of $\mathbf{P}$ and $\mathbf{1}$ is the associated right eigenvector since $\mathbf{P}$ is stochastic so that $\mathbf{P1} = \mathbf{1}$. The associated PF left eigenvector $\boldsymbol{\pi}$ is such that

$$\boldsymbol{\pi}^\top = \boldsymbol{\pi}^\top \mathbf{P}, \quad \boldsymbol{\pi}^\top \mathbf{1} = 1,$$

which is strictly positive componentwise; viz. $\boldsymbol{\pi}$ is the stationary distribution of $\mathbf{P}$; see (2.15). Recall that these results have been obtained in a more general setting in Theorem 2.8.

At this point, we provide several illustrative examples.

Example 2.9 A one-dimensional random walk is a Markov chain on a countable state space in which in a single transition a particle either does not change the state or moves to one of the adjacent states. Suppose that there are two boundaries at 0 and N, $0 < N < \infty$, and that the particle starts from a nonboundary state. The transition matrix of the *finite* random walk $\{X_n\}$ has the form

$$\mathbf{P} = \begin{pmatrix} r_0 & p_0 & 0 & 0 & \cdots & 0 \\ q_1 & r_1 & p_1 & 0 & \cdots & 0 \\ 0 & q_2 & r_2 & p_2 & \cdots & 0 \\ \vdots & & \ddots & \ddots & \ddots & \vdots \\ 0 & \cdots & 0 & q_{N-1} & r_{N-1} & p_{N-1} \\ 0 & \cdots & 0 & 0 & q_N & r_N \end{pmatrix}, \qquad (2.32)$$

where the $q_i > 0$ are the downward transition probabilities, the $r_i \geq 0$ are the self-transition probabilities, and the $p_i > 0$ are the upward transition probabilities. Note that $q_0 = p_N = 0$ because of the boundaries. The probability r_0 describes the boundary property of state 0, while r_N describes the boundary property of state N. If $r_0 = 1$ then state 0 is *absorbing* as in the gambler's ruin problem (see Example 2.3). If $r_0 = 0$ so that $p_0 = 1$, then state 0 is called *reflecting*. In the case where $0 < r_0 < 1$, we call state 0 *retaining*.

Suppose that not all the r_i are zero, so that the finite random walk is ergodic. It is easily shown that the stationary distribution $\tilde{\pi} = (\tilde{\pi}_i)$ is given by

$$\tilde{\pi}_i = \frac{\pi_i}{\sum_{i=0}^{N} \pi_i}, \qquad i \in \mathcal{N},$$

where π_i are defined by (2.20).

Example 2.10 The *Ehrenfest urn model* is a classical mathematical description of diffusion through a permeable membrane. Suppose two urns contain a total of $2a$ balls. An urn, A say, contains k balls and the other, B say, contains the remaining $2a - k$ balls. A ball is selected at random (all selections are equally likely) from the total of the $2a$ balls and moves to the other urn (cf. Exercise 1.8). Let X_n be the number of balls in urn A at the nth stage. Each selection generates a transition of the process. The process $\{X_n\}$ is a finite random walk on the state space $\{0, 1, \cdots, 2a-1, 2a\}$ with transition probabilities

$$p_i = \frac{2a - i}{2a}, \qquad r_i = 0, \qquad q_i = \frac{i}{2a}.$$

The process $\{X_n\}$ is governed by a restoring force that is proportional to the distance from position a. Note that $p_0 = q_{2a} = 1$. Hence the states 0 and $2a$ act as reflecting boundaries. Note that the Markov chain $\{X_n\}$ is periodic with period 2.

Example 2.11 Let $Y_1, Y_2, \cdots$ be IID integer-valued random variables. We consider the partial-sum process defined by $X_n = \sum_{i=1}^{n} Y_i$ with two boundaries at 0 and N, $0 < N < \infty$. Namely, as in Example 2.9, let

$$f(x) = \begin{cases} 0, & x < 0, \\ x, & 0 \le x \le N, \\ N, & x > N, \end{cases}$$

and define

$$X_{n+1} = f(X_n + Y_{n+1}), \qquad n = 0, 1, \cdots.$$

That is, when $X_n + Y_{n+1} < 0$ or $> N$, the process is forced back to state 0 or state N respectively. Hence, the boundaries act as retaining boundaries. The process $\{X_n\}$ is a homogeneous Markov chain on the state space $\mathcal{N} = \{0, 1, \cdots, N\}$ with transition probabilities

$$p_{ij} = \begin{cases} A_{-i}, & j = 0, \\ a_{j-i}, & j = 1, \cdots, N-1, \\ \overline{A}_{N-i}, & j = N, \end{cases}$$

where $a_k \equiv P[Y = k]$, $k = 0, \pm 1, \cdots$, and

$$A_k \equiv P[Y \le k] = \sum_{m=-\infty}^{k} a_m, \quad \overline{A}_k \equiv P[Y \ge k] = \sum_{m=k}^{\infty} a_m.$$

The transition matrix is given by

$$\mathbf{P} = \begin{pmatrix} A_0 & a_1 & a_2 & \cdots & a_{N-1} & \overline{A}_N \\ A_{-1} & a_0 & a_1 & \cdots & a_{N-2} & \overline{A}_{N-1} \\ \vdots & \vdots & \vdots & \ddots & \vdots & \vdots \\ A_{-N} & a_{1-N} & a_{2-N} & \cdots & a_{-1} & \overline{A}_0 \end{pmatrix}. \tag{2.33}$$

Note that the diagonal elements of $\mathbf{P}$, except the two boundaries, are $a_0 = P[Y = 0]$, that those above (below, respectively) are a_1 (a_{-1}), and so on. This is so because $\{X_n\}$ is spatially homogeneous with two boundaries. A sufficient condition for aperiodicity is $a_0 > 0$.

Example 2.12 As in Example 1.3, let U_1, $U_2, \cdots$ be IID positive integer-valued random variables representing, for example, the lifetimes of successively replaced systems. Replacements are assumed to be instantaneous. Suppose that the system is replaced at age $N+1$ even if the actual lifetime is longer than $N+1$. This replacement policy, called *age replacement*, is common in practice to prevent unpredicted failures. To describe the situation, let $\widehat{U}_n = \min\{U_n, N+1\}$ and consider the partial-sum process $T_n = \sum_{i=1}^{n} \widehat{U}_i$ with $T_0 \equiv 0$. Note that the $\widehat{U}_n$ are IID random variables. Hence the process $\{X_n\}$ defined by

$$X_n = n - T_k, \quad T_k \le n < T_{k+1},$$

is the discrete-time age process associated with $\{\widehat{U}_n\}$. Since replacements are instantaneous, X_n cannot take the value $N+1$. Therefore, the age process $\{X_n\}$ is a homogeneous Markov chain with state space $\{0, 1, \cdots, N\}$. For $i \ne N$, the transition probabilities are given by

$$p_{ij} = \begin{cases} h_i, & j = 0, \\ 1 - h_i, & j = i+1, \\ 0, & \text{otherwise}, \end{cases}$$

where $h_k = P[U = k | U \ge k]$, $k = 1, 2, \cdots$. For $i = N$, replacement takes place at the next time epoch so that $p_{N0} = 1$. Summarizing, the transition matrix is given by

$$\mathbf{P} = \begin{pmatrix} h_1 & 1-h_1 & 0 & \cdots & 0 \\ h_2 & 0 & 1-h_2 & \cdots & 0 \\ \vdots & \vdots & \vdots & \ddots & \vdots \\ h_N & 0 & 0 & \cdots & 1-h_N \\ 1 & 0 & 0 & \cdots & 0 \end{pmatrix}. \tag{2.34}$$

The conditional probability h_i is given by

$$h_i = \frac{P[U = i]}{P[U \geq i]}, \quad i = 1, 2, \cdots,$$

meaning that, conditional on survival up to age $i-1$, the probability of system failure occurring at age i is h_i. In this regard, the sequence of conditional probabilities h_i is called the *hazard rate function* in reliability theory. If $h_i < 1$, $i = 1, \cdots, N$, then the Markov chain $\{X_n\}$ is irreducible. It is aperiodic if h_1 is nonzero.

The stationary distribution $\boldsymbol{\pi} = (\pi_i)$ satisfies

$$\pi_i = \pi_{i-1}(1 - h_i), \quad i = 1, \cdots, N,$$

and $\sum_{i=0}^{N} \pi_i = 1$. Note that adding the above equations yields

$$1 - \pi_0 = 1 - \pi_N - \sum_{i=1}^{N} \pi_{i-1} h_i,$$

i.e. the remaining equation in $\boldsymbol{\pi}^\top = \boldsymbol{\pi}^\top \mathbf{P}$.

Example 2.13 Consider the GI/M/1 queue given in Example 1.4, but now with a waiting room of capacity $N-1$, $N \geq 1$, so that a customer finding N customers upon arrival cannot enter the system and leaves never to return. Let X_n be the queue size just before the nth arrival. Then, $\{X_n\}$ is a homogeneous Markov chain with state space $\{0, 1, \cdots, N\}$ and transition matrix

$$\mathbf{P} = \begin{pmatrix} \overline{A}_1 & a_0 & 0 & \cdots & 0 \\ \overline{A}_2 & a_1 & a_0 & \cdots & 0 \\ \vdots & \vdots & \vdots & \ddots & \vdots \\ \overline{A}_N & a_{N-1} & a_{N-2} & \cdots & a_0 \\ \overline{A}_N & a_{N-1} & a_{N-2} & \cdots & a_0 \end{pmatrix}, \tag{2.35}$$

where a_k is given by (1.8), which is the probability that the number of potential departures between interarrivals of successive customers is k, and $\overline{A}_k = \sum_{i=k}^{\infty} a_i$, $k = 1, \cdots, N$. Note that the last two rows of $\mathbf{P}$ in (2.35) are identical. The Markov chain $\{X_n\}$ is irreducible and aperiodic as long as $0 < a_0 < 1$.

Example 2.14 Consider in turn an M/G/1 queue with a waiting room of finite capacity N, $N \geq 0$, so that a customer finding $N+1$ customers upon arrival leaves the system never to return (cf. Example 1.5). Let X_n be the number of customers in the system just after the nth departure. Then $\{X_n\}$ is a homogeneous Markov chain with state space $\{0, 1, \cdots, N\}$ and transition matrix

$$\mathbf{P} = \begin{pmatrix} b_0 & b_1 & b_2 & \cdots & b_{N-1} & \overline{B}_N \\ b_0 & b_1 & b_2 & \cdots & b_{N-1} & \overline{B}_N \\ 0 & b_0 & b_1 & \cdots & b_{N-2} & \overline{B}_{N-1} \\ \vdots & \vdots & \ddots & \ddots & \vdots & \vdots \\ 0 & 0 & \cdots & b_0 & b_1 & \overline{B}_2 \\ 0 & 0 & \cdots & 0 & b_0 & \overline{B}_1 \end{pmatrix}. \tag{2.36}$$

Here b_k denotes the probability that the number of arrivals during a service completion is k and $\overline{B}_k = \sum_{i=k}^{\infty} b_i$. It is interesting to note that if the states of $\mathbf{P}$ in (2.36) were relabeled so that $0 \to N$, $1 \to N-1$, $\cdots$, $N \to 0$, then the resulting transition matrix would coincide with the transition matrix given by (2.35).

Suppose $0 < b_0 < 1$. The invariant vector $\boldsymbol{\pi} = (\pi_i)$ satisfies

$$\pi_i = b_i \pi_0 + \sum_{k=1}^{i+1} b_{i-k+1} \pi_k, \quad i = 0, 1, \cdots, N-1.$$

Hence, setting $\pi_0 = 1$, solving the above equation in terms of π_i, $i = 1, \cdots, N$, and then normalizing them to sum to unity yields the stationary distribution.

We have seen that if a finite Markov chain is ergodic then the stationary distribution is given by the PF left eigenvector, normed to sum to unity, of its regular transition matrix. Moreover, in this case, we can state the rate of convergence to stationarity explicitly. Let $\{X_n\}$ be an ergodic Markov chain with regular transition matrix $\mathbf{P}$, which we denote by

$$\mathbf{P} = \mathbf{1}\boldsymbol{\pi}^{\mathsf{T}} + \boldsymbol{\Delta}.$$

Note that $\boldsymbol{\Delta}\mathbf{1} = \mathbf{0}$ and $\boldsymbol{\pi}^{\mathsf{T}}\boldsymbol{\Delta} = \mathbf{0}^{\mathsf{T}}$. It follows that

$$\mathbf{P}^n = \mathbf{1}\boldsymbol{\pi}^{\mathsf{T}} + \boldsymbol{\Delta}^n, \quad n = 1, 2, \cdots. \tag{2.37}$$

Let λ_i be the eigenvalues of $\mathbf{P}$ such that $1 = \lambda_0 > |\lambda_1| \geq |\lambda_j|$ for $j \neq 0, 1$. From Theorem A.2, $\boldsymbol{\Delta}^n$ converges to the zero matrix $\mathbf{O}$ as $n \to \infty$ at the rate $|\lambda_1|$. Hence $\mathbf{P}^n$ converges to $\mathbf{1}\boldsymbol{\pi}^{\mathsf{T}}$ as $n \to \infty$ geometrically fast at the same rate. The value $|\lambda_1|$ plays the role of the *rate of convergence* to stationarity and is called the *decay parameter* of the Markov chain. In this regard, the *relaxation time* defined by

$$T_{\mathrm{REL}}(\mathbf{P}) = \frac{1}{1 - |\lambda_1|} \tag{2.38}$$

is often used as a measure of convergence to stationarity. Note that $0 \leq$

$|\lambda_1| < 1$. The larger the value of $|\lambda_1|$, the longer the relaxation time. We shall return to the rate of convergence to stationarity and the relaxation time in Section 2.6.

Example 2.15 Consider a two-state Markov chain on the state space $\{0, 1\}$ with transition matrix

$$\mathbf{P} = \begin{pmatrix} 1-a & a \\ b & 1-b \end{pmatrix}, \quad 0 < a,\ b < 1. \tag{2.39}$$

This Markov chain can be viewed as a special case of the random walk given in Example 2.9 with two states, where the two states play retaining boundaries. The stationary distribution is given by

$$\boldsymbol{\pi}^{\top} = \left(\frac{b}{a+b}, \frac{a}{a+b} \right),$$

from which

$$\boldsymbol{\Delta} = \mathbf{P} - \mathbf{1}\boldsymbol{\pi}^{\top} = \frac{(1-a-b)}{a+b} \begin{pmatrix} a & -a \\ -b & b \end{pmatrix}.$$

It is readily confirmed that $\boldsymbol{\Delta}\mathbf{1} = \mathbf{0}$ and $\boldsymbol{\pi}^{\top}\boldsymbol{\Delta} = \mathbf{0}^{\top}$. The n-step transition matrix is

$$\mathbf{P}^n = \frac{1}{a+b} \begin{pmatrix} b & a \\ b & a \end{pmatrix} + \frac{(1-a-b)^n}{a+b} \begin{pmatrix} a & -a \\ -b & b \end{pmatrix}, \quad n = 1, 2, \cdots.$$

Hence the decay parameter of this Markov chain is $|1-a-b|$. The relaxation time is given by

$$T_{\mathrm{REL}}(\mathbf{P}) = \begin{cases} \dfrac{1}{a+b}, & a+b \le 1, \\[2ex] \dfrac{1}{2-a-b}, & a+b > 1. \end{cases}$$

In fact, it is an easy exercise to show that $\lambda_1 = 1 - a - b$.

The next theorem summarizes the above discussion.

Theorem 2.12 *Let $\{X_n\}$ be an ergodic Markov chain with regular transition matrix* $\mathbf{P}$. *Then:*

(i) *There exists a unique stationary distribution $\boldsymbol{\pi} = (\pi_i)$, strictly positive componentwise, which is the PF left eigenvector of* $\mathbf{P}$;

(ii) *$p_{ij}(n)$ converges to π_j as $n \to \infty$ geometrically fast for all $i,\ j \in \mathcal{N}$ at the rate $|\lambda_1|$, where λ_1 is the eigenvalue of* $\mathbf{P}$ *that is largest in magnitude other than unity.*

Recall that the stationary distribution is a probability vector satisfying (2.15) for a given stochastic matrix $\mathbf{P}$. Conversely, suppose that a stationary distribution $\boldsymbol{\pi}$ is given and we want to know which stochastic matrix will produce $\boldsymbol{\pi}$ as the stationary distribution. This problem is known as the *inverse problem* of the stationary equation. To clarify the problem, suppose that $\boldsymbol{\pi}^{\mathsf{T}} = (\pi, 1-\pi)$ is given, where $0 \le \pi \le 1$. Let $\mathbf{P}$ be as in (2.39), where $0 \le a,\ b \le 1$. Equation (2.18) implies that

$$a\pi = b(1-\pi).$$

Hence, given the probability vector $\boldsymbol{\pi}$, any 2×2 matrix satisfying the above relation with $0 \le a,\ b \le 1$ has $\boldsymbol{\pi}$ as its stationary distribution. The inverse problem was first discussed by Karr (1978).

2.5 Time-reversible Markov chains

For a Markov chain $\{X_n\}$, let $A_j(n)$ be the number of visits to state j up to time n, and let $D_j(n)$ be the number of departures from state j up to time n. When there are successive visits to state j, we count both a visit and a departure each time. Then, because visits and departures alternate, one has

$$|A_j(n) - D_j(n)| \le 1, \quad n = 1, 2, \cdots. \tag{2.40}$$

Divide (2.40) by n and let $n \to \infty$. Since $\lim_{n\to\infty} A_j(n)/n$, if it exists, is the long-run average of the number of visits to state j per unit of time, it represents the *transition rate* into state j. Similarly, $\lim_{n\to\infty} D_j(n)/n$, if it exists, represents the transition rate out of state j. It follows from (2.40) that, for each j, we have

$$\text{transition rate out of state } j = \text{transition rate into state } j. \tag{2.41}$$

Let $\{X_n\}$ be an ergodic Markov chain with state space $\mathcal{N}$ and transition matrix $\mathbf{P} = (p_{ij})$. Note that, by definition, the transition probability p_{ij} is the ratio of transitions into state j to transitions out of state i. Since the stationary probability π_i represents the long-run frequency of being in state i, by the basic limit theorem (Theorem 2.7), the quantity $\sum_i \pi_i p_{ij}$ is the transition rate into state j and the quantity $\sum_i \pi_j p_{ji} = \pi_j$ is the transition rate out of state j, which by (2.41) must be identical. Hence, the stationary equation (2.15) is just a particular instance of (2.41).

Similarly, transition rates across arbitrary boundaries must be the same. That is, suppose that we partition the state space $\mathcal{N}$ into two disjoint subsets, A and A^c, say. By the same arguments as are given above, it must hold for any such partition that

$$\sum_{j\in A}\sum_{i\in A^c} \pi_j p_{ji} = \sum_{j\in A^c}\sum_{i\in A} \pi_j p_{ji}. \tag{2.42}$$

The left-hand side of (2.42) describes the transition rate from A to A^c, while the right-hand side describes the transition rate from A^c to A. Equation (2.42) is called a *full balance equation*. When A is a singleton, (2.42) reduces to

$$\pi_j(1-p_{jj}) = \sum_{i \neq j} \pi_i p_{ij}, \quad j \in \mathcal{N},$$

which agrees with the stationary equation (2.15).

Suppose that for a pair i and j we have

$$\pi_i p_{ij} = \pi_j p_{ji}, \quad i, j \in \mathcal{N}, \tag{2.43}$$

the left-hand side of which describes the transition rate from state i to state j, while the right-hand side means the transition rate from j to i. Equation (2.43) is called a *detailed balance equation.* If detailed balance holds for every pair of states, then full balance (2.42) holds for any partition. But detailed balance implies more, as we will see shortly.

For an ergodic Markov chain $\{X_n\}$ with state space $\mathcal{N}$ and transition matrix $\mathbf{P}$, let $\boldsymbol{\pi}_{\mathrm{D}}$ be the diagonal matrix whose diagonal elements are the stationary probabilities π_i, $i \in \mathcal{N}$. From (2.25), the dual of $\mathbf{P}$ is given by

$$\mathbf{P}_{\mathrm{R}} = \boldsymbol{\pi}_{\mathrm{D}}^{-1}\mathbf{P}^{\top}\boldsymbol{\pi}_{\mathrm{D}}.$$

When the state space is finite, it is easily seen that if $\mathbf{P}$ is regular then the dual $\mathbf{P}_{\mathrm{R}}$ is also regular. In fact, we have

$$\mathbf{P}_{\mathrm{R}}^k = (\boldsymbol{\pi}_{\mathrm{D}}^{-1}\mathbf{P}^{\top}\boldsymbol{\pi}_{\mathrm{D}})^k = \boldsymbol{\pi}_{\mathrm{D}}^{-1}(\mathbf{P}^k)^{\top}\boldsymbol{\pi}_{\mathrm{D}}, \quad k = 1, 2, \cdots, \tag{2.44}$$

from which the claim follows (see Lemma 2.9).

Suppose that detailed balance (2.43) holds for any pair of states. The resulting relations show that the matrix $\boldsymbol{\pi}_{\mathrm{D}}\mathbf{P}$ is symmetric, i.e.,

$$\boldsymbol{\pi}_{\mathrm{D}}\mathbf{P} = (\boldsymbol{\pi}_{\mathrm{D}}\mathbf{P})^{\top} = \mathbf{P}^{\top}\boldsymbol{\pi}_{\mathrm{D}},$$

whence

$$\mathbf{P} = \boldsymbol{\pi}_{\mathrm{D}}^{-1}\mathbf{P}^{\top}\boldsymbol{\pi}_{\mathrm{D}} = \mathbf{P}_{\mathrm{R}}. \tag{2.45}$$

The transition matrix $\mathbf{P}$ and its dual $\mathbf{P}_{\mathrm{R}}$ are identical if and only if the detailed balance equation holds for every pair of states.

Definition 2.8 An ergodic Markov chain is said to be *reversible in time* *if its governing transition matrix and its dual are identical.*

A necessary condition for time reversibility is that if $p_{ij} > 0$, then $p_{ji} > 0$, and vice versa. Hence, for example, a periodic Markov chain cannot be reversible in time (see Theorem 2.6 for the canonical form of a periodic transition matrix).

The term 'time reversible' is justified by the following fact. From (2.45), we have

$$\mathbf{P}^n = \mathbf{P}_{\mathrm{R}}^n = \boldsymbol{\pi}_{\mathrm{D}}^{-1}(\mathbf{P}^{\top})^n\boldsymbol{\pi}_{\mathrm{D}}, \quad n = 0, 1, 2, \cdots,$$

so that

$$\boldsymbol{\pi}_{\mathrm{D}}\mathbf{P}^n = (\boldsymbol{\pi}_{\mathrm{D}}\mathbf{P}^n)^{\top}, \quad n = 0, 1, 2, \cdots.$$

The (i,j)th component of $\pi_D \mathbf{P}^n$ is equal to

$$\pi_i \, p_{ij}(n) = P[X_m = i, \, X_{m+n} = j]$$

if the chain $\{X_n\}$ is stationary. Hence, if (2.45) holds, then we have

$$P[X_m = i, \, X_{m+n} = j] = P[X_m = j, \, X_{m+n} = i],$$

so dividing both sides by $\pi_i = P[X_m = i] = P[X_{m+n} = i]$ yields

$$P[X_{m+n} = j | X_m = i] = P[X_m = j | X_{m+n} = i],$$

i.e. reversibility in time.

The concept of time reversibility is heavily exploited by Kelly (1979). In the remainder of this section, we see the importance of this notion in the theory of Markov chains. The next theorem is known as *Kolmogorov's criterion* for time reversibility.

Theorem 2.13 *An ergodic Markov chain is reversible in time if and only if*

$$p_{i,i_1} p_{i_1,i_2} \cdots p_{i_k,i} = p_{i,i_k} p_{i_k,i_{k-1}} \cdots p_{i_1,i} \tag{2.46}$$

for all $k \geq 2$ and all states $i, \; i_1, \cdots, \; i_k$.

Proof. Suppose that detailed balance (2.43) holds for every pair of states. Then, for any states $i, \; i_1, \cdots, \; i_k$, we have

$$\pi_i p_{i,i_1} \pi_{i_1} p_{i_1,i_2} \cdots \pi_{i_k} p_{i_k,i} = \pi_{i_1} p_{i_1,i} \, \pi_{i_2} p_{i_2,i_1} \cdots \pi_i p_{i,i_k}.$$

Hence, dividing both sides by $\pi_i \pi_{i_1} \cdots \pi_{i_k} > 0$ yields (2.46). Conversely, summing (2.46) over all states $i_1, \cdots, i_{k-1}$ yields

$$p_{i,i_k}(k) \, p_{i_k,i} = p_{i,i_k} \, p_{i_k,i}(k),$$

which, on letting $k \to \infty$, converges to (2.43). □

Corollary 2.8 *For an ergodic Markov chain with transition matrix $\mathbf{P} = (p_{ij})$, suppose $p_{ij} > 0$ for all i and j. Then, the Markov chain is reversible in time if and only if, for any fixed state i,*

$$p_{ij} p_{jk} p_{ki} = p_{ik} p_{kj} p_{ji}$$

for all j and k.

Proof. To prove sufficiency, let $\pi_j = p_{ij}/p_{ji}$. Then, using $p_{ij} p_{jk} p_{ki} = p_{ik} p_{kj} p_{ji}$, we have

$$\sum_j \pi_j p_{jk} = \sum_j \frac{p_{ij} p_{jk}}{p_{ji}} = \sum_j \frac{p_{ik} p_{kj}}{p_{ki}} = \frac{p_{ik}}{p_{ki}} = \pi_k.$$

Hence, $\pi = (\pi_j)$ must be proportional to the stationary distribution. On the other hand,

$$\pi_j p_{jk} = \frac{p_{ij} p_{jk}}{p_{ji}} = \frac{p_{ik} p_{kj}}{p_{ki}} = \pi_k p_{kj}$$

for all j and k, whence the result. □

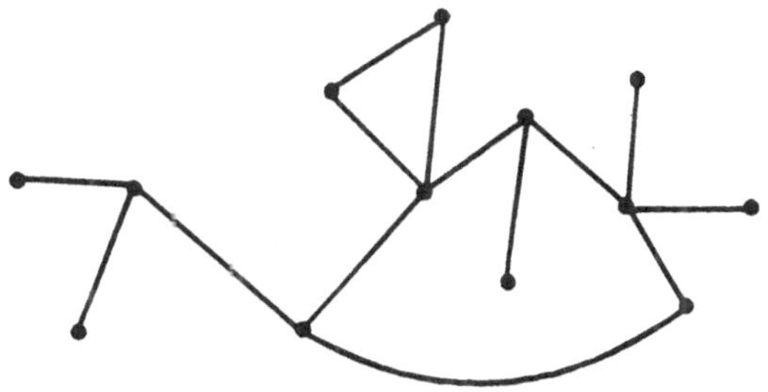

Figure 2.4 *A simple connected graph.*

We provide some examples of time-reversible Markov chains.

Example 2.16 Consider the two-state Markov chain given in Example 2.15. The stationary distribution is

$$\pi^{\mathsf{T}} = \left(\frac{b}{a+b}, \frac{a}{a+b}\right),$$

from which

$$\pi_1 p_{10} = \frac{a}{a+b} b = \frac{b}{a+b} a = \pi_0 p_{01},$$

so that the two-state Markov chain is reversible in time.

Example 2.17 Let $\{X_n\}$ be the random walk considered in Example 2.6. For π_i defined by (2.20), suppose that $\sum_i \pi_i < \infty$, i.e., the random walk is ergodic. By definition (2.20), we have

$$\pi_j p_{j,j-1} = \pi_j q_j = \pi_{j-1} p_{j-1} = \pi_{j-1} p_{j-1,j}, \quad j = 1, 2, \cdots,$$

whence any ergodic random walk is reversible in time. Recall that the two-state Markov chain is a random walk.

Example 2.18 Let $G = (\mathcal{N}, E)$ be an undirected graph with vertex set $\mathcal{N}$ and edge set E. Suppose that G is connected and simple, i.e., G has no loop or multiple edges (see Figure 2.4). A process starts from an initial vertex i_0 and thereafter proceeds by choosing a neighboring vertex with uniform probability, where a neighboring state is a vertex connected by an edge. Thus, if b_i denotes the degree of vertex i, then

$$p_{ij} = \begin{cases} b_i^{-1}, & \{i,j\} \in E, \\ 0, & \text{otherwise}. \end{cases}$$

The process is a Markov chain with state space $\mathcal{N}$ and is called a *random walk on a graph G*. Since the graph is connected, the Markov chain is irreducible. If the chain is aperiodic, then the stationary probabilities are given by

$$\pi_i = \frac{b_i}{2|E|}, \quad i \in \mathcal{N},$$

where $|E|$ denotes the number of edges in E. To confirm this, we check that $\sum_i \pi_i = 1$ and that they satisfy the stationarity equation

$$\sum_i \pi_i p_{ij} = \sum_{i:\{i,j\}\in E} \frac{1}{2|E|} = \frac{b_j}{2|E|} = \pi_j, \quad j \in \mathcal{N}.$$

Note that

$$\pi_i p_{ij} = \pi_j p_{ji} = \begin{cases} \dfrac{1}{2|E|}, & \{i,j\} \in E, \\ 0, & \text{otherwise.} \end{cases}$$

Hence, if the random walk on graph G is aperiodic, then it is reversible in time. A sufficient condition for aperiodicity is that there exists an odd cycle. For more information, see Diaconis (1988) or Diaconis and Stroock (1991), and references therein.

Example 2.19 Suppose that a transition matrix itself is symmetric, i.e., $p_{ij} = p_{ji}$ for all i and j. Then, since $\sum_j p_{ij} = \sum_j p_{ji} = 1$, the transition matrix is doubly stochastic (see Exercise 2.14). The stationary distribution of a finite, doubly stochastic matrix is uniform, whence the corresponding finite Markov chain is reversible in time.

We now turn to a finite, time-reversible Markov chain. Let $\{X_n\}$ be an ergodic Markov chain with state space $\mathcal{N} = \{0, 1, \cdots, N\}$ and transition matrix $\mathbf{P}$. The stationary distribution is denoted by $\boldsymbol{\pi} = (\pi_i)$. Let $\boldsymbol{\pi}_D^{1/2}$ and $\boldsymbol{\pi}_D^{-1/2}$ be the diagonal matrices whose diagonal elements are $\pi_i^{1/2}$ and $\pi_i^{-1/2}$, $i \in \mathcal{N}$, respectively. Since pre- and post-multiplication of a symmetric matrix by another symmetric matrix preserves symmetry, it follows that $\boldsymbol{\pi}_D\mathbf{P}$is symmetric if and only if

$$\boldsymbol{\pi}_D^{-1/2}\boldsymbol{\pi}_D\mathbf{P}\boldsymbol{\pi}_D^{-1/2} = \boldsymbol{\pi}_D^{1/2}\mathbf{P}\boldsymbol{\pi}_D^{-1/2}$$

is symmetric. Hence, the Markov chain $\{X_n\}$ is reversible in time if and only if

$$(\boldsymbol{\pi}_D^{1/2}\mathbf{P}\boldsymbol{\pi}_D^{-1/2})^n = \boldsymbol{\pi}_D^{1/2}\mathbf{P}^n\boldsymbol{\pi}_D^{-1/2}, \quad n = 0, 1, \cdots, \tag{2.47}$$

are all symmetric. Now, let λ_j, $j = 0, 1, \cdots, N$, be the eigenvalues of the symmetric matrix $\boldsymbol{\pi}_D^{1/2}\mathbf{P}\boldsymbol{\pi}_D^{-1/2}$. Since $\boldsymbol{\pi}_D^{1/2}\mathbf{P}\boldsymbol{\pi}_D^{-1/2}$ is a similarity transform, the eigenvalues of $\boldsymbol{\pi}_D^{1/2}\mathbf{P}\boldsymbol{\pi}_D^{-1/2}$ are the same as those of $\mathbf{P}$ (see, e.g., Noble and Daniel, Section 8.4, 1977). Therefore, time reversibility ensures that the eigenvalues of $\mathbf{P}$ are all real. See Keilson (1979) for more information.

In what follows, we assume without loss of generality that

$$1 = \lambda_0 > |\lambda_1| \geq |\lambda_2| \geq \cdots \geq |\lambda_N|.$$

The spectral decomposition of the symmetric matrix $\boldsymbol{\pi}_D^{1/2}\mathbf{P}\boldsymbol{\pi}_D^{-1/2}$ is given by

$$\boldsymbol{\pi}_D^{1/2}\mathbf{P}\boldsymbol{\pi}_D^{-1/2} = \sum_{j=0}^{N} \lambda_j\, \mathbf{x}_j \mathbf{x}_j^{\top}, \tag{2.48}$$

where each $\mathbf{x}_j$ is the eigenvector associated with eigenvalue λ_j such that the system $\{\mathbf{x}_j\}$ is orthonormal, i.e., $\mathbf{x}_i^{\top}\mathbf{x}_j = \delta_{ij}$. From (2.47) and (2.48), we have

$$\mathbf{P}^n = \sum_{j=0}^{N} \lambda_j^n\, \mathbf{v}_j \mathbf{u}_j^{\top}, \quad n = 0, 1, \cdots,$$

where

$$\mathbf{u}_j = \boldsymbol{\pi}_D^{1/2}\mathbf{x}_j, \quad \mathbf{v}_j = \boldsymbol{\pi}_D^{-1/2}\mathbf{x}_j; \quad j = 0, 1, \cdots, N.$$

Note that $\mathbf{u}_0 = \boldsymbol{\pi}$ and $\mathbf{v}_0 = \mathbf{1}$. Thus,

$$\mathbf{P}^n = \mathbf{1}\boldsymbol{\pi}^{\top} + \sum_{j=1}^{N} \lambda_j^n\, \mathbf{v}_j \mathbf{u}_j^{\top}, \quad n = 0, 1, \cdots. \tag{2.49}$$

Note that $\mathbf{u}_i^{\top}\mathbf{v}_j = \delta_{ij}$. Such a system $\{\mathbf{u}_j, \mathbf{v}_j\}$ is called *biorthonormal.* Therefore, the transition matrix of any time-reversible Markov chain has a *spectral decomposition* (2.49) in terms of the biorthonormal system $\{\mathbf{u}_j, \mathbf{v}_j\}$. Representation (2.49) is useful for computing n-step transition probabilities for large n. The significance of representation (2.49) will become clearer later.

From (2.49), if the initial distribution is $\boldsymbol{\alpha}$, then the state distribution at time n is given by

$$\boldsymbol{\pi}(n) = \boldsymbol{\pi} + \sum_{j=1}^{N} \lambda_j^n (\boldsymbol{\alpha}^{\top}\mathbf{v}_j)\mathbf{u}_j, \quad n = 0, 1, \cdots, \tag{2.50}$$

which converges to the stationary distribution $\boldsymbol{\pi}$ as $n \to \infty$, since $|\lambda_j| < 1$ for $j = 1, \cdots, N$. Moreover, it is readily seen from (2.49) that

$$p_{ii}(n) = \pi_i + \sum_{j=1}^{N} \lambda_j^n x_{ji}^2, \quad n = 0, 1, \cdots, \tag{2.51}$$

where x_{ji} denotes the ith component of vector $\mathbf{x}_j$. Thus, if $\lambda_j \geq 0$ for all j, then the transition probability $p_{ii}(n)$ converges to π_i from above as $n \to \infty$. In general, $p_{ii}(2n)$ has this property provided that the Markov chain is reversible in time.

Example 2.20 For the two-state Markov chain given in Example 2.15, we have

$$\pi_D^{1/2}\mathbf{P}\pi_D^{-1/2} = \begin{pmatrix} 1-a & \sqrt{ab} \\ \sqrt{ab} & 1-b \end{pmatrix},$$

which is symmetric. The eigenvalues of $\mathbf{P}$ are 1 and $1-a-b$. The spectral decomposition of $\mathbf{P}$ is given by

$$\mathbf{P}^n = \frac{1}{a+b}\begin{pmatrix} b & a \\ b & a \end{pmatrix} + \frac{(1-a-b)^n}{a+b}\begin{pmatrix} a & -a \\ -b & b \end{pmatrix}, \quad n = 0, 1, \cdots,$$

which coincides with the preceding result.

2.6 The rate of convergence to stationarity

Let $\{X_n\}$ be an ergodic Markov chain with state space $\mathcal{N} = \{0, 1, \cdots, N\}$ and transition matrix $\mathbf{P} = (p_{ij})$. The state distribution at time n is given by

$$\boldsymbol{\pi}^\top(n) = \boldsymbol{\alpha}^\top \mathbf{P}^n, \quad n = 0, 1, \cdots,$$

where $\boldsymbol{\alpha}$ is the initial distribution. Hence, in principle, the time-dependent distribution, called the *transient behavior*, of the Markov chain $\{X_n\}$ is determined by matrix multiplications. However, when the state space is finite but large, computational difficulties can arise. For example, computing $\mathbf{P}^n$ for large n is not a trivial task. In fact, the time complexity for calculating $\mathbf{P}^n$ is $O(N^{2.81}\log n)$. The reader is referred to, e.g., Aho, Hopcroft and Ullman (1974) for the algorithm and its time complexity. Hence, it is of interest to determine when one can use the stationary distribution $\boldsymbol{\pi} = (\pi_i)$ to approximate $\boldsymbol{\pi}(n)$ for large n, since $\mathbf{P}^n$ converges to $\mathbf{1}\boldsymbol{\pi}^\top$ as $n \to \infty$. Note that the time complexity for solving the stationary equation (2.15) is $O(N^{2.81})$. It is thus important in practice to know how rapidly the state distribution approaches the stationary distribution.

Recall from Theorem 2.12 that $p_{ij}(n)$ converges to π_j as $n \to \infty$ at the rate of the decay parameter $|\lambda_1|$, where λ_1 is the (possibly complex) eigenvalue of $\mathbf{P}$ that is largest in magnitude except for the unit eigenvalue. It turns out that λ_1 is the eigenvalue of $\mathbf{P} - \mathbf{1}\boldsymbol{\pi}^\top$ that is largest in magnitude, and therefore a standard method such as the power method may be applied to obtain $|\lambda_1|$. See, e.g., Golub and Van Loan (page 351, 1989) for numerical methods. However, there may be numerical difficulties if N is large. It is therefore of practical importance to obtain an upper bound on the decay parameter of an ergodic Markov chain. Recall that the relaxation time T_{REL} defined in (2.38) involves the decay parameter. An upper bound on $|\lambda_1|$ turns out to be an upper bound on the relaxation time. In this section, we consider finite, ergodic Markov chains and investigate various measures of convergence to stationarity. These measures will then be compared with the decay parameter $|\lambda_1|$.

For a stochastic matrix $\mathbf{P} = (p_{ij})$, define

$$\tau(\mathbf{P}) = \frac{1}{2} \max_{i,j \in \mathcal{N}} \sum_{k=0}^{N} |p_{ik} - p_{jk}|, \tag{2.52}$$

i.e. the *coefficient of ergodicity*, which satisfies the following properties:

(a) $0 \leq \tau(\mathbf{P}) \leq 1$,

(b) $\tau(\mathbf{P}) = 0$ if and only if $\mathbf{P} = \mathbf{1}\boldsymbol{\pi}^\top$,

(c) $\tau(\mathbf{P}_1\mathbf{P}_2) \leq \tau(\mathbf{P}_1)\tau(\mathbf{P}_2)$.

It is obvious from the definition that properties (a) and (b) hold. Property (c) will be proved later. From property (b), the coefficient $\tau(\mathbf{P}^n)$ converges to 0 as $n \to \infty$ if the Markov chain is ergodic.

In the following, we write $||\mathbf{x}||_1 = \sum_i |x_i|$, i.e. the ℓ_1-norm of $\mathbf{x} = (x_i)$. A preliminary lemma is needed whose proof may be found in Seneta (page 63, 1981).

Lemma 2.10 *Let* $\mathbf{x}$ *be a real vector such that* $\mathbf{x} \neq \mathbf{0}$ *and* $\mathbf{x}^\top\mathbf{1} = 0$. *Then there exists a set of real numbers* $\eta_{ij} \geq 0$ *such that*

$$\mathbf{x} = \sum_{i,j} \frac{\eta_{ij}}{2}(\boldsymbol{\delta}_i - \boldsymbol{\delta}_j), \quad \sum_{i,j} \eta_{ij} = ||\mathbf{x}||_1,$$

where $\boldsymbol{\delta}_i$ *denotes the ith unit vector, i.e., the jth component is* 1 *if* $j = i$ *and* 0 *otherwise.*

The next theorem bounds the decay parameter $|\lambda_1|$ from above. The proof is taken from Seneta (1981). See Diaconis and Stroock (1991) for another bound on $|\lambda_1|$ for time-reversible Markov chains, and Fill (1991) for a corresponding result for nonreversible Markov chains.

Theorem 2.14 *Let* λ_1 *be the eigenvalue that is largest in magnitude other than the unit eigenvalue of a transition matrix* $\mathbf{P} = (p_{ij})$. *Then* $|\lambda_1| \leq \tau(\mathbf{P})$.

Proof. Let $\mathbf{x}$ be as in Lemma 2.10. Then, for any *complex* vector $\mathbf{z} = (z_i)$, we have

$$|\mathbf{x}^\top\mathbf{z}| \leq \sum_{i,j} \frac{\eta_{ij}}{2}|z_i - z_j| \leq \frac{1}{2} f(\mathbf{z})\, ||\mathbf{x}||_1, \tag{2.53}$$

where $f(\mathbf{z}) = \max_{i,j} |z_i - z_j|$. Consider

$$f(\mathbf{P}\mathbf{z}) = \max_{i,j} \left| \sum_k (p_{ik} - p_{jk}) z_k \right|.$$

Denote the ith row vector of $\mathbf{P}$ by $\mathbf{p}_i^\top = (p_{ik})$, and write $\mathbf{x}_{ij} = \mathbf{p}_i - \mathbf{p}_j$ so that $\mathbf{x}_{ij}^\top\mathbf{1} = 0$. If $\mathbf{x}_{ij} = \mathbf{0}$ for all i and j then $\mathbf{P} = \mathbf{1}\boldsymbol{\pi}^\top$, whose eigenvalues are 0 and 1. Hence the theorem holds for this case. So, suppose $\mathbf{x}_{ij} \neq \mathbf{0}$ for some i and j. For any complex vector $\mathbf{z}$, it follows from (2.53) that

$$|\mathbf{x}_{ij}^\top\mathbf{z}| \leq \frac{1}{2} f(\mathbf{z})\, ||\mathbf{x}_{ij}||_1,$$

whence

$$f(\mathbf{Pz}) = \max_{i,j} |\mathbf{x}_{ij}^\top \mathbf{z}| \le f(\mathbf{z}) \frac{1}{2} \max_{i,j} \|\mathbf{x}_{ij}\|_1 = f(\mathbf{z})\tau(\mathbf{P}).$$

Let λ be any eigenvalue of $\mathbf{P}$ other than unity and let $\mathbf{z}$ be the associated eigenvector. It follows that

$$|\lambda| f(\mathbf{z}) = f(\lambda \mathbf{z}) = f(\mathbf{Pz}) \le f(\mathbf{z})\tau(\mathbf{P}).$$

Since $\lambda \neq 1$ implies $f(\mathbf{z}) \neq 0$, the theorem follows. □

For $\mathbf{x}$ given in Lemma 2.10, we have

$$\|(\mathbf{x}^\top \mathbf{P})^\top\|_1 \le \sum_{i,j} \frac{\eta_{ij}}{2} \|\mathbf{p}_i - \mathbf{p}_j\|_1 \le \|\mathbf{x}\|_1 \tau(\mathbf{P}),$$

from which

$$\tau(\mathbf{P}) \ge \sup_{\mathbf{x}} \frac{\|\mathbf{P}^\top \mathbf{x}\|_1}{\|\mathbf{x}\|_1}.$$

But, letting i and j be such that $\tau(\mathbf{P}) = \|\mathbf{p}_i - \mathbf{p}_j\|_1/2$ and defining $\mathbf{x} = \boldsymbol{\delta}_i - \boldsymbol{\delta}_j$, we have $\|(\mathbf{x}^\top \mathbf{P})^\top\|_1 = \|\mathbf{p}_i - \mathbf{p}_j\|_1$ so that $\tau(\mathbf{P}) = \|\mathbf{P}^\top \mathbf{x}\|_1/\|\mathbf{x}\|_1$. It follows that

$$\tau(\mathbf{P}) = \sup_{\mathbf{x} \neq 0,\, \mathbf{x}^\top \mathbf{1} = 0} \frac{\|\mathbf{P}^\top \mathbf{x}\|_1}{\|\mathbf{x}\|_1}, \tag{2.54}$$

whence the coefficient of ergodicity satisfies the inequality

$$\tau(\mathbf{P}_1 \mathbf{P}_2) \le \tau(\mathbf{P}_1)\tau(\mathbf{P}_2),$$

as stated earlier in property (c). Therefore, for any transition matrix $\mathbf{P}$, we have

$$\tau(\mathbf{P}^n) \le \tau^n(\mathbf{P}), \quad n = 0, 1, \cdots.$$

Noting that, for any real pair x and y,

$$|x - y| = (x + y) - 2\min\{x, y\} = 2\max\{x, y\} - (x + y),$$

the coefficient of ergodicity can also be written as

$$\begin{aligned} \tau(\mathbf{P}) &= 1 - \min_{i,j \in \mathcal{N}} \sum_{k=0}^{N} \min\{p_{ik}, p_{jk}\} \\ &= \max_{i,j \in \mathcal{N}} \sum_{k=0}^{N} \max\{p_{ik}, p_{jk}\} - 1. \end{aligned}$$

Example 2.21 For the two-state Markov chain given in Example 2.15, the coefficient of ergodicity is given by $\tau(\mathbf{P}) = |1 - a - b|$, coincident with the decay parameter $|\lambda_1|$ (see Example 2.20). For the random walk of Example 2.9 with state space $\{0, 1, 2, 3\}$, however, we have

$$\sum_{k=0}^{3} \min\{p_{0k}, p_{3k}\} = 0,$$

from which $\tau(\mathbf{P}) = 1$. Hence, Theorem 2.14 above produces a trivial bound $|\lambda_1| \le 1$.

If $\mathbf{P}$ is sparse, as for the case of random walks, then the coefficient of ergodicity is often equal to 1. For such cases, we may need to compute the power $\mathbf{P}^m$ so that $\mathbf{P}^m$ has fewer zero components. If $\mathbf{P}^m$ has no zero components then it necessarily holds that $\tau(\mathbf{P}^m) < 1$. Therefore,

$$|\lambda_1| \le \tau^{1/m}(\mathbf{P}^m) < 1.$$

When $\mathbf{P}$ is regular, we have the following. For the proof, see Paz (1971).

Lemma 2.11 *Let* $\mathbf{P}$ *be any regular transition matrix defined on the state space* $\mathcal{N} = \{0, 1, \cdots, N\}$, *and let* $\gamma = N(N+1)/2$. *Then, for any* $n \ge \gamma$, $\mathbf{P}^n$ *has no zero components.*

Another useful application of the coefficient of ergodicity is the study of the sensitivity of finite Markov chains under perturbation. Let $\mathbf{P}$ be a regular transition matrix with stationary distribution $\boldsymbol{\pi}$. Of interest is how much the stationary distribution changes from $\boldsymbol{\pi}$ if we perturb $\mathbf{P}$ to another regular transition matrix $\overline{\mathbf{P}} = \mathbf{P} + \mathbf{C}$. Let $\overline{\boldsymbol{\pi}}$ be the stationary distribution of $\overline{\mathbf{P}}$ and define

$$\mathbf{A}^\sharp = (\mathbf{I} - \mathbf{P} + \mathbf{1}\boldsymbol{\pi}^\top)^{-1} - \mathbf{1}\boldsymbol{\pi}^\top. \tag{2.55}$$

The inverse exists since $\mathbf{P}$ is regular. It is readily seen that

$$(\overline{\boldsymbol{\pi}} - \boldsymbol{\pi})^\top (\mathbf{I} - \mathbf{P} + \mathbf{1}\boldsymbol{\pi}^\top) = \overline{\boldsymbol{\pi}}^\top \mathbf{C},$$

from which, and since $\mathbf{C}\mathbf{1} = \mathbf{0}$, we have

$$\overline{\boldsymbol{\pi}}^\top - \boldsymbol{\pi}^\top = \overline{\boldsymbol{\pi}}^\top \mathbf{C}\mathbf{A}^\sharp. \tag{2.56}$$

For matrix $\mathbf{A} = (a_{ij})$, we define the matrix ℓ_1-norm and ℓ_∞-norm by

$$||\mathbf{A}||_1 = \sup_{||\mathbf{x}||_1 = 1} ||\mathbf{A}\mathbf{x}||_1, \quad ||\mathbf{A}||_\infty = \sup_{||\mathbf{x}||_\infty = 1} ||\mathbf{A}\mathbf{x}||_\infty$$

respectively. It is well known that $||\mathbf{A}||_1 = \max_j \sum_i |a_{ij}|$ is the maximal absolute column sum, while $||\mathbf{A}||_\infty = \max_i \sum_j |a_{ij}|$ is the maximal absolute row sum. It follows that

$$||\mathbf{A}||_1 = ||\mathbf{A}^\top||_\infty.$$

Now, from (2.56) and (2.54),

$$||\overline{\boldsymbol{\pi}} - \boldsymbol{\pi}||_1 = ||(\overline{\boldsymbol{\pi}}^\top \mathbf{C})^\top||_1 \, \frac{||(\overline{\boldsymbol{\pi}}^\top \mathbf{C}\mathbf{A}^\sharp)^\top||_1}{||(\overline{\boldsymbol{\pi}}^\top \mathbf{C})^\top||_1} \le ||\mathbf{C}^\top||_1 \, \tau(\mathbf{A}^\sharp),$$

so that, although $\mathbf{A}^\sharp$ may not be stochastic, we have

$$\frac{||\overline{\boldsymbol{\pi}} - \boldsymbol{\pi}||_1}{||\mathbf{C}||_\infty} \le \tau(\mathbf{A}^\sharp).$$

Meyer (1994) considered the question of whether the closeness of the non-unit eigenvalues of $\mathbf{P}$ to unity provides complete information about the relative sensitivity of $\mathbf{P}$. Seneta (1993) confirmed this by deriving the inequalities

$$\frac{1}{\min_{1\le j\le N}|1-\lambda_j|} \le \tau(\mathbf{A}^\sharp) \le \frac{n}{\min_{1\le j\le N}|1-\lambda_j|},$$

where the λ_j are the eigenvalues of $\mathbf{P}$ (see Exercise 2.21).

The term 'coefficient of ergodicity' comes from the next result, which is due to Seneta (1991).

Theorem 2.15 *For any regular* $\mathbf{P}$ *defined on the state space* $\{0, 1, \cdots, N\}$, *let* $\gamma = N(N+1)/2$. *Then we have*

$$\|\mathbf{P}^n - \mathbf{1}\boldsymbol{\pi}^\top\|_\infty \le C(\tau^{1/\gamma}(\mathbf{P}^\gamma))^n, \quad n \ge \gamma,$$

where C *is a constant independent of* n.

Proof. Note that $\boldsymbol{\pi}^\top\mathbf{P}^n = \boldsymbol{\pi}^\top$ for any n. Hence, by definition,

$$\|\mathbf{P}^n - \mathbf{1}\boldsymbol{\pi}^\top\|_\infty = \sup_{\|\mathbf{x}\|_1=1} \|(\mathbf{x}^\top(\mathbf{I}-\mathbf{1}\boldsymbol{\pi}^\top)\mathbf{P}^n)^\top\|_1.$$

Since $\mathbf{x}^\top(\mathbf{I}-\mathbf{1}\boldsymbol{\pi}^\top)\mathbf{1} = 0$, one has, for $\mathbf{x} \ne \boldsymbol{\pi}$,

$$\begin{aligned}\|(\mathbf{x}^\top(\mathbf{I}-\mathbf{1}\boldsymbol{\pi}^\top)\mathbf{P}^n)^\top\|_1 &= \|(\mathbf{x}^\top(\mathbf{I}-\mathbf{1}\boldsymbol{\pi}^\top))^\top\|_1 \frac{\|(\mathbf{x}^\top(\mathbf{I}-\mathbf{1}\boldsymbol{\pi}^\top)\mathbf{P}^n)^\top\|_1}{\|(\mathbf{x}^\top(\mathbf{I}-\mathbf{1}\boldsymbol{\pi}^\top))^\top\|_1} \\ &\le \|\mathbf{x}-(\mathbf{x}^\top\mathbf{1})\boldsymbol{\pi}\|_1\,\tau(\mathbf{P}^n),\end{aligned}$$

where the inequality follows from (2.54). For $n \ge \gamma$, write $n = m\gamma + r$ with $0 \le r < \gamma$. Then, from property (c) of the coefficient τ, we have

$$\tau(\mathbf{P}^n) \le \tau^m(\mathbf{P}^\gamma)\tau(\mathbf{P}^r) = \frac{\tau(\mathbf{P}^r)}{(\tau^{1/\gamma}(\mathbf{P}^\gamma))^r}(\tau^{1/\gamma}(\mathbf{P}^\gamma))^n,$$

completing the proof. □

Another way of studying the rate of convergence is *coupling*. Let $\{X_n\}$ be an ergodic Markov chain with state space $\mathcal{N}$ and transition matrix $\mathbf{P}$. Recall that, for any initial distribution $\boldsymbol{\alpha}$, the state distribution $\boldsymbol{\alpha}^\top\mathbf{P}^n$ converges to its stationary distribution $\boldsymbol{\pi}$ as $n \to \infty$. The idea of coupling is as follows. Introduce a parallel process $\{\widehat{X}_n\}$, independent of $\{X_n\}$, with transition matrix $\mathbf{P}$ and initial distribution $\boldsymbol{\pi}$ so that $\{\widehat{X}_n\}$ is stationary. Define $\{Y_n\}$ by

$$Y_n = \begin{cases} X_n, & n < T, \\ \widehat{X}_n, & n \ge T, \end{cases}$$

where

$$T = \inf\{n : X_n = \widehat{X}_n\}.$$

That is, the two Markov chains are coupled at T, called the *coupling time*. Due to the strong Markov property, we have

$$P[X_n = j,\, n \ge T] = P[\widehat{X}_n = j,\, n \ge T], \quad j \in \mathcal{N}.$$

It follows that

$$\begin{aligned}|P[X_n = j] - \pi_j| &= |P[X_n = j] - P[\widehat{X}_n = j]| \\ &= |P[X_n = j,\, n < T] - P[\widehat{X}_n = j,\, n < T]| \\ &\leq P[X_n = j,\, n < T] + P[\widehat{X}_n = j,\, n < T],\end{aligned}$$

whence

$$\|(\boldsymbol{\alpha}^\top \mathbf{P}^n)^\top - \boldsymbol{\pi}\|_1 \leq 2P[T > n].$$

Thus, if we find the tail probabilities of the coupling time T, then we can obtain an immediate bound on the rate of convergence. Fortunately, we can in fact do so and there is a huge literature on coupling. See e.g., Aldous (1983) and Lindvall (1992), and references therein.

There are many recent papers on the rate of convergence. Among them, Spieksma (1993) has identified upper bounds on the rate of convergence for some specific queueing models, and Lund and Tweedie (1996) treat rates of convergence for monotone processes. Meyn and Tweedie (1993) focus on the topic of geometric ergodicity. Many further references on these subjects can be found therein. See also Rosenthal (1995) for recent results.

Now let $\{X_n\}$ be a finite, ergodic Markov chain with transition matrix $\mathbf{P}$ and let $\boldsymbol{\pi} = (\pi_i)$ denote its stationary distribution. A stationary Markov chain $\{\widehat{X}_n\}$ corresponding to the ergodic Markov chain $\{X_n\}$ is defined as above. The *correlation coefficient* of $\mathbf{P}$ is defined by

$$d(\mathbf{P}) = \sup_{f,g} \mathrm{Corr}[f(\widehat{X}_0),\, g(\widehat{X}_1)], \tag{2.57}$$

where $\mathbf{f} = (f(i))$ and $\mathbf{g} = (g(i))$ are real-valued functions on the finite state space $\mathcal{N}$, and the supremum is taken over all such real functions. Here Corr denotes the correlation operator

$$\mathrm{Corr}[X, Y] = \frac{E[XY] - E[X]E[Y]}{\sqrt{V[X]V[Y]}}$$

for random variables X and Y. Note that

$$E[f(\widehat{X}_0)] = \sum_{i \in \mathcal{N}} f(i)\pi_i = \mathbf{f}^\top \boldsymbol{\pi}, \quad E[g(\widehat{X}_1)] = \mathbf{g}^\top \boldsymbol{\pi},$$

and that

$$E[f^2(\widehat{X}_0)] = \sum_{i \in \mathcal{N}} f^2(i)\pi_i = \mathbf{f}^\top \boldsymbol{\pi}_{\mathrm{D}} \mathbf{f}, \quad E[g^2(\widehat{X}_1)] = \mathbf{g}^\top \boldsymbol{\pi}_{\mathrm{D}} \mathbf{g},$$

where $\boldsymbol{\pi}_{\mathrm{D}}$ is the diagonal matrix with diagonal elements π_i. Hence the variances are given by

$$V[f(\widehat{X}_0)] = \mathbf{f}^\top \boldsymbol{\pi}_{\mathrm{D}} \mathbf{f} - (\mathbf{f}^\top \boldsymbol{\pi})^2, \quad V[g(\widehat{X}_1)] = \mathbf{g}^\top \boldsymbol{\pi}_{\mathrm{D}} \mathbf{g} - (\mathbf{g}^\top \boldsymbol{\pi})^2.$$

Also,

$$\begin{aligned} E[f(\widehat{X}_0)g(\widehat{X}_1)] &= \sum_{i,j\in\mathcal{N}} f(i)g(j)P[\widehat{X}_0=i,\,\widehat{X}_1=j] \\ &= \sum_{i,j\in\mathcal{N}} f(i)g(j)\pi_i p_{ij} \\ &= \mathbf{f}^\top \boldsymbol{\pi}_{\mathrm{D}}\mathbf{P}\,\mathbf{g}. \end{aligned}$$

Since $\mathbf{f}^\top\boldsymbol{\pi} = \mathbf{f}^\top\boldsymbol{\pi}_{\mathrm{D}}\mathbf{1}$, it follows that

$$d(\mathbf{P}) = \sup_{\mathbf{f},\mathbf{g}} \frac{\mathbf{f}^\top\boldsymbol{\pi}_{\mathrm{D}}(\mathbf{P}-\mathbf{1}\boldsymbol{\pi}^\top)\mathbf{g}}{\sqrt{\mathbf{f}^\top\boldsymbol{\pi}_{\mathrm{D}}\mathbf{f}-(\mathbf{f}^\top\boldsymbol{\pi})^2}\,\sqrt{\mathbf{g}^\top\boldsymbol{\pi}_{\mathrm{D}}\mathbf{g}-(\mathbf{g}^\top\boldsymbol{\pi})^2}}. \tag{2.58}$$

For the stationary distribution $\boldsymbol{\pi} = (\pi_i)$, let $\boldsymbol{\pi}_{\mathrm{D}}^{1/2}$ and $\boldsymbol{\pi}_{\mathrm{D}}^{-1/2}$ be the diagonal matrices with diagonal elements $\pi_i^{1/2}$ and $\pi_i^{-1/2}$ respectively. Write

$$\widetilde{\mathbf{P}} \equiv \boldsymbol{\pi}_{\mathrm{D}}^{1/2}\mathbf{P}\boldsymbol{\pi}_{\mathrm{D}}^{-1/2}. \tag{2.59}$$

Let $\mathbf{x} = \boldsymbol{\pi}_{\mathrm{D}}^{1/2}\mathbf{f}$ and $\mathbf{y} = \boldsymbol{\pi}_{\mathrm{D}}^{1/2}\mathbf{g}$. Then (2.58) can be written as

$$d(\mathbf{P}) = \sup_{\mathbf{x},\mathbf{y}} \frac{\mathbf{x}^\top(\widetilde{\mathbf{P}}-\sqrt{\boldsymbol{\pi}}\sqrt{\boldsymbol{\pi}}^\top)\mathbf{y}}{\sqrt{\mathbf{x}^\top\mathbf{x}-(\mathbf{x}^\top\sqrt{\boldsymbol{\pi}})^2}\,\sqrt{\mathbf{y}^\top\mathbf{y}-(\mathbf{y}^\top\sqrt{\boldsymbol{\pi}})^2}},$$

where $\sqrt{\boldsymbol{\pi}} = \boldsymbol{\pi}_{\mathrm{D}}^{1/2}\mathbf{1}$ is the column vector with components $\pi_i^{1/2}$. Let W be a subspace of R^{N+1} that is orthogonal to $\sqrt{\boldsymbol{\pi}}$, i.e.,

$$W = \{\mathbf{y} : \mathbf{y}^\top\sqrt{\boldsymbol{\pi}} = 0\}.$$

Any vector $\mathbf{x}$ on R^{N+1} can be decomposed as $\mathbf{x} = x_0\sqrt{\boldsymbol{\pi}} + \widetilde{\mathbf{x}}$, with $x_0 \in R$ and $\widetilde{\mathbf{x}} \in W$. Also, writing $\mathbf{P} = \mathbf{1}\boldsymbol{\pi}^\top + \boldsymbol{\Delta}$, we have

$$\widetilde{\mathbf{P}} = \sqrt{\boldsymbol{\pi}}\sqrt{\boldsymbol{\pi}}^\top + \widetilde{\boldsymbol{\Delta}}; \quad \widetilde{\boldsymbol{\Delta}} = \boldsymbol{\pi}_{\mathrm{D}}^{1/2}\boldsymbol{\Delta}\boldsymbol{\pi}_{\mathrm{D}}^{-1/2}. \tag{2.60}$$

Note that $\widetilde{\boldsymbol{\Delta}}$ is orthogonal to $\sqrt{\boldsymbol{\pi}}$. Since

$$\sqrt{\boldsymbol{\pi}}^\top\sqrt{\boldsymbol{\pi}} = \sum_i \sqrt{\pi_i}^2 = \sum_i \pi_i = 1,$$

we have

$$\mathbf{x}^\top\mathbf{x} = x_0^2 + \widetilde{\mathbf{x}}^\top\widetilde{\mathbf{x}} \quad\text{and}\quad (\mathbf{x}^\top\sqrt{\boldsymbol{\pi}})^2 = x_0^2.$$

Hence, writing $\mathbf{y} = y_0\sqrt{\boldsymbol{\pi}} + \widetilde{\mathbf{y}}$, it follows that

$$d(\mathbf{P}) = \sup_{\widetilde{\mathbf{x}},\widetilde{\mathbf{y}}} \frac{\widetilde{\mathbf{x}}^\top\widetilde{\boldsymbol{\Delta}}\widetilde{\mathbf{y}}}{\|\widetilde{\mathbf{x}}\|_2\,\|\widetilde{\mathbf{y}}\|_2},$$

where $\|\mathbf{x}\|_2 = \left(\sum_i x_i^2\right)^{1/2}$ denotes the ℓ_2-norm of $\mathbf{x} = (x_i)$. Therefore, one arrives at the expression

$$d(\mathbf{P}) = \sup_{\mathbf{x},\mathbf{y}\in W} \frac{\mathbf{x}^{\top}\widetilde{\mathbf{P}}\,\mathbf{y}}{\|\mathbf{x}\|_2\,\|\mathbf{y}\|_2}. \tag{2.61}$$

It is obvious that

(a) $0 \le d(\mathbf{P}) \le 1$,

(b) $d(\mathbf{P}) = 0$ if and only if $\mathbf{P} = \mathbf{1}\boldsymbol{\pi}^{\top}$,

as for the coefficient of ergodicity τ. Property (a) follows from definition (2.57) and (b) from (2.61). Note that property (c) of τ need not hold for the correlation coefficient.

Theorem 2.16 *Suppose that* $\widetilde{\mathbf{P}}$ *in* (2.59) *is symmetric. Then* $d(\mathbf{P}) = |\lambda_1|$, *where* λ_1 *is the eigenvalue of* $\mathbf{P}$ *that is largest in magnitude other than unity. In words, the correlation coefficient coincides with the decay parameter.*

Proof. Since $\widetilde{\mathbf{P}}$ is symmetric, we have the spectral decomposition

$$\widetilde{\mathbf{P}} = \sqrt{\boldsymbol{\pi}}\sqrt{\boldsymbol{\pi}}^{\top} + \sum_{j=1}^{N} \lambda_j\, \mathbf{z}_j \mathbf{z}_j^{\top},$$

where the λ_j, $j = 1, \cdots, N$, are the eigenvalues of $\mathbf{P}$, other than unity, and the $\mathbf{z}_j$ are the associated eigenvectors. Note that the system $\{\mathbf{z}_1, \cdots, \mathbf{z}_N\}$ is an orthonormal basis of W. Hence, any vector $\mathbf{x} \in W$ can be written as $\mathbf{x} = \sum_{j=1}^{N} \alpha_j \mathbf{z}_j$ with some real numbers α_j. Writing $\mathbf{y} = \sum_{j=1}^{N} \beta_j \mathbf{z}_j$, it follows that

$$\frac{\mathbf{x}^{\top}\widetilde{\mathbf{P}}\,\mathbf{y}}{\|\mathbf{x}\|_2\,\|\mathbf{y}\|_2} = \frac{\sum_{j=1}^{N} \alpha_j \beta_j \lambda_j}{\sqrt{\sum_j \alpha_j^2}\sqrt{\sum_j \beta_j^2}}.$$

Since $\sum_j |\alpha_j \beta_j| \le \sqrt{\sum_j \alpha_j^2}\sqrt{\sum_j \beta_j^2}$, we have

$$|\lambda_1| - \frac{\sum_{j=1}^{N} |\alpha_j \beta_j \lambda_j|}{\sqrt{\sum_j \alpha_j^2}\sqrt{\sum_j \beta_j^2}} \ge \frac{\sum_{j=1}^{N} |\alpha_j \beta_j|(|\lambda_1| - |\lambda_j|)}{\sqrt{\sum_j \alpha_j^2}\sqrt{\sum_j \beta_j^2}} \ge 0.$$

Therefore, $|\lambda_1| \ge d(\mathbf{P})$. On the other hand, let $\mathbf{x} = \mathbf{z}_1$, and let $\mathbf{y} = \mathbf{z}_1$ if $\lambda_1 \ge 0$ and $\mathbf{y} = -\mathbf{z}_1$ if $\lambda_1 < 0$. Then

$$d(\mathbf{P}) \ge \frac{\mathbf{x}^{\top}\widetilde{\mathbf{P}}\,\mathbf{y}}{\|\mathbf{x}\|_2\,\|\mathbf{y}\|_2} = |\lambda_1|.$$

Combining the above completes the proof. □

The next corollary shows the importance of the relaxation time

$$T_{\mathrm{REL}}(\mathbf{P}) = \frac{1}{1 - |\lambda_1|}$$

for time-reversible Markov chains.

Corollary 2.9 *Let $\{X_n\}$ be a finite, ergodic Markov chain with transition matrix* $\mathbf{P}$. *If $\{X_n\}$ is reversible in time, then*

$$T_{\mathrm{REL}}(\mathbf{P}) = \sup_{f,g} \sum_{n=0}^{\infty} \mathrm{Corr}[f(\widehat{X}_0), g(\widehat{X}_n)],$$

where the supremum is taken over all real-valued functions f and g on $\mathcal{N}$.

Proof. First note that

$$\sup_{f,g} \sum_{n=0}^{\infty} \mathrm{Corr}[f(\widehat{X}_0), g(\widehat{X}_n)] \le \sum_{n=0}^{\infty} d(\mathbf{P}^n).$$

Since $\{X_n\}$ is reversible in time, $\widetilde{\mathbf{P}}^n$ is symmetric so that $d(\mathbf{P}^n) = |\lambda_1|^n$, by Theorem 2.16. Therefore,

$$T_{\mathrm{REL}}(\mathbf{P}) = \frac{1}{1-|\lambda_1|} \ge \sup_{f,g} \sum_{n=0}^{\infty} \mathrm{Corr}[f(\widehat{X}_0), g(\widehat{X}_n)].$$

On the other hand, as in the proof of Theorem 2.16, we have

$$\widetilde{\mathbf{P}}^n = \sqrt{\boldsymbol{\pi}}\sqrt{\boldsymbol{\pi}}^{\top} + \sum_{j=1}^{N} \lambda_j^n \mathbf{z}_j \mathbf{z}_j^{\top}.$$

Let $\mathbf{f} = \boldsymbol{\pi}_{\mathrm{D}}^{-1/2}\mathbf{z}_1$, and let $\mathbf{g} = \boldsymbol{\pi}_{\mathrm{D}}^{-1/2}\mathbf{z}_1$ if $\lambda_1 \ge 0$ and $\mathbf{g} = -\boldsymbol{\pi}_{\mathrm{D}}^{-1/2}\mathbf{z}_1$ if $\lambda_1 < 0$. Then

$$\sum_{n=0}^{\infty} \mathrm{Corr}[f(\widehat{X}_0), g(\widehat{X}_n)] = \sum_{n=0}^{\infty} |\lambda_1|^n = T_{\mathrm{REL}}(\mathbf{P}),$$

whence the result. □

For any real matrix $\mathbf{A}$, define the matrix ℓ_2-norm by

$$||\mathbf{A}||_2 = \sup_{||\mathbf{x}||_2=1} ||\mathbf{A}\mathbf{x}||_2 = \sup_{||\mathbf{x}||_2=1} \sqrt{\mathbf{x}^{\top}\mathbf{A}^{\top}\mathbf{A}\mathbf{x}}.$$

This matrix norm is related to the largest singular value of the matrix. $\mathbf{A}^{\top}\mathbf{A}$ is symmetric and positive semidefinite since $\mathbf{x}^{\top}\mathbf{A}^{\top}\mathbf{A}\mathbf{x} = ||\mathbf{A}\mathbf{x}||_2^2 \ge 0$ for any real $\mathbf{x}$. Hence the eigenvalues of $\mathbf{A}^{\top}\mathbf{A}$ are all nonnegative. The singular values of $\mathbf{A}$ are defined to be the squared nonzero eigenvalues of $\mathbf{A}^{\top}\mathbf{A}$. Following the same proof as the proof of Theorem 2.16, we have

$$||\mathbf{A}||_2 = \rho(\mathbf{A}^{\top}\mathbf{A}),$$

where $\rho(\mathbf{A}^{\top}\mathbf{A})$ denotes the largest eigenvalue of $\mathbf{A}^{\top}\mathbf{A}$. Note that the nonzero eigenvalues of $\mathbf{A}^{\top}\mathbf{A}$ coincide with those of $\mathbf{A}\,\mathbf{A}^{\top}$. Hence, $||\mathbf{A}||_2 = ||\mathbf{A}^{\top}||_2$.

For an ergodic Markov chain with transition matrix $\mathbf{P}$, let $\mathbf{P}_{\mathrm{R}}$ be the dual of $\mathbf{P}$ (see Definition 2.7). Define

$$\mathbf{P}_1^* = \mathbf{P}\,\mathbf{P}_R \quad \text{and} \quad \mathbf{P}_2^* = \mathbf{P}_R\mathbf{P}. \tag{2.62}$$

The matrices $\mathbf{P}_i^*$, $i = 1, 2$, are stochastic and have the same stationary distribution $\boldsymbol{\pi}$, i.e.,

$$\boldsymbol{\pi}^\top \mathbf{P}_i^* = \boldsymbol{\pi}^\top, \quad i = 1, 2.$$

By definition, we have

$$\widetilde{\mathbf{P}}_1^* \equiv \boldsymbol{\pi}_D^{1/2}\mathbf{P}_1^*\boldsymbol{\pi}_D^{-1/2} = \boldsymbol{\pi}_D^{1/2}\mathbf{P}\boldsymbol{\pi}_D^{-1}\mathbf{P}^\top\boldsymbol{\pi}_D\boldsymbol{\pi}_D^{-1/2} = \widetilde{\mathbf{P}}\,\widetilde{\mathbf{P}}^\top.$$

Similarly, $\widetilde{\mathbf{P}}_2^* = \widetilde{\mathbf{P}}^\top\widetilde{\mathbf{P}}$. From (2.60) and since $\widetilde{\boldsymbol{\Delta}}\sqrt{\boldsymbol{\pi}} = \mathbf{0}$, it follows that

$$\widetilde{\mathbf{P}}_1^* = \sqrt{\boldsymbol{\pi}}\sqrt{\boldsymbol{\pi}}^\top + \widetilde{\boldsymbol{\Delta}}\,\widetilde{\boldsymbol{\Delta}}^\top \quad \text{and} \quad \widetilde{\mathbf{P}}_2^* = \sqrt{\boldsymbol{\pi}}\sqrt{\boldsymbol{\pi}}^\top + \widetilde{\boldsymbol{\Delta}}^\top\widetilde{\boldsymbol{\Delta}}.$$

Hence each $\widetilde{\mathbf{P}}_i^*$ is symmetric and a Markov chain with transition matrix $\mathbf{P}_i^*$ is reversible in time. The next result is due to Kijima (1989a).

Theorem 2.17 *For any regular* $\mathbf{P}$*, we have*

$$|\lambda_1| \le \sqrt{d(\mathbf{P}_i^*)} = \|\widetilde{\boldsymbol{\Delta}}\|_2, \quad i = 1, 2.$$

Moreover, if $\mathbf{P}_1^* = \mathbf{P}_2^*$*, then* $|\lambda_1| = \|\widetilde{\boldsymbol{\Delta}}\|_2$.

Proof. Since the $\widetilde{\mathbf{P}}_i^*$ are symmetric and since $\|\widetilde{\boldsymbol{\Delta}}\|_2 = \|\widetilde{\boldsymbol{\Delta}}^\top\|_2$, we have $\|\widetilde{\boldsymbol{\Delta}}\|_2^2 = d(\mathbf{P}_i^*)$ for $i = 1, 2$. Now, let $\mathbf{x}$ be such that*

$$\lambda_1\mathbf{x} = \widetilde{\boldsymbol{\Delta}}\mathbf{x} \quad \text{and} \quad \|\mathbf{x}\|_2 = 1.$$

Then $|\lambda_1| = \|\widetilde{\boldsymbol{\Delta}}\mathbf{x}\|_2$ so that

$$|\lambda_1| \le \sup_{\|\mathbf{x}\|_2=1} \|\widetilde{\boldsymbol{\Delta}}\mathbf{x}\|_2 = \|\widetilde{\boldsymbol{\Delta}}\|_2.$$

If $\widetilde{\mathbf{P}}_1^* = \widetilde{\mathbf{P}}_2^*$ then $\widetilde{\boldsymbol{\Delta}}\widetilde{\boldsymbol{\Delta}}^\top = \widetilde{\boldsymbol{\Delta}}^\top\widetilde{\boldsymbol{\Delta}}$; viz. $\widetilde{\boldsymbol{\Delta}}$ is a normal matrix. The standard spectral theory for normal matrices then leads to the result (see, e.g., Noble and Daniel, Section 9.3, 1977). □

Aldous (1988) considered the correlation coefficient $d(\mathbf{P}^n)$ as a measure of dependence between $\widehat{X}_n$ and $\widehat{X}_0$, which may be informative for the finite time behavior of the Markov chain $\{X_n\}$. To bound $d(\mathbf{P}^n)$ from above, the matrix norm is again useful.

Theorem 2.18 *For any regular* $\mathbf{P}$*, we have*

$$d(\mathbf{P}^n) \le \|\widetilde{\boldsymbol{\Delta}}\|_2^n, \quad n = 0, 1, \cdots.$$

Proof. Note that

$$\widetilde{\mathbf{P}}^n = \sqrt{\boldsymbol{\pi}}\sqrt{\boldsymbol{\pi}}^\top + \widetilde{\boldsymbol{\Delta}}^n, \quad n = 0, 1, \cdots.$$

* If $\mathbf{x} = (x_i)$ is a complex vector, the ℓ_2-norm is defined by $\|\mathbf{x}\|_2 = \left(\sum_i |x_i|^2\right)^{1/2}$.

Hence, letting $\mathbf{x}$ and $\mathbf{y}$ be such that

$$d(\mathbf{P}^n) = \mathbf{x}^\top \widetilde{\boldsymbol{\Delta}}^n \mathbf{y} \quad \text{and} \quad \|\mathbf{x}\|_2 = \|\mathbf{y}\|_2 = 1,$$

one obtains

$$\begin{aligned} d(\mathbf{P}^n) &= \mathbf{x}^\top (\widetilde{\boldsymbol{\Delta}}^n \mathbf{y}) \\ &\le \|\mathbf{x}\|_2 \, \|\widetilde{\boldsymbol{\Delta}}^n \mathbf{y}\|_2 \\ &\le \sup_{\|\mathbf{y}\|_2 = 1} \|\widetilde{\boldsymbol{\Delta}}^n \mathbf{y}\|_2 \\ &= \|\widetilde{\boldsymbol{\Delta}}^n\|_2. \end{aligned}$$

Since

$$\|\widetilde{\boldsymbol{\Delta}}^n\|_2 = \sup \frac{\|\widetilde{\boldsymbol{\Delta}}\mathbf{x}\|_2}{\|\mathbf{x}\|_2} \frac{\|\widetilde{\boldsymbol{\Delta}}^{n-1} \widetilde{\boldsymbol{\Delta}}\mathbf{x}\|_2}{\|\widetilde{\boldsymbol{\Delta}}\mathbf{x}\|_2} \le \|\widetilde{\boldsymbol{\Delta}}\|_2 \, \|\widetilde{\boldsymbol{\Delta}}^{n-1}\|_2,$$

the result follows at once. □

So far, our main concern has been the decay parameter of a finite, ergodic Markov chain $\{X_n\}$ with transition matrix $\mathbf{P}$. Recall that the decay parameter is independent of states. On the other hand, we have

$$\mathbf{P}^n - \mathbf{1}\boldsymbol{\pi}^\top = \boldsymbol{\Delta}^n, \quad n = 1, 2, \cdots,$$

and $\mathbf{P}^n \to \mathbf{1}\boldsymbol{\pi}^\top$ as $n \to \infty$, provided that $\mathbf{P}$ is regular. Define the matrix $\mathbf{Z} = (z_{ij})$ by

$$\mathbf{Z} \equiv \mathbf{I} + \sum_{n=1}^{\infty} \{\mathbf{P}^n - \mathbf{1}\boldsymbol{\pi}^\top\} = (\mathbf{I} - \{\mathbf{P} - \mathbf{1}\boldsymbol{\pi}^\top\})^{-1}. \tag{2.63}$$

If the component z_{ij} is large in magnitude, this means that $p_{ij}(n)$ converges slowly to the limiting probability π_j. Therefore, $\mathbf{Z}$ can be considered to represent the information regarding the speed of convergence to stationarity. The matrix $\mathbf{Z}$ is usually called the *fundamental matrix* of the ergodic Markov chain $\{X_n\}$. The fundamental matrix arises quite often in the study of Markov chains. In particular, it provides a tool for evaluating mean first passage times (see Kemeny and Snell, 1960; and Iosifescu, 1980). Some identities which the fundamental matrix (2.63) satisfies are provided in Exercise 2.22.

Another method of quantifying nonstationarity is *separation*:

$$s_i(n) = \max_{j \in \mathcal{N}} \left\{ 1 - \frac{p_{ij}(n)}{\pi_j} \right\} = 1 - \min_{j \in \mathcal{N}} \frac{p_{ij}(n)}{\pi_j}, \quad n = 0, 1, \cdots, \tag{2.64}$$

given by Aldous and Diaconis (1987). Note the absence of an absolute value sign. It is easily seen that the separation is an upper bound on the *variation distance*

$$\|\mathbf{p}_i(n) - \boldsymbol{\pi}\| = \max_{A \subset \mathcal{N}} \left| \sum_{j \in A} \{p_{ij}(n) - \pi_j\} \right|, \quad n = 0, 1, \cdots, \tag{2.65}$$

where $\mathbf{p}_i(n) = (p_{ij}(n))$. See Diaconis and Fill (1990) for details. Note that the computation of separation, as well as the variation distance, is in general as difficult as that of the transition probabilities. Hence, it is of interest to bound separation from above. We will consider this problem in Chapter 3.

2.7 Absorbing Markov chains and their applications

Consider an *absorbing Markov chain* $\{X_n\}$ defined on $\mathcal{N}$ with k absorbing states, $k \geq 1$. Renumbering the states, the transition matrix can be written in the canonical form as

$$\mathbf{P} = \begin{pmatrix} \mathbf{I} & \mathbf{O} \\ \mathbf{R} & \mathbf{T} \end{pmatrix}, \tag{2.66}$$

where $\mathbf{I}$ denotes the identity matrix of order k. The submatrix $\mathbf{T}$ is square and corresponds to nonabsorbing states. It is assumed that nonabsorbing states communicate with each other and are transient. Hence, the matrix $\mathbf{T}$ is strictly substochastic, i.e., $\mathbf{T} \geq \mathbf{O}$ and $\mathbf{T1} \leq \mathbf{1}$ with at least one strict inequality. We denote by $\mathcal{A}$ the set of absorbing states ($\mathcal{A}^c$ is the set of transient states). The matrix $\mathbf{R}$ is nonnegative and nonzero but need not be square. Throughout this section, we assume that $\mathbf{T}$ is finite and primitive (see Definition A.1). A strictly substochastic, primitive matrix has a PF eigenvalue strictly less than unity. The PF eigenvalue is positive, simple and largest in magnitude (see Theorem A.1).

From (2.66), we have

$$\mathbf{P}^n = \begin{pmatrix} \mathbf{I} & \mathbf{O} \\ \mathbf{R}_n & \mathbf{T}^n \end{pmatrix}, \quad n = 1, 2, \cdots, \tag{2.67}$$

where

$$\mathbf{R}_n = (\mathbf{I} + \mathbf{T} + \cdots + \mathbf{T}^{n-1})\mathbf{R};$$

cf. (2.31). Under the assumptions stated above, we have $\lim_{n\to\infty} \mathbf{T}^n = \mathbf{O}$. It follows from Lemma A.1 that the inverse $(\mathbf{I} - \mathbf{T})^{-1}$ exists and is given by

$$\mathbf{N} \equiv (\mathbf{I} - \mathbf{T})^{-1} = \mathbf{I} + \mathbf{T} + \mathbf{T}^2 + \cdots; \tag{2.68}$$

cf. (2.8). The matrix $\mathbf{N} = (n_{ij})$ is called the *fundamental matrix* of the absorbing Markov chain $\{X_n\}$. See (2.63) for the fundamental matrix of an ergodic Markov chain. Recall that if $\mathbf{T}$ is primitive then there is some positive integer k such that $\mathbf{T}^n$ has no zero components for all $n \geq k$. Hence, under the assumptions, the fundamental matrix $\mathbf{N}$ is positive componentwise.

Let T_j be the random variable representing a time at which absorption at state $j \in \mathcal{A}$ occurs. If the Markov chain never reaches state j, we write $T_j = \infty$. Let

$$a_{ij} \equiv P_i[T_j < \infty], \quad i \in \mathcal{A}^c, \ j \in \mathcal{A},$$

and define $\mathbf{A} = (a_{ij})$. The quantity a_{ij} is the probability that, starting from state $i \in \mathcal{A}^c$, absorption occurs at state $j \in \mathcal{A}$. It is readily seen that

$$\mathbf{A} = \mathbf{NR}, \tag{2.69}$$

since, from (2.67) and (2.68),

$$\lim_{n\to\infty} \mathbf{P}^n = \begin{pmatrix} \mathbf{I} & \mathbf{O} \\ \mathbf{NR} & \mathbf{O} \end{pmatrix}. \tag{2.70}$$

Note that $\mathbf{R1} = (\mathbf{I} - \mathbf{T})\mathbf{1}$ since $\mathbf{P1} = \mathbf{1}$. It follows that

$$\mathbf{A1} = (\mathbf{I} - \mathbf{T})^{-1}\mathbf{R1} = \mathbf{1},$$

whence the Markov chain is eventually absorbed at one of the absorbing states. Since $\mathbf{N}$ has no zero components, $a_{ij} > 0$ if and only if the jth column of $\mathbf{R}$ is nonzero.

Alternatively, (2.69) can be obtained as follows. From a first step analysis, it is easy to see that

$$\begin{aligned} a_{ij} &= \sum_{k\in\mathcal{N}} P_i[X_1 = k,\, T_j < \infty] \\ &= P_i[X_1 = j,\, T_j < \infty] + \sum_{k\in\mathcal{A}^c} P_i[X_1 = k,\, T_j < \infty] \\ &= p_{ij} + \sum_{k\in\mathcal{A}^c} p_{ik} P_k[T_j < \infty], \end{aligned}$$

whence

$$a_{ij} = p_{ij} + \sum_{k\in\mathcal{A}^c} p_{ik} a_{kj}, \quad i \in \mathcal{A}^c, \; j \in \mathcal{A}. \tag{2.71}$$

The above equations are written in matrix form as

$$\mathbf{A} = \mathbf{R} + \mathbf{TA},$$

which, together with (2.68), leads to (2.69).

The fundamental matrix $\mathbf{N}$ plays a central role in the study of absorbing Markov chains. It appears in many expressions of quantities of interest. For example, from (2.2), (2.3) and (2.68), it is obvious that, for $i,\; j \in \mathcal{A}^c$,

$$(f_{ij}^*) = (f_{ij}^*(1)) = (\mathbf{N} - \mathbf{I})\mathbf{n}_{\mathrm{D}}^{-1},$$

where $\mathbf{n}_{\mathrm{D}}$ denotes the diagonal matrix whose diagonal entries are n_{jj}. Also, as in Section 2.1, let N_j represent the number of visits to state $j \in \mathcal{A}^c$ before absorption occurs. Define

$$\widehat{n}_{ij} = E_i[N_j], \quad i,\; j \in \mathcal{A}^c.$$

Recall that

$$\widehat{n}_{ij} = \sum_{n=0}^{\infty} p_{ij}(n) = \delta_{ij} + \sum_{n=1}^{\infty} p_{ij}(n).$$

The Chapman–Kolmogorov equation,

$$p_{ij}(n) = \sum_{k \in \mathcal{A}^c} p_{ik}\, p_{kj}(n-1), \qquad n = 1, 2, \cdots,$$

then yields

$$\begin{aligned} \widehat{n}_{ij} &= \delta_{ij} + \sum_{n=1}^{\infty} \sum_{k \in \mathcal{A}^c} p_{ik}\, p_{kj}(n-1) \\ &= \delta_{ij} + \sum_{k \in \mathcal{A}^c} \sum_{n=1}^{\infty} p_{ik}\, p_{kj}(n-1), \end{aligned}$$

where the interchange of the summations is allowed by Fubini's theorem. It follows that

$$\widehat{n}_{ij} = \delta_{ij} + \sum_{k \in \mathcal{A}^c} p_{ik} \widehat{n}_{kj}, \qquad i,\ j \in \mathcal{A}^c.$$

Hence, writing $\widehat{\mathbf{N}} = (\widehat{n}_{ij})$, we have

$$\widehat{\mathbf{N}} = \mathbf{I} + \mathbf{T}\widehat{\mathbf{N}},$$

which leads to the identity

$$\widehat{\mathbf{N}} = (\mathbf{I} - \mathbf{T})^{-1} = \mathbf{N}.$$

That is, the expected number of visits to state $j \in \mathcal{A}^c$ before absorption, starting from state $i \in \mathcal{A}^c$, is the (i,j)th component of the fundamental matrix $\mathbf{N}$. Other examples that demonstrate the usefulness of the fundamental matrix in absorbing Markov chains may be found in, e.g., Kemeny and Snell (1960) and Iosifescu (1980).

Example 2.22 Consider the gambler's ruin problem given in Example 2.3, where the transition matrix is

$$\mathbf{P} = \begin{pmatrix} 1 & 0 & 0 & 0 & \cdots & 0 \\ q_1 & r_1 & p_1 & 0 & \cdots & 0 \\ 0 & q_2 & r_2 & p_2 & \cdots & 0 \\ \vdots & & \ddots & \ddots & \ddots & \vdots \\ 0 & \cdots & 0 & q_{N-1} & r_{N-1} & p_{N-1} \\ 0 & \cdots & 0 & 0 & 0 & 1 \end{pmatrix},$$

with $p_i q_i > 0$ for all $i = 1, \cdots, N-1$. After renumbering the states, the canonical form (2.66) becomes

$$\mathbf{T} = \begin{pmatrix} r_1 & p_1 & 0 & \cdots & 0 \\ q_2 & r_2 & p_2 & \cdots & 0 \\ \vdots & \ddots & \ddots & \ddots & \vdots \\ 0 & \cdots & q_{N-2} & r_{N-2} & p_{N-2} \\ 0 & \cdots & 0 & q_{N-1} & r_{N-1} \end{pmatrix}; \quad \mathbf{R} = \begin{pmatrix} q_1 & 0 \\ 0 & 0 \\ \vdots & \vdots \\ 0 & 0 \\ 0 & p_{N-1} \end{pmatrix}.$$

The fundamental matrix is then given by

$$\mathbf{N} = \begin{pmatrix} 1-r_1 & -p_1 & 0 & \cdots & 0 \\ -q_2 & 1-r_2 & -p_2 & \cdots & 0 \\ \vdots & \ddots & \ddots & \ddots & \vdots \\ 0 & \cdots & -q_{N-2} & 1-r_{N-2} & -p_{N-2} \\ 0 & \cdots & 0 & -q_{N-1} & 1-r_{N-1} \end{pmatrix}^{-1}.$$

Write $a_i = a_{i0}$ so that a_i represents player A's ruin, with the initial fortune i. In order to obtain a_i, it is easier to consider (2.71) rather than the matrix inverse. Now the a_i satisfy the recursive relationship

$$a_i = p_i a_{i+1} + r_i a_i + q_i a_{i-1}, \quad i = 1, \cdots, N-1, \tag{2.72}$$

where $a_0 = 1$ and $a_N = 0$. Let

$$\rho_i = \frac{q_1 \cdots q_i}{p_1 \cdots p_i}, \quad i = 1, \cdots, N-1.$$

Noting $r_i = 1 - p_i - q_i$, it is not difficult to show that the general solution to (2.72) is given by

$$a_i = \frac{\rho_i + \cdots + \rho_{N-1}}{1 + \rho_1 + \cdots + \rho_{N-1}}, \quad i = 1, \cdots, N-1.$$

Example 2.23 Consider an absorbing Markov chain defined on the state space $\mathcal{N} = \{0, 1, \cdots, N\}$, with absorbing states 0 and N both accessible from the intermediate states. Let $\mathbf{P} = (p_{ij})$ be the transition matrix with the property that $p_{ij} = 0$ for $j \geq i+2$. Since the Markov chain can then move only to the neighboring state to the right, such a Markov chain is called *skip-free to the right*. If $p_{ij} = 0$ for $j \leq i-2$, it is called *skip-free to the left*. If a Markov chain is skip-free both to the left and to the right, it is a random walk. The following result is taken from Kijima (1993a).

Suppose that the Markov chain is skip-free to the right and let r_i be the probability of absorption at state 0 prior to state N when it starts from state i. From (2.71), we have

$$r_i = \sum_{j=0}^{i+1} p_{ij} r_j, \quad i = 1, \cdots, N-1, \tag{2.73}$$

where $r_0 = 1$ and $r_N = 0$. Since $r_i > 0$, $i \neq N$, we can define $\chi_i = r_i / r_{i-1}$ for $i = 1, 2, \cdots, N$. It follows from (2.73) that

$$1 = p_{i,i+1}\, \chi_{i+1} + p_{ii} + \sum_{j=0}^{i-1} \frac{p_{ij}}{\prod_{k=j+1}^{i} \chi_k}, \quad i = 1, \cdots, N-1.$$

The accessibility assumption ensures that $p_{i,i+1} > 0$ so that

$$\chi_{i+1} = \frac{1}{p_{i,i+1}} \left(1 - p_{ii} - \sum_{j=0}^{i-1} \frac{p_{ij}}{\prod_{k=j+1}^{i} \chi_k} \right), \quad i = 1, \cdots, N-1. \quad (2.74)$$

Note that $\chi_i > 0$ for $i = 1, \cdots, N-1$ and $\chi_N = 0$ since $r_N = 0$. Now choose $\chi_1(r) = r$ arbitrary, where $0 < r < 1$, and generate $\chi_i(r)$ for $i = 2, \cdots, N$ successively by

$$\chi_{i+1}(r) = \frac{1}{p_{i,i+1}} \left(1 - p_{ii} - \sum_{j=0}^{i-1} \frac{p_{ij}}{\prod_{k=j+1}^{i} \chi_k(r)} \right). \quad (2.75)$$

Since the absorption probabilities satisfying (2.73) are unique, if we find r such that $0 < r < 1$, $\chi_i(r) > 0$ for $i = 2, \cdots, N-1$ and $\chi_N(r) = 0$, then this r must be the absorption probability r_1 that we seek. This follows at once from the definition of $\chi_i(r)$, (2.74) and the fact that $\chi_1(r_1) = r_1/r_0$ with $r_0 = 1$. The remainder of the ruin probabilities are determined by

$$r_{i+1} = \chi_{i+1}(r_1) r_i = \cdots = \prod_{j=1}^{i+1} \chi_j(r_1), \quad i = 1, \cdots, N-2.$$

We note that $\chi_i(r)$, $i = 1, \cdots, N$, is strictly increasing in r, $0 < r < 1$. This fact is the key to finding $r = r_1$ numerically and is readily proved by induction. Suppose r is such that either $\chi_i(r) \leq 0$ for some $i = 2, \cdots, N-1$, or $\chi_i(r) > 0$ for $i = 2, \cdots, N-1$ and $\chi_N(r) < 0$. Then, since $\chi_i(r)$ for $i = 1, \cdots, N-1$ must be positive and $\chi_N(r)$ must be zero, this value of r is too small to be the correct value r_1 due to the monotonicity of $\chi_i(r)$ with respect to r. On the other hand, if r is such that $\chi_i(r) > 0$ for $i = 2, \cdots, N$, then $r > r_1$. Hence, we have the following bisection search algorithm to find r_1, which provides an estimate of r_1 with an error less than some prespecified $\varepsilon > 0$.

Algorithm

Step 1 $L \leftarrow 0$ and $R \leftarrow 1$.

Step 2 If $R - L < \varepsilon$, then $r_1 \leftarrow (L + R)/2$ and terminate. Otherwise, $r \leftarrow (L + R)/2$ and $i \leftarrow 1$.

Step 3 Calculate $\chi_{i+1}(r)$ by (2.75). If $\chi_{i+1}(r) \leq 0$ then $L \leftarrow r$ and go to Step 2.

Step 4 $i \leftarrow i + 1$. If $i < N - 1$ then go to Step 3.

Step 5 Calculate $\chi_N(r)$ by (2.75). If $\chi_N(r) < 0$ (> 0) then $L \leftarrow r$ ($R \leftarrow r$, respectively) and go to Step 2. If $\chi_N(r) = 0$ then $r_1 \leftarrow r$ and terminate.

If the Markov chain is skip-free to the left, we consider the probability of absorption at N rather than at 0, and use a similar algorithm to find the absorption probability.

We now turn our attention to the first passage times of an irreducible Markov chain with state space $\mathcal{N} = \{0, 1, \cdots, N\}$. Let T_j denote the first passage time to state $j \in \mathcal{N}$ of an irreducible Markov chain $\{X_n\}$. For $i \neq j$, let

$$f_{ij}(n) = P_i[T_j = n], \quad n = 1, 2, \cdots.$$

For the study of the first-passage-time distribution $f_{ij}(n)$ of $\{X_n\}$, we consider an absorbing Markov chain constructed from the original Markov chain $\{X_n\}$ by making state j absorbing. In what follows, renumbering the states, we assume $j = 0$ without loss of generality. Then, the absorbing Markov chain has the transition matrix

$$\mathbf{P} = \begin{pmatrix} 1 & \mathbf{0}^{\top} \\ \mathbf{r} & \mathbf{T} \end{pmatrix}, \tag{2.76}$$

if the original transition matrix is given by

$$\begin{pmatrix} p_{00} & \mathbf{p}_0^{\top} \\ \mathbf{r} & \mathbf{T} \end{pmatrix},$$

where $\mathbf{p}_0^{\top} = (p_{0i})$. Note that $\mathbf{r} = \mathbf{1} - \mathbf{T1}$. If the original Markov chain $\{X_n\}$ is irreducible then the absorbing state 0 is reached directly from some state so that $\mathbf{r}$ is nonnegative and nonzero, whence $\mathbf{T}$ is strictly substochastic.

Let $p_{ij}(n)$ denote the n-step transition probabilities of the absorbing Markov chain with transition matrix $\mathbf{P}$ given in (2.76). Then the following key observation holds:

$$p_{i0}(n) = P_i[T_0 \leq n], \quad n = 1, 2, \cdots,$$

where T_0 is the first passage time to the absorbing state 0. From (2.67), we have

$$\mathbf{P}^n = \begin{pmatrix} 1 & \mathbf{0}^{\top} \\ \mathbf{r}_n & \mathbf{T}^n \end{pmatrix}, \quad n = 1, 2, \cdots,$$

where

$$\mathbf{r}_n = (\mathbf{I} + \mathbf{T} + \cdots + \mathbf{T}^{n-1})\mathbf{r}.$$

Let $\mathbf{f}(n) = (f_{i0}(n))$. Since $f_{i0}(n) = p_{i0}(n) - p_{i0}(n-1)$ for $n = 1, 2, \cdots$, it follows that

$$\mathbf{f}(n) = \mathbf{r}_n - \mathbf{r}_{n-1} = \mathbf{T}^{n-1}\mathbf{r}, \quad n = 1, 2, \cdots.$$

Now define the generating function

$$f_i^*(z) = \sum_{n=1}^{\infty} f_{i0}(n) z^n, \quad |z| < 1; \quad i = 1, \cdots, N.$$

Let $\mathbf{f}^*(z)$ be the column vector with components $f_i^*(z)$. The proof of the next theorem is left to the reader (see Exercise 2.27).

Theorem 2.19 *The vector* $\mathbf{f}^*(z) = (f_i^*(z))$ *of the generating functions is given by*

$$\mathbf{f}^*(z) = z(\mathbf{I} - z\mathbf{T})^{-1}\mathbf{r}, \quad |z| < 1. \tag{2.77}$$

When $\mathbf{T}$ is irreducible and strictly substochastic (so that $\mathbf{r}$ is nonzero), we have

$$(P_i[T_0 < \infty]) = \mathbf{f}^*(1) = (\mathbf{I} - \mathbf{T})^{-1}\mathbf{r} = \mathbf{1},$$

that is, the first-passage-time distribution $f_{i0}(n)$ is not defective for every state i. Note that the inverse $(\mathbf{I} - \mathbf{T})^{-1}$ exists under the irreducibility assumption. Also,

$$\begin{aligned} & (\mathbf{I} - (z+h)\mathbf{T})^{-1} - (\mathbf{I} - z\mathbf{T})^{-1} \\ = \ & (\mathbf{I} - (z+h)\mathbf{T})^{-1}(\mathbf{I} - z\mathbf{T})(\mathbf{I} - z\mathbf{T})^{-1} \\ & \qquad -(\mathbf{I} - (z+h)\mathbf{T})^{-1}(\mathbf{I} - (z+h)\mathbf{T})(\mathbf{I} - z\mathbf{T})^{-1} \\ = \ & (\mathbf{I} - (z+h)\mathbf{T})^{-1}[\mathbf{I} - z\mathbf{T} - \mathbf{I} + (z+h)\mathbf{T}](\mathbf{I} - z\mathbf{T})^{-1} \\ = \ & h(\mathbf{I} - (z+h)\mathbf{T})^{-1}\mathbf{T}(\mathbf{I} - z\mathbf{T})^{-1}. \end{aligned}$$

Hence, differentiation of (2.77) with respect to z yields

$$\mathbf{f}^{*\prime}(z) = (\mathbf{I} - z\mathbf{T})^{-1}\mathbf{r} + z(\mathbf{I} - z\mathbf{T})^{-1}\mathbf{T}(\mathbf{I} - z\mathbf{T})^{-1}\mathbf{r}, \quad |z| < 1. \tag{2.78}$$

It follows that

$$\begin{aligned} \mathbf{f}^{*\prime}(1) &= (\mathbf{I} - \mathbf{T})^{-1}\mathbf{r} + (\mathbf{I} - \mathbf{T})^{-1}\mathbf{T}(\mathbf{I} - \mathbf{T})^{-1}\mathbf{r} \\ &= (\mathbf{I} - \mathbf{T})^{-2}\mathbf{r} \\ &= (\mathbf{I} - \mathbf{T})^{-1}\mathbf{1}. \end{aligned}$$

Recalling $\mathbf{N} = (\mathbf{I} - \mathbf{T})^{-1}$, the fundamental matrix (2.68), we thus have

$$(E_i[T_0]) = \mathbf{N1}.$$

Higher order moments of T_0 can be obtained by repeated differentiation of (2.78).

Now, suppose that the original Markov chain $\{X_n\}$ is ergodic and reversible in time, i.e., (2.43) holds for all $i, j \in \mathcal{N}$. In the modified absorbing Markov chain, since $\mathbf{T}$ is a submatrix of the original transition matrix, it is clear that the set of equations

$$\pi_i p_{ij} = \pi_j p_{ji}, \quad i, j = 1, \cdots, N$$

still holds, where $\boldsymbol{\pi} = (\pi_i)$ is the stationary distribution of the original Markov chain $\{X_n\}$. Hence $\boldsymbol{\pi}_\mathrm{D}\mathbf{T}$ is symmetric and so is $\boldsymbol{\pi}_\mathrm{D}^{1/2}\mathbf{T}\boldsymbol{\pi}_\mathrm{D}^{-1/2}$. It follows that $\boldsymbol{\pi}_\mathrm{D}^{1/2}\mathbf{T}\boldsymbol{\pi}_\mathrm{D}^{-1/2}$ admits the spectral decomposition

$$\boldsymbol{\pi}_\mathrm{D}^{1/2}\mathbf{T}\boldsymbol{\pi}_\mathrm{D}^{-1/2} = \sum_{j=1}^{N} \lambda_j \, \mathbf{x}_j \mathbf{x}_j^\mathsf{T},$$

where the λ_j are the eigenvalues of $\mathbf{T}$, which are real, such that

$$1 > |\lambda_1| > |\lambda_2| \geq \cdots \geq |\lambda_N|,$$

and the $\mathbf{x}_j$ are the associated orthonormal eigenvectors. Since, then,

$$\mathbf{T}^n = \sum_{j=1}^{N} \lambda_j^n \mathbf{v}_j \mathbf{u}_j^\top, \quad n = 0, 1, \cdots,$$

where $\mathbf{u}_j = \boldsymbol{\pi}_D^{1/2}\mathbf{x}_j$ and $\mathbf{v}_j = \boldsymbol{\pi}_D^{-1/2}\mathbf{x}_j$, it is readily seen that

$$(\mathbf{I} - z\mathbf{T})^{-1} = \sum_{n=0}^{\infty} z^n \mathbf{T}^n = \sum_{j=1}^{N} \frac{1}{1 - \lambda_j z} \mathbf{v}_j \mathbf{u}_j^\top, \quad |z| < 1.$$

Therefore, from (2.77), we have

$$\mathbf{f}^*(z) = \sum_{j=1}^{N} \frac{z}{1 - \lambda_j z} (\mathbf{u}_j^\top \mathbf{r}) \mathbf{v}_j, \quad |z| < 1. \tag{2.79}$$

Note that, if $\lambda > 0$, then

$$\frac{(1-\lambda)z}{1 - \lambda z}, \quad |z| < 1,$$

is the generating function of the geometric distribution

$$g(n; \lambda) = (1 - \lambda)\lambda^{n-1}, \quad n = 1, 2, \cdots.$$

Hence, if $\lambda_j > 0$ for all $j = 1, \cdots, N$, then we conclude that the first-passage-time distribution $f_{i0}(n)$ is given by

$$f_{i0}(n) = \sum_{j=1}^{N} \beta_{ij} (1 - \lambda_j) \lambda_j^{n-1} \quad n = 1, 2, \cdots; \quad \beta_{ij} = \frac{(\boldsymbol{\delta}_i^\top \mathbf{v}_j)(\mathbf{u}_j^\top \mathbf{r})}{1 - \lambda_j}.$$

Note that the β_{ij} may be negative. If they are all positive, then $f_{i0}(n)$ is a mixture of geometric distributions.

Example 2.24 Let $\{X_n\}$ be a one-dimensional random walk with transition matrix

$$\mathbf{P} = \begin{pmatrix} r_0 & p_0 & 0 & 0 & \cdots & 0 \\ q_1 & r_1 & p_1 & 0 & \cdots & 0 \\ 0 & q_2 & r_2 & p_2 & \cdots & 0 \\ \vdots & & \ddots & \ddots & \ddots & \vdots \\ 0 & \cdots & 0 & q_{N-1} & r_{N-1} & p_{N-1} \\ 0 & \cdots & 0 & 0 & q_N & r_N \end{pmatrix},$$

where $p_i q_{i+1} > 0$ for all $i = 0, 1, \cdots, N-1$, and at least one of r_i, $i = 1, 2, \cdots, N$ is positive. These assumptions ensure that the random walk is ergodic (see Example 2.9). In this example we consider the first passage time T_0. For this purpose, we need to define

$$\mathbf{T} = \begin{pmatrix} r_1 & p_1 & 0 & \cdots & 0 \\ q_2 & r_2 & p_2 & \cdots & 0 \\ \vdots & \ddots & \ddots & \ddots & \vdots \\ 0 & \cdots & q_{N-1} & r_{N-1} & p_{N-1} \\ 0 & \cdots & 0 & q_N & r_N \end{pmatrix}; \quad \mathbf{r} = \begin{pmatrix} q_1 \\ 0 \\ \vdots \\ 0 \\ 0 \end{pmatrix}.$$

Since $\mathbf{T}$ is nonnegative and tridiagonal, if λ_j denote the eigenvalues of $\mathbf{T}$, then the λ_j are real and simple. Defining $\mathbf{x}_j = (x_{ij})$, (2.79) leads to

$$f_1^*(z) = \sum_{j=1}^{N} \frac{z}{1 - \lambda_j z} x_{1j}^2 q_1, \quad |z| < 1.$$

Therefore, if $\lambda_j > 0$ for all $j = 1, 2, \cdots, N$, then the first-passage-time distribution $f_{10}(n)$ is given by a mixture of geometric distributions. A necessary and sufficient condition for the λ_js to be positive is that the symmetrized matrix $\boldsymbol{\pi}_\mathbf{D}^{1/2}\mathbf{T}\boldsymbol{\pi}_\mathbf{D}^{-1/2}$ is positive definite.

On the other hand, from (2.77) and the form of $\mathbf{r}$, we have

$$f_N^*(z) = z\, q_1[(\mathbf{I} - z\mathbf{T})^{-1}]_{N1}, \quad |z| < 1,$$

where $[\mathbf{A}]_{ij}$ means the (i,j)th component of matrix $\mathbf{A}$. Let $\widetilde{a}_{ij}$ denote the (i,j)th cofactor of the matrix $(\mathbf{I} - z\mathbf{T})$. Then, from a well-known result in linear algebra (see, e.g., Noble and Daniel, page 208, 1977), it follows that

$$f_N^*(z) = z\, q_1 \frac{\widetilde{a}_{1N}}{\det(\mathbf{I} - z\mathbf{T})}, \quad |z| < 1,$$

where $\det(\mathbf{A})$ is the determinant of matrix $\mathbf{A}$. Note that, since $\mathbf{T}$ is strictly substochastic, one has $\det(\mathbf{I} - z\mathbf{T}) \neq 0$ for $|z| < 1$. Now, the tridiagonality of $(\mathbf{I} - z\mathbf{T})$ yields

$$\widetilde{a}_{1N} = (-1)^{N+1} \prod_{k=2}^{N} (-q_k z) = z^{N-1} \prod_{k=2}^{N} q_k,$$

from which we obtain

$$f_N^*(z) = z^N \frac{\prod_{k=1}^{N} q_k}{\det(\mathbf{I} - z\mathbf{T})}, \quad |z| < 1.$$

By induction, we can prove (try!) that $\prod_{k=1}^{N} q_k = \det(\mathbf{I} - \mathbf{T})$, so that

$$\prod_{k=1}^{N} q_k = \prod_{k=1}^{N} (1 - \lambda_k).$$

Also,

$$\det(\mathbf{I} - z\mathbf{T}) = \prod_{k=1}^{N}(1-\lambda_k z), \quad |z| < 1.$$

It follows that

$$f_N^*(z) = \prod_{k=1}^{N} \frac{(1-\lambda_k)z}{1-\lambda_k z}, \quad |z| < 1.$$

Hence, if $\lambda_j > 0$ for all j, the first-passage-time distribution $f_{N0}(n)$ is given by a convolution of geometric distributions. See Sumita and Masuda (1985) and Masuda (1988) for a different proof of these two results.

Let $\mathbf{T}$ be a finite, strictly substochastic matrix and let $\boldsymbol{\alpha}$ be a probability vector of the same size. For the pair $(\boldsymbol{\alpha}, \mathbf{T})$, define

$$f(n) = \boldsymbol{\alpha}^{\top}\mathbf{T}^{n-1}\mathbf{r}, \quad n = 1, 2, \cdots, \tag{2.80}$$

where $\mathbf{r} = \mathbf{1} - \mathbf{T1} = (\mathbf{I} - \mathbf{T})\mathbf{1}$ is nonnegative and nonzero. Then the distribution $f(n)$ is not defective since

$$\sum_{n=1}^{\infty} f(n) = \boldsymbol{\alpha}^{\top} \sum_{n=0}^{\infty} \mathbf{T}^n \mathbf{r} = \boldsymbol{\alpha}^{\top}(\mathbf{I} - \mathbf{T})^{-1}\mathbf{r} = \boldsymbol{\alpha}^{\top}\mathbf{1} = 1.$$

Defining $\overline{F}(n) = \sum_{k=n}^{\infty} f(k)$, $n = 1, 2, \cdots$, we have

$$\overline{F}(n) = \boldsymbol{\alpha}^{\top}\mathbf{T}^{n-1}\mathbf{1}, \quad n = 1, 2, \cdots, \tag{2.81}$$

since

$$\overline{F}(n) = \boldsymbol{\alpha}^{\top}\mathbf{T}^{n-1}(\mathbf{I} - \mathbf{T})^{-1}\mathbf{r}.$$

The distribution $f(n)$ generated by the pair $(\boldsymbol{\alpha}, \mathbf{T})$, as in (2.80), is called a *phase-type distribution* (see Neuts, 1981, for details). Recalling (2.76), the phase-type distribution $f(n)$ can be interpreted as the first-passage-time distribution to the absorbing state 0 with the initial distribution $\boldsymbol{\alpha}$ supported on $\{1, \cdots, N\}$.

2.8 Lossy Markov chains

As in Chung (1967), we introduce transition probabilities with taboo states or, more briefly, *taboo transition probabilities*. For a Markov chain $\{X_n\}$ with state space $\mathcal{N}$ ($\mathcal{N}$ can be denumerable), let H be an arbitrary subset of $\mathcal{N}$ and define

$${}_Hp_{ij}(n) = P_i[X_n = j;\ X_k \notin H,\ 0 < k < n], \quad n = 1, 2, \cdots. \tag{2.82}$$

In words, the taboo transition probability ${}_Hp_{ij}(n)$ is the transition probability, starting from state i, of entering state j at the nth step under the restriction of no visits to states in H in-between (exclusive of both ends). The set H is called the *taboo set* and the states in it are called *taboo states*.

H may be empty; in this case it is omitted from the notation and the taboo transition probabilities are just the ordinary transition probabilities. If H consists of a single state k, we shall denote the taboo transition probability by ${}_kp_{ij}(n)$. If the taboo set consists of the union of a set H and a state k (which may belong to H), the corresponding taboo transition probability is denoted by ${}_{k,H}p_{ij}(n)$. We note that in our previous notation,

$$f_{ij}(n) = {}_jp_{ij}(n), \quad n = 0, 1, \cdots,$$

i.e. the first-passage-time distribution. We define ${}_Hp_{ij}(0) = \delta_{ij}$ if $i \notin H$ and ${}_Hp_{ij}(0) = 0$ if $i \in H$. Note also that ${}_Hp_{ij}(1) = p_{ij}(1)$.

We first provide *basic decomposition formulas* for the taboo transition probabilities. The proof is left to the reader (see Exercise 2.28).

Theorem 2.20 *For $k \notin H$, we have*

$$ {}_Hp_{ij}(n) = {}_{k,H}p_{ij}(n) + \sum_{m=1}^{n-1} {}_{k,H}p_{ik}(m)\, {}_Hp_{kj}(n-m) \tag{2.83}$$

and

$$ {}_Hp_{ij}(n) = {}_{k,H}p_{ij}(n) + \sum_{m=1}^{n-1} {}_Hp_{ik}(m)\, {}_{k,H}p_{kj}(n-m) \tag{2.84}$$

for all $n = 1, 2, \cdots$.

Formulas (2.83) and (2.84) are called, respectively, the *first entrance decomposition* and the *last exit decomposition*. An application of the basic decomposition formulas is given next.

Example 2.25 Consider the one-sided random walk given in Example 2.6, where the state space is given by $\mathcal{N} = \{0, 1, \cdots\}$. Let T_j be the first passage time to state j. For $i < j$, define

$$f_{ij}(n) = P_i[T_j = n], \quad n = 1, 2, \cdots.$$

Then, from the last-exit-decomposition formula (2.84), with $H = \{j\}$ and $k = i$, we have

$$f_{ij}(n) = \sum_{m=0}^{n-1} {}_jp_{ii}(m)\, {}_if_{ij}(n-m), \quad n = 1, 2, \cdots; \quad i < j,$$

where ${}_kf_{ij}(n)$ denotes the first-passage-time distribution from state i to state j with taboo state k. Because of the skip-free property of the random walk, we see that

$$ {}_if_{ij}(n) = {}_if_{i+1,j}(n-1)\, p_i, \quad n = 1, 2, \cdots; \quad i < j,$$

whence

$$f_{ij}(n) = p_i \sum_{m=0}^{n-1} {}_jp_{ii}(m)\, {}_if_{i+1,j}(n-m-1), \quad n = 1, 2, \cdots. \tag{2.85}$$

For $|z| < 1$, let $f^*_{ij}(z) = \sum_{n=1}^{\infty} f_{ij}(n) z^n$ and let ${}_k p^*_{ij}(z) = \sum_{n=0}^{\infty} {}_k p_{ij}(n) z^n$. The generating function ${}_k f^*_{ij}(z)$ is defined similarly. Then, from (2.85), it can be readily shown that

$$f^*_{ij}(z) = z\, p_{i\,j} p^*_{ii}(z)\, {}_i f^*_{i+1,j}(z), \quad |z| < 1.$$

It follows that

$$f^*_{ij}(z) = \left(z\, p_{i\,i} f^*_{i+1,j}(1)\, {}_j p^*_{ii}(z)\right) \left(\frac{{}_i f^*_{i+1,j}(z)}{{}_i f^*_{i+1,j}(1)} \right), \quad |z| < 1. \tag{2.86}$$

Note that the term in the first parentheses of the right-hand side of (2.86) corresponds to the upward last exit time of state i given no visit to state j (denoted by $T^j_{i,i+1}$), whereas the term in the second parentheses corresponds to the subsequent first passage time to state j from state $i+1$ without returning to state i (denoted by ${}^i T_{i+1,j}$). Then (2.86) states that

$$T_{ij} \stackrel{\mathrm{d}}{=} T^j_{i,i+1} + {}^i T_{i+1,j},$$

where the two random variables $T^j_{i,i+1}$ and ${}^i T_{i+1,j}$ are independent of each other and $\stackrel{\mathrm{d}}{=}$ stands for equality in distribution. This result is a discrete counterpart of Keilson (1981).

Let $\mathbf{P}$ be the transition matrix of $\{X_n\}$ and let $\mathbf{T}$ be the submatrix of $\mathbf{P}$ for the states corresponding to $\mathcal{N} \setminus H$. In other words, $\mathbf{T}$ is obtained from $\mathbf{P}$ by deleting all the columns and rows corresponding to states in H, whence $\mathbf{T}$ is substochastic. If $\{X_n\}$ is irreducible and H is nonempty, then $\mathbf{T}$ is strictly substochastic. It is easy to see that, for $i,\ j \notin H$,

$$\mathbf{T}^n = ({}_H p_{ij}(n)), \quad n = 1, 2, \cdots. \tag{2.87}$$

It follows that

$${}_H p_{ij}(m+n) = \sum_{k \notin H} {}_H p_{ik}(m)\, {}_H p_{kj}(n), \quad i,\ j \in \mathcal{N} \setminus H,$$

i.e. the Chapman–Kolmogorov equation for taboo transition probabilities. The *taboo state probabilities* ${}_H \pi_j(n)$ are given by

$${}_H \boldsymbol{\pi}(n) = \boldsymbol{\alpha}^{\mathsf{T}} \mathbf{T}^n, \quad n = 0, 1, 2, \cdots, \tag{2.88}$$

where ${}_H \boldsymbol{\pi}(n) = ({}_H \pi_j(n))$, called the *taboo state distribution*, and $\boldsymbol{\alpha}$ is the initial distribution restricted on $\mathcal{N} \setminus H$.

Conversely, as in Theorem 1.2, any substochastic matrix and probability vector whose indices run over $\mathcal{N} \setminus H$ together determine a Markov chain with the taboo set H (H cannot be observed) on a probability space, since only the substochastic matrix $\mathbf{T}$ is relevant to the taboo state distribution. In this light, a Markov chain governed by a strictly substochastic transition matrix is called a *lossy Markov chain* on the state space $\mathcal{N} \setminus H$ (see Keilson, 1979). In the remainder of this section, we shall consider a lossy Markov chain $\{X_n\}$ with transition matrix $\mathbf{T}$ and initial distribution $\boldsymbol{\alpha}$. Recall that $\mathbf{T}$ is strictly substochastic and $\boldsymbol{\alpha}$ is a probability vector. The transition probabilities of $\{X_n\}$, which are the taboo transition probabilities in the former context, will be denoted by $p_{ij}(n)$, where $\mathbf{T}^n = (p_{ij}(n))$, unless confusion could occur.

Recall from Section 2.2 that accessibility $i \to j$ is the notion of the existence of a nonnegative integer n such that $p_{ij}(n) > 0$. Hence, the concept of communication, and hence equivalence classes for lossy Markov chains, can be defined in the same manner through transition probabilities $p_{ij}(n)$. This in turn implies that the notion of irreducibility of lossy Markov chains is the same as that for ordinary Markov chains. The notion of periodicity is unchanged for lossy Markov chains, too. A finite, strictly substochastic matrix $\mathbf{T}$ is called *regular* or *primitive* if there is some positive integer n such that $\mathbf{T}^n$ has no zero components (see Definition A.1).

Let $\{X_n\}$ be a lossy Markov chain with state space $\mathcal{N}\setminus H = \{0, 1, \cdots, N\}$ and transition matrix $\mathbf{T}$. Suppose that the strictly substochastic matrix $\mathbf{T}$ is regular. Then the Perron–Frobenius theorem (Theorem A.1) shows that there exists a unique probability vector $\mathbf{u} = (u_i)$, which is positive componentwise, such that

$$\gamma \mathbf{u}^{\mathsf{T}} = \mathbf{u}^{\mathsf{T}}\mathbf{T}, \quad \mathbf{u}^{\mathsf{T}}\mathbf{1} = 1, \tag{2.89}$$

where γ denotes the PF eigenvalue of $\mathbf{T}$. Furthermore, there exists a unique vector positive componentwise such that

$$\gamma \mathbf{v} = \mathbf{T}\mathbf{v}, \quad \mathbf{u}^{\mathsf{T}}\mathbf{v} = 1. \tag{2.90}$$

Let $\boldsymbol{\Delta}$ be such that

$$\mathbf{T} = \gamma\, \mathbf{v}\mathbf{u}^{\mathsf{T}} + \boldsymbol{\Delta}.$$

Note that $\boldsymbol{\Delta}\mathbf{v} = \mathbf{0}$ and $\mathbf{u}^{\mathsf{T}}\boldsymbol{\Delta} = \mathbf{0}^{\mathsf{T}}$. It follows that

$$\mathbf{T}^n = \gamma^n\, \mathbf{v}\mathbf{u}^{\mathsf{T}} + \boldsymbol{\Delta}^n, \quad n = 1, 2, \cdots. \tag{2.91}$$

Recall that the spectral radius of $\boldsymbol{\Delta}$ is strictly less than γ. Hence, $\boldsymbol{\Delta}^n/\gamma^n \to \mathbf{O}$ as $n \to \infty$ (see Theorem A.2).

The next result is termed the *ratio limit theorem*. See Orey (Chapter 3, 1971) for more information.

Theorem 2.21 *Let $\{X_n\}$ be a finite, lossy Markov chain with regular transition matrix $\mathbf{T} = (p_{ij})$. Then*

$$\lim_{n\to\infty} \frac{p_{ij}(n)}{p_{ik}(n)} = \frac{u_j}{u_k}, \quad i,\ j,\ k \in \mathcal{N}\setminus H,$$

where $\mathbf{u} = (u_i)$ is the PF left eigenvector of $\mathbf{T}$.

Proof. Defining $\mathbf{v} = (v_i)$, Theorem A.2 implies that

$$p_{ij}(n) = \gamma^n v_i u_j + o(\gamma^n),$$

where $o(\gamma^n)$ means that $\lim_{n\to\infty} o(\gamma^n)/\gamma^n = 0$. It follows that

$$\lim_{n\to\infty} \frac{p_{ij}(n)}{p_{ik}(n)} = \lim_{n\to\infty} \frac{\gamma^n v_i u_j + o(\gamma^n)}{\gamma^n v_i u_k + o(\gamma^n)} = \frac{u_j}{u_k},$$

as desired. □

The ratio limit theorem holds in a more general setting. Let $\mathbf{T} = (p_{ij})$ be irreducible, aperiodic and substochastic. Define the generating function

$$p_{ij}^*(z) = \sum_{n=0}^{\infty} p_{ij}(n) z^n$$

for all real z for which the series converges, and let

$$R_{ij} = \sup\{z \geq 0 : p_{ij}^*(z) < \infty\}$$

be the radius of convergence of $p_{ij}^*(z)$. It is well known that the $p_{ij}^*(z)$ have a common convergence radius $R = R_{ij}$ and converge or diverge together (see, e.g., Seneta, page 200, 1981). If $p_{ij}^*(R) < \infty$, then $\mathbf{T}$ is called *R-transient*, while it is called *R-recurrent* otherwise. Note that, since $\mathbf{T}$ is substochastic, we must have $R \geq 1$. Let

$$f_{ii}^*(z) = \sum_{n=0}^{\infty} f_{ii}(n) z^n = \sum_{n=0}^{\infty} {}_i p_{ii}(n) z^n$$

for all real z for which the series converges. The generating function converges in at least $|z| < R$. It is readily seen that $f_{ii}^*(R) = 1$ for all i if $\mathbf{T}$ is R-recurrent and $f_{ii}^*(R) < 1$ for all i if it is R-transient. Further, let

$$\mu_i(R) = f_{ii}^{*\prime}(R-).$$

For the case of R-recurrence, the $\mu_i(R)$ all converge or diverge together, thereby further classifying R-recurrence. Namely, $\mathbf{T}$ is called *R-positive* if $\mu_i(R) < \infty$ and *R-null* if $\mu_i(R) = \infty$. These classifications should be compared with those given in Section 2.2. It is readily verified that any finite, regular substochastic matrix $\mathbf{T}$ is R-positive with $R = \gamma^{-1} \geq 1$, where γ is the PF eigenvalue of $\mathbf{T}$.

The following result, called R-theory of Markov chains, was largely developed by Tweedie (1974a, 1974b). See also Nummelin (1984) for more information about the theory.

Definition 2.9 (i) A nonnegative, nonzero vector $\mathbf{x} = (x_i)$ is said to be a *β-subinvariant measure* of $\mathbf{T} = (p_{ij})$ if

$$x_i \geq \beta \sum_j x_j p_{ji}.$$

If the inequality is replaced by an equality, $\mathbf{x}$ is called a *β-invariant measure*.

(ii) A nonnegative, nonzero vector $\mathbf{y} = (y_i)$ is said to be a *β-subinvariant vector* of $\mathbf{T} = (p_{ij})$ if

$$y_i \geq \beta \sum_j p_{ij} y_j .$$

If the inequality is replaced by an equality, $\mathbf{y}$ is called a *β-invariant vector.*

Recall from Definition 2.5 that a 1-subinvariant (1-invariant) measure is simply called subinvariant (invariant, respectively). The vector $\mathbf{1}$ is a 1-subinvariant vector of any substochastic matrix. Note that, since the same proof as in Lemma 2.7 can be applied, if $\mathbf{x}$ is a β-subinvariant measure of $\mathbf{T}$ then it is strictly positive componentwise.

The next result is due to Vere-Jones (1962) and should be compared with Theorem 2.11(i). The proof is omitted.

Theorem 2.22 *For an R-recurrent matrix* $\mathbf{T}$*, an R-invariant measure always exists which is positive componentwise and unique up to constant multiples. An R-subinvariant measure which is not invariant exists if and only if* $\mathbf{T}$ *is R-transient.*

For a nonnegative matrix $\mathbf{T}$, consider its transpose $\mathbf{T}^\top$. It is readily shown that the common convergence radius of $\mathbf{T}^\top$ is the same as that of $\mathbf{T}$, and the subinvariant vector properties of $\mathbf{T}$ can be determined from the subinvariant measure properties of $\mathbf{T}^\top$. Therefore, we conclude that, for an R-recurrent matrix $\mathbf{T}$, an R-invariant vector always exists and is positive componentwise.

Suppose that $\mathbf{x} = (x_i)$ is a β-invariant measure of $\mathbf{T} = (p_{ij})$. Then, as in (2.25), we define

$$\mathbf{T}_\mathrm{R} = \beta \mathbf{x}_\mathrm{D}^{-1} \mathbf{T}^\top \mathbf{x}_\mathrm{D},$$

where $\mathbf{x}_\mathrm{D}$ is the diagonal matrix with diagonal components x_i. It is easily seen that the matrix $\mathbf{T}_\mathrm{R}$ is stochastic. Suppose, further, that $\mathbf{T}$ admits a β-invariant vector $\mathbf{y} = (y_i)$ and that $\sum_i x_i y_i < \infty$. Put $z_i = x_i y_i / \sum_i x_i y_i$ so that $\sum_i z_i = 1$. The positive vector $\mathbf{z} = (z_i)$ is given by $\mathbf{z} = C \mathbf{y}^\top \mathbf{x}_\mathrm{D}$, where $C^{-1} = \mathbf{y}^\top \mathbf{x}_\mathrm{D} \mathbf{1} = \mathbf{y}^\top \mathbf{x}$. Since

$$\mathbf{y}^\top \mathbf{x}_\mathrm{D} \mathbf{T}_\mathrm{R} = \beta (\mathbf{T}\mathbf{y})^\top \mathbf{x}_\mathrm{D} = \mathbf{y}^\top \mathbf{x}_\mathrm{D},$$

the probability vector $\mathbf{z}$ is invariant with respect to $\mathbf{T}_\mathrm{R}$. It follows from Theorem 2.8 that $\mathbf{T}_\mathrm{R} = (p_{ij}^R)$ is positive recurrent and

$$\lim_{n\to\infty} p_{ij}^R(n) = \lim_{n\to\infty} \frac{\beta^n p_{ji}(n) x_j}{x_i} \neq 0,$$

where $\mathbf{T}_\mathrm{R}^n = (p_{ij}^R(n))$. By the definition of the radius of convergence R, it must follow that $\beta = R$ and that $\mathbf{T}$ is R-positive; for, otherwise, $R^n p_{ji}(n) \to 0$ as $n \to \infty$. In fact, we can prove the following (see Exercise 2.30). The next result, which is taken from Seneta (1981), is parallel to Theorem 2.8.

Theorem 2.23 *Suppose* $\mathbf{x} = (x_i)$ *is a* β*-invariant measure and* $\mathbf{y} = (y_i)$ *is a* β*-invariant vector of* $\mathbf{T}$. *Then* $\mathbf{T}$ *is* R*-positive if* $\sum_i x_i y_i < \infty$, *in which case* $\beta = R$ *and both* $\mathbf{x}$ *and* $\mathbf{y}$ *are unique up to constant multiples.*

Conversely, if $\mathbf{T}$ *is* R*-positive and* $\mathbf{x}$ *and* $\mathbf{y}$ *are, respectively, an invariant measure and an invariant vector of* $\mathbf{T}$, *then* $\sum_i x_i y_i < \infty$.

Suppose $\mathbf{T}$ is R-positive. Let $\mathbf{u} = (u_i)$ be an R-invariant measure of $\mathbf{T}$, and define

$$\mathbf{T}_R = R\mathbf{u}_D^{-1}\mathbf{T}^\top \mathbf{u}_D. \tag{2.92}$$

Let $\mathbf{v} = (v_i)$ be an R-invariant vector of $\mathbf{T}$ such that $\sum_i u_i v_i = 1$. Write $\mathbf{d} = (d_i)$, where $d_i = u_i v_i$. We have seen that

$$\lim_{n\to\infty} p_{ji}^R(n) = \lim_{n\to\infty} \frac{R^n p_{ij}(n)\, u_i}{u_j} = d_i,$$

so that

$$R^n p_{ij}(n) \to u_j v_i \quad \text{as} \quad n \to \infty. \tag{2.93}$$

Therefore the ratio limit theorem (Theorem 2.21) holds.

For a lossy Markov chain $\{X_n\}$ with state space $\mathcal{N} \setminus H$ and transition matrix $\mathbf{T}$, let $\boldsymbol{\alpha}$ be the initial distribution concentrated on $\mathcal{N} \setminus H$. For $j \notin H$, consider the conditional probabilities

$$q_j(n) \equiv P_\alpha[X_n = j | X_k \notin H;\ k = 0, 1, \cdots, n], \quad n = 0, 1, \cdots.$$

Denoting the taboo state distribution (2.88) of the lossy Markov chain by $\boldsymbol{\pi}(n) = (\pi_i(n))$, it follows that

$$q_j(n) = \frac{\pi_j(n)}{\sum_{j\notin H} \pi_j(n)}, \quad n = 0, 1, \cdots; \quad j \notin H. \tag{2.94}$$

In matrix notation, (2.94) can be written as

$$\mathbf{q}^\top(n) = \frac{\boldsymbol{\pi}^\top(n)}{\boldsymbol{\pi}^\top(n)\mathbf{1}} = \frac{\boldsymbol{\alpha}^\top \mathbf{T}^n}{\boldsymbol{\alpha}^\top \mathbf{T}^n \mathbf{1}}, \quad n = 0, 1, \cdots,$$

where $\mathbf{q}(n) = (q_j(n))$. It follows that

$$\mathbf{q}^\top(n+1) = \frac{\mathbf{q}^\top(n)\mathbf{T}}{\mathbf{q}^\top(n)\mathbf{T}\mathbf{1}}, \quad n = 0, 1, \cdots. \tag{2.95}$$

Let $\mathbf{q}$ be the initial distribution of the lossy Markov chain $\{X_n\}$ and suppose that

$$\mathbf{q}^\top = \frac{\mathbf{q}^\top \mathbf{T}}{\mathbf{q}^\top \mathbf{T}\mathbf{1}}. \tag{2.96}$$

That is, $\mathbf{q}$ is a probability vector with the property that, starting with $\mathbf{q}$, the conditional distribution at time 1 given that the Markov chain is in $\mathcal{N} \setminus H$ is equal to the initial distribution $\mathbf{q}$. By induction, we have

$$\mathbf{q}^\top = \frac{\mathbf{q}^\top \mathbf{T}^n}{\mathbf{q}^\top \mathbf{T}^n \mathbf{1}}, \quad n = 0, 1, \cdots,$$

so that the conditional process is strictly stationary. For a substochastic matrix $\mathbf{T}$, a probability vector $\mathbf{q} = (q_i)$ satisfying (2.96) is called a *quasi-stationary distribution* of $\mathbf{T}$. From (2.96), we have

$$\gamma \mathbf{q}^\top = \mathbf{q}^\top \mathbf{T}; \quad \mathbf{q}^\top \mathbf{1} = 1, \tag{2.97}$$

where $\gamma = \mathbf{q}^\top \mathbf{T} \mathbf{1}$. If $\mathbf{T}$ is R-positive, Theorems 2.22 and 2.23 imply that $R = \gamma^{-1}$ and $\mathbf{q}$ is a positive R-invariant measure of $\mathbf{T}$ and if, in addition, the invariant measure is summable, then the normalized measure is the unique quasi-stationary distribution. If $\mathbf{T}$ is finite and regular, then this is of course a consequence of the Perron–Frobenius theorem, in which case γ is the PF eigenvalue and $\mathbf{q}$ is the associated PF left eigenvector. It should be noted that, even if $\mathbf{T}$ is R-positive, the quasi-stationary distribution may not exist since not every R-invariant measure of $\mathbf{T}$ is summable ($\mathbf{q}^\top \mathbf{v}$ is *always* convergent; see Theorem 2.23). Note also that, when $\mathbf{T}$ is stochastic so that $\mathbf{T1} = \mathbf{1}$, (2.97) coincides with the stationary equation (2.15). In this regard, we shall call (2.97) the *quasi-stationary equation*.

We next turn our attention to the limit of conditional probability (2.94) as $n \to \infty$. In the case where $\boldsymbol{\alpha} = \boldsymbol{\delta}_i$, $i \notin H$, we denote the conditional probability by $q_{ij}(t)$ rather than by $q_j(t)$.

Theorem 2.24 *Suppose* $\mathbf{T}$ *is R-positive. Let* $\mathbf{u} = (u_i)$ *and* $\mathbf{v} = (v_i)$ *be, respectively, an R-invariant measure and an R-invariant vector of* $\mathbf{T}$ *such that* $\sum_i u_i v_i = 1$. *If* $\mathbf{u}$ *is summable so that* $\sum_i u_i = 1$, *then*

$$\lim_{n\to\infty} q_{ij}(n) = u_j, \quad i,\ j \in \mathcal{N} \setminus H,$$

independently of the initial state. For an initial distribution $\boldsymbol{\alpha}$ *concentrated on* $\mathcal{N} \setminus H$, *if either* $\boldsymbol{\alpha}$ *is dominated by some multiple of* $\mathbf{u}$ *or* $\sum_i \alpha_i v_i < \infty$ *and* $\mathbf{v}$ *is bounded away from zero, then*

$$\lim_{n\to\infty} q_j(n) = u_j, \quad j \in \mathcal{N} \setminus H.$$

Proof. From (2.94) and (2.88), we have

$$q_{ij}(n) = \frac{p_{ij}(n)}{\sum_{k\notin H} p_{ik}(n)} = \frac{R^n p_{ij}(n)}{\sum_{k\notin H} R^n p_{ik}(n)}, \quad n = 0, 1, \cdots.$$

We know from (2.93) that $R^n p_{ij}(n)$ converges to $u_j v_i$ as $n \to \infty$. Let $\mathbf{T}_\mathrm{R} = (p_{ij}^R)$ be defined as in (2.92). Then, since $\lim_{n\to\infty} p_{ki}^R(n) = d_i$, where $d_i = u_i v_i$, we have as $n \to \infty$

$$\sum_{k\notin H} R^n p_{ik}(n) = \frac{1}{u_i} \sum_{k\notin H} u_k\, p_{ki}^R(n) \quad \to \quad \frac{1}{u_i} \sum_{k\notin H} u_k d_i = v_i,$$

by the dominated convergence theorem (Lemma 2.6). The proof of the general case may be found in Seneta and Vere-Jones (1966). □

From Theorem 2.24, the conditional distribution $\mathbf{q}(n)$ with some initial distribution converges to a limiting distribution $\mathbf{u}$ as $n \to \infty$. The distribution $\mathbf{u}$ is positive componentwise, independent of the initial distribution. It follows from (2.95) that $\mathbf{u}$ coincides with the quasi-stationary distribution $\mathbf{q}$, which is unique from Theorem 2.23 (see Seneta and Vere-Jones, 1966, for more detailed discussions). In this regard, $\mathbf{u}$ is often called the ***quasi-limiting distribution***, if it exists, of the lossy Markov chain $\{X_n\}$, which can be computed by solving the quasi-stationary equation (2.97).

An interpretation of the quasi-limiting distribution is as follows. For a Markov chain $\{X_n\}$ defined on $\mathcal{N}$, let H be a subset of $\mathcal{N}$. Let T be the first passage time of $\{X_n\}$ into the set H. That is, $T = \inf\{n : X_n \in H\}$, with the understanding that the infimum of the empty set is ∞. Suppose that the Markov chain $\{X_n\}$ enters the set H in a finite time with probability one. The time T may, however, be sufficiently long to allow $\{X_n\}$ to settle down to a quasi-statistical equilibrium within this period. When this happens, one is interested in the limiting conditional probabilities

$$q_j = \lim_{n\to\infty} P_\alpha[X_n = j|T > n], \quad j \in \mathcal{N} \setminus H, \tag{2.98}$$

since q_j may well closely approximate the conditional steady state probabilities. Note that the event $\{T > n\}$ is equivalent to the event $\{X_k \notin H,\ k = 0, 1, \cdots, n\}$, so that

$$q_j(n) = P_\alpha[X_n = j|T > n], \quad n = 0, 1, \cdots.$$

Hence, we will consider a lossy Markov chain with the taboo set H to derive the limiting conditional probabilities (2.98). Recall that the R-positivity assumption guarantees the existence of an ergodic dual Markov chain (2.92). See Kijima (1993b) and Kesten (1995) for non-R-positive cases.

The quasi-limiting distribution provides particularly useful information if the time to visit the set H is substantially longer than that to approach to the limiting conditional distribution. For, in that case, the process relaxes to the quasi-limiting regime after a relatively short time, and then, after a much longer period, a visit to H will eventually occur. Interesting examples of this phenomenon can be found in the chemistry literature. See Dambrine and Moreau (1981), Parsons and Pollett (1987), and references therein. When $\mathcal{N} \setminus H$ is finite, the speed of convergence to quasi-stationarity can be characterized by the gap between the PF eigenvalue γ and the spectral radius of $\boldsymbol{\Delta}$ in (2.91). See van Doorn (1991) for the case of denumerable birth–death processes.

Example 2.26 Consider the GI/M/1 queue described in Example 1.4. Let $\{X_n\}$ be the embedded Markov chain with transition matrix $\mathbf{P}$ given in (1.28). Suppose that the embedded chain $\{X_n\}$ is observed up until period t and suppose that $X_n \le N$ for all $n \le t$. Suppose, further, that we observe a statistical equilibrium in the system within this period. Then the conditional steady state probabilities can be approximated by the quasi-limiting distribution of the strictly substochastic matrix

$$\mathbf{T} = \begin{pmatrix} \overline{A}_1 & a_0 & 0 & \cdots & 0 \\ \overline{A}_2 & a_1 & a_0 & \cdots & 0 \\ \vdots & \vdots & \vdots & \ddots & \vdots \\ \overline{A}_N & a_{N-1} & a_{N-2} & \cdots & a_0 \\ \overline{A}_{N+1} & a_N & a_{N-1} & \cdots & a_1 \end{pmatrix},$$

where $\mathcal{N} \setminus H = \{0, 1, \cdots, N\}$. The quasi-stationary equation can be written componentwise as

$$\gamma q_i = \sum_{k=0}^{N-i+1} a_k q_{i-1+k}, \quad i = N, \cdots, 1; \quad \gamma q_0 = \sum_{k=0}^{N} \overline{A}_{k+1} q_k.$$

Let $\chi_i = q_{i-1}/q_i$ for $i = N, \cdots, 1$ (cf. Exercise 2.31). It follows that

$$\gamma = a_0 \chi_i + a_1 + \sum_{k=2}^{N-i+1} \frac{a_k}{\prod_{j=i+1}^{i+k-1} \chi_j}, \quad i = N, \cdots, 1,$$

and

$$(\gamma - \overline{A}_1)\chi_1 = \overline{A}_2 + \sum_{k=2}^{N} \frac{\overline{A}_{k+1}}{\prod_{j=2}^{k} \chi_j},$$

where the empty sum is understood to be zero. As in Example 2.23, choose r arbitrary such that $0 < r < 1$, and generate $\chi_i(r)$ in reverse order by

$$\chi_i(r) = \frac{1}{a_0} \left(r - a_1 - \sum_{k=2}^{N-i+1} \frac{a_k}{\prod_{j=i+1}^{i+k-1} \chi_j(r)} \right), \quad i = N, \cdots, 1. \tag{2.99}$$

If $\chi_i(r) > 0$ for all i and if the equation

$$(r - \overline{A}_1)\chi_1(r) = \overline{A}_2 + \sum_{k=2}^{N} \frac{\overline{A}_{k+1}}{\prod_{j=2}^{k} \chi_j(r)} \tag{2.100}$$

is satisfied, then the quasi-stationary equation (2.97) holds after a normalization, in which case this r coincides with the PF eigenvalue γ of $\mathbf{T}$. The quasi-stationary distribution $\mathbf{q} = (q_i)$ is then obtained via (2.99), the relation $q_{i-1} = q_i \chi_i(r)$ and a normalization. To this end, the key fact is that each $\chi_i(r)$ is strictly increasing in r. This can be readily verified by an induction argument using (2.99). Therefore, a bisection search algorithm similar to the one given in Example 2.23 can be developed for finding the PF eigenvalue γ.

Motivated by (2.98), of further interest is the *doubly limiting conditional probabilities*

$$d_j = \lim_{n \to \infty} \lim_{m \to \infty} P_\alpha[X_n = j | T > n + m], \quad j \in \mathcal{N} \setminus H,$$

where T denotes the first passage time to the set H. Define

$$d_j(n,m) = P_\alpha[X_n = j | T > n+m]$$

and let $\mathbf{d}(n,m) = (d_j(n,m))$. Then, for $n,\ m \geq 0$, we have

$$d_j(n,m) = \frac{P_\alpha[X_n = j,\, T > m+n]}{P_\alpha[T > n+m]}. \tag{2.101}$$

In order to obtain the limiting conditional distribution, we again consider a lossy Markov chain governed by a substochastic matrix $\mathbf{T}$ restricted on $\mathcal{N} \setminus H$. We write $d_{ij}(n,m)$ for $d_j(n,m)$ if $\boldsymbol{\alpha} = \boldsymbol{\delta}_i,\ i \notin H$.

Theorem 2.25 *Suppose* $\mathbf{T}$ *is R-positive. Let* $\mathbf{u} = (u_i)$ *and* $\mathbf{v} = (v_i)$ *be, respectively, an R-invariant measure and an R-invariant vector of* $\mathbf{T}$ *such that* $\sum_i u_i v_i = 1$. *If* $\mathbf{u}$ *is summable so that* $\sum_i u_i = 1$, *then*

$$\lim_{n\to\infty} \lim_{m\to\infty} d_{ij}(n,m) = d_j > 0, \quad j \in \mathcal{N} \setminus H,$$

independently of the initial state, where $d_j = u_j v_j$.

Proof. From (2.101), we have

$$d_{ij}(n,m) = \frac{p_{ij}(n) P_j[T > m]}{P_i[T > n+m]} = \frac{p_{ij}(n) \sum_{k \notin H} R^m p_{jk}(m)}{\sum_{k \notin H} R^m p_{ik}(n+m)}, \quad n = 0, 1, \cdots,$$

where $\mathbf{T}^n = (p_{ij}(n))$. In the proof of Theorem 2.24, we have shown that

$$\lim_{m\to\infty} \sum_{k \notin H} R^m p_{jk}(m) = v_j.$$

Similarly, as $m \to \infty$,

$$\sum_{k \notin H} R^m p_{ik}(n+m) = R^{-n} \sum_{k \notin H} R^{n+m} p_{ik}(n+m) \quad \to \quad R^{-n} v_i.$$

It follows that

$$\lim_{m\to\infty} d_{ij}(n,m) = \frac{R^n p_{ij}(n) v_j}{v_i}, \quad n = 0, 1, \cdots.$$

Since $\lim_{n\to\infty} R^n p_{ij}(n) = u_j v_i$, the result follows. □

Recall that $\mathbf{d} = (d_j)$ is the limiting distribution of an ergodic Markov chain with transition matrix $\mathbf{T}_\mathrm{R}$ given by (2.92). Another result related to the doubly limiting conditional distribution is given in Exercise 2.33. Ziedins (1987) used the doubly limiting conditional distribution to analyze one-dimensional circuit-switched networks.

As stated earlier, in order to obtain the quasi-stationary distribution, we have to solve the quasi-stationary equation (2.97). Comparing the quasi-stationary equation with the ordinary stationary equation (2.15), the difficulty in the former case arises from the task of finding γ (the PF eigenvalue for the finite case). That is, for any stochastic matrix, it is known *a priori* that $\gamma = 1$, while γ in (2.97) is in general not easy to obtain. Hence, the quasi-stationary distribution is much more difficult to compute than the ordinary stationary counterpart. When the lossy Markov chain is skip-free, there is a simple numerical algorithm based on a bisection search to solve the quasi-stationary equation (2.97), as in Example 2.26. See also Kijima and Makimoto (1992) for an algorithm to compute quasi-stationary distributions in some finite queueing systems. We will discuss some bounds on the quasi-stationary distribution in Chapter 3.

2.9 Exercises

Exercise 2.1 For a Markov chain with the transition matrix given in Exercise 1.13, obtain the first-passage-time distribution $f_{ij}(n)$.

Exercise 2.2 Verify the identity (2.3).

Exercise 2.3 Prove Theorem 2.2 by mimicking the proof of Theorem 1.3.

Exercise 2.4 Consider a Markov chain with N states. Prove that if state j can be reached from state $i \neq j$, then it can be reached in N steps or less.

Exercise 2.5 Let $X_1, X_2, \cdots$ be independent Poisson random variables, each having mean 1, and define $S_n = \sum_{i=1}^{n} X_i$. Applying the central limit theorem to S_n for large n (assume that the central limit theorem can be applied), show that

$$P[S_n = n] \approx \frac{1}{\sqrt{2\pi n}},$$

from which we can deduce Stirling's formula

$$n! \approx n^n e^{-n} \sqrt{2\pi n}$$

(Ross, 1989).

Exercise 2.6 Let $Y_1, Y_2, \cdots$ be IID random variables such that

$$P[Y = 1] = p = 1 - P[Y = -1], \quad 0 < p < 1.$$

Prove that the random walk considered in Example 2.1 is transient for $p \neq 1/2$ by the strong law of large numbers.

Exercise 2.7 Prove Lemma 2.3 and transitivity so as to confirm the fact that $\leftrightarrow$ is an equivalence relation.

Exercise 2.8 For a Markov chain, suppose that $p_{jj}(n)$ converges to π_j, say, as $n \to \infty$. Using the generating function $p_{jj}^*(z) = \sum_{n=0}^{\infty} p_{jj}(n) z^n$, prove that $\pi_j = \mu_j^{-1}$, where μ_j is the mean return time to state j.

Exercise 2.9 Consider the M/G/1 queue given in Example 1.13. For $\mathbf{P}$ defined in (1.29), let $\boldsymbol{\pi} = (\pi_i)$ satisfy the equation $\boldsymbol{\pi}^\top = \boldsymbol{\pi}^\top \mathbf{P}$. Write the equation componentwise, and then derive the generating function $g(z) = \sum_{n=0}^{\infty} \pi_n z^n$ of $\boldsymbol{\pi}$.

Exercise 2.10 If a Markov chain $\{X_n\}$ is recurrent, then, for any initial distributions $\boldsymbol{\alpha}$ and $\boldsymbol{\beta}$, we have

$$\lim_{n\to\infty} \|\boldsymbol{\pi}_\alpha(n) - \boldsymbol{\pi}_\beta(n)\|_1 = 0,$$

where $\boldsymbol{\pi}_\alpha(n)$ denotes the state distribution at time n when the initial distribution $\boldsymbol{\alpha}$ is used. Using this fact, prove that if $\{X_n\}$ is null recurrent then $p_{ij}(n) \to 0$ as $n \to \infty$ (Lindvall, 1992).

Exercise 2.11 Prove (2.28) for the periodic case. Also, prove that $p_{jj}(nd)$ converges to d/μ_j as $n \to \infty$, where $d \geq 2$ is the period of the Markov chain.

Exercise 2.12 Consider a deck with N cards labeled by the numbers $1, 2, \cdots, N$ from the top card to the bottom. The deck is mixed by repeated single shuffles carried out independently of each other and according to the same principle each time. A single shuffle consists of inserting the top card at a random place with probability $1/N$ for each of the N possibilities. (Note that a replacement at the top is one of them.) Suppose $N = 3$. Formulate an appropriate Markov chain and show that the deck is well mixed after sufficiently many shuffles (Aldous, 1983).

Exercise 2.13 Let $\mathbf{P}$ be given in (2.30) and suppose that $\mathbf{Q}$ is aperiodic. Let $\widetilde{\boldsymbol{\pi}}^\top = (\boldsymbol{\pi}^\top, \mathbf{0}^\top)$, where $\boldsymbol{\pi}$ is the stationary distribution of $\mathbf{Q}$. Then, prove that $\mathbf{P}^n$ converges to $\mathbf{1}\widetilde{\boldsymbol{\pi}}^\top$ as $n \to \infty$.

Exercise 2.14 Let $\mathbf{P} = (p_{ij})$ be a stochastic matrix, i.e., the row sums of $\mathbf{P}$ are all unity. Suppose, further, that the column sums of $\mathbf{P}$ are all unity. That is, $\sum_i p_{ij} = 1$ for all j. Such a matrix is called *doubly stochastic.* Prove that the stationary distribution of any finite, doubly stochastic matrix is a uniform distribution on the state space.

Exercise 2.15 In the Ehrenfest urn model described in Example 2.10, let $m_n = E[X_n]$. Show that

$$m_{n+1} = 1 + \left(1 - \frac{1}{a}\right) m_n$$

and so

$$m_n = a + \left(\frac{a-1}{a}\right)^n (m_0 - a).$$

Exercise 2.16 Suppose that the transition matrix $\mathbf{P}$ is given by (2.35). Derive the stationary distribution.

Exercise 2.17 Let X_n be the sum of n independent rolls of a fair die. By defining an appropriate Markov chain, find

$$\lim_{n\to\infty} P[X_n \text{ is a multiple of } 13]$$

(Karlin and Taylor, 1975).

Exercise 2.18 Let $\{X_n\}$ be an ergodic Markov chain with limiting probabilities π_i. Define the process $\{Y_n\}$ by $Y_n = (X_{n-1}, X_n)$, i.e. a two-dimensional process. Is $\{Y_n\}$ Markovian? If so, find the limiting probabilities of $\{Y_n\}$.

Exercise 2.19 Consider the age process $\{X_n\}$ of a successively replaced system described in Example 2.12, but now without planned age replacement. That is, letting $U_1, U_2, \cdots$ be IID positive integer-valued random variables, we have

$$X_n = n - T_k, \quad T_k \leq n < T_{k+1},$$

where $T_n = \sum_{i=1}^{n} U_i$ with $T_0 \equiv 0$. Then X_n represents the age of the system in operation. The process $\{X_n\}$ is a Markov chain with transition matrix

$$\mathbf{P} = \begin{pmatrix} h_1 & 1-h_1 & 0 & 0 & \cdots \\ h_2 & 0 & 1-h_2 & 0 & \cdots \\ h_3 & 0 & 0 & 1-h_3 & \cdots \\ \vdots & \vdots & \vdots & \ddots & \ddots \end{pmatrix},$$

where $h_i \equiv P[U = i | U \geq i]$, i.e. the hazard rate function of U. Suppose that $E[U] < \infty$ and let $\pi_i = P[U \geq i] / \sum_i P[U \geq i]$, $i = 1, 2, \cdots$. Prove that $\boldsymbol{\pi} = (\pi_i)$ satisfies $\sum_i \pi_i = 1$ and is the stationary distribution of the age process $\{X_n\}$.

Exercise 2.20 For a Markov chain with state space $\mathcal{Z}_+$ and transition matrix $\mathbf{P} = (p_{ij})$, the *truncated Markov chain* $\{\overline{X}_n\}$ on the state space $\mathcal{N} \subset \mathcal{Z}_+$ is defined by a Markov chain with the transition matrix $\overline{\mathbf{P}} = (\overline{p}_{ij})$, where

$$\overline{p}_{ij} = \frac{p_{ij}}{\sum_{j \in \mathcal{N}} p_{ij}}, \quad i, j \in \mathcal{N}.$$

Suppose that $\{X_n\}$ is reversible in time and that the truncated Markov chain is irreducible and aperiodic. Identify the stationary distribution of the truncated Markov chain and show that it is reversible in time.

Exercise 2.21 Let $\mathbf{A}^\sharp$ be as defined in (2.55). Prove that the eigenvalues of $\mathbf{A}^\sharp$ are 0 and $(1 - \lambda_j)^{-1}$, $j = 1, \cdots, N$, where the λ_j are the eigenvalues of $\mathbf{P}$ other than unity (Seneta, 1993).

Exercise 2.22 Let $\mathbf{Z}$ be the fundamental matrix defined in (2.63). Prove the identities

$$\mathbf{PZ} = \mathbf{ZP}, \quad \boldsymbol{\pi}^\top \mathbf{Z} = \boldsymbol{\pi}^\top, \quad \mathbf{Z1} = \mathbf{1}, \quad \mathbf{I} - \mathbf{Z} = \mathbf{1}\boldsymbol{\pi}^\top - \mathbf{PZ},$$

where $\boldsymbol{\pi}$ is the stationary distribution of $\mathbf{P}$. Also, show that the fundamental matrix of the time-reversal Markov chain is given by $\mathbf{Z}_\mathrm{R} = \boldsymbol{\pi}_\mathrm{D}^{-1} \mathbf{Z}^\top \boldsymbol{\pi}_\mathrm{D}$.

Exercise 2.23 In the gambler's ruin problem given in Example 2.22, let

$$T = \min\{n \geq 0 : X_n = 0 \text{ or } X_n = N\}$$

and define $v_i = E_i[T]$, the mean hitting time to states 0 or N. Show that

$$v_i = 1 + q_i v_{i-1} + r_i v_i + p_i v_{i+1}, \quad i = 1, \cdots, N-1,$$

where $v_0 = v_N = 0$.

Exercise 2.24 For a finite, irreducible Markov chain with state space $\mathcal{N} = \{0, 1, \cdots, N\}$, let x_i be the probability that, starting from state i, the Markov chain visits state N before state 0. Obtain a set of equations that the x_i satisfy.

Exercise 2.25 Let $\{X_n\}$ be an absorbing Markov chain with state space $\mathcal{N} = \{0, 1, \cdots, N\}$, where 0 and N are absorbing states, both accessible from any intermediate state. Let μ_k be the expected time until absorption with the initial state $k \neq 0$, N. For the transition matrix $\mathbf{P}$ of $\{X_n\}$, we construct a stochastic matrix $\overline{\mathbf{P}} = (\overline{p}_{ij})$ such that $\overline{p}_{0k} = \overline{p}_{Nk} = 1$ and leave the other rows unchanged. If the transient states of $\{X_n\}$ communicate with each other, then $\overline{\mathbf{P}}$ is clearly regular. Let $\boldsymbol{\pi} = (\pi_i)$ be the stationary distribution of $\overline{\mathbf{P}}$. Prove that

$$\mu_k = \frac{1}{\pi_0 + \pi_N} - 1$$

for $k \neq 0$, N (Karlin and Taylor, 1975).

Exercise 2.26 Let $\{X_n\}$ be an irreducible Markov chain with state space $\mathcal{Z}_+$ and consider an absorbing Markov chain constructed by making state 0 of $\{X_n\}$ absorbing. Let $\mathbf{P} = (p_{ij})$ denote the transition matrix of the absorbing Markov chain. Prove that if we have a bounded, nonconstant solution $\{y_i\}$ for

$$\sum_{j=0}^{\infty} p_{ij} y_j = y_i, \quad i = 0, 1, \cdots,$$

then the original Markov chain $\{X_n\}$ is transient.

Exercise 2.27 Using the recurrence relation (2.1), prove Theorem 2.19.

Exercise 2.28 Prove Theorem 2.20 by decomposing the events regarding when the Markov chain visits state k.

Exercise 2.29 For a recurrent Markov chain with state space $\mathcal{Z}_+$, let ${}_k p_{ij}(n)$ be the taboo transition probabilities. Let $v_0 = 1$ and

$$v_i = \sum_{n=0}^{\infty} {}_0 p_{0i}(n), \quad i = 1, 2, \cdots.$$

Show that $\{v_n\}$ is a solution of the system of equations

$$v_i = \sum_{j=0}^{\infty} v_j p_{ji}, \quad i \in \mathcal{Z}_+.$$

Exercise 2.30 Suppose that $\mathbf{T}$ is R-positive and that $\mathbf{x}$ and $\mathbf{y}$ are, respectively, an R-invariant measure and an R-invariant vector of $\mathbf{T}$. By considering the stochastic matrix $\mathbf{P} = R\mathbf{x}_{\mathrm{D}}^{-1}\mathbf{T}^{\top}\mathbf{x}_{\mathrm{D}}$, prove that $\mathbf{x}^{\top}\mathbf{y}$ is convergent.

Exercise 2.31 In Example 2.26, let $\chi_i = q_i/q_{i-1}$. Obtain $\chi_i(r)$ similarly to (2.99) and (2.100), and show that each $\chi_i(r)$ is strictly decreasing in r. Using this fact, develop a bisection algorithm to find the PF eigenvalue.

Exercise 2.32 Let $\mathbf{T}$ be substochastic and R-positive with $R \geq 1$. Prove that either $R > 1$, or $R = 1$ and the matrix is stochastic (Seneta and Vere-Jones, 1966).

Exercise 2.33 For a lossy Markov chain $\{X_n\}$ with transition matrix $\mathbf{T}$, let $N_j(n)$ be the number of visits to state j before time n. Prove that, as $n \to \infty$,

$$E_i\left[\frac{N_j(n)}{n}\middle| T > n\right] \quad \to \quad d_j,$$

i.e. the doubly limiting conditional probability, under the conditions of Theorem 2.25 (Darroch and Seneta, 1965).

Exercise 2.34 Consider a discrete-time process on a finite state space $\mathcal{N}$. After observing the state of the process, an action is chosen where the set of possible actions, which we denote by $\mathcal{A}$, is assumed to be finite. If the process is in state i at time n and action a is chosen, then the next state of the process is determined according to the transition probabilities $p_{ij}(a)$. Namely, if we let X_n be the state of the process at time n and let A_n be the action chosen at time n, then we have

$$P[X_{n+1} = j | X_0, A_0, \cdots, X_n = i, A_n = a] = p_{ij}(a).$$

Let $R(i, a)$ be the reward earned whenever action a is chosen in state i. For $0 < r < 1$, we want to determine a *policy* so as to maximize the expected total discounted reward $E[\sum_{n=0}^{\infty} r^n R(X_n, A_n)]$. Note that not only X_n but also A_n are random variables. In this setting, the process $\{X_n\}$ is called a *Markov decision process*. See, e.g., Puterman (1994) for details.

(i) For a given policy, let y_{ja} denote the expected discounted time that the process is in state j and action a is chosen, i.e.,

$$y_{ja} = E\left[\sum_{n=0}^{\infty} r^n \mathrm{I}_{\{X_n = j,\, A_n = a\}}\right],$$

where I_A denotes the indicator function of event A. Show that

$$\sum_a y_{ja} = \alpha_j + r\sum_i \sum_a y_{ia}\, p_{ij}(a); \quad \sum_j \sum_a y_{ja} = \frac{1}{1-r},$$

where $\boldsymbol{\alpha} = (\alpha_i)$ is the initial distribution.

(ii) Argue that y_{ja} can be interpreted as the expected discounted time mentioned above when the initial distribution is $\boldsymbol{\alpha}$ and the policy is such that, in state i, action a is chosen with probability $y_{ia}/\sum_i y_{ia}$. Hence, an optimal policy can be obtained by first solving the linear program

$$\begin{aligned}
\text{maximize} \quad & \sum_j \sum_a y_{ja} R(j,a) \\
\text{subject to} \quad & \sum_j \sum_a y_{ja} = \frac{1}{1-r}, \\
& \sum_a y_{ja} = \alpha_j + r \sum_i \sum_a y_{ia}\, p_{ij}(a), \\
& y_{ja} \geq 0, \quad j \in \mathcal{N}, \ a \in \mathcal{A},
\end{aligned}$$

and then defining the policy that, in state i, action a is chosen with probability

$$\beta_i^*(a) = \frac{y_{ia}^*}{\sum_i y_{ia}^*},$$

where the y_{ia}^* are the solutions of the linear program (Ross, 1989).

3

Monotone Markov chains

In this chapter, we consider the monotonicity properties of discrete-time Markov chains where each monotonicity is characterized in terms of transition matrices. A Markov chain $\{X_n\}$ is said to be increasing (decreasing, respectively) if $X_{n+1} \succ X_n$ ($X_n \succ X_{n+1}$) for all $n = 0, 1, \cdots$, where $\succ$ denotes an ordering relation in some stochastic sense, and in either case we call $\{X_n\}$ *internally monotone*, or monotone for short. An *external monotonicity* is such that, for two Markov chains $\{X_n\}$ and $\{Y_n\}$, we have $X_n \succ Y_n$ for all n. Monotonicity properties are important both theoretically and practically because they lead to a variety of structural insights. In particular, they are a basic tool for deriving many useful inequalities in Markov chains for stochastic modeling.

To clarify the importance of monotonicity properties, we consider the following two examples.

Example 3.1 Consider an ergodic Markov chain $\{X_n\}$ on the denumerable state space $\mathcal{Z}_+$; we wish to calculate the stationary distribution, which is usually very difficult to obtain. To approximate the stationary distribution, it is common to truncate the state space so that we have a finite ergodic Markov chain $\{Y_n\}$ whose stationary distribution can be calculated with ease. Suppose it is known that, for some n, X_n is greater (smaller, respectively) than Y_n in some stochastic sense, $X_n \succ Y_n$ ($Y_n \succ X_n$), say, and the inequality is preserved for $n+1$. Then we have $X_m \succ Y_m$ ($Y_m \succ X_m$) for all $m \geq n$, and so the stationary distribution of $\{X_n\}$ is bounded from below (above) by the stationary distribution of $\{Y_n\}$.

Example 3.2 Consider an M/G/1 queue with arrival rate λ and service time distribution $G(t)$. Supposing that the mean service time μ^{-1} is finite, we denote the residual distribution of $G(t)$ by

$$G_R(t) = \mu \int_0^t \{1 - G(x)\}dx, \quad x \geq 0.$$

Let $X_1, X_2, \cdots$ be IID random variables with distribution function $G_R(t)$, and let N be a geometrically distributed random variable with parameter $\rho \equiv \lambda/\mu < 1$ which is independent of $\{X_n\}$. It is well known (see, e.g., Kleinrock, page 200, 1975) that the generic stationary waiting time for the M/G/1 queue is given by $W = \sum_{i=1}^{N} X_i$. Hence, in general, computation of the waiting-time distribution requires inversion of the Laplace transform (see Appendix B.2)

$$\phi_W(s) = \frac{1-\rho}{1-\rho\,\phi_X(s)},$$

or a similar transformation. On the other hand, if the service time is exponentially distributed, then we have $G_R(t) = 1 - e^{-\mu t}$, and so

$$P[W \le t] = 1 - \rho\, e^{-\mu(1-\rho)t}, \quad t \ge 0.$$

Suppose $\{Y_n\}$ is another sequence of independent, exponentially distributed random variables with parameter μ' which is also independent of N. Assume that $X_n \succ Y_n$. Defining $W_Y = \sum_{i=1}^{N} Y_i$, if one can show that $W \succ W_Y$, then we have useful information on the waiting-time distribution such as

$$P[W \le t] \le P[W_Y \le t] = 1 - \rho' e^{-\mu'(1-\rho')t}, \quad t \ge 0,$$

where $\rho' = \lambda/\mu' < 1$. As we shall see, this claim can be proved by constructing two appropriate Markov chains and comparing them in some stochastic sense.

Throughout this chapter, 'increasing' means 'nondecreasing', while 'decreasing' means 'nonincreasing.' To begin with, we provide some basic information used throughout this chapter. Throughout we shall denote the state space of a Markov chain $\{X_n\}$ under consideration by $\mathcal{N}$. As usual, we assume that $\mathcal{N} = \{0, 1, \cdots, N\}$ for the finite case and $\mathcal{N} = \{0, 1, \cdots\}$ for the denumerable case.

3.1 Preliminaries

Suppose, firstly, that the state space $\mathcal{N}$ is finite. Then we introduce the matrix

$$\mathbf{U} = \begin{pmatrix} 1 & 0 & 0 & \cdots & 0 \\ 1 & 1 & 0 & \cdots & 0 \\ 1 & 1 & 1 & \cdots & 0 \\ \vdots & \vdots & \vdots & \ddots & \vdots \\ 1 & 1 & 1 & \cdots & 1 \end{pmatrix} \tag{3.1}$$

defined on the state space $\mathcal{N}$. Clearly, $\mathbf{U}$ has inverse

$$\mathbf{U}^{-1} = \begin{pmatrix} 1 & 0 & 0 & \cdots & 0 \\ -1 & 1 & 0 & \cdots & 0 \\ 0 & -1 & 1 & \cdots & 0 \\ \vdots & & \ddots & \ddots & \vdots \\ 0 & \cdots & 0 & -1 & 1 \end{pmatrix}. \tag{3.2}$$

The matrix $\mathbf{U}$ was first introduced in Keilson and Kester (1977). Also, as in Kijima (1990a), we define

$$\mathbf{V} = \mathbf{U}^{\top} = \begin{pmatrix} 1 & 1 & 1 & \cdots & 1 \\ 0 & 1 & 1 & \cdots & 1 \\ 0 & 0 & 1 & \cdots & 1 \\ \vdots & \vdots & \vdots & \ddots & \vdots \\ 0 & 0 & 0 & \cdots & 1 \end{pmatrix}. \tag{3.3}$$

The inverse of $\mathbf{V}$ exists and is given by

$$\mathbf{V}^{-1} = (\mathbf{U}^{-1})^{\top} = \begin{pmatrix} 1 & -1 & 0 & \cdots & 0 \\ 0 & 1 & -1 & \cdots & 0 \\ \vdots & & \ddots & \ddots & \vdots \\ 0 & \cdots & 0 & 1 & -1 \\ 0 & \cdots & 0 & 0 & 1 \end{pmatrix}. \tag{3.4}$$

Secondly, when the state space is denumerably infinite, we need to take care in matrix multiplication (see Appendix A.3). Recall that, for two infinite matrices $\mathbf{A} = (a_{ij})$ and $\mathbf{B} = (b_{ij})$, the matrix product $\mathbf{C} = (c_{ij}) = \mathbf{AB}$ is well defined if

$$|c_{ij}| = \left| \sum_{k=0}^{\infty} a_{ik} b_{kj} \right| < \infty$$

for all $i,\ j \in \mathcal{N}$. Let

$$\mathbf{U} = \begin{pmatrix} 1 & 0 & 0 & \cdots \\ 1 & 1 & 0 & \cdots \\ 1 & 1 & 1 & \cdots \\ \vdots & \vdots & \vdots & \ddots \end{pmatrix} \tag{3.5}$$

and

$$\mathbf{U}^{-1} = \begin{pmatrix} 1 & 0 & 0 & 0 & \cdots \\ -1 & 1 & 0 & 0 & \cdots \\ 0 & -1 & 1 & 0 & \cdots \\ 0 & 0 & -1 & 1 & \cdots \\ \vdots & \vdots & \vdots & \ddots & \ddots \end{pmatrix}. \tag{3.6}$$

(We use the same notation as in the finite case.) It is easily seen that the products $\mathbf{U}\,\mathbf{U}^{-1}$ and $\mathbf{U}^{-1}\mathbf{U}$ are well defined and $\mathbf{U}\,\mathbf{U}^{-1} = \mathbf{U}^{-1}\mathbf{U} = \mathbf{I}$, the identity matrix of infinite order. Similarly, let

$$\mathbf{V} = \mathbf{U}^{\top} = \begin{pmatrix} 1 & 1 & 1 & \cdots \\ 0 & 1 & 1 & \cdots \\ 0 & 0 & 1 & \cdots \\ \vdots & \vdots & \vdots & \ddots \end{pmatrix} \tag{3.7}$$

and

$$\mathbf{V}^{-1} = (\mathbf{U}^{-1})^{\top} = \begin{pmatrix} 1 & -1 & 0 & 0 & \cdots \\ 0 & 1 & -1 & 0 & \cdots \\ 0 & 0 & 1 & -1 & \cdots \\ 0 & 0 & 0 & 1 & \cdots \\ \vdots & \vdots & \vdots & \vdots & \ddots \end{pmatrix}. \tag{3.8}$$

Again, we have $\mathbf{V}\,\mathbf{V}^{-1} = \mathbf{V}^{-1}\mathbf{V} = \mathbf{I}$.

In what follows, we shall denote the (i,j)th component of matrix $\mathbf{A}$ by $[\mathbf{A}]_{ij}$, that is, if $\mathbf{A} = (a_{ij})$ then $[\mathbf{A}]_{ij} = a_{ij}$. We also write

$$A_{ij} \equiv \sum_{k=0}^{j} a_{ik}, \quad \overline{A}_{ij} \equiv \sum_{k=j}^{\infty} a_{ik},$$

whenever these sums exist. Note that $\mathbf{AV} = (A_{ij})$ and $\mathbf{AU} = (\overline{A}_{ij})$. If $\mathbf{AU}$ is well defined, i.e., if the $\overline{A}_{ij}$ exist for all $j \in \mathcal{N}$, then

$$\lim_{m\to\infty} \overline{A}_{im} = 0, \quad i \in \mathcal{N}.$$

Lemma 3.1 *Let* $\mathbf{A} = (a_{ij})$ *be an infinite matrix.*

(i) *Suppose that* $\mathbf{AU}$ *is well defined. Then*

$$(\mathbf{AU})\mathbf{U}^{-1} = \mathbf{A}$$

and

$$\mathbf{U}^{-1}(\mathbf{AU}) = (\mathbf{U}^{-1}\mathbf{A})\mathbf{U} = \mathbf{U}^{-1}\mathbf{AU}.$$

(ii) *Suppose* $\lim_{j\to\infty} a_{ij} = 0$. *Then*

$$(\mathbf{AU}^{-1})\mathbf{U} = \mathbf{A}.$$

Proof. To prove the lemma, we make use of Theorem A.9. For any j, we have $[\mathbf{U}^{-1}]_{nj} = 0$ for all $n > j+1$. Hence

$$\sum_{k=0}^{\infty} a_{ik} \sum_{n=m}^{\infty} [\mathbf{U}]_{kn}\,[\mathbf{U}^{-1}]_{nj} = 0$$

for sufficiently large m, and the first part of assertion (i) follows from

Theorem A.9(i). For the second part, let $j < m$. Then

$$\sum_{k=0}^{\infty}[\mathbf{U}^{-1}]_{ik}\sum_{n=m}^{\infty}a_{kn}\,[\mathbf{U}]_{nj} = \overline{A}_{im} - \overline{A}_{i-1,m}.$$

Since $\lim_{m\to\infty}\overline{A}_{im} = 0$ for all $i \in \mathcal{N}$, the second part follows.

(ii) If $j < m$ then

$$\sum_{n=m}^{\infty}[\mathbf{U}^{-1}]_{kn}\,[\mathbf{U}]_{nj} = \begin{cases} 1, & k = m, \\ 0, & \text{otherwise,} \end{cases}$$

whence we have

$$\sum_{k=0}^{\infty}a_{ik}\sum_{n=m}^{\infty}[\mathbf{U}^{-1}]_{kn}\,[\mathbf{U}]_{nj} = a_{im}.$$

Since, by assumption, $\lim_{m\to\infty}a_{im} = 0$, assertion (ii) is proved. □

Lemma 3.2 *For two infinite matrices* $\mathbf{A} = (a_{ij})$ *and* $\mathbf{B} = (b_{ij})$, *suppose that* $\mathbf{AU}$ *and* $\mathbf{AB}$ *are well defined. If* b_{ij} *is bounded then*

$$(\mathbf{AU})(\mathbf{U}^{-1}\mathbf{B}) = \mathbf{AB}$$

and

$$(\mathbf{U}^{-1}\mathbf{AU})(\mathbf{U}^{-1}\mathbf{B}) = \mathbf{U}^{-1}(\mathbf{AB}).$$

Proof. Since $(\mathbf{AU})\mathbf{U}^{-1} = \mathbf{A}$ from Lemma 3.1(i), we need to prove that

$$(\mathbf{AU})(\mathbf{U}^{-1}\mathbf{B}) = ((\mathbf{AU})\mathbf{U}^{-1})\mathbf{B}$$

for the first part. To this end, we have

$$\sum_{k=m}^{\infty}\overline{A}_{ik}\,[\mathbf{U}^{-1}]_{kn} = \begin{cases} \overline{A}_{in} - \overline{A}_{i,n+1}, & m \le n, \\ -\overline{A}_{i,n+1}, & m = n+1, \\ 0, & \text{otherwise.} \end{cases}$$

Since $\overline{A}_{in} - \overline{A}_{i,n+1} = a_{in}$, it follows that

$$\sum_{n=0}^{\infty}b_{nj}\sum_{k=m}^{\infty}\overline{A}_{ik}\,[\mathbf{U}^{-1}]_{kn} = -b_{m-1,j}\,\overline{A}_{im} + \sum_{n=m}^{\infty}a_{in}b_{nj}.$$

Now, since $\mathbf{AU}$ and $\mathbf{AB}$ are well defined, we have

$$\lim_{m\to\infty}\overline{A}_{im} = \lim_{m\to\infty}\sum_{n=m}^{\infty}a_{in}b_{nj} = 0.$$

The result follows at once from Theorem A.9(ii) since b_{ij} is bounded.

Alternatively, we have, for any $N > 0$,

$$\sum_{n=0}^{N}\overline{A}_{in}(b_{nj} - b_{n-1,j}) = \overline{A}_{iN}b_{Nj} + \sum_{n=0}^{N-1}a_{in}b_{nj}.$$

Under the assumption, letting $N \to \infty$ yields

$$\sum_{n=0}^{\infty} \overline{A}_{in}(b_{nj} - b_{n-1,j}) = \sum_{n=0}^{\infty} a_{in} b_{nj}.$$

Hence $(\mathbf{AU})(\mathbf{U}^{-1}\mathbf{B}) = \mathbf{AB}$. The second part follows from Lemma 3.1(i) and the first part of this lemma. □

Turning to the matrix $\mathbf{V}$, we have the following. The proof of the next lemma is similar to those of Lemmas 3.1 and 3.2 and is omitted. Note that the matrix $\mathbf{AV} = (A_{ij})$ is always well defined.

Lemma 3.3 (i) *Let* $\mathbf{A} = (a_{ij})$ *be an infinite matrix. We have*

$$(\mathbf{AV})\mathbf{V}^{-1} = (\mathbf{AV}^{-1})\mathbf{V} = \mathbf{A}$$

and

$$\mathbf{V}^{-1}(\mathbf{AV}) = (\mathbf{V}^{-1}\mathbf{A})\mathbf{V} = \mathbf{V}^{-1}\mathbf{AV}.$$

(ii) *For two infinite matrices* $\mathbf{A} = (a_{ij})$ *and* $\mathbf{B} = (b_{ij})$, *suppose* $\mathbf{AB}$ *is well defined. If* A_{ij} *is bounded and* $\lim_{m\to\infty} b_{mj} = 0$ *for all* j *then*

$$(\mathbf{AV})(\mathbf{V}^{-1}\mathbf{B}) = \mathbf{AB}$$

and

$$(\mathbf{V}^{-1}\mathbf{AV})(\mathbf{V}^{-1}\mathbf{B}) = \mathbf{V}^{-1}(\mathbf{AB}).$$

Our next concern is the total positivity properties of nonnegative matrices. The theory of total positivity is summarized in Appendix C. The next definition consists of restatements of Definitions C.1 and C.2 in terms of nonnegative matrices.

Definition 3.1 (i) A nonnegative matrix $\mathbf{A}$ on $\mathcal{N} \times \mathcal{N}$ is called *totally positive of order* n, $n = 2, 3, \cdots$, denoted by $\mathbf{A} \in \mathrm{TP}_n$, if, for all $k \leq n$, all the $k \times k$ minors of $\mathbf{A}$ are nonnegative.

(ii) A nonnegative vector $\mathbf{f} = (f_i)$ on $\mathcal{N}$ is said to be a *Pólya frequency of order* n, denoted by $\mathbf{f} \in \mathrm{PF}_n$, if the matrix $\mathbf{F} = (f_{i-j})$ is TP_n.*

The next result describes the structural properties of TP_2 matrices. The proof may be found in Keilson and Kester (1978).

Lemma 3.4 *Let matrix* $\mathbf{A} = (a_{ij})$ *have no null rows or columns.*

(i) *Suppose* $\mathbf{A}$ *is* TP_2. *If* $a_{ij} = 0$ *then either* $a_{k\ell} = 0$ *for* $k \leq i$ *and* $\ell \geq j$, *or* $a_{k\ell} = 0$ *for* $k \geq i$ *and* $\ell \leq j$.

(ii) $\mathbf{A}$ *is* TP_2 *if and only if all its local* 2×2 *minors are nonnegative, i.e.,*

$$a_{ij}\, a_{i+1,j+1} \geq a_{i,j+1}\, a_{i+1,j}$$

for all $i,\ j \in \mathcal{N}$.

* Throughout this book, we assume that $f_i = 0$ if $i < 0$.

We note that a nonnegative vector $\mathbf{f} = (f_i)$ is PF_2 if and only if the matrix

$$\begin{pmatrix} f_0 & f_1 & f_2 & \cdots \\ 0 & f_0 & f_1 & \cdots \end{pmatrix}$$

is TP_2, that is, if and only if

$$f_{i+1}^2 \geq f_i f_{i+2}, \quad i \in \mathcal{N}.$$

The next result is a restatement of the *basic composition formula* (Theorem C.1). For the sake of completeness, we give a proof.

Theorem 3.1 *Suppose that both* $\mathbf{A} = (a_{ij})$ *and* $\mathbf{B} = (b_{ij})$ *are* TP_2 *matrices. If* $\mathbf{AB}$ *is well defined, then it is* TP_2.

Proof. Let $\mathbf{C} = (c_{ij}) = \mathbf{AB}$. Since $c_{ij} = \sum_k a_{ik} b_{kj}$, we have

$$\begin{aligned}
& c_{ij} c_{i+1,j+1} - c_{i+1,j} c_{i,j+1} \\
&= \left(\sum_k a_{ik} b_{kj}\right)\left(\sum_\ell a_{i+1,\ell} b_{\ell,j+1}\right) - \left(\sum_k a_{i+1,k} b_{kj}\right)\left(\sum_\ell a_{i\ell} b_{\ell,j+1}\right) \\
&= \sum_{k,\ell} (a_{ik} a_{i+1,\ell} b_{kj} b_{\ell,j+1} - a_{i+1,k} a_{i\ell} b_{kj} b_{\ell,j+1}) \\
&= \sum_{k<\ell} (a_{ik} a_{i+1,\ell} - a_{i+1,k} a_{i\ell}) b_{kj} b_{\ell,j+1} \\
&\qquad + \sum_{k>\ell} (a_{ik} a_{i+1,\ell} - a_{i+1,k} a_{i\ell}) b_{kj} b_{\ell,j+1} \\
&= \sum_{k<\ell} (a_{ik} a_{i+1,\ell} - a_{i+1,k} a_{i\ell}) b_{kj} b_{\ell,j+1} \\
&\qquad + \sum_{\ell>k} (a_{i\ell} a_{i+1,k} - a_{i+1,\ell} a_{ik}) b_{\ell j} b_{k,j+1} \\
&= \sum_{k<\ell} (a_{ik} a_{i+1,\ell} - a_{i+1,k} a_{i\ell})(b_{kj} b_{\ell,j+1} - b_{k,j+1} b_{\ell j}) \geq 0,
\end{aligned}$$

where the inequality follows from the TP_2 properties of $\mathbf{A}$ and $\mathbf{B}$. The general 2×2 minors can similarly be proved to be nonnegative. □

It is easily seen that the matrices $\mathbf{U}$ and $\mathbf{V}$ given in (3.5) and (3.7), respectively, are TP_2 (the same is true for the finite case). Hence, from Theorem 3.1, if $\mathbf{P} \in \mathrm{TP}_2$ and $\mathbf{PU}$ is well defined, then $\mathbf{PU} \in \mathrm{TP}_2$. For example, if $\mathbf{P} \in \mathrm{TP}_2$ is stochastic, then $\mathbf{PU}$ is well defined and $\mathbf{PU} \in \mathrm{TP}_2$. Moreover, this case ensures that $\mathbf{U}^{-1}\mathbf{PU} \geq \mathbf{O}$, since $[\mathbf{PU}]_{i0} = 1$ for all i and $\mathbf{PU} \in \mathrm{TP}_2$ implies

$$[\mathbf{PU}]_{ij} = [\mathbf{PU}]_{i0}\,[\mathbf{PU}]_{ij} \geq [\mathbf{PU}]_{i+1,0}\,[\mathbf{PU}]_{i+1,j} = [\mathbf{PU}]_{i+1,j}$$

for all i and j. Recall from Lemma 3.1(i) that $\mathbf{U}^{-1}(\mathbf{PU}) = \mathbf{U}^{-1}\mathbf{PU}$ whenever $\mathbf{PU}$ is well defined. We have thus obtained the next result.

Corollary 3.1 *If* $\mathbf{P} \in \mathrm{TP}_2$ *then* $\mathbf{PU}$, $\mathbf{UP}$, $\mathbf{PV}$ *and* $\mathbf{VP}$ *are all* TP_2, *whenever they are well defined. Moreover, if* $\mathbf{P}$ *is stochastic then each of* $\mathbf{PU} \in \mathrm{TP}_2$ *or* $\mathbf{PV} \in \mathrm{TP}_2$ *implies both* $\mathbf{U}^{-1}\mathbf{PU} \geq \mathbf{O}$ *and* $\mathbf{V}^{-1}\mathbf{PV} \geq \mathbf{O}$.

The next theorem is due to Shanthikumar (1988) and plays an essential role in what follows. The result will be referred to as the *composition law.*

Theorem 3.2 *Let* $\mathbf{A} = (a_{ij})$ *be a nonnegative matrix with the properties that* $\mathbf{AU}$ *is well defined and has no null rows or columns. Let* $\mathbf{B} = (b_{ij})$ *be a nonnegative matrix such that* $\mathbf{BU}$ *is well defined,* $\overline{B}_{ij}$ *is bounded and has no null rows or columns. Further, let* $\mathbf{C} = (c_{ij}) = \mathbf{AB}$ *and suppose that* $\mathbf{C}$ *as well as* $\mathbf{A}(\mathbf{BU})$ *is well defined. If* $\mathbf{AU}$ *and* $\mathbf{BU}$ *are* TP_2 *and* $\mathbf{U}^{-1}\mathbf{BU} \geq \mathbf{O}$, *then* $\mathbf{CU}$ *is* TP_2.

Proof. First note that Theorem A.8(i) ensures that $\mathbf{CU} = \mathbf{A}(\mathbf{BU})$, i.e.,

$$\overline{C}_{ij} = \sum_{k=0}^{\infty} a_{ik}\overline{B}_{kj}, \quad i,\ j \in \mathcal{N}.$$

Also, we have $\overline{B}_{ij} \geq \overline{B}_{i-1,j}$ for all $i \geq 0$ with $\overline{B}_{-1,j} = 0$ since $\mathbf{U}^{-1}\mathbf{BU} \geq \mathbf{O}$. Now, for any m and $N > m$, we have

$$\begin{aligned}\sum_{k=m}^{N} a_{ik}\overline{B}_{kj} &= \sum_{k=m}^{N} (\overline{A}_{ik} - \overline{A}_{i,k+1})\overline{B}_{kj} \\ &= \overline{A}_{im}\overline{B}_{mj} + \sum_{k=m+1}^{N} \overline{A}_{ik}(\overline{B}_{kj} - \overline{B}_{k-1,j}) - \overline{A}_{i,N+1}\overline{B}_{Nj}.\end{aligned}$$

Under the assumptions, the limit exists as $N \to \infty$ and is given by

$$\sum_{k=m}^{\infty} a_{ik}\overline{B}_{kj} = \overline{A}_{im}\overline{B}_{mj} + \sum_{k=m+1}^{\infty} \overline{A}_{ik}(\overline{B}_{kj} - \overline{B}_{k-1,j}). \tag{3.9}$$

Let $j < j'$. It follows that

$$\begin{aligned}\overline{C}_{ij'} &= \sum_{m=0}^{\infty} a_{im}\overline{B}_{mj'} \\ &= \sum_{m=0}^{\infty} \frac{\overline{B}_{mj'}}{\overline{B}_{mj}} a_{im}\overline{B}_{mj} \\ &= \frac{\overline{B}_{0j'}}{\overline{B}_{0j}} \sum_{m=0}^{\infty} a_{im}\overline{B}_{mj} + \sum_{m=1}^{\infty} \left(\frac{\overline{B}_{mj'}}{\overline{B}_{mj}} - \frac{\overline{B}_{m-1,j'}}{\overline{B}_{m-1,j}}\right) \sum_{k=m}^{\infty} a_{ik}\overline{B}_{kj} \\ &= \frac{\overline{B}_{0j'}}{\overline{B}_{0j}} \overline{C}_{ij} + \sum_{m=1}^{\infty} \left(\frac{\overline{B}_{mj'}}{\overline{B}_{mj}} - \frac{\overline{B}_{m-1,j'}}{\overline{B}_{m-1,j}}\right) \sum_{k=m}^{\infty} a_{ik}\overline{B}_{kj},\end{aligned} \tag{3.10}$$

where the convention $0/0 = 0$ is used. Here the second equality is valid since **B** is nonnegative so that $\overline{B}_{ij} = 0$ implies $\overline{B}_{ij'} = 0$ for $j' > j$. Indeed since the matrix $\mathbf{BU} = (\overline{B}_{ij})$ has no null rows or columns and is TP_2, if $\overline{B}_{ij} = 0$, then $\overline{B}_{m\ell} = 0$ for $m \le i$ and $\ell \ge j$, from Lemma 3.4(i). Hence, the TP_2 property of $(\overline{B}_{ij})$ can be expressed as

$$1 \ge \frac{\overline{B}_{mj'}}{\overline{B}_{mj}} \ge \frac{\overline{B}_{m-1,j'}}{\overline{B}_{m-1,j}} \ge 0, \quad j < j',$$

which, together with the fact that $\lim_{m\to\infty} \sum_{k=m}^{\infty} a_{ik}\overline{B}_{kj} = 0$, ensures the validity of the third equation in (3.10). Now let $i < i'$. From (3.10), we see that

$$\overline{C}_{ij}\overline{C}_{i'j'} \ge \overline{C}_{ij'}\overline{C}_{i'j}$$

if and only if

$$\sum_{m=1}^{\infty}\left(\frac{\overline{B}_{mj'}}{\overline{B}_{mj}} - \frac{\overline{B}_{m-1,j'}}{\overline{B}_{m-1,j}}\right)\overline{C}_{ij}\sum_{k=m}^{\infty} a_{i'k}\overline{B}_{kj}$$

$$\ge \sum_{m=1}^{\infty}\left(\frac{\overline{B}_{mj'}}{\overline{B}_{mj}} - \frac{\overline{B}_{m-1,j'}}{\overline{B}_{m-1,j}}\right)\overline{C}_{i'j}\sum_{k=m}^{\infty} a_{ik}\overline{B}_{kj}.$$

But, since $\mathbf{BU} = (\overline{B}_{ij})$ is TP_2, the above inequality holds if

$$\sum_{\ell=0}^{\infty} a_{i\ell}\overline{B}_{\ell j}\sum_{k=m}^{\infty} a_{i'k}\overline{B}_{kj} \ge \sum_{\ell=0}^{\infty} a_{i'\ell}\overline{B}_{\ell j}\sum_{k=m}^{\infty} a_{ik}\overline{B}_{kj}$$

or, equivalently,

$$\sum_{\ell=0}^{m-1} a_{i\ell}\overline{B}_{\ell j}\sum_{k=m}^{\infty} a_{i'k}\overline{B}_{kj} \ge \sum_{\ell=0}^{m-1} a_{i'\ell}\overline{B}_{\ell j}\sum_{k=m}^{\infty} a_{ik}\overline{B}_{kj} \tag{3.11}$$

for all $m = 1, 2, \cdots$. To prove (3.11), we note that

$$\overline{A}_{i'm}\overline{A}_{i\ell} \ge \overline{A}_{i'\ell}\overline{A}_{im}, \quad m > \ell,$$

since the matrix $\mathbf{AU} = (\overline{A}_{ij})$ is TP_2. It follows from (3.9) that

$$\begin{aligned}
&\overline{A}_{i'm}\sum_{k=m}^{\infty} a_{ik}\overline{B}_{kj} \\
&= \overline{A}_{i'm}\overline{A}_{im}\overline{B}_{mj} + \sum_{k=m+1}^{\infty} \overline{A}_{i'm}\overline{A}_{ik}(\overline{B}_{kj} - \overline{B}_{k-1,j}) \\
&\le \overline{A}_{im}\overline{A}_{i'm}\overline{B}_{mj} + \sum_{k=m+1}^{\infty} \overline{A}_{im}\overline{A}_{i'k}(\overline{B}_{kj} - \overline{B}_{k-1,j}) \\
&= \overline{A}_{im}\sum_{k=m}^{\infty} a_{i'k}\overline{B}_{kj}.
\end{aligned}$$

On the other hand, we have

$$\sum_{\ell=0}^{m-1} a_{i\ell}\overline{B}_{\ell j} = \overline{A}_{i0}\overline{B}_{0j} + \sum_{\ell=1}^{m-1} \overline{A}_{i\ell}(\overline{B}_{\ell j} - \overline{B}_{\ell-1,j}) - \overline{A}_{im}\overline{B}_{m-1,j}. \tag{3.12}$$

It follows that

$$\begin{aligned} & \overline{A}_{i'm} \sum_{\ell=0}^{m-1} a_{i\ell}\overline{B}_{\ell j} \\ = \; & \overline{A}_{i'm}\overline{A}_{i0}\overline{B}_{0j} + \sum_{\ell=0}^{m-1} \overline{A}_{i'm}\overline{A}_{i\ell}(\overline{B}_{\ell j} - \overline{B}_{\ell-1,j}) - \overline{A}_{i'm}\overline{A}_{im}\overline{B}_{m-1,j} \\ \geq \; & \overline{A}_{im}\overline{A}_{i'0}\overline{B}_{0j} + \sum_{\ell=0}^{m-1} \overline{A}_{im}\overline{A}_{i'\ell}(\overline{B}_{\ell j} - \overline{B}_{\ell-1,j}) - \overline{A}_{im}\overline{A}_{i'm}\overline{B}_{m-1,j} \\ = \; & \overline{A}_{im} \sum_{\ell=0}^{m-1} a_{i'\ell}\overline{B}_{\ell j}. \end{aligned}$$

Suppose that $\overline{A}_{im} > 0$. Then $\overline{A}_{i'm} > 0$ for $i' > i$, since otherwise $\overline{A}_{im} = 0$, from Lemma 3.4(i). In this case, combining the above two inequalities yields (3.11). If $\overline{A}_{im} = 0$ then, again from Lemma 3.4(i), we have $\overline{A}_{ik} = 0$ for $k \geq m$. It follows from (3.9) that $\sum_{k=m}^{\infty} a_{ik}\overline{B}_{kj} = 0$ so that (3.11) always holds. This proves the theorem. □

The next composition law is dual to Theorem 3.2.

Theorem 3.3 *Let* $\mathbf{A} = (a_{ij})$ *be a nonnegative matrix such that* $\mathbf{AV}$ *has no null rows or columns, and that the limit of* A_{ij} *exists as* $j \to \infty$, A_i, *say, for all* i. *Let* $\mathbf{B} = (b_{ij})$ *be a nonnegative matrix such that* $\mathbf{BV}$ *has no null rows or columns. Further, let* $\mathbf{C} = (c_{ij}) = \mathbf{AB}$ *and suppose that* $\mathbf{C}$ *and* $\mathbf{A}(\mathbf{BV})$ *are well defined. If* $\mathbf{AV}$ *and* $\mathbf{BV}$ *are* TP_2 *and* $\mathbf{V}^{-1}\mathbf{BV} \geq \mathbf{O}$ *then* $\mathbf{CV}$ *is* TP_2.

Proof. Let $j' < j$. Since $\mathbf{B} = (b_{ij})$ is nonnegative, $B_{ij} = 0$ implies $B_{ij'} = 0$. Hence, the TP_2 property of matrix $\mathbf{BV} = (B_{ij})$ implies that if $B_{ij} = 0$ then $B_{m\ell} = 0$ for $m \geq i$ and $\ell \leq j$, from Lemma 3.4(i). It follows that the TP_2 property may be expressed as

$$1 \geq \frac{B_{mj'}}{B_{mj}} \geq \frac{B_{m+1,j'}}{B_{m+1,j}} \geq 0, \quad j > j',$$

with the convention $0/0 = 0$. Also, since $\mathbf{V}^{-1}\mathbf{BV} \geq \mathbf{O}$, we have $B_{kj} \geq B_{k+1,j}$ so that $B_j \equiv \lim_{k\to\infty} B_{kj}$ exists and is nonnegative for all j. Now, from Theorem A.8(i), we immediately conclude that $(\mathbf{AB})\mathbf{V} = \mathbf{A}(\mathbf{BV})$, i.e.,

$$C_{ij} \equiv \sum_{k=0}^{j} c_{ik} = \sum_{k=0}^{\infty} a_{ik}B_{kj}, \quad i, j \in \mathcal{N}.$$

For any m and $N > m$, we have

$$\sum_{k=m}^{N} a_{ik}B_{kj} = \sum_{k=m}^{N}(A_{ik} - A_{i,k-1})B_{kj}$$
$$= A_{iN}B_{Nj} + \sum_{k=m}^{N-1} A_{ik}(B_{kj} - B_{k+1,j}) - A_{i,m-1}B_{mj},$$

where $A_{i,-1} = 0$. Under the assumptions, the limit exists as $N \to \infty$ and is given by

$$\sum_{k=m}^{\infty} a_{ik}B_{kj} = A_iB_j + \sum_{k=m}^{\infty} A_{ik}(B_{kj} - B_{k+1,j}) - A_{i,m-1}B_{mj}.$$

On the other hand,

$$\sum_{k=0}^{N} a_{ik}B_{kj} = A_{iN}B_{Nj} + \sum_{k=0}^{N-1} A_{ik}(B_{kj} - B_{k+1,j}).$$

The rest of the proof is similar to that of Theorem 3.2 and is omitted (see Exercise 3.1). □

An immediate consequence of the composition law is the following.

Corollary 3.2 *Let* $\mathbf{A} = (a_{ij})$ *be a nonnegative matrix.*

(i) *Suppose that* $\mathbf{A}^n\mathbf{U}$, $n = 1, 2, \cdots$, *are well defined and have no null rows or columns. Further, suppose that* $\mathbf{AU} = (\overline{A}_{ij})$ *is bounded. If* $\mathbf{AU}$ *is* TP_2 *and* $\mathbf{U}^{-1}\mathbf{AU} \geq \mathbf{O}$ *then* $\mathbf{A}^n\mathbf{U} \in \mathrm{TP}_2$ *for all* $n = 1, 2, \cdots$.

(ii) *Suppose that* $\mathbf{A}^n\mathbf{V}$, $n = 1, 2, \cdots$, *are well defined and have no null rows or columns. Further, suppose that the limit* $\lim_{j\to\infty} A_{ij}$ *exists for every* i *where* $\mathbf{AV} = (A_{ij})$. *If* $\mathbf{AV}$ *is* TP_2 *and* $\mathbf{V}^{-1}\mathbf{AV} \geq \mathbf{O}$ *then* $\mathbf{A}^n\mathbf{V} \in \mathrm{TP}_2$ *for all* $n = 1, 2, \cdots$.

When the state space is finite, the composition laws (Theorems 3.2 and 3.3) can be stated as follows.

Corollary 3.3 *Let* $\mathbf{A}$ *and* $\mathbf{B}$ *be finite, nonnegative matrices. Let* $\mathbf{C} = \mathbf{AB}$.

(i) *Suppose that* $\mathbf{AU}$ *and* $\mathbf{BU}$ *have no null rows or columns. If* $\mathbf{AU}$ *and* $\mathbf{BU}$ *are* TP_2 *and* $\mathbf{U}^{-1}\mathbf{BU} \geq \mathbf{O}$ *then* $\mathbf{CU}$ *is* TP_2.

(ii) *Suppose that* $\mathbf{AV}$ *and* $\mathbf{BV}$ *have no null rows or columns. If* $\mathbf{AV}$ *and* $\mathbf{BV}$ *are* TP_2 *and* $\mathbf{V}^{-1}\mathbf{BV} \geq \mathbf{O}$ *then* $\mathbf{CV}$ *is* TP_2.

We now turn to the unimodality of nonnegative vectors.

Definition 3.2 Let $\mathbf{a} = (a_i)$ be a nonnegative vector defined on $\mathcal{N}$.

(i) $\mathbf{a}$ is said to be *unimodal* if there is some $i_0 \in \mathcal{N}$ such that a_j is increasing in $j \leq i_0$ and a_j is decreasing in $j \geq i_0$. The index i_0 is called the *mode* of $\mathbf{a}$.

(ii) If a_j is decreasing (increasing, respectively), then $\mathbf{a}$ is said to be *decreasing* (*increasing*). In either case, $\mathbf{a}$ is called *monotone*.

(iii) If $\mathbf{a}$ is unimodal but not monotone, then it is called *strictly unimodal.*

The next definition is a restatement of that of the counting function $S(f)$ given in (C.6).

Definition 3.3 Let $\mathbf{a} = (a_n)$ be a real vector on the ordered set $\mathcal{N}$. The *sign sequence* of $\mathbf{a}$ is the sequence of $+1$s or -1s generated by $\text{sgn}(a_n)$ as n runs through $\mathcal{N}$, where zeros are neglected. The function $S(\mathbf{a})$ counts the sign changes in the sign sequence of $\mathbf{a}$.

The unimodality and monotonicity of nonnegative vectors can be stated in terms of matrices $\mathbf{U}^{-1}$ and $\mathbf{V}^{-1}$. The proof of the next result is left to the reader (Exercise 3.2).

Theorem 3.4 *Let* $\mathbf{a}$ *be a nonnegative vector on* $\mathcal{N}$.

(i) $\mathbf{a}$ *is increasing (decreasing, respectively) if and only if*

$$\mathbf{U}^{-1}\mathbf{a} \geq \mathbf{0} \quad (\mathbf{V}^{-1}\mathbf{a} \geq \mathbf{0}).$$

(ii) $\mathbf{a}$ *is either strictly unimodal or decreasing if and only if* $S(\mathbf{U}^{-1}\mathbf{a}) = 1$ *and the sign changes from* $+1$ *to* -1.

(iii) *Suppose that* $\mathcal{N}$ *is denumerably infinite;* $\mathbf{a}$ *is strictly unimodal if and only if* $S(\mathbf{V}^{-1}\mathbf{a}) = 1$ *and the sign changes from* -1 *to* $+1$. *In the finite case,* $\mathbf{a}$ *is either strictly unimodal or increasing if and only if* $S(\mathbf{V}^{-1}\mathbf{a}) = 1$ *and the sign changes from* -1 *to* $+1$.

3.2 Distribution classes of interest

This section introduces important distribution classes of discrete random variables. Let X be a discrete random variable on $\mathcal{N}$ with probability vector $\mathbf{a} = (a_i)$, where $a_i = P[X = i]$ and $\sum_{i \in \mathcal{N}} a_i = 1$. Throughout this section, we assume that $\mathcal{N} = \mathcal{Z}_+$. If the support of X is finite, i.e., there is some N such that $\sum_{i=0}^{N} a_i = 1$, then we put $a_i = 0$ for all $i > N$. For distribution classes of continuous random variables, we refer to Barlow and Proschan (1975).

First, recall that the *hazard rate function* of X is defined by

$$h_i = \frac{a_i}{\overline{A}_i}; \quad \overline{A}_i \equiv \sum_{k=i}^{\infty} a_k, \quad i \in \mathcal{N}, \tag{3.13}$$

whenever $\overline{A}_i > 0$. Note that $\overline{A}_i = 0$ implies $\overline{A}_j = 0$ for all $j > i$. An interpretation of the hazard rate function is

$$h_i = \frac{P[X = i]}{P[X \geq i]} = P[X = i | X \geq i], \quad i \in \mathcal{N},$$

that is, h_i is the probability that, conditional on survival, the lifetime of X is equal to i.

Definition 3.4 A discrete random variable X with probability vector $\mathbf{a} = (a_i)$ is called the *increasing hazard rate*, denoted by $X \in$ IHR or $\mathbf{a} \in$ IHR, if

$$\overline{A}_{i+1}^2 \geq \overline{A}_i \, \overline{A}_{i+2}, \quad i \in \mathcal{N}.$$

If the inequality is reversed, X is called the *decreasing hazard rate* and is denoted by $X \in$ DHR or $\mathbf{a} \in$ DHR.

We note that $X \in$ IHR if and only if the vector $\mathbf{a}^\top \mathbf{U} = (\overline{A}_i)$ is PF$_2$. The property $X \in$ DHR can be stated in matrix form as

$$\begin{pmatrix} \overline{A}_0 & \overline{A}_1 & \overline{A}_2 & \cdots \\ \overline{A}_1 & \overline{A}_2 & \overline{A}_3 & \cdots \end{pmatrix} \quad \text{is TP}_2.$$

Moreover, since $\overline{A}_i = 0$ implies $\overline{A}_j = 0$ for $j > i$, $X \in$ IHR if and only if

$$\frac{\overline{A}_{i+1}}{\overline{A}_i} \geq \frac{\overline{A}_{i+2}}{\overline{A}_{i+1}}, \quad i \in \mathcal{N},$$

with the convention $0/0 = 0$. Since

$$\frac{\overline{A}_{i-1}}{\overline{A}_i} = 1 - \frac{a_i}{\overline{A}_i} = 1 - h_i,$$

provided that $\overline{A}_i > 0$, $X \in$ IHR if and only if the hazard rate function of X is increasing on the support of X. Similarly, it is easily seen that $X \in$ DHR if and only if $\overline{A}_{i+1}/\overline{A}_i$ is increasing in i, that is, if and only if the hazard rate function of X is decreasing.* These observations justify the terms IHR and DHR in Definition 3.4.

We next define the *reversed hazard rate function* of X by

$$r_i = \frac{a_i}{A_i}; \quad A_i \equiv \sum_{k=0}^{i} a_k, \quad i \in \mathcal{N}, \tag{3.14}$$

whenever $A_i > 0$. Note that $A_i = 0$ implies $A_j = 0$ for all $j < i$. An interpretation of the reversed hazard rate function is

$$r_i = \frac{P[X = i]}{P[X \leq i]} = P[X = i | X \leq i], \quad i \in \mathcal{N}.$$

The term 'reversed hazard rate' is taken from Shanthikumar, Yamazaki and Sakasegawa (1991).

Definition 3.5 A discrete random variable X with probability vector $\mathbf{a} =$

* If $\overline{A}_N = 0$ for some N, then the DHR property implies $\overline{A}_i = 0$ for all $i \leq N$, which is impossible. Hence, in order for X to be DHR, $\overline{A}_i > 0$ for all $i \in \mathcal{N}$.

(a_i) is called the *decreasing reversed hazard rate*, denoted by $X \in$ DRHR or $\mathbf{a} \in$ DRHR, if

$$A_{i+1}^2 \geq A_i \, A_{i+2}, \quad i \in \mathcal{N}.$$

Observe that $X \in$ DRHR if and only if the vector $\mathbf{a}^\top \mathbf{V} = (A_i)$ is PF_2. Moreover, since $A_i = 0$ implies $A_j = 0$ for $j < i$, $X \in$ DRHR if and only if

$$\frac{A_i}{A_{i+1}} \leq \frac{A_{i+1}}{A_{i+2}}, \quad i \in \mathcal{N},$$

with the convention $0/0 = 0$. Since

$$\frac{A_i}{A_{i+1}} = 1 - \frac{a_{i+1}}{A_{i+1}} = 1 - r_{i+1},$$

provided that $A_{i+1} > 0$, $X \in$ DRHR if and only if the reversed hazard rate function of X is decreasing on the support of X (recall that $r_0 = 1$ if $a_0 > 0$). It should be noted that any discrete random variable with infinite support cannot have an increasing reversed hazard rate (see Exercise 3.4). To see this, suppose $A_0 = a_0 > 0$ for simplicity. Since $1 - r_i = A_{i-1}/A_i$, we have

$$A_n = A_0 \prod_{i=1}^{n} (1 - r_i)^{-1}, \quad n \in \mathcal{N}.$$

Hence if r_i is increasing in $i \geq 1$, there exists some n such that $A_n > 1$, unless $r_i \equiv 0$.

Finally, the *likelihood ratio function* of X is defined by

$$\ell_i = \frac{a_{i+1}}{a_i}, \quad i \in \mathcal{N}, \tag{3.15}$$

whenever $a_i > 0$.

Definition 3.6 A discrete random variable X with probability vector $\mathbf{a} = (a_i)$ is called the *decreasing likelihood ratio*, denoted by $X \in$ DLR or $\mathbf{a} \in$ DLR, if

$$a_{i+1}^2 \geq a_i \, a_{i+2}, \quad i \in \mathcal{N}.$$

If the inequality is reversed, X is called the *increasing likelihood ratio* and is denoted by $X \in$ ILR or $\mathbf{a} \in$ ILR.

We note that $X \in$ DLR if and only if the vector $\mathbf{a} = (a_i)$ itself is PF_2. The property $X \in$ ILR can be stated in matrix form as

$$\begin{pmatrix} a_0 & a_1 & a_2 & \cdots \\ a_1 & a_2 & a_3 & \cdots \end{pmatrix} \quad \text{is } \text{TP}_2.$$

Now suppose for simplicity that $0 < a_0 < 1$. Then, if $X \in$ DLR, $a_i = 0$ implies $a_j = 0$ for all $j > i$. Hence $X \in$ DLR implies

$$\frac{a_{i+1}}{a_i} = \ell_i \geq \ell_{i+1} = \frac{a_{i+2}}{a_{i+1}}, \quad i \in \mathcal{N},$$

with the convention $0/0 = 0$, that is, the likelihood ratio function of X is decreasing. Similarly, if $X \in$ ILR then $a_i > 0$ for all $i \in \mathcal{N}$ and the likelihood ratio function of X is increasing.

We have seen that the three classes DLR, IHR and DRHR are characterized in terms of the PF_2 properties of nonnegative vectors. Because these characterizations play a very important role in what follows, we repeat them as a theorem.

Theorem 3.5 *Let X be a discrete random variable with probability vector* $\mathbf{a}$. *Then,*

(i) $X \in$ DLR *if and only if the vector* $\mathbf{a}$ *itself is* PF_2.

(ii) $X \in$ IHR *if and only if the vector* $\mathbf{a}^\top \mathbf{U}$ *is* PF_2.

(iii) $X \in$ DRHR *if and only if the vector* $\mathbf{a}^\top \mathbf{V}$ *is* PF_2.

A relation between DHR random variables and DRHR random variables is given next.

Lemma 3.5 *Any discrete* DHR *random variable is* DRHR.

Proof. For $\mathbf{a} = (a_i)$, if $\mathbf{a} \in$ DHR then

$$0 \le a_{i+1}^2 \le (a_i - a_{i+1})\overline{A}_{i+2},$$

so that a_i is necessarily decreasing in i. Note that $\mathbf{a} \in$ DRHR if and only if

$$a_{i+1}^2 \ge (a_{i+2} - a_{i+1})A_i,$$

which is satisfied if a_i is decreasing in i. □

Let X and Y be mutually independent random variables on $\mathcal{N}$ with probability vectors $\mathbf{a} = (a_i)$ and $\mathbf{b} = (b_i)$ respectively. Consider the random variable $Z = X + Y$ and let $\mathbf{c} = (c_i)$ be the probability vector of Z. The probability vector $\mathbf{c}$ is given by

$$c_i = \sum_{k=0}^{i} a_k b_{i-k}, \quad i \in \mathcal{N}. \tag{3.16}$$

This operation is called *discrete convolution.*

Definition 3.7 Let X be a discrete random variable on $\mathcal{N}$ with probability vector $\mathbf{a}$.

(i) X is called *unimodal* if the probability vector $\mathbf{a}$ is unimodal. If $\mathbf{a}$ is unimodal but not monotone then X is called *strictly unimodal.*

(ii) X is called *strongly unimodal*, denoted by $X \in$ SU or $\mathbf{a} \in$ SU, if $X + Y$ is unimodal for any discrete random variable Y which is unimodal and independent of X.

We note that, by taking $Y = 0$ almost surely in Definition 3.7(ii), $X \in$ SU implies that X is indeed unimodal. The SU class of random variables is identical to the DLR class. Here we prove the implication that any DLR random variable is strongly unimodal. For the converse, see Keilson and Gerber (1971).

Lemma 3.6 *Any* DLR *random variable is* SU.

Proof. For probability vector $\mathbf{a} = (a_i)$, define the upper triangular matrix

$$\mathbf{A} = \begin{pmatrix} a_0 & a_1 & a_2 & \cdots \\ 0 & a_0 & a_1 & \cdots \\ 0 & 0 & a_0 & \cdots \\ \vdots & \vdots & \vdots & \ddots \end{pmatrix}. \tag{3.17}$$

The convolution operation (3.16) is then written in matrix form as

$$\mathbf{c}^\top = \mathbf{b}^\top \mathbf{A}.$$

It is not difficult to check that all the conditions of Lemma 3.2 are satisfied (in the transposed matrices). Hence

$$\mathbf{c}^\top \mathbf{V}^{-1} = (\mathbf{b}^\top \mathbf{V}^{-1})(\mathbf{VAV}^{-1}). \tag{3.18}$$

Suppose that $\mathbf{b}$ is unimodal (either decreasing or strictly unimodal). Then $S(\mathbf{b}^\top \mathbf{V}^{-1}) = 1$ and the sign changes from $+1$ to -1 when the index runs through $\mathcal{N}$ (see Theorem 3.4). Note that $\mathbf{VAV}^{-1} = \mathbf{A}$. If $\mathbf{a} \in$ DLR so that $\mathbf{A} \in \mathrm{TP}_2$, then the VDP (variation diminishing property, Theorem C.4) applied to (3.18) reveals that the vector $\mathbf{c}^\top \mathbf{V}^{-1}$ changes sign at most once when the index runs through $\mathcal{N}$, and, if it does, it changes from $+1$ to -1. It is impossible that the sign will remain the same. Therefore, $\mathbf{c}$ is unimodal. Since the unimodal vector $\mathbf{b}$ is arbitrary, this implies that $\mathbf{a}$ is strongly unimodal. □

We next examine closure properties of distribution classes under discrete convolution.

Theorem 3.6 *Let X and Y be mutually independent random variables on $\mathcal{N}$ with probability vectors $\mathbf{a} = (a_i)$ and $\mathbf{b} = (b_i)$ respectively.*

(i) *If X, $Y \in$ DLR then $X + Y \in$ DLR.*

(ii) *If X, $Y \in$ IHR then $X + Y \in$ IHR.*

(iii) *If X, $Y \in$ DRHR then $X + Y \in$ DRHR.*

Proof. Let $\mathbf{c} = (c_i)$ be the probability vector of $X + Y$. From (3.16), we have

$$\mathbf{C} = \mathbf{AB},$$

where each matrix is upper triangular, as defined in (3.17).

(i) Note that X, $Y \in$ DLR if and only if both $\mathbf{A}$ and $\mathbf{B}$ are TP_2. The basic composition formula (Theorem 3.1) then shows that $\mathbf{C}$ is TP_2, whence $X + Y \in$ DLR.

(ii) From Theorem 3.5(ii), X, $Y \in$ IHR if and only if $\mathbf{AU}$, $\mathbf{BU} \in \mathrm{TP}_2$. It is easily seen that $\mathbf{U}^{-1}\mathbf{BU} \geq \mathbf{O}$. Also, it is readily verified that all the conditions of Theorem 3.2 are satisfied. Hence $\mathbf{CU} \in \mathrm{TP}_2$ so that $X+Y \in$ IHR. Part (iii) can be proved similarly using Theorem 3.5(iii) and Theorem 3.3. □

We note that the DHR class of discrete random variables is not closed under convolution (see, e.g., Barlow and Proschan, page 101, 1975). However, the broader class DRHR $\supset$ DHR (see Lemma 3.5) is closed under this operation. This makes the DRHR class of random variables important in applied probability because addition of independent random variables is frequently encountered. For the DRHR class of continuous random variables, see Kijima (1989b), where the term 'PF_2 random variable' is used.

Another distribution class of interest in this chapter is the following. The next definition is a restatement of Definition B.3 in particular for discrete random variables. For $0 < \alpha < 1$, the probability vector $\mathbf{b} = (b_i)$ given by

$$b_i = (1-\alpha)\alpha^i, \quad i = 0, 1, \cdots, \tag{3.19}$$

is called geometrically distributed.

Definition 3.8 A discrete random variable X with probability vector $\mathbf{a}$ is called *completely monotone*, denoted by $X \in$ CM or $\mathbf{a} \in$ CM, if $\mathbf{a}$ is a mixture of geometric distributions; that is, if $\mathbf{a} = (a_i)$ is given by

$$a_i = \int_0^1 (1-\alpha)\alpha^i d\mu(\alpha), \quad i = 0, 1, \cdots,$$

where μ is a probability measure on $(0, 1)$.

Geometric distributions play a special role among discrete distributions. Let X be a geometrically distributed random variable with probability vector $\mathbf{b} = (b_i)$. It is easy to see that the hazard rate function, as well as the likelihood ratio function, is constant. A geometrically distributed random variable X is then called the *constant hazard rate*, and is denoted by $X \in$ CHR or $\mathbf{b} \in$ CHR. The converse is also true. That is, $X \in$ CHR if and only if the discrete random variable X is geometrically distributed. On the other hand, the reversed hazard rate function of X is given by

$$r_i = \frac{(1-\alpha)\alpha^i}{1-\alpha^{i+1}}, \quad i = 0, 1, \cdots.$$

Since r_i is decreasing, the geometric distribution has the decreasing reversed hazard rate function (DRHR). Recall that CHR = IHR $\cap$ DHR $\subset$ DRHR.

Definition 3.9 Let X be a discrete random variable with probability vector $\mathbf{a} = (a_i)$.

(i) X is called the *increasing hazard rate average*, denoted by $X \in$ IHRA or $\mathbf{a} \in$ IHRA, if

$$(\overline{A}_i)^{i+1} \geq (\overline{A}_{i+1})^i, \quad i \in \mathcal{N}.$$

If the above inequality is reversed, X is called the *decreasing hazard rate average* and is denoted by $X \in$ DHRA or $\mathbf{a} \in$ DHRA.

(ii) X is called *new better than used* and is denoted by $X \in$ NBU or $\mathbf{a} \in$ NBU, if

$$\overline{A}_{i+j} \leq \overline{A}_i \overline{A}_j, \quad i,\ j \in \mathcal{N}.$$

If the inequality is reversed, X is called *new worse than used* and is denoted by $X \in$ NWU or $\mathbf{a} \in$ NWU.

Since $1 - h_i = \overline{A}_{i+1}/\overline{A}_i$ whenever $\overline{A}_i > 0$, it follows that

$$\overline{A}_n = \prod_{i=0}^{n-1}(1 - h_i), \quad n \in \mathcal{N},$$

provided that $\overline{A}_n > 0$. Hence $X \in$ IHRA if and only if

$$-\frac{1}{n}\log \overline{A}_n = -\frac{1}{n}\sum_{i=0}^{n-1}\log(1 - h_i)$$

is increasing in n on the support of X. Note that $-\log(1 - h_i) \approx h_i$ if h_i is sufficiently small. In Definition 3.9(i), X is called IHRA (DHRA, respectively) because the average of the hazard rate function is increasing (decreasing). On the other hand, the NBU (NWU) property is written as

$$P[X \geq i] = \overline{A}_i \geq \ (\leq)\ \frac{\overline{A}_{i+j}}{\overline{A}_j} = P[X \geq i + j | X \geq j], \quad j = 1, 2, \cdots,$$

with the convention $0/0 = 0$. The left-hand side in the above inequality denotes the survival probability of a new unit, whereas the right-hand side is the survival probability of an old unit with age j. The term 'new better (worse) than used' is thus justified.

According to Barlow and Proschan (Sections 4.4 and 6.5, 1975), the IHRA and NBU classes of random variables are closed under convolution (the closure property of the IHR class is proved in Theorem 3.6(ii)), while they are not closed under mixture. In contrast, the DHR, DHRA and NWU classes are closed under mixture, while they are not closed under convolution. See Barlow and Proschan (1975) for more details and other distribution classes of interest in reliability.

In the next theorem, we list the inclusive relations between the distribution classes defined so far. Figure 3.1 depicts the relationship.

Theorem 3.7 *In the distribution classes of discrete random variables, we have:*

(i) DLR $\subset$ IHR $\cap$ DRHR; IHR $\subset$ IHRA $\subset$ NBU;

(ii) CM $\subset$ ILR $\subset$ DHR $\subset$ DHRA $\subset$ NWU; DHR $\subset$ DRHR.

Proof. The fact that IHR $\subset$ IHRA follows from the above discussion. Let

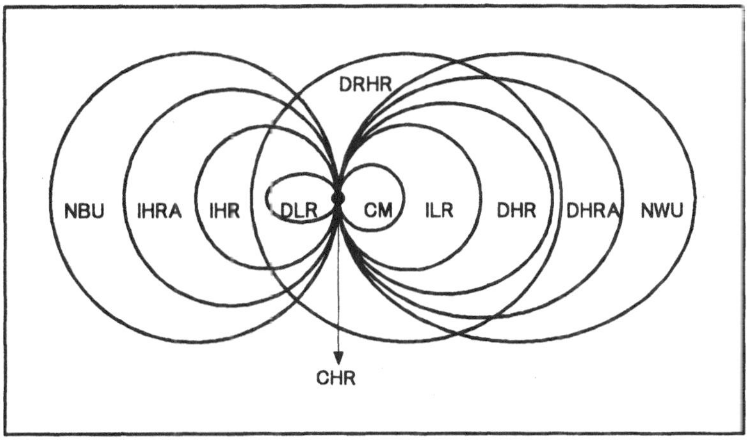

Figure 3.1 *The relationship between distribution classes.*

X be a discrete random variable with probability vector $\mathbf{a} = (a_i)$ and let $\mathbf{A}$ be the upper triangular matrix defined in (3.17). Suppose $X \in$ DLR so that $\mathbf{A}$ is TP_2. Then, from Corollary 3.1, $\mathbf{AU}$, as well as $\mathbf{AV}$, is TP_2. Hence $X \in$ IHR $\cap$ DRHR. Next, let $i > j$. If $X \in$ IHRA then

$$(\overline{A}_j)^i \geq (\overline{A}_i)^j \quad \text{and} \quad (\overline{A}_i)^{i+j} \geq (\overline{A}_{i+j})^i.$$

Combining, we have $(\overline{A}_j\overline{A}_i)^i \geq (\overline{A}_{i+j})^i$ so that $X \in$ NBU. The proof of part (ii) is left to the reader (see Exercise 3.5). □

Finally, we provide some illustrative examples.

Example 3.3 Let X and Y be mutually independent, geometrically distributed random variables with parameter α, $0 < \alpha < 1$. A straightforward computation shows that

$$P[X + Y = i] = (1 - \alpha)^2(i + 1)\alpha^i, \quad i = 0, 1, \cdots.$$

The likelihood ratio function of $X + Y$ is given by

$$\ell_i = \frac{i+2}{i+1}\alpha, \quad i = 0, 1, \cdots,$$

which is decreasing in i and converges to α. Hence $X + Y \in$ DLR. On the other hand, consider a geometric distribution with parameter β where $0 < \beta < \alpha$. Let $0 < p < 1$. The probability vector $\mathbf{c} = (c_i)$ defined by

$$c_i = p(1 - \alpha)\alpha^i + (1 - p)(1 - \beta)\beta^i, \quad i = 0, 1, \cdots,$$

is a mixture of geometric distributions and hence CM. It is readily verified that the likelihood ratio function is increasing in i and converges to α for any $\beta < \alpha$.

Example 3.4 (Uniform distribution) Let X be a *uniformly distributed* discrete random variable with support $\{0, 1, \cdots, n\}$, i.e., its probability vector $\mathbf{a} = (a_i)$ is given by

$$a_i = \frac{1}{n+1}, \quad i = 0, 1, \cdots, n.$$

Writing $a_m = 0$ for $m > n$, the likelihood ratio function is given by

$$\ell_i = \begin{cases} 1, & i = 0, 1, \cdots, n-1, \\ 0, & i = n, \end{cases}$$

which is decreasing in i, so that the uniform distribution is DLR and hence, from Theorem 3.7(i), it is IHR and DRHR simultaneously. In fact, the hazard rate and reversed hazard rate functions are given by

$$h_i = \frac{1}{n+1-i}; \quad r_i = \frac{1}{i+1}, \quad i = 0, 1, \cdots, n,$$

respectively, which are increasing and decreasing on the support of X.

Example 3.5 (Binomial distribution) For $0 < p < 1$, the probability vector $\mathbf{a} = (a_i)$ with

$$a_i = {}_nC_i\, p^i (1-p)^{n-i}, \quad i = 0, 1, \cdots, n,$$

is called a *binomial distribution*, where ${}_nC_i = n!/i!\,(n-i)!$ is the binomial coefficient. The quantity a_i is the $(i+1)$th term in the binomial expansion of $(p + (1-p))^n$. Clearly, a_i is the probability that there are exactly i successes in n repeated Bernoulli trials with probability p of success. The likelihood ratio function is given by

$$\ell_i = \frac{p}{1-p} \frac{n-i}{i+1}, \quad i = 0, 1, \cdots, n,$$

which is decreasing in i. Hence binomial distributions are DLR.

Example 3.6 (Negative binomial distribution) For $m > 0$ and $0 < p < 1$, the probability vector $\mathbf{a} = (a_i)$ with

$$a_i = (1-p)^m \, {}_{m+i-1}C_i\, p^i, \quad i = 0, 1, \cdots,$$

is called a *negative binomial distribution*. Noting ${}_{m+i-1}C_i = (-1)^i \, {}_{-m}C_i$, the quantity $a_i/(1-p)^m$ is the $(i+1)$th term in the (generalized) binomial expansion of $(1-p)^{-m}$. When $m = 1$, this special case reduces to a geometric distribution. The likelihood ratio function is

$$\ell_i = (1-p)\frac{m+i}{i+1}, \quad i = 0, 1, \cdots.$$

For $0 < m < 1$, the likelihood ratio function is increasing in i while it is decreasing in i for the case where $m > 1$. Hence negative binomial distributions are ILR if $0 < m < 1$ while they are DLR if $m > 1$.

Example 3.7 (Poisson distribution) For $\lambda > 0$, the probability vector $\mathbf{a} = (a_i)$ with

$$a_i = \frac{\lambda^i}{i!} e^{-\lambda}, \quad i = 0, 1, \cdots,$$

is called a *Poisson distribution*. It is well known that the Poisson distribution can be obtained from a binomial distribution by a suitable limiting argument (see, e.g., Neuts, 1973). The likelihood ratio function is given by

$$\ell_i = \frac{\lambda}{i+1}, \quad i = 0, 1, \cdots,$$

which is decreasing in i. Hence Poisson distributions are DLR.

3.3 Stochastic ordering relations

In this section, we define some important stochastic ordering relations for discrete random variables defined on $\mathcal{N} = \mathcal{Z}_+$. For other stochastic ordering relations of interest in stochastic modeling, see Stoyan (1983), Shaked and Shanthikumar (1994) and Szekli (1995). Throughout this section, if $\mathbf{a} = (a_i)$ is a probability vector, we write $\overline{A}_n = \sum_{k \geq n} a_k$ and $A_n = \sum_{k \leq n} a_k$, as before. Also, the convention $0/0 = 0$ is used.

Definition 3.10 Let X and Y be discrete random variables with probability vectors $\mathbf{a} = (a_i)$ and $\mathbf{b} = (b_i)$ respectively.

(i) X is said to be greater than Y in the sense of *likelihood ratio ordering*, denoted by $X \geq_{\text{lr}} Y$ or $\mathbf{a} \geq_{\text{lr}} \mathbf{b}$, if

$$a_i b_j \geq a_j b_i \quad \text{for all } i > j.$$

(ii) X is said to be greater than Y in the sense of *hazard rate ordering*, denoted by $X \geq_{\text{hr}} Y$ or $\mathbf{a} \geq_{\text{hr}} \mathbf{b}$, if

$$\overline{A}_i \overline{B}_j \geq \overline{A}_j \overline{B}_i \quad \text{for all } i > j.$$

(iii) X is said to be greater than Y in the sense of *reversed hazard rate ordering*, denoted by $X \geq_{\text{rh}} Y$ or $\mathbf{a} \geq_{\text{rh}} \mathbf{b}$, if

$$A_i B_j \geq A_j B_i \quad \text{for all } i > j.$$

(iv) X is said to be greater than Y in the sense of *stochastic ordering* or simply *stochastically greater* than Y, denoted by $X \geq_{\text{st}} Y$ or $\mathbf{a} \geq_{\text{st}} \mathbf{b}$, if

$$\overline{A}_i \geq \overline{B}_i \text{ (or, equivalently, } A_i \leq B_i) \quad \text{for all } i.$$

The reason why we use the terms 'likelihood ratio ordering', 'hazard rate ordering' and 'reversed hazard rate ordering' is justified in Exercise 3.7.

Also, these ordering relations can be more succinctly expressed using the triangular matrices **U** and **V**:

$$X \geq_{\rm lr} Y \quad \text{if and only if} \quad \begin{pmatrix} \mathbf{b}^\top \\ \mathbf{a}^\top \end{pmatrix} \quad \text{is TP}_2, \tag{3.20}$$

$$X \geq_{\rm hr} Y \quad \text{if and only if} \quad \begin{pmatrix} \mathbf{b}^\top \\ \mathbf{a}^\top \end{pmatrix} \mathbf{U} \quad \text{is TP}_2, \tag{3.21}$$

and

$$X \geq_{\rm rh} Y \quad \text{if and only if} \quad \begin{pmatrix} \mathbf{b}^\top \\ \mathbf{a}^\top \end{pmatrix} \mathbf{V} \quad \text{is TP}_2. \tag{3.22}$$

See Keilson and Sumita (1982), where they use different terminologies. For the ordering relation $X \geq_{\rm st} Y$, we have

$$X \geq_{\rm st} Y \quad \text{if and only if} \quad \mathbf{a}^\top \mathbf{U} \geq \mathbf{b}^\top \mathbf{U} \ (\text{or } \mathbf{a}^\top \mathbf{V} \leq \mathbf{b}^\top \mathbf{V}). \tag{3.23}$$

The equivalence (3.23) is first observed by Keilson and Kester (1977).

Another characterization of the stochastic ordering $\geq_{\rm st}$ is the following.* For a vector $\mathbf{f} = (f_i)$, we sometimes write $f(i)$ for f_i. Let

$$\mathcal{F}_{\rm st} = \{\mathbf{f} = (f_i) : f_i \geq f_{i-1} \text{ for all } i\}, \tag{3.24}$$

i.e., $\mathcal{F}_{\rm st}$ is the class of increasing vectors.

Theorem 3.8 *Let X and Y be random variables with probability vectors $\mathbf{a} = (a_i)$ and $\mathbf{b} = (b_i)$ respectively. Then, $X \geq_{\rm st} Y$ if and only if*

$$E[f(X)] \geq E[f(Y)] \quad \textit{for all} \quad \mathbf{f} \in \mathcal{F}_{\rm st}$$

for which the expectations exist.

Proof. Suppose that $X \geq_{\rm st} Y$. For $\mathbf{f} \in \mathcal{F}_{\rm st}$, by considering the vector $\mathbf{f} - f_0\mathbf{1}$, we can assume without loss of generality that $\mathbf{f} \geq \mathbf{0}$. Now, if $E[f(X)]$ exists then, since $\mathbf{f}$ is increasing, we have

$$0 \leq \overline{A}_{N+1} f_N \leq \sum_{i=N+1}^{\infty} a_i f_i \to 0 \quad \text{as} \quad N \to \infty.$$

On the other hand, defining $f_{-1} = 0$, we have

$$\begin{aligned} \sum_{i=0}^{N} a_i f_i &= \sum_{i=0}^{N} (\overline{A}_i - \overline{A}_{i+1}) f_i \\ &= \sum_{i=0}^{N} \overline{A}_i (f_i - f_{i-1}) - \overline{A}_{N+1} f_N. \end{aligned}$$

* It can be shown that $X \geq_{\rm st} Y$ if and only if two random variables $\widehat{X}$ and $\widehat{Y}$ exist, defined on some probability space, such that $\widehat{X} \stackrel{\rm d}{=} X$, $\widehat{Y} \stackrel{\rm d}{=} Y$ and $\widehat{X} \geq \widehat{Y}$ almost surely. Although we do not use this property in this book, this characterization is very useful in applications.

It follows that

$$E[f(X)] = \sum_{i=0}^{\infty} \overline{A}_i (f_i - f_{i-1}).$$

Hence, if both $E[f(X)]$ and $E[f(Y)]$ exist, then

$$\begin{aligned} E[f(X) - f(Y)] &= \sum_{i=0}^{\infty} (\overline{A}_i - \overline{B}_i)(f_i - f_{i-1}) \\ &= \sum_{i=1}^{\infty} (\overline{A}_i - \overline{B}_i)(f_i - f_{i-1}), \end{aligned}$$

where the fact $\overline{A}_0 = \overline{B}_0 = 1$ is employed. Since $X \geq_{\text{st}} Y$ implies $\overline{A}_i \geq \overline{B}_i$ and since $\mathbf{f}$ is increasing, we have $E[f(X)] \geq E[f(Y)]$.

Conversely, let $\mathbf{f} = (f_i)$ be such that

$$f_j = \begin{cases} 1, & j \geq i, \\ 0, & j < i. \end{cases}$$

Then $\mathbf{f} \in \mathcal{F}_{\text{st}}$, $E[f(X)] = \overline{A}_i$ and $E[f(Y)] = \overline{B}_i$, whence $\overline{A}_i \geq \overline{B}_i$, by assumption. This inequality holds for all $i \in \mathcal{N}$. Hence $X \geq_{\text{st}} Y$ and the theorem is proved. □

It should be noted that these ordering relations are *partial* and need not be *total*. That is, not all random variables satisfy the binary relation $\succ$, where $\succ$ denotes each of $\geq_{\text{lr}}$, $\geq_{\text{hr}}$, $\geq_{\text{rh}}$, or $\geq_{\text{st}}$. However, they satisfy the properties that $X \succ X$ (reflexivity); if $X \succ Y$ and $Y \succ Z$ then $X \succ Y$ (transitivity); and if $X \succ Y$ and $Y \succ X$ then $X \stackrel{\text{d}}{=} Y$ (anti-symmetry). Also, when discussing the ordering relation $X \succ Y$, the random variables X and Y need not be independent since we are only interested in their marginal distributions.

Theorem 3.9 *For two discrete random variables X and Y, $X \geq_{\text{lr}} Y$ implies both $X \geq_{\text{hr}} Y$ and $X \geq_{\text{rh}} Y$. Each of $X \geq_{\text{hr}} Y$ or $X \geq_{\text{rh}} Y$ implies $X \geq_{\text{st}} Y$.*

Proof. Let $\mathbf{a}$ and $\mathbf{b}$ be the probability vectors of X and Y respectively. Suppose $X \geq_{\text{lr}} Y$, i.e., (3.20) holds. Since $\mathbf{U}$ and $\mathbf{V}$ are TP_2, both $\begin{pmatrix} \mathbf{b}^\top \\ \mathbf{a}^\top \end{pmatrix} \mathbf{U}$ and $\begin{pmatrix} \mathbf{b}^\top \\ \mathbf{a}^\top \end{pmatrix} \mathbf{V}$ are TP_2, whence $X \geq_{\text{hr}} Y$ as well as $X \geq_{\text{rh}} Y$ respectively. If, in turn, (3.21) holds then Corollary 3.1 implies that

$$\begin{pmatrix} 1 & 0 \\ -1 & 1 \end{pmatrix} \begin{pmatrix} \mathbf{b}^\top \\ \mathbf{a}^\top \end{pmatrix} \mathbf{U} = \begin{pmatrix} \mathbf{b}^\top \mathbf{U} \\ \mathbf{a}^\top \mathbf{U} - \mathbf{b}^\top \mathbf{U} \end{pmatrix} \geq \mathbf{O},$$

so that (3.23) holds. Hence $X \geq_{\rm hr} Y$ implies $X \geq_{\rm st} Y$. That $X \geq_{\rm rh} Y$ implies $X \geq_{\rm st} Y$ can be proved analogously. □

The ordering relation $X \geq_{\rm hr} Y$ does not necessarily imply $X \geq_{\rm rh} Y$ and vice versa, as the next example shows.

Example 3.8 ($X \geq_{\rm hr} Y$ does not imply $X \geq_{\rm rh} Y$.) Let X and Y be random variables on $\{0, 1, 2\}$ with probability vectors $\mathbf{a}^\top = (1/3,\ 1/3,\ 1/3)$ and $\mathbf{b}^\top = (1/3,\ 2/5,\ 4/15)$ respectively. The hazard rate function of X is $(h_0^X,\ h_1^X,\ h_2^X) = (1/3,\ 1/2,\ 1)$ and that of Y is $(h_0^Y,\ h_1^Y,\ h_2^Y) = (1/3,\ 3/5,\ 1)$, so that $X \geq_{\rm hr} Y$ (see Exercise 3.7). On the other hand, the reversed hazard rate function of X is $(r_0^X,\ r_1^X,\ r_2^X) = (1,\ 1/2,\ 1/3)$ and that of Y is $(r_0^Y,\ r_1^Y,\ r_2^Y) = (1,\ 6/11,\ 4/15)$. Hence, X and Y are not ordered in the sense of reversed hazard rate ordering.

Stoyan (1983) proposed the following properties that all stochastic ordering relations are desired to possess. Let $\succ$ denote some stochastic ordering. For random variables X, Y and X_n, Y_n, $n = 1, 2, \cdots$:

(C) $X \succ Y$ implies $X + Z \succ Y + Z$ for any random variable Z independent of X and Y.

(R) For real numbers a and b with $a \geq b$, one has $a \succ b$.

(M) $X \succ Y$ implies $aX \succ aY$ for any $a > 0$.

(W) If X_n and Y_n converge to X and Y, respectively, in distribution as $n \to \infty$ and if $X_n \succ Y_n$ for every n. then $X \succ Y$.

(E) $X \succ Y$ implies $E[X] \geq E[Y]$.

Property (C) is called the *convolution property*. We note that any real number can be considered as a *degenerate random variable* which is independent of any other random variables. Hence, property (C) reduces to

(C′) $X \succ Y$ implies $X + c \succ Y + c$ for any real number c.

Property (R) is called the *real number property*, (M) the *multiplication property*, (W) the *weak convergence property*, and (E) the *expectation property*. It can be shown (see Stoyan, page 3, 1983) that any stochastic ordering relation having properties (C), (R), (M), and (W) must have property (E).

The next result is well known and the proof is omitted (see Exercise 3.11).

Lemma 3.7 *The stochastic ordering $\geq_{\rm st}$ satisfies all the properties stated above. Moreover, it is closed under mixtures of distributions.*

In contrast to the stochastic ordering $\geq_{\rm st}$, the other orderings given in Definition 3.10 do not satisfy the convolution property (C), but they do satisfy the other properties.

Example 3.9 (The likelihood ratio ordering $\geq_{\rm lr}$ does not satisfy the convolution property.) Let X and Y be geometrically distributed with parameters α and β, respectively, where $\alpha > \beta$. Since the likelihood ratio functions are given by $\ell_i^X = \alpha$ and $\ell_i^Y = \beta$, respectively, it follows that $X \geq_{\rm lr} Y$ (see

Exercise 3.7). Now, let Z be another random variable independent of X and Y and let its probability vector $\mathbf{c} = (c_i)$ be given by

$$c_0 = \gamma, \quad c_1 = 0, \quad c_2 = 1 - \gamma; \quad c_i = 0, \quad i \geq 3.$$

Denoting the probability vector of $X + Z$ by (p_i), simple computation of (3.16) yields

$$p_0 = (1-\alpha)\gamma, \quad p_1 = (1-\alpha)\alpha\gamma, \quad p_2 = (1-\alpha)(1-\gamma+\alpha^2\gamma).$$

Similarly, denoting the probability vector of $Y + Z$ by (q_i), one has

$$q_0 = (1-\beta)\gamma, \quad q_1 = (1-\beta)\beta\gamma, \quad q_2 = (1-\beta)(1-\gamma+\beta^2\gamma).$$

Note that $\ell_1^{X+Z} < \ell_1^{Y+Z}$ if and only if

$$\frac{1-\gamma+\alpha^2\gamma}{\alpha\gamma} < \frac{1-\gamma+\beta^2\gamma}{\beta\gamma}.$$

Hence, if $1 - \gamma - \alpha\beta\gamma > 0$, we have $\ell_1^{X+Z} < \ell_1^{Y+Z}$ so that $X + Z \geq_{\mathrm{lr}} Y + Z$ is not true.

We next provide a sufficient condition under which the convolution property (C) holds true for each of $\geq_{\mathrm{lr}}$, $\geq_{\mathrm{hr}}$ and $\geq_{\mathrm{rh}}$.

Theorem 3.10 *Let Z be independent of X and Y.*

(i) *Suppose $Z \in$ DLR. If $X \geq_{\mathrm{lr}} Y$ then $X + Z \geq_{\mathrm{lr}} Y + Z$.*

(ii) *Suppose $Z \in$ IHR. If $X \geq_{\mathrm{hr}} Y$ then $X + Z \geq_{\mathrm{hr}} Y + Z$.*

(iii) *Suppose $Z \in$ DRHR. If $X \geq_{\mathrm{rh}} Y$ then $X + Z \geq_{\mathrm{rh}} Y + Z$.*

Proof. Let $\mathbf{a}$ and $\mathbf{b}$ be the probability vectors of X and Y respectively. Let $\mathbf{c}$ be the probability vector of Z, and define $\mathbf{C}$ to be the upper triangular matrix given in (3.17). It follows that the probability vectors of $X + Z$ and $Y + Z$ are given by $\mathbf{a}^\top\mathbf{C}$ and $\mathbf{b}^\top\mathbf{C}$ respectively.

(i) Note that $X + Z \geq_{\mathrm{lr}} Y + Z$ if and only if

$$\begin{pmatrix} \mathbf{b}^\top\mathbf{C} \\ \mathbf{a}^\top\mathbf{C} \end{pmatrix} = \begin{pmatrix} \mathbf{b}^\top \\ \mathbf{a}^\top \end{pmatrix}\mathbf{C} \quad \text{is } \mathrm{TP}_2.$$

$X \geq_{\mathrm{lr}} Y$ if and only if $\begin{pmatrix} \mathbf{b}^\top \\ \mathbf{a}^\top \end{pmatrix}$ is TP_2, and $Z \in$ DLR if and only if $\mathbf{C}$ is TP_2. Hence, by the basic composition formula (Theorem 3.1), $X \geq_{\mathrm{lr}} Y$ and $Z \in$ DLR together imply $X + Z \geq_{\mathrm{lr}} Y + Z$.

(ii) We need to prove

$$\begin{pmatrix} \mathbf{b}^\top\mathbf{CU} \\ \mathbf{a}^\top\mathbf{CU} \end{pmatrix} = \begin{pmatrix} \mathbf{b}^\top \\ \mathbf{a}^\top \end{pmatrix}\mathbf{CU} \quad \text{is } \mathrm{TP}_2.$$

Under the assumptions, $\begin{pmatrix} \mathbf{b}^\top \\ \mathbf{a}^\top \end{pmatrix}\mathbf{U}$ and $\mathbf{CU}$ are TP_2. Also $\mathbf{U}^{-1}(\mathbf{CU}) =$

$\mathbf{C} \geq \mathbf{O}$. The composition law (Theorem 3.2) then ensures the desired conclusion. The proof of part (iii) is similar. □

Corollary 3.4 *Let $\succ$ denote $\geq_{\rm lr}$ ($\geq_{\rm hr}$ or $\geq_{\rm rh}$, respectively). Suppose that random vectors (X_i, Y_i), $i = 1, \cdots, n$, are mutually independent. Moreover, suppose that either condition* (i) *or condition* (ii) *holds:*

(i) *$X_i \succ Y_i$ and either $X_i \in$ DLR (IHR or DRHR) or $Y_i \in$ DLR (IHR or DRHR) for all i;*

(ii) *There exist random variables Z_i, $i = 1, \cdots, n$, such that $X_i \succ Z_i \succ Y_i$ and $Z_i \in$ DLR (IHR or DRHR) for all i.*

Then, for each n, we have

$$\sum_{i=1}^{n} X_i \succ \sum_{i=1}^{n} Y_i.$$

The ordering relations given in Definition 3.10 have the following characterizations. For matrix $\mathbf{G} = (g_{ij})$, we sometimes write $g(i,j)$ for g_{ij}. Also, we write

$$\widetilde{g}_{ij} = g_{ij} - g_{ji}, \quad i, j \in \mathcal{N}.$$

Note that $\widetilde{g}_{ii} = 0$ and $\widetilde{g}_{ij} = -\widetilde{g}_{ji}$ for all $i, j \in \mathcal{N}$. We introduce the following matrix classes:

$$\mathcal{G}_{\rm lr} = \{\mathbf{G} : \widetilde{g}_{ij} \geq 0 \text{ for } i > j\};$$

$$\mathcal{G}_{\rm hr} = \{\mathbf{G} : \widetilde{g}_{ij} \geq \widetilde{g}_{i-1,j} \text{ for } i > j\};$$

$$\mathcal{G}_{\rm rh} = \{\mathbf{G} : \widetilde{g}_{ij} \geq \widetilde{g}_{i-1,j} \text{ for } i < j\};$$

$$\mathcal{G}_{\rm st} = \{\mathbf{G} : \widetilde{g}_{ij} \geq \widetilde{g}_{i-1,j} \text{ for all } i\}.$$

It is obvious that

$$\mathcal{G}_{\rm lr} \supset \mathcal{G}_{\rm hr} \supset \mathcal{G}_{\rm st} \quad \text{and} \quad \mathcal{G}_{\rm lr} \supset \mathcal{G}_{\rm rh} \supset \mathcal{G}_{\rm st}.$$

The next *bivariate characterization* is due to Shanthikumar and Yao (1991).

Theorem 3.11 *Let X and Y be mutually independent random variables with probability vectors $\mathbf{a} = (a_i)$ and $\mathbf{b} = (b_i)$ respectively. Let $\mathbf{G} = (g_{ij})$.*

(i) *$X \geq_{\rm lr} Y$ if and only if $E[\widetilde{g}(X,Y)] \geq 0$ for all $\mathbf{G} \in \mathcal{G}_{\rm lr}$ for which the expectation exists.*

(ii) *$X \geq_{\rm hr} Y$ if and only if $E[\widetilde{g}(X,Y)] \geq 0$ for all $\mathbf{G} \in \mathcal{G}_{\rm hr}$ for which the expectation exists.*

(iii) *$X \geq_{\rm rh} Y$ if and only if $E[\widetilde{g}(X,Y)] \geq 0$ for all $\mathbf{G} \in \mathcal{G}_{\rm rh}$ for which the expectation exists.*

(iv) *$X \geq_{\rm st} Y$ if and only if $E[\widetilde{g}(X,Y)] \geq 0$ for all $\mathbf{G} \in \mathcal{G}_{\rm st}$ for which the expectation exists.*

Proof. Noting $\tilde{g}_{ij} = -\tilde{g}_{ji}$, we have

$$\begin{aligned} E[\tilde{g}(X,Y)] &= \sum_{i>j} a_i b_j \tilde{g}_{ij} + \sum_{i<j} a_i b_j \tilde{g}_{ij} \\ &= \sum_{i>j} a_i b_j \tilde{g}_{ij} - \sum_{j<i} a_j b_i \tilde{g}_{ij} \\ &= \sum_{i>j} (a_i b_j - a_j b_i)\tilde{g}_{ij}. \end{aligned} \tag{3.25}$$

(i) Suppose $X \geq_{\rm lr} Y$. Then, by definition, $a_i b_j \geq a_j b_i$ for $i > j$, whence $X \geq_{\rm lr} Y$ implies $E[\tilde{g}(X,Y)] \geq 0$ for all $\mathbf{G} \in \mathcal{G}_{\rm lr}$. To prove the converse, let k and ℓ be any integers such that $k > \ell$. Define $g_{ij} = 1$ for $i = k$ and $j = \ell$, and $g_{ij} = 0$ otherwise. It is obvious that $\mathbf{G} = (g_{ij}) \in \mathcal{G}_{\rm lr}$. Then

$$0 \leq E[\tilde{g}(X,Y)] = \sum_{i,j} a_i b_j \tilde{g}_{ij} = a_k b_\ell - a_\ell b_k, \quad k > \ell.$$

Since k and ℓ are arbitrary, we conclude that $X \geq_{\rm lr} Y$.

(ii) Note that, for any $N > 0$,

$$\begin{aligned} \sum_{i=0}^{N}\sum_{j=0}^{i-1} a_i b_j \tilde{g}_{ij} &= \sum_{i=0}^{N}\sum_{j=0}^{i-1} \overline{A}_i b_j \tilde{g}_{ij} - \sum_{i=0}^{N}\sum_{j=0}^{i-1} \overline{A}_{i+1} b_j \tilde{g}_{ij} \\ &= \sum_{i=1}^{N}\sum_{j=0}^{i-1} \overline{A}_i b_j (\tilde{g}_{ij} - \tilde{g}_{i-1,j}) - \overline{A}_{N+1} \sum_{j=0}^{N-1} b_j \tilde{g}_{Nj}, \end{aligned}$$

where we have made use of the fact that $\tilde{g}_{ii} = 0$. It is not difficult to show that if $E[\tilde{g}(X,Y)]$ exists then $\overline{A}_{N+1} \sum_{j=0}^{N-1} b_j \tilde{g}_{Nj} \to 0$ as $N \to \infty$. It follows that

$$\sum_{i>j} a_i b_j \tilde{g}_{ij} = \sum_{i>j} \overline{A}_i b_j (\tilde{g}_{ij} - \tilde{g}_{i-1,j})$$

and hence

$$E[\tilde{g}(X,Y)] = \sum_{i>j} (\overline{A}_i b_j - a_j \overline{B}_i)(\tilde{g}_{ij} - \tilde{g}_{i-1,j}).$$

Now, if $X \geq_{\rm hr} Y$ then $\overline{A}_{i+1}\overline{B}_i \geq \overline{A}_i \overline{B}_{i+1}$ so that

$$a_i \overline{B}_i = \overline{A}_i \overline{B}_i - \overline{A}_{i+1}\overline{B}_i \leq \overline{A}_i \overline{B}_i - \overline{A}_i \overline{B}_{i+1} = \overline{A}_i b_i.$$

Also, since $\overline{A}_i \overline{B}_j \geq \overline{A}_j \overline{B}_i$ for $i > j$, we have

$$b_j \overline{A}_i \overline{B}_j \geq b_j \overline{A}_j \overline{B}_i \geq a_j \overline{B}_j \overline{B}_i, \quad i > j,$$

whence $\overline{A}_i b_j \geq a_j \overline{B}_i$ for $i > j$. Therefore $X \geq_{\rm hr} Y$ implies $E[\tilde{g}(X,Y)] \geq 0$ for all $\mathbf{G} \in \mathcal{G}_{\rm hr}$. To prove the converse, let $k > \ell$ and define $g_{ij} = 1$ for $i \geq k$ and $j \geq \ell$ and $g_{ij} = 0$ otherwise. Obviously, $\mathbf{G} \in \mathcal{G}_{\rm hr}$ so that

$$0 \leq E[\tilde{g}(X,Y)] = \overline{A}_k \overline{B}_\ell - \overline{A}_\ell \overline{B}_k, \quad k > \ell,$$

whence $X \geq_{\rm hr} Y$. The proof of part (iii) is similar and is left to the reader (see Exercise 3.13).

(iv) Note that

$$E[\widetilde{g}(X,Y)] = \sum_{i>j} \overline{A}_i b_j (\widetilde{g}_{ij} - \widetilde{g}_{i-1,j}) - \sum_{i<j} A_i b_j (\widetilde{g}_{i+1,j} - \widetilde{g}_{ij}). \qquad (3.26)$$

Let Y^* be a random variable independent of Y with the same probability vector $\mathbf{b} = (b_i)$ as Y. Then, from (3.25),

$$E[\widetilde{g}(Y^*,Y)] = \sum_{i>j} (b_i b_j - b_j b_i)\widetilde{g}_{ij} = 0,$$

while, from (3.26),

$$E[\widetilde{g}(Y^*,Y)] = \sum_{i>j} \overline{B}_i b_j (\widetilde{g}_{ij} - \widetilde{g}_{i-1,j}) - \sum_{i<j} B_i b_j (\widetilde{g}_{i+1,j} - \widetilde{g}_{ij}).$$

If $X \geq_{\rm st} Y$, we have $\overline{A}_i \geq \overline{B}_i$ and $A_i \leq B_i$ for all i. It follows that

$$E[\widetilde{g}(X,Y)] \geq E[\widetilde{g}(Y^*,Y)] = 0$$

for all $\mathbf{G} \in \mathcal{G}_{\rm st}$. The converse is readily proved by defining $g_{ij} = 1$ for $i \geq j$ and $g_{ij} = 0$ otherwise. □

The bivariate characterization given in Theorem 3.11 can be used as follows. See Kijima and Ohnishi (1996) for details.

Example 3.10 (Stochastic Hardy–Littlewood–Pólya inequality) Suppose that $\mathbf{f} = (f_i) \in \mathcal{F}_{\rm st}$ and $\mathbf{G} = (g_{ij}) \in \mathcal{G}_{\rm lr}$. Then, it is easy to see that $\mathbf{F} = (f(g_{ij})) \in \mathcal{G}_{\rm lr}$. Hence, if X and Y are mutually independent such that $X \geq_{\rm lr} Y$, then Theorem 3.11(i) implies

$$E[f(g(X,Y))] \geq E[f(g(Y,X))] \quad \text{for all } \mathbf{f} \in \mathcal{F}_{\rm st},$$

so that $g(X,Y) \geq_{\rm st} g(Y,X)$, from Theorem 3.8.

Now consider $g_{ij} = ai + bj$ with $a \geq b$. Then $\widetilde{g}_{ij} = (a-b)(i-j) \geq 0$ for $i > j$, whence $\mathbf{G} = (g_{ij}) \in \mathcal{G}_{\rm lr}$. It follows that

$$aX + bY \geq_{\rm st} bX + aY, \quad a \geq b.$$

Using pairwise comparisons, the above stochastic ordering relation can be generalized to the stochastic Hardy–Littlewood–Pólya inequality

$$\sum_{i=1}^n a_i X_i \geq_{\rm st} \sum_{i=1}^n a_{[i]} X_i \geq_{\rm st} \sum_{i=1}^n a_{n-i+1} X_i,$$

where X_i are mutually independent random variables such that $X_1 \geq_{\rm lr} \cdots \geq_{\rm lr} X_n$, a_i are real numbers such that $a_1 \geq \cdots \geq a_n$, and $[1], \cdots, [n]$ denote any permutation of the integers $1, \cdots, n$.

3.4 Monotone Markov chains

In studying the transient behavior of a Markov chain, it is very helpful to know a relation between the state distributions in time. For example, if we know that the state distributions are monotone in time in some sense, this information ensures the existence of the limiting distribution and provides us with many useful inequalities. In this section, we define some important notions of stochastic monotonicity in Markov chains. Throughout this section, we assume that the Markov chain under consideration is irreducible. We shall denote the Markov chain by $\{X_n\}$ and its transition matrix by $\mathbf{P}$, unless specified otherwise. Note that, by irreducibility, the matrix $\mathbf{P}$ has no null columns.

Definition 3.11 Let $\{X_n\}$ be an irreducible Markov chain with transition matrix $\mathbf{P}$.

(i) $\{X_n\}$ is said to be *monotone in the sense of likelihood ratio ordering*, denoted by $\{X_n\} \in \mathcal{M}_{\text{lr}}$ or $\mathbf{P} \in \mathcal{M}_{\text{lr}}$, if the transition matrix $\mathbf{P}$ itself is TP_2.

(ii) $\{X_n\}$ is said to be *monotone in the sense of hazard rate ordering*, denoted by $\{X_n\} \in \mathcal{M}_{\text{hr}}$ or $\mathbf{P} \in \mathcal{M}_{\text{hr}}$, if $\mathbf{PU}$ is TP_2.

(iii) $\{X_n\}$ is said to be *monotone in the sense of reversed hazard rate ordering*, denoted by $\{X_n\} \in \mathcal{M}_{\text{rh}}$ or $\mathbf{P} \in \mathcal{M}_{\text{rh}}$, if $\mathbf{PV}$ is TP_2.

(iv) $\{X_n\}$ is said to be *monotone in the sense of stochastic ordering*, or simply *stochastically monotone* for short, denoted by $\{X_n\} \in \mathcal{M}_{\text{st}}$ or $\mathbf{P} \in \mathcal{M}_{\text{st}}$, if $\mathbf{U}^{-1}\mathbf{PU} \geq \mathbf{O}$ or, equivalently, $\mathbf{V}^{-1}\mathbf{PV} \geq \mathbf{O}$.

Recall that $\mathbf{P} \in \text{TP}_2$ implies that both $\mathbf{PU}$ and $\mathbf{PV}$ are TP_2, each of which in turn implies that $\mathbf{U}^{-1}\mathbf{PU} \geq \mathbf{O}$ (see Corollary 3.1). Hence, $\mathbf{P} \in \mathcal{M}_{\text{lr}}$ implies both $\mathbf{P} \in \mathcal{M}_{\text{hr}}$ and $\mathbf{P} \in \mathcal{M}_{\text{rh}}$, and each of these implies $\mathbf{P} \in \mathcal{M}_{\text{st}}$.

Lemma 3.8 *Let $\mathcal{M}$ denote each of $\mathcal{M}_{\text{lr}}$, $\mathcal{M}_{\text{hr}}$, $\mathcal{M}_{\text{rh}}$ or $\mathcal{M}_{\text{st}}$. Then, $\mathbf{P} \in \mathcal{M}$ if and only if $\mathbf{P}^n \in \mathcal{M}$ for all $n = 1, 2, \cdots$.*

Proof. If $\mathcal{M} = \mathcal{M}_{\text{lr}}$, the result is a consequence of the basic composition formula (Theorem 3.1). The results for $\mathcal{M}_{\text{hr}}$ and $\mathcal{M}_{\text{rh}}$ are due to a combination of Corollaries 3.1 and 3.2. For $\mathcal{M} = \mathcal{M}_{\text{st}}$, we invoke Lemma 3.2 to conclude that

$$(\mathbf{U}^{-1}\mathbf{PU})(\mathbf{U}^{-1}(\mathbf{PU})) = \mathbf{U}^{-1}(\mathbf{P}(\mathbf{PU})).$$

But, Theorem A.8(i) implies $\mathbf{P}(\mathbf{PU}) = \mathbf{P}^2\mathbf{U}$ so that $\mathbf{U}^{-1}\mathbf{P}^2\mathbf{U} \geq \mathbf{O}$ and the result follows by an induction argument. □

Theorem 3.12 (i) *$\mathbf{P} \in \mathcal{M}_{\text{lr}}$ if and only if $\mathbf{a}^\top\mathbf{P} \geq_{\text{lr}} \mathbf{b}^\top\mathbf{P}$ for all probability vectors $\mathbf{a}$ and $\mathbf{b}$ such that $\mathbf{a} \geq_{\text{lr}} \mathbf{b}$.*

(ii) *$\mathbf{P} \in \mathcal{M}_{\text{hr}}$ if and only if $\mathbf{a}^\top\mathbf{P} \geq_{\text{hr}} \mathbf{b}^\top\mathbf{P}$ for all probability vectors $\mathbf{a}$ and $\mathbf{b}$ such that $\mathbf{a} \geq_{\text{hr}} \mathbf{b}$.*

(iii) $\mathbf{P} \in \mathcal{M}_{\mathrm{rh}}$ *if and only if* $\mathbf{a}^\top \mathbf{P} \geq_{\mathrm{rh}} \mathbf{b}^\top \mathbf{P}$ *for all probability vectors* $\mathbf{a}$ *and* $\mathbf{b}$ *such that* $\mathbf{a} \geq_{\mathrm{rh}} \mathbf{b}$.

(iv) $\mathbf{P} \in \mathcal{M}_{\mathrm{st}}$ *if and only if* $\mathbf{a}^\top \mathbf{P} \geq_{\mathrm{st}} \mathbf{b}^\top \mathbf{P}$ *for all probability vectors* $\mathbf{a}$ *and* $\mathbf{b}$ *such that* $\mathbf{a} \geq_{\mathrm{st}} \mathbf{b}$.

Proof. Let $\mathbf{p}_i^\top$ denote the ith row (probability) vector of $\mathbf{P}$.

(i) From (3.20), $\mathbf{a}^\top \mathbf{P} \geq_{\mathrm{lr}} \mathbf{b}^\top \mathbf{P}$ if and only if

$$\begin{pmatrix} \mathbf{b}^\top \mathbf{P} \\ \mathbf{a}^\top \mathbf{P} \end{pmatrix} = \begin{pmatrix} \mathbf{b}^\top \\ \mathbf{a}^\top \end{pmatrix} \mathbf{P} \quad \text{is } \mathrm{TP}_2.$$

Hence, if $\mathbf{P} \in \mathrm{TP}_2$ and $\begin{pmatrix} \mathbf{b}^\top \\ \mathbf{a}^\top \end{pmatrix}$ is TP_2, then $\begin{pmatrix} \mathbf{b}^\top \\ \mathbf{a}^\top \end{pmatrix} \mathbf{P}$ is TP_2, from Theorem 3.1. Conversely, let $\mathbf{a} = \boldsymbol{\delta}_i$ and $\mathbf{b} = \boldsymbol{\delta}_j$ with $i > j$. Then, clearly, $\begin{pmatrix} \mathbf{b}^\top \\ \mathbf{a}^\top \end{pmatrix}$ is TP_2. Since $\mathbf{a}^\top \mathbf{P} = \mathbf{p}_i^\top$ and $\mathbf{b}^\top \mathbf{P} = \mathbf{p}_j^\top$ so that $\mathbf{p}_i \geq_{\mathrm{lr}} \mathbf{p}_j$, it follows that $\begin{pmatrix} \mathbf{p}_j^\top \\ \mathbf{p}_i^\top \end{pmatrix}$ is TP_2 for all $i > j$, implying that $\mathbf{P} \in \mathrm{TP}_2$.

(ii) From (3.21), $\mathbf{a}^\top \mathbf{P} \geq_{\mathrm{hr}} \mathbf{b}^\top \mathbf{P}$ if and only if

$$\begin{pmatrix} \mathbf{b}^\top \mathbf{P} \\ \mathbf{a}^\top \mathbf{P} \end{pmatrix} \mathbf{U} = \begin{pmatrix} \mathbf{b}^\top \\ \mathbf{a}^\top \end{pmatrix} \mathbf{PU} \quad \text{is } \mathrm{TP}_2.$$

Hence, if $\mathbf{PU} \in \mathrm{TP}_2$ and $\begin{pmatrix} \mathbf{b}^\top \\ \mathbf{a}^\top \end{pmatrix} \mathbf{U}$ is TP_2, then $\begin{pmatrix} \mathbf{b}^\top \\ \mathbf{a}^\top \end{pmatrix} \mathbf{PU}$ is TP_2, from Theorem 3.2. For the converse, we define $\mathbf{a}$ and $\mathbf{b}$ as in the proof of part (i). Then, $\mathbf{a} \geq_{\mathrm{hr}} \mathbf{b}$ and $\mathbf{p}_i \geq_{\mathrm{hr}} \mathbf{p}_j$. Since $\begin{pmatrix} \mathbf{p}_j^\top \\ \mathbf{p}_i^\top \end{pmatrix} \mathbf{U}$ is TP_2 for all $i > j$, we conclude that $\mathbf{PU}$ is TP_2. Assertion (iii) can be proved similarly using Theorem 3.3.

(iv) From (3.23) and Lemma 3.2, $\mathbf{a}^\top \mathbf{P} \geq_{\mathrm{st}} \mathbf{b}^\top \mathbf{P}$ if and only if

$$\mathbf{a}^\top \mathbf{PU} - \mathbf{b}^\top \mathbf{PU} = (\mathbf{a}^\top \mathbf{U} - \mathbf{b}^\top \mathbf{U}) \mathbf{U}^{-1} \mathbf{PU} \geq \mathbf{O},$$

which holds true if $\mathbf{U}^{-1} \mathbf{PU} \geq \mathbf{O}$ and $\mathbf{a}^\top \mathbf{U} \geq \mathbf{b}^\top \mathbf{U}$. Conversely, let $\mathbf{a}$ and $\mathbf{b}$ be defined as above, with $i = j + 1$. Then $\mathbf{p}_{i+1}^\top \mathbf{U} \geq \mathbf{p}_i^\top \mathbf{U}$ so that $\mathbf{U}^{-1} \mathbf{PU} \geq \mathbf{O}$. □

Let $\mathbf{p}_i^\top$ denote the ith row (probability) vector of $\mathbf{P}$. We saw in the proof of Theorem 3.12 that $\mathbf{P} \in \mathcal{M}_{\mathrm{lr}}$ if and only if $\mathbf{p}_i$ is increasing in i in the sense of $\geq_{\mathrm{lr}}$. The same is true for the other orderings. This observation leads to the following restatement of Theorem 3.12.

Corollary 3.5 *Let* $\mathbf{p}_i^\top$ *denote the ith row vector of transition matrix* $\mathbf{P}$.

(i) $\mathbf{P} \in \mathcal{M}_{\mathrm{lr}}$ *if and only if* $\mathbf{p}_{i+1} \geq_{\mathrm{lr}} \mathbf{p}_i$ *for all* i.

(ii) $\mathbf{P} \in \mathcal{M}_{\mathrm{hr}}$ *if and only if* $\mathbf{p}_{i+1} \geq_{\mathrm{hr}} \mathbf{p}_i$ *for all* i.

(iii) $\mathbf{P} \in \mathcal{M}_{\mathrm{rh}}$ *if and only if* $\mathbf{p}_{i+1} \geq_{\mathrm{rh}} \mathbf{p}_i$ *for all* i.

(iv) $\mathbf{P} \in \mathcal{M}_{\mathrm{st}}$ *if and only if* $\mathbf{p}_{i+1} \geq_{\mathrm{st}} \mathbf{p}_i$ *for all* i.

As is apparent from Corollary 3.5, Definition 3.11 implicitly assumes that the state space of the Markov chain $\{X_n\}$ is a totally ordered set. It is possible to extend the notions in Definition 3.11 to Markov chains with partially ordered state spaces. See, e.g., Kamae, Krengel and O'Brien (1977) or Shaked and Shanthikumar (1987) for such extensions.

Before proceeding, we provide some examples of monotone Markov chains.

Example 3.11 Let $\{X_n\}$ be a finite Markov chain with transition matrix

$$\mathbf{P} = \begin{pmatrix} A_0 & a_1 & a_2 & \cdots & \overline{A}_N \\ A_{-1} & a_0 & a_1 & \cdots & \overline{A}_{N-1} \\ \vdots & \vdots & \vdots & \ddots & \vdots \\ A_{-N} & a_{1-N} & a_{2-N} & \cdots & \overline{A}_0 \end{pmatrix}$$

given by (2.33). Suppose that the probability vector $\mathbf{a} = (a_i)$ is DLR, i.e., a_{i+1}/a_i is decreasing in i with the convention $0/0 = 0$. Then, it is readily seen that $\mathbf{P}$ is TP_2 so that $\mathbf{P} \in \mathcal{M}_{\mathrm{lr}}$. Also,

$$\mathbf{PU} = \begin{pmatrix} 1 & \overline{A}_1 & \overline{A}_2 & \cdots & \overline{A}_N \\ 1 & \overline{A}_0 & \overline{A}_1 & \cdots & \overline{A}_{N-1} \\ \vdots & \vdots & \vdots & \ddots & \vdots \\ 1 & \overline{A}_{1-N} & \overline{A}_{2-N} & \cdots & \overline{A}_0 \end{pmatrix}.$$

Hence, if $\mathbf{a}$ is IHR, i.e., $\overline{A}_{i+1}/\overline{A}_i$ is decreasing in i, then $\mathbf{PU}$ is TP_2 so that $\mathbf{P} \in \mathcal{M}_{\mathrm{hr}}$. Similarly, if $\mathbf{a}$ is DRHR, then $\mathbf{P} \in \mathcal{M}_{\mathrm{rh}}$. Finally, since

$$\mathbf{U}^{-1}\mathbf{PU} = \begin{pmatrix} 1 & \overline{A}_1 & \overline{A}_2 & \cdots & \overline{A}_N \\ 0 & a_0 & a_1 & \cdots & a_{N-1} \\ \vdots & \vdots & \vdots & \ddots & \vdots \\ 0 & a_{1-N} & a_{2-N} & \cdots & a_0 \end{pmatrix},$$

such a Markov chain is *always* stochastically monotone. See Keilson and Kester (1977) for other spatially homogeneous Markov chains.

Example 3.12 Let $\{X_n\}$ be a finite random walk with transition matrix

$$\mathbf{P} = \begin{pmatrix} r_0 & p_0 & 0 & 0 & \cdots & 0 \\ q_1 & r_1 & p_1 & 0 & \cdots & 0 \\ 0 & q_2 & r_2 & p_2 & \cdots & 0 \\ \vdots & & \ddots & \ddots & \ddots & \vdots \\ 0 & \cdots & 0 & q_{N-1} & r_{N-1} & p_{N-1} \\ 0 & \cdots & 0 & 0 & q_N & r_N \end{pmatrix};$$

see (2.32). Because of the zero structure, $\mathbf{P}$ is TP_2 if and only if

$$r_i r_{i+1} \geq p_i q_{i+1}, \quad i = 0, \cdots, N-1.$$

A sufficient condition for this is $r_i \geq 0.5$ for all i. On the other hand,

$$\mathbf{PU} = \begin{pmatrix} 1 & p_0 & 0 & 0 & \cdots & 0 \\ 1 & 1-q_1 & p_1 & 0 & \cdots & 0 \\ 1 & 1 & 1-q_2 & p_2 & \cdots & 0 \\ \vdots & \vdots & \ddots & \ddots & \ddots & \vdots \\ 1 & 1 & \cdots & 1 & 1-q_{N-1} & p_{N-1} \\ 1 & 1 & \cdots & 1 & 1 & 1-q_N \end{pmatrix}.$$

Hence, $\mathbf{PU} \in \mathrm{TP}_2$ if and only if

$$(1-q_i)(1-q_{i+1}) \geq p_i, \quad i = 0, \cdots, N-1,$$

where $q_0 = 0$. It is readily verified that a sufficient condition for this is $\mathbf{P} \in \mathrm{TP}_2$. Also, $\mathbf{P} \in \mathcal{M}_{\mathrm{st}}$ if and only if $p_i + q_{i+1} \leq 1$ for all i.

Example 3.13 Consider the age process $\{X_n\}$ of the successively replaced system described in Exercise 2.19. The process $\{X_n\}$ is a Markov chain with transition matrix

$$\mathbf{P} = \begin{pmatrix} h_1 & 1-h_1 & 0 & 0 & \cdots \\ h_2 & 0 & 1-h_2 & 0 & \cdots \\ h_3 & 0 & 0 & 1-h_3 & \cdots \\ \vdots & \vdots & \vdots & \ddots & \ddots \end{pmatrix}. \tag{3.27}$$

Note that, because of the zero structure, $\mathbf{P}$ itself cannot be TP_2. However, since

$$\mathbf{PU} = \begin{pmatrix} 1 & 1-h_1 & 0 & 0 & \cdots \\ 1 & 1-h_2 & 1-h_2 & 0 & \cdots \\ 1 & 1-h_3 & 1-h_3 & 1-h_3 & \cdots \\ \vdots & \vdots & \vdots & \vdots & \ddots \end{pmatrix},$$

if the hazard rate function h_i is decreasing in i, $\mathbf{PU}$ is TP_2 so that $\mathbf{P} \in \mathcal{M}_{\mathrm{hr}}$. It should be noted that $\mathbf{PV}$ cannot be TP_2.

Example 3.14 Let X_n be the number of customers in an M/G/1 queue just after the nth departure, where the arrival rate is λ and the service time distribution is $G(t)$. The process $\{X_n\}$ is a Markov chain with transition matrix

$$\mathbf{P} = \begin{pmatrix} b_0 & b_1 & b_2 & b_3 & \cdots \\ b_0 & b_1 & b_2 & b_3 & \cdots \\ 0 & b_0 & b_1 & b_2 & \cdots \\ 0 & 0 & b_0 & b_1 & \cdots \\ \vdots & \vdots & \vdots & \ddots & \ddots \end{pmatrix};$$

see (1.29). The components b_n are given by

$$b_n = \frac{1}{\lambda} \int_0^\infty p(n+1,t)\, dG(t), \quad n = 0, 1, \cdots,$$

where $p(n,t)$ denotes the density function of the *Erlang distribution* of order n, i.e.,

$$p(n,t) = \lambda \frac{(\lambda t)^{n-1}}{(n-1)!} e^{-\lambda t}, \quad t \geq 0; \quad n = 1, 2, \cdots,$$

and $p(0,t) = \delta(t)$, the Dirac delta function, meaning that $\delta(t) = 0$ for all $t > 0$ and $\int_0^\eta \delta(t)dt = 1$ for any $\eta > 0$. It is readily seen that $\mathbf{P} \in \mathcal{M}_{\rm st}$, as in Example 3.11. In this example, we shall prove that if the service time distribution is (continuous) IHR then $\mathbf{P} \in \mathcal{M}_{\rm hr}$.

Differentiating $p(n,t)$ with respect to t yields

$$p'(n+1,t) = \lambda\, p(n,t) - \lambda\, p(n+1,t), \quad n = 1, 2, \cdots.$$

It follows from integration by parts that

$$\overline{B}_n \equiv \sum_{i=n}^{\infty} b_i = \int_0^\infty p(n,t)\overline{G}(t)dt, \quad n = 0, 1, \cdots,$$

where $\overline{G}(t) = 1 - G(t)$. It is not difficult to derive the Laplace transform

$$\int_0^\infty e^{-st} p(n,t)dt = \left(\frac{\lambda}{s+\lambda}\right)^n, \quad n = 0, 1, \cdots; \quad \mathrm{Re}\,(s) > -\lambda,$$

from which we have

$$p(n,t) * p(m,t) = p(n+m,t), \quad n,\ m = 0, 1, \cdots,$$

where $*$ denotes the convolution operator with respect to t (see Appendix B.2). It follows that

$$\overline{B}_{n+m} = \int_0^\infty p(n,u) \int_0^\infty p(m,t)\overline{G}(t+u)dt\, du, \quad n,\ m = 0, 1, \cdots.$$

Define

$$\phi(m,u) = \int_0^\infty p(m,t)\overline{G}(t+u)dt, \quad m = 0, 1, \cdots; \quad u \geq 0.$$

Suppose $\overline{G}(x+h)/\overline{G}(x)$ is decreasing in x for any $h > 0$ or, equivalently,

$$\overline{G}(t_1+u_1)\,\overline{G}(t_2+u_2) \leq \overline{G}(t_1+u_2)\,\overline{G}(t_2+u_1), \quad t_1 < t_2, \ u_1 < u_2.$$

Such a continuous distribution is called the *increasing hazard rate* (IHR; see Definition 3.4 for the discrete case). Since $p(m,t)$ is TP_2 in $m = 0, 1, \cdots$ and $t \geq 0$ (see Exercise 3.16), the basic composition formula (Theorem C.1) implies that

$$\phi(m_1,u_1)\,\phi(m_2,u_2) \leq \phi(m_1,u_2)\,\phi(m_2,u_1), \quad m_1 < m_2, \ u_1 < u_2.$$

Another application of the basic composition formula then shows that

$$\overline{B}_{n_1+m_1}\overline{B}_{n_2+m_2} \le \overline{B}_{n_1+m_2}\overline{B}_{n_2+m_1}, \quad n_1 < n_2, \ m_1 < m_2,$$

so that the vector $(\overline{B}_n)$ is PF_2 (see Karlin, 1968, for more details). It follows that if the service time distribution is IHR then $\mathbf{P} \in \mathcal{M}_{\rm hr}$.

The reason why we call Markov chains given in Definition 3.11 *monotone* is given by the next theorem.

Theorem 3.13 *Let $\{X_n\}$ be an irreducible Markov chain with transition matrix* $\mathbf{P}$.

(i) *Suppose $\mathbf{P} \in \mathcal{M}_{\rm lr}$. If $X_0 \le_{\rm lr} X_1$ ($X_0 \ge_{\rm lr} X_1$, respectively) then $X_n \le_{\rm lr} X_{n+1}$ ($X_n \ge_{\rm lr} X_{n+1}$) for all $n = 0, 1, \cdots$, i.e., $\{X_n\}$ is monotonically increasing (decreasing) in the sense of likelihood ratio ordering.*

(ii) *Suppose $\mathbf{P} \in \mathcal{M}_{\rm hr}$. If $X_0 \le_{\rm hr} X_1$ ($X_0 \ge_{\rm hr} X_1$) then $X_n \le_{\rm hr} X_{n+1}$ ($X_n \ge_{\rm hr} X_{n+1}$) for all $n = 0, 1, \cdots$, i.e., $\{X_n\}$ is monotonically increasing (decreasing) in the sense of hazard rate ordering.*

(iii) *Suppose $\mathbf{P} \in \mathcal{M}_{\rm rh}$. If $X_0 \le_{\rm rh} X_1$ ($X_0 \ge_{\rm rh} X_1$) then $X_n \le_{\rm rh} X_{n+1}$ ($X_n \ge_{\rm rh} X_{n+1}$) for all $n = 0, 1, \cdots$, i.e., $\{X_n\}$ is monotonically increasing (decreasing) in the sense of reversed hazard rate ordering.*

(iv) *Suppose $\mathbf{P} \in \mathcal{M}_{\rm st}$. If $X_0 \le_{\rm st} X_1$ ($X_0 \ge_{\rm st} X_1$) then $X_n \le_{\rm st} X_{n+1}$ ($X_n \ge_{\rm st} X_{n+1}$) for all $n = 0, 1, \cdots$, i.e., $\{X_n\}$ is monotonically increasing (decreasing) in the sense of stochastic ordering or, simply, stochastically increasing (decreasing).*

Proof. Let $\boldsymbol{\pi}_n$ denote the state distribution of $\{X_n\}$ at time n. Recall that

$$\boldsymbol{\pi}_{n+1}^{\top} = \boldsymbol{\pi}_n^{\top}\mathbf{P}, \quad n = 0, 1, \cdots. \tag{3.28}$$

(i) From (3.28), we have

$$\begin{pmatrix} \boldsymbol{\pi}_n^{\top} \\ \boldsymbol{\pi}_{n+1}^{\top} \end{pmatrix} = \begin{pmatrix} \boldsymbol{\pi}_{n-1}^{\top} \\ \boldsymbol{\pi}_n^{\top} \end{pmatrix} \mathbf{P}, \quad n = 1, 2, \cdots. \tag{3.29}$$

According to (3.20), $X_0 \le_{\rm lr} X_1$ if and only if $\begin{pmatrix} \boldsymbol{\pi}_0^{\top} \\ \boldsymbol{\pi}_1^{\top} \end{pmatrix}$ is TP_2. Therefore, if $\mathbf{P}$ is TP_2, Theorem 3.1 applied to (3.29) shows that $\begin{pmatrix} \boldsymbol{\pi}_1^{\top} \\ \boldsymbol{\pi}_2^{\top} \end{pmatrix}$ is TP_2 and hence $X_1 \le_{\rm lr} X_2$. If $X_0 \ge_{\rm lr} X_1$, we consider

$$\begin{pmatrix} \boldsymbol{\pi}_{n+1}^{\top} \\ \boldsymbol{\pi}_n^{\top} \end{pmatrix} = \begin{pmatrix} \boldsymbol{\pi}_n^{\top} \\ \boldsymbol{\pi}_{n-1}^{\top} \end{pmatrix} \mathbf{P}, \quad n = 1, 2, \cdots.$$

A similar argument then leads to $X_1 \ge_{\rm lr} X_2$. Assertion (i) now follows by an induction argument.

(ii) From (3.29), we have

$$\begin{pmatrix} \pi_n^\top \\ \pi_{n+1}^\top \end{pmatrix} \mathbf{U} = \begin{pmatrix} \pi_{n-1}^\top \\ \pi_n^\top \end{pmatrix} \mathbf{PU}, \quad n = 1, 2, \cdots.$$

Recall that $X_0 \leq_{\rm hr} X_1$ if and only if $\begin{pmatrix} \pi_0^\top \\ \pi_1^\top \end{pmatrix} \mathbf{U}$ is TP_2. Therefore, if $\mathbf{PU}$ is TP_2, then Theorem 3.2 shows that $\begin{pmatrix} \pi_1^\top \\ \pi_2^\top \end{pmatrix} \mathbf{U}$ is also TP_2 and hence $X_1 \leq_{\rm hr} X_2$. The rest of the proof is similar to that of assertion (i). The proof of assertion (iii) is analogous.

(iv) From (3.28) and Lemma 3.2, we have

$$\pi_{n+1}^\top \mathbf{U} - \pi_n^\top \mathbf{U} = (\pi_n^\top \mathbf{U} - \pi_{n-1}^\top \mathbf{U})(\mathbf{U}^{-1}\mathbf{PU}), \quad n = 1, 2, \cdots.$$

Hence, if $\pi_0^\top \mathbf{U} \leq \pi_1^\top \mathbf{U}$ and $\mathbf{U}^{-1}\mathbf{PU} \geq \mathbf{O}$, then $\pi_1^\top \mathbf{U} \leq \pi_2^\top \mathbf{U}$ so that $X_1 \leq_{\rm st} X_2$. The rest of the proof is similar to that of assertion (i). □

As we saw in Example 3.11, if $\mathbf{P}$ is stochastic then $\mathbf{U}^{-1}\mathbf{PU}$ is of the form

$$\mathbf{U}^{-1}\mathbf{PU} = \begin{pmatrix} 1 & \mathbf{a}^\top \\ 0 & \mathbf{A} \end{pmatrix}$$

and, hence, $\mathbf{P} \in \mathcal{M}_{\rm st}$ if and only if $\mathbf{A} \geq \mathbf{O}$. For the case of finite state space, since $\mathbf{U}^{-1}\mathbf{PU}$ is a similarity transform, the eigenvalues of $\mathbf{P}$ other than unity are the same as those of $\mathbf{A}$. Hence, if the transition matrix $\mathbf{P}$ is stochastically monotone, the Perron–Frobenius eigenvalue of $\mathbf{A}$ coincides with the second largest eigenvalue of $\mathbf{P}$ in magnitude, which must be nonnegative (see Keilson and Kester, 1977). Therefore, for example, the relaxation time (2.38) of a stochastically monotone Markov chain is given by

$$T_{\rm REL}(\mathbf{P}) = \frac{1}{1 - \lambda(\mathbf{A})},$$

where $\lambda(\mathbf{A})$ denotes the PF eigenvalue of $\mathbf{A}$.

3.5 Unimodality of transition probabilities

In the remainder of this chapter, we provide several applications of monotone Markov chains. We shall consider finite Markov chains unless stated otherwise. The results obtained in this section can be generalized to the denumerable case with no difficulty. Let $\{X_n\}$ be a Markov chain with state space $\mathcal{N} = \{0, 1, \cdots, N\}$ and transition matrix $\mathbf{P}$. This section studies the monotonicity and unimodality of transition probabilities

$$p_{ij}(n) = P_i[X_n = j], \quad n = 0, 1, \cdots; \quad i, j \in \mathcal{N}.$$

We shall investigate the properties of $p_{ij}(n)$ with respect to $n = 0, 1, \cdots$. For this purpose, let $\mathbf{Q}(n) = \mathbf{P}^{n+1} - \mathbf{P}^n$ for $n = 0, 1, \cdots$, where $\mathbf{P}^n = (p_{ij}(n))$. It is easily seen that

$$\mathbf{Q}(n+1) = \mathbf{Q}(n)\mathbf{P} = \mathbf{P}\mathbf{Q}(n), \quad n = 0, 1, \cdots. \tag{3.30}$$

Let $\mathbf{q}_i^\top(n)$ denote the ith row vector of $\mathbf{Q}(n)$ and let $\mathbf{p}_i^\top$ be the ith row vector of $\mathbf{P}$. From (3.30), we have

$$\mathbf{q}_i^\top(n+1) = \mathbf{q}_i^\top(n)\mathbf{P}, \quad n = 0, 1, \cdots; \quad i \in \mathcal{N}, \tag{3.31}$$

where $\mathbf{q}_i^\top(0) = \mathbf{p}_i^\top - \boldsymbol{\delta}_i^\top$.

As before, let $\mathcal{S}$ be the class of stochastic matrices and let $\mathcal{M}_{\mathrm{st}}$ be the class of stochastically monotone matrices. From Definition 3.11(iv),

$$\mathcal{M}_{\mathrm{st}} = \{\mathbf{P} \in \mathcal{S} : \mathbf{U}^{-1}\mathbf{P}\mathbf{U} \geq \mathbf{O}\} = \{\mathbf{P} \in \mathcal{S} : \mathbf{V}^{-1}\mathbf{P}\mathbf{V} \geq \mathbf{O}\}.$$

First we discuss the monotonicity of $p_{00}(n)$ with respect to n. See Karlin (1964) for related results.

Theorem 3.14 *Suppose* $\mathbf{P} \in \mathcal{M}_{\mathrm{st}}$. *Then* $p_{00}(n)$ *is decreasing and* $p_{0N}(n)$ *is increasing in* n.

Proof. It is easy to see that $\mathbf{q}_0^\top(0)\mathbf{V} \leq \mathbf{0}^\top$ and $\mathbf{q}_0^\top(0)\mathbf{U} \geq \mathbf{0}^\top$. Also, from (3.31),

$$\mathbf{q}_0^\top(n+1)\mathbf{V} = \mathbf{q}_0^\top(n)\mathbf{V}(\mathbf{V}^{-1}\mathbf{P}\mathbf{V}).$$

Since $\mathbf{P} \in \mathcal{M}_{\mathrm{st}}$ so that $\mathbf{V}^{-1}\mathbf{P}\mathbf{V} \geq \mathbf{O}$, we have $\mathbf{q}_0^\top(n)\mathbf{V} \leq \mathbf{0}^\top$ for all n, by induction. Thus

$$[\mathbf{q}_0^\top(n)\mathbf{V}]_0 = p_{00}(n+1) - p_{00}(n) \leq 0,$$

i.e., $p_{00}(n)$ is decreasing in n. Similarly,

$$[\mathbf{q}_0^\top(n)\mathbf{U}]_N = p_{0N}(n+1) - p_{0N}(n) \geq 0,$$

so that $p_{0N}(n)$ is increasing in n. $\square$

The proof of Theorem 3.14 reveals that, if $\mathbf{P} \in \mathcal{M}_{\mathrm{st}}$, then

$$\sum_{j=0}^{i} p_{0j}(n+1) \leq \sum_{j=0}^{i} p_{0j}(n), \quad n = 0, 1, \cdots,$$

for all $i \in \mathcal{N}$. Also, if $\mathbf{P} \in \mathcal{M}_{\mathrm{st}}$, then $p_{N0}(n)$ is increasing and $p_{NN}(n)$ is decreasing in n. Note that, even when $\mathbf{P} \in \mathcal{M}_{\mathrm{st}}$, the transition probability $p_{ii}(n)$, $i \neq 0, N$, need not be monotonically decreasing (cf. the time-reversible case in Section 2.5).

Higher order extensions of $\mathcal{M}_{\mathrm{st}}$ are of importance for unimodality. We define two classes of stochastic matrices by means of the TP_2 property, namely,

$$\mathcal{M}_2^U = \{\mathbf{P} \in \mathcal{S} : \mathbf{U}^{-1}\mathbf{P}\mathbf{U} \in \mathrm{TP}_2\}; \quad \mathcal{M}_2^V = \{\mathbf{P} \in \mathcal{S} : \mathbf{V}^{-1}\mathbf{P}\mathbf{V} \in \mathrm{TP}_2\},$$

see Keilson and Kester (1978). It should be noted that $\mathcal{M}_2^U$ and $\mathcal{M}_2^V$ are in general distinct. Also, since $\mathbf{PU} = \mathbf{U}(\mathbf{U}^{-1}\mathbf{PU})$, $\mathbf{P} \in \mathcal{M}_2^U$ implies that $\mathbf{PU} \in TP_2$. Similarly, if $\mathbf{P} \in \mathcal{M}_2^V$ then $\mathbf{PV}$ is TP_2. We shall write $\mathcal{M}_2 = \mathcal{M}_2^U \cup \mathcal{M}_2^V$. The next result is due to Kijima (1990a).

Theorem 3.15 *Suppose* $\mathbf{P} \in \mathcal{M}_2^V$ ($\mathbf{P} \in \mathcal{M}_2^U$, *respectively*). *If in addition* $\mathbf{PU}$ ($\mathbf{PV}$) *is* TP_2 *then* $p_{0j}(n)$ *is unimodal in* n *for any* $j \in \mathcal{N}$.

Proof. Let $\mathbf{r}^\top(n) = -\mathbf{q}_0^\top(n)\mathbf{V}$. We saw in the proof of Theorem 3.14 that $\mathbf{r}(n) \geq \mathbf{0}$. From (3.31), we have

$$\begin{pmatrix} \mathbf{r}^\top(n+1) \\ \mathbf{r}^\top(n+2) \end{pmatrix} = \begin{pmatrix} \mathbf{r}^\top(n) \\ \mathbf{r}^\top(n+1) \end{pmatrix} \mathbf{V}^{-1}\mathbf{PV}, \quad n = 0, 1, \cdots. \tag{3.32}$$

Suppose that $\begin{pmatrix} \mathbf{r}^\top(0) \\ \mathbf{r}^\top(1) \end{pmatrix}$ is TP_2. Since $\mathbf{V}^{-1}\mathbf{PV} \in TP_2$, it follows from the basic composition formula (Theorem 3.1) that $\begin{pmatrix} \mathbf{r}^\top(n) \\ \mathbf{r}^\top(n+1) \end{pmatrix}$ is TP_2 for all n by induction. Thus, the matrix $\mathbf{R}$ with $\mathbf{r}^\top(n)$ as its nth row vector is TP_2. Consider, then, the vector $-\mathbf{V}^{-1}\boldsymbol{\delta}_j$ for $j \neq 0$. It is easy to verify that $S(-\mathbf{V}^{-1}\boldsymbol{\delta}_j) = 1$ and the sign changes from $+1$ to -1. The variation diminishing property (VDP, Theorem C.4) of the TP_2 matrix $\mathbf{R}$ then guarantees that $S(-\mathbf{R}\mathbf{V}^{-1}\boldsymbol{\delta}_j) \leq 1$, and if $S(-\mathbf{R}\mathbf{V}^{-1}\boldsymbol{\delta}_j) = 1$ then the sign must change from $+1$ to -1. Since

$$-\mathbf{R}\mathbf{V}^{-1}\boldsymbol{\delta}_j = (q_{0j}(0),\ q_{0j}(1),\ \cdots)^\top,$$

one concludes that $p_{0j}(n)$ is unimodal with respect to n. It remains to show that $\begin{pmatrix} \mathbf{r}^\top(0) \\ \mathbf{r}^\top(1) \end{pmatrix}$ is TP_2. By definition, we have

$$\mathbf{r}^\top(0) = \mathbf{1}^\top - \mathbf{p}_0^\top\mathbf{V}; \quad \mathbf{r}^\top(1) = \mathbf{p}_0^\top\mathbf{V} - \mathbf{p}_0^\top(2)\mathbf{V}.$$

Since $\mathbf{r}(n) \geq \mathbf{0}$, it suffices to verify that

$$(1 - P_{0j})(P_{0,j+1} - P_{0,j+1}(2)) - (1 - P_{0,j+1})(P_{0j} - P_{0j}(2)) \geq 0$$

for $j = 0, \cdots, N-1$, where $P_{0j}(n) = \sum_{k=0}^{j} p_{0k}(n)$, or, equivalently, that

$$\overline{P}_{0,j+1}\overline{P}_{0,j+2}(2) - \overline{P}_{0,j+2}\overline{P}_{0,j+1}(2) \geq 0, \quad j = 0, \cdots, N-1,$$

where $\overline{P}_{0j}(n) = \sum_{k=j}^{N} p_{0k}(n)$. This is so if $\mathbf{PU} \in TP_2$, since

$$\begin{pmatrix} \mathbf{p}_0^\top \\ \mathbf{p}_0^\top(2) \end{pmatrix} \mathbf{U} = \begin{pmatrix} \boldsymbol{\delta}_0^\top \\ \mathbf{p}_0^\top \end{pmatrix} \mathbf{PU}$$

and $\begin{pmatrix} \boldsymbol{\delta}_0^\top \\ \mathbf{p}_0^\top \end{pmatrix}\mathbf{U}$ is TP_2. The other case can be proved similarly. □

Theorem 3.16 *Suppose that* $\mathbf{P} \in \mathcal{M}_2$. *Then* $p_{i0}(n)$ *is unimodal in* n *for every* $i \in \mathcal{N}$.

Proof. It is readily seen that, for $i \neq 0, N$, we have $S(\mathbf{q}_i^\top(0)\mathbf{V}) = 1$, and the sign changes from $+1$ to -1. Suppose $\mathbf{P} \in \mathcal{M}_2^V$. By induction, it is shown that

$$S(\mathbf{q}_i^\top(n)\mathbf{V}) \leq 1$$

and, if it is 1, the sign changes from $+1$ to -1. Once $S(\mathbf{q}_i^\top(n)\mathbf{V}) = 0$ for some n, the sign of $\mathbf{q}_i^\top(m)\mathbf{V}$ for $m \geq n$ remains the same, since $\mathbf{V}^{-1}\mathbf{P}\mathbf{V} \geq \mathbf{O}$ under the assumption. Note that $[\mathbf{q}_i^\top(n)\mathbf{V}]_0 = q_{i0}(n)$. Thus, the sign of the vector $(q_{i0}(n))$ with respect to n changes from $+1$ to -1 or it remains $+1$. Hence $p_{i0}(n)$ is unimodal in n. The other case follows similarly. □

We now turn our attention to the unimodality of transition probabilities $p_{ij}(n)$ with respect to i or j. For this purpose, let $\widehat{\mathbf{p}}_j(n) = (p_{ij}(n))$, $j \in \mathcal{N}$, denote the jth column vector of $\mathbf{P}^n$. Then

$$\widehat{\mathbf{p}}_j(n) = \mathbf{P}\,\widehat{\mathbf{p}}_j(n-1) = \mathbf{P}^n\,\widehat{\mathbf{p}}_j(0), \quad n = 0, 1, \cdots.$$

Note that

$$\mathbf{U}^{-1}\mathbf{P}^n\mathbf{U} = (\mathbf{U}^{-1}\mathbf{P}\mathbf{U})^n, \quad \mathbf{V}^{-1}\mathbf{P}^n\mathbf{V} = (\mathbf{V}^{-1}\mathbf{P}\mathbf{V})^n.$$

Also if $\mathbf{P} \in \mathcal{M}_2$ then $\mathbf{P}^n \in \mathcal{M}_2$, by Theorem 3.1. The next result can be proved with the aid of Theorem 3.4. The proof is left to the reader (see Exercise 3.18).

Lemma 3.9 (i) *Let* $\mathbf{P} \in \mathcal{M}_{\mathrm{st}}$. *Then,* $p_{i0}(n)$ *is decreasing and* $p_{iN}(n)$ *is increasing in* i *for every* n.

(ii) *Let* $\mathbf{P} \in \mathcal{M}_2$. *Then* $p_{ij}(n)$, $j \neq 0, N$, *is unimodal in* i *for every* n.

For an ergodic Markov chain $\{X_n\}$ with transition matrix $\mathbf{P}$, let $\boldsymbol{\pi} = (\pi_i)$ be the stationary distribution, i.e., $\boldsymbol{\pi}^\top\mathbf{P} = \boldsymbol{\pi}^\top$ and $\boldsymbol{\pi}^\top\mathbf{1} = 1$. Suppose that the Markov chain is reversible in time. This means that

$$\pi_i\, p_{ij}(n) = \pi_j\, p_{ji}(n), \quad n = 1, 2, \cdots; \quad i, j \in \mathcal{N}. \tag{3.33}$$

Under the condition of Lemma 3.9(i), we know that $p_{j0}(n)$ is decreasing and $p_{jN}(n)$ is increasing in j. It follows from (3.33) that $p_{0j}(n)/\pi_j$ is decreasing and $p_{Nj}(n)/\pi_j$ is increasing in j for every n. Similarly, under the condition of Lemma 3.9(ii), if $\{X_n\}$ is reversible in time, $p_{ij}(n)/\pi_j$ is unimodal in j for every n. Now, let $\boldsymbol{\pi}(n) = (\pi_i(n))$ denote the state distribution of $\{X_n\}$ at time n, i.e., $\boldsymbol{\pi}^\top(n) = \boldsymbol{\alpha}^\top\mathbf{P}^n$, where $\boldsymbol{\alpha} = (\alpha_i)$ is the initial distribution. If $\{X_n\}$ is reversible in time, we have from (3.33) that

$$\frac{\pi_i(n)}{\pi_i} = \sum_j p_{ij}(n)\,\frac{\alpha_j}{\pi_j}, \quad n = 0, 1, \cdots.$$

Hence, if $\mathbf{P} \in \mathcal{M}_{\mathrm{st}}$ and if α_j/π_j is decreasing (increasing, respectively) in

j, then $\pi_i(n)/\pi_i$ is decreasing (increasing) in i for every n. Also, if $\mathbf{P} \in \mathcal{M}_2$ and if α_j/π_j is strictly unimodal, i.e., unimodal but not monotone, then $\pi_i(n)/\pi_i$ is unimodal (not necessarily strictly) in i for every n. We summarize these results in the next theorem (see also Keilson and Kester, 1977).

Theorem 3.17 *For a time-reversible Markov chain, let* $\boldsymbol{\alpha} = (\alpha_i)$ *be the initial distribution,* $\mathbf{P}$ *the transition matrix, and* $\boldsymbol{\pi} = (\pi_i)$ *the stationary distribution.*

(i) *Suppose* α_i/π_i *is decreasing (increasing, respectively) in* i. *If* $\mathbf{P} \in \mathcal{M}_{\mathrm{st}}$ *then* $\pi_i(n)/\pi_i$ *is decreasing (increasing) in* i *for every* n.

(ii) *Suppose* α_i/π_i *is unimodal in* i. *If* $\mathbf{P} \in \mathcal{M}_2$ *then* $\pi_i(n)/\pi_i$ *is unimodal in* i *for every* n.

Taking $\boldsymbol{\alpha} = \boldsymbol{\delta}_i$ in Theorem 3.17, the next result follows immediately. Note that $\boldsymbol{\delta}_0$ ($\boldsymbol{\delta}_N$, respectively) is decreasing (increasing), while $\boldsymbol{\delta}_i$, $i \neq 0, N$, is strictly unimodal.

Corollary 3.6 *For a time-reversible Markov chain with transition matrix* $\mathbf{P} = (p_{ij})$ *and stationary distribution* $\boldsymbol{\pi} = (\pi_i)$:

(i) *If* $\mathbf{P} \in \mathcal{M}_{\mathrm{st}}$ *then, for every* $j \in \mathcal{N}$,

$$\frac{p_{0j}(n)}{\pi_j} \geq \frac{p_{0N}(n)}{\pi_N} \quad \text{and} \quad \frac{p_{Nj}(n)}{\pi_j} \geq \frac{p_{N0}(n)}{\pi_0}.$$

(ii) *If* $\mathbf{P} \in \mathcal{M}_2$ *then*

$$\frac{p_{ij}(n)}{\pi_j} \geq \min\left\{\frac{p_{i0}(n)}{\pi_0}, \frac{p_{iN}(n)}{\pi_N}\right\}, \quad i, j \in \mathcal{N}.$$

For an ergodic Markov chain $\{X_n\}$, recall from (2.64) that the separation of $\{X_n\}$ is given by

$$s_i(n) = 1 - \min_{j \in \mathcal{N}} \frac{p_{ij}(n)}{\pi_j}, \quad n = 0, 1, \cdots; \quad i \in \mathcal{N}.$$

Hence, if the Markov chain is reversible in time and $\mathbf{P} \in \mathcal{M}_{\mathrm{st}}$, then

$$s_0(n) = 1 - \frac{p_{0N}(n)}{\pi_N} \quad \text{and} \quad s_N(n) = 1 - \frac{p_{N0}(n)}{\pi_0}, \tag{3.34}$$

while if $\mathbf{P} \in \mathcal{M}_2$ then

$$s_i(n) = \max\left\{1 - \frac{p_{i0}(n)}{\pi_0}, 1 - \frac{p_{iN}(n)}{\pi_N}\right\}, \quad i \in \mathcal{N}. \tag{3.35}$$

In the following, we investigate the separation of an ergodic (not necessarily time-reversible) Markov chain.

For the stationary distribution $\boldsymbol{\pi} = (\pi_i)$, let $\boldsymbol{\pi}_{\mathrm{D}}$ denote the diagonal matrix whose diagonal elements are π_i. Let $\mathbf{P}_{\mathrm{R}} = \boldsymbol{\pi}_{\mathrm{D}}^{-1}\mathbf{P}^{\top}\boldsymbol{\pi}_{\mathrm{D}}$ be the dual of $\mathbf{P}$. The dual $\mathbf{P}_\mathrm{R}$ is a stochastic matrix and has the same stationary distribution $\boldsymbol{\pi}$. We define

$$\mathcal{M}_{\mathrm{st}}^{\mathrm{R}} = \{\mathbf{P} \in \mathcal{S} : \mathbf{U}^{-1}\mathbf{P}_\mathrm{R}\mathbf{U} \geq \mathbf{O}\} = \{\mathbf{P} \in \mathcal{S} : \mathbf{V}^{-1}\mathbf{P}_\mathrm{R}\mathbf{V} \geq \mathbf{O}\}.$$

That is, $\mathcal{M}_{\mathrm{st}}^{\mathrm{R}}$ is the class of stochastic matrices whose duals are stochastically monotone. In general, $\mathbf{P} \neq \mathbf{P}_\mathrm{R}$ so that $\mathcal{M}_{\mathrm{st}}$ and $\mathcal{M}_{\mathrm{st}}^{\mathrm{R}}$ are not identical. When the Markov chain is reversible in time, then $\mathbf{P} = \mathbf{P}_\mathrm{R}$ and $\mathbf{P} \in \mathcal{M}_{\mathrm{st}}$ is equivalent to $\mathbf{P} \in \mathcal{M}_{\mathrm{st}}^{\mathrm{R}}$.

Let $\boldsymbol{\alpha} = (\alpha_i)$ be a probability vector on $\mathcal{N}$, i.e., $\alpha_i \geq 0$ and $\sum_i \alpha_i = 1$. Recall from Theorem 3.4 that $\boldsymbol{\alpha}$ is decreasing (increasing, respectively) if and only if $\boldsymbol{\alpha}^\top \mathbf{U}^{-1} \geq \mathbf{0}^\top$ ($\boldsymbol{\alpha}^\top \mathbf{V}^{-1} \geq \mathbf{0}^\top$), whereas $\boldsymbol{\alpha}$ is strictly unimodal if and only if the sign of $\boldsymbol{\alpha}^\top \mathbf{U}^{-1}$ ($\boldsymbol{\alpha}^\top \mathbf{V}^{-1}$) changes exactly once from -1 to $+1$ (from $+1$ to -1) when the index i traverses from 0 to N. Let $\boldsymbol{\delta}_i$ denote the ith unit vector having a point mass at $i \in \mathcal{N}$. Then,

$$\frac{p_{ij}(n)}{\pi_j} = [\boldsymbol{\delta}_i^\top \mathbf{P}^n \boldsymbol{\pi}_\mathrm{D}^{-1}]_j, \quad n = 0, 1, \cdots.$$

Note that $\boldsymbol{\delta}_0^\top \boldsymbol{\pi}_\mathrm{D}^{-1}$ and $\boldsymbol{\delta}_N^\top \boldsymbol{\pi}_\mathrm{D}^{-1}$ are monotone while $\boldsymbol{\delta}_i^\top \boldsymbol{\pi}_\mathrm{D}^{-1}$, $i \neq 0, N$, are strictly unimodal. In the remainder of this section, we denote the state distribution by $\boldsymbol{\pi}_\alpha(n) = (\pi_{\alpha j}(n))$ and separation by $s_\alpha(n)$ when the initial distribution is $\boldsymbol{\alpha}$. If $\boldsymbol{\alpha} = \boldsymbol{\delta}_i$ in particular, we write $\boldsymbol{\pi}_i(n)$ and $s_i(n)$, as before. The following results are taken from Kijima (1994). See Diaconis and Fill (1990) for more general results.

Let $\boldsymbol{\alpha}$ be the initial distribution such that $\boldsymbol{\alpha}^\top \boldsymbol{\pi}_\mathrm{D}^{-1}$ is decreasing, i.e., $\boldsymbol{\alpha}^\top \boldsymbol{\pi}_\mathrm{D}^{-1} \mathbf{U}^{-1} \geq \mathbf{0}^\top$. Suppose $\mathbf{P} \in \mathcal{M}_{\mathrm{st}}^{\mathrm{R}}$. Then

$$\boldsymbol{\pi}_\alpha^\top(1)\boldsymbol{\pi}_\mathrm{D}^{-1}\mathbf{U}^{-1} = \boldsymbol{\alpha}^\top \mathbf{P} \boldsymbol{\pi}_\mathrm{D}^{-1}\mathbf{U}^{-1} = \boldsymbol{\alpha}^\top \boldsymbol{\pi}_\mathrm{D}^{-1}\mathbf{U}^{-1}(\mathbf{V}^{-1}\mathbf{P}_\mathrm{R}\mathbf{V})^\top \geq \mathbf{0}^\top.$$

In general, since

$$\boldsymbol{\pi}_\alpha^\top(n)\boldsymbol{\pi}_\mathrm{D}^{-1}\mathbf{U}^{-1} = \boldsymbol{\pi}_\alpha^\top(n-1)\boldsymbol{\pi}_\mathrm{D}^{-1}\mathbf{U}^{-1}(\mathbf{V}^{-1}\mathbf{P}_\mathrm{R}\mathbf{V})^\top, \tag{3.36}$$

we have $\boldsymbol{\pi}_\alpha^\top(n)\boldsymbol{\pi}_\mathrm{D}^{-1}\mathbf{U}^{-1} \geq \mathbf{0}^\top$ for all n by induction. Hence, if $\boldsymbol{\alpha}^\top \boldsymbol{\pi}_\mathrm{D}^{-1}$ is decreasing and $\mathbf{P} \in \mathcal{M}_{\mathrm{st}}^{\mathrm{R}}$, then $\boldsymbol{\pi}_\alpha^\top(n)\boldsymbol{\pi}_\mathrm{D}^{-1}$ is decreasing for all n. It follows that the minimum of $\pi_{\alpha j}(n)/\pi_j$ is attained at $j = N$. This proves the next theorem, which should be compared with the time-reversible case (3.34).

Theorem 3.18 *Suppose that* $\mathbf{P} \in \mathcal{M}_{\mathrm{st}}^{\mathrm{R}}$. *If* $\boldsymbol{\alpha}^\top \boldsymbol{\pi}_\mathrm{D}^{-1}$ *is decreasing, then the separation* $s_\alpha(n)$ *for the Markov chain is given by*

$$s_\alpha(n) = 1 - \frac{\pi_{\alpha N}(n)}{\pi_N}, \quad n = 0, 1, \cdots,$$

where $\boldsymbol{\alpha}$ *is the initial distribution and* $\boldsymbol{\pi} = (\pi_i)$ *denotes the stationary distribution.*

Let $\boldsymbol{\Pi}_\mathrm{D}$ be the diagonal matrix whose ith diagonal component is given by $\sum_{j=0}^{i} \pi_j$, $i \in \mathcal{N}$. For $\mathbf{P} \in \mathcal{M}_{\mathrm{st}}^{\mathrm{R}}$, define

$$\widehat{\mathbf{P}} = \boldsymbol{\Pi}_{\mathrm{D}}^{-1}(\mathbf{V}^{-1}\mathbf{P}_{\mathrm{R}}\mathbf{V})^{\mathsf{T}}\boldsymbol{\Pi}_{\mathrm{D}}. \tag{3.37}$$

Writing $\mathbf{C} = \boldsymbol{\Pi}_{\mathrm{D}}^{-1}\mathbf{U}\boldsymbol{\pi}_{\mathrm{D}}$, $\widehat{\mathbf{P}}$ can be expressed as $\widehat{\mathbf{P}} = \mathbf{C}\mathbf{P}\mathbf{C}^{-1}$. It is not difficult to see that if $\mathbf{P} \in \mathcal{M}_{\mathrm{st}}^{\mathrm{R}}$ then $\widehat{\mathbf{P}}$ is a stochastic matrix with absorbing state N. Note that

$$\frac{\pi_{\alpha N}(n)}{\pi_N} = [\boldsymbol{\pi}_\alpha^{\mathsf{T}}(n)\boldsymbol{\pi}_{\mathrm{D}}^{-1}]_N = [\boldsymbol{\pi}_\alpha^{\mathsf{T}}(n)\boldsymbol{\pi}_{\mathrm{D}}^{-1}\mathbf{U}^{-1}\boldsymbol{\Pi}_{\mathrm{D}}]_N = [\boldsymbol{\alpha}^{\mathsf{T}}\mathbf{P}^n\mathbf{C}^{-1}]_N.$$

It follows that

$$\frac{\pi_{\alpha N}(n)}{\pi_N} = [\boldsymbol{\alpha}^{\mathsf{T}}\mathbf{C}^{-1}\widehat{\mathbf{P}}^n]_N = [\widehat{\boldsymbol{\alpha}}^{\mathsf{T}}\widehat{\mathbf{P}}^n]_N, \tag{3.38}$$

where

$$\widehat{\boldsymbol{\alpha}}^{\mathsf{T}} = \boldsymbol{\alpha}^{\mathsf{T}}\mathbf{C}^{-1} = \boldsymbol{\alpha}^{\mathsf{T}}\boldsymbol{\pi}_{\mathrm{D}}^{-1}\mathbf{U}^{-1}\boldsymbol{\Pi}_{\mathrm{D}}. \tag{3.39}$$

It should be noted that if $\boldsymbol{\alpha}^{\mathsf{T}}\boldsymbol{\pi}_{\mathrm{D}}^{-1}$ is decreasing, i.e., $\boldsymbol{\alpha}^{\mathsf{T}}\boldsymbol{\pi}_{\mathrm{D}}^{-1}\mathbf{U}^{-1} \geq \mathbf{0}^{\mathsf{T}}$, then $\widehat{\boldsymbol{\alpha}}$ is a probability vector, since $\widehat{\boldsymbol{\alpha}} \geq \mathbf{0}$ and, from (3.39),

$$\widehat{\boldsymbol{\alpha}}^{\mathsf{T}}\mathbf{1} = \boldsymbol{\alpha}^{\mathsf{T}}\boldsymbol{\pi}_{\mathrm{D}}^{-1}\mathbf{U}^{-1}\boldsymbol{\Pi}_{\mathrm{D}}\mathbf{1} = \boldsymbol{\alpha}^{\mathsf{T}}\boldsymbol{\pi}_{\mathrm{D}}^{-1}\boldsymbol{\pi} = \boldsymbol{\alpha}^{\mathsf{T}}\mathbf{1} = 1.$$

Hence, letting $\widehat{\xi}$ be the first passage time to the absorbing state N of an absorbing Markov chain $\{\widehat{X}_n\}$ with initial distribution $\widehat{\boldsymbol{\alpha}}$ and transition matrix $\widehat{\mathbf{P}}$, it follows from (3.38) that

$$\frac{\pi_{\alpha N}(n)}{\pi_N} = P_{\widehat{\alpha}}[\widehat{\xi} \leq n], \quad n = 0, 1, \cdots.$$

The key fact here is that the state N is absorbing in $\{\widehat{X}_n\}$. It follows from Theorem 3.18 that the separation of the original Markov chain $\{X_n\}$ is characterized as

$$s_\alpha(n) = P_{\widehat{\alpha}}[\widehat{\xi} > n], \quad n = 0, 1, \cdots, \tag{3.40}$$

which is a result from Diaconis and Fill (1990). We note that, for the initial distribution $\boldsymbol{\alpha}$ such that $\boldsymbol{\alpha}^{\mathsf{T}}\boldsymbol{\pi}_{\mathrm{D}}^{-1}$ is increasing, i.e., $\boldsymbol{\alpha}^{\mathsf{T}}\boldsymbol{\pi}_{\mathrm{D}}^{-1}\mathbf{V}^{-1} \geq \mathbf{0}^{\mathsf{T}}$, we have the same results by reversing the order of the states (see Exercise 3.19).

In order to consider the strictly unimodal case, we introduce

$$\mathcal{M}_2^{\mathrm{R}} = \{\mathbf{P} \in \mathcal{S} : \mathbf{U}^{-1}\mathbf{P}_{\mathrm{R}}\mathbf{U} \in \mathrm{TP}_2 \text{ or } \mathbf{V}^{-1}\mathbf{P}_{\mathrm{R}}\mathbf{V} \in \mathrm{TP}_2\}.$$

Let $\boldsymbol{\alpha}$ be the initial distribution such that $\boldsymbol{\alpha}^{\mathsf{T}}\boldsymbol{\pi}_{\mathrm{D}}^{-1}$ is strictly unimodal and suppose that $\mathbf{P} \in \mathcal{M}_2^{\mathrm{R}}$. Suppose that $\mathbf{V}^{-1}\mathbf{P}_{\mathrm{R}}\mathbf{V}$ is TP_2. Since $\boldsymbol{\alpha}^{\mathsf{T}}\boldsymbol{\pi}_{\mathrm{D}}^{-1}$ is strictly unimodal, the sign of $\boldsymbol{\alpha}^{\mathsf{T}}\boldsymbol{\pi}_{\mathrm{D}}^{-1}\mathbf{U}^{-1}$ changes from -1 to $+1$ exactly once. Now, applying the VDP of the TP_2 matrix $\mathbf{V}^{-1}\mathbf{P}_{\mathrm{R}}\mathbf{V}$ to (3.36), an induction argument shows that, for all n, the sign of $\boldsymbol{\pi}_\alpha^{\mathsf{T}}(n)\boldsymbol{\pi}_{\mathrm{D}}^{-1}\mathbf{U}^{-1}$ changes at most once and, if it does change once, the sign changes from -1 to $+1$. Hence, $\boldsymbol{\pi}_\alpha^{\mathsf{T}}(n)\boldsymbol{\pi}_{\mathrm{D}}^{-1}$ is either monotone or strictly unimodal. If $\mathbf{U}^{-1}\mathbf{P}_{\mathrm{R}}\mathbf{U}$ is

TP_2, then we consider $\boldsymbol{\pi}_\alpha^\top(n)\boldsymbol{\pi}_D^{-1}\mathbf{V}^{-1}$, and a similar argument to the above leads to the same conclusion. It follows that the minimum of $\pi_{\alpha j}(n)/\pi_j$ is attained at either $j = 0$ or $j = N$. The next result thus follows; cf. the time-reversible case (3.35).

Theorem 3.19 *Suppose that* $\mathbf{P} \in \mathcal{M}_2^R$. *If* $\boldsymbol{\alpha}^\top\boldsymbol{\pi}_D^{-1}$ *is strictly unimodal, then the separation for the Markov chain* $\{X_n\}$ *is given by*

$$s_\alpha(n) = \max\left\{1 - \frac{\pi_{\alpha 0}(n)}{\pi_0},\ 1 - \frac{\pi_{\alpha N}(n)}{\pi_N}\right\}, \quad n = 0, 1, \cdots.$$

Since $\mathcal{M}_2^R \subset \mathcal{M}_{st}^R$, the absorbing stochastic matrix $\widehat{\mathbf{P}}$ of (3.37) is still well defined under the assumptions of Theorem 3.19. For $\boldsymbol{\alpha} = \boldsymbol{\delta}_i$, $i \neq 0, N$, it is not difficult to show from (3.38) and (3.39) that

$$\frac{p_{iN}(n)}{\pi_N} = \frac{\sum_{j=1}^{i}\pi_j}{\pi_i}[\boldsymbol{\delta}_i^\top\widehat{\mathbf{P}}^n]_N - \frac{\sum_{j=1}^{i-1}\pi_j}{\pi_i}[\boldsymbol{\delta}_{i-1}^\top\widehat{\mathbf{P}}^n]_N. \tag{3.41}$$

Note that if $\mathbf{V}^{-1}\mathbf{P}_R\mathbf{V}$ is TP_2 then so is $\widehat{\mathbf{P}}$ from (3.37). Hence, $\widehat{\mathbf{P}}\mathbf{U}$ is TP_2 so that $\widehat{\mathbf{P}} \in \mathcal{M}_{st}^R$. Since $\boldsymbol{\delta}_i^\top\mathbf{U} \geq \boldsymbol{\delta}_{i-1}^\top\mathbf{U}$, it follows that $\left[\boldsymbol{\delta}_i^\top\widehat{\mathbf{P}}^n\right]_N \geq \left[\boldsymbol{\delta}_{i-1}^\top\widehat{\mathbf{P}}^n\right]_N$ for all n. Hence, from (3.41), one has

$$\frac{p_{iN}(n)}{\pi_N} \geq [\boldsymbol{\delta}_i^\top\widehat{\mathbf{P}}^n]_N, \quad n = 0, 1, \cdots.$$

Let $\widehat{\xi}$ be the time of absorption at state N for an absorbing Markov chain with initial distribution $\boldsymbol{\delta}_i$ and transition matrix $\widehat{\mathbf{P}}$. As before, it then follows that

$$\frac{p_{iN}(n)}{\pi_N} \geq P_i[\widehat{\xi} \leq n], \quad n = 0, 1, \cdots.$$

Similarly, $p_{i0}(n)/\pi_0$ can be bounded from below using the first passage time, $\widehat{\xi}^*$ say, of another absorbing Markov chain. Hence, from Theorem 3.19, the separation is bounded from above as follows:

$$s_i(n) \leq \max\{P_i[\widehat{\xi} > n],\ P_i[\widehat{\xi}^* > n]\}, \quad n = 0, 1, \cdots.$$

3.6 First-passage-time distributions

Let $\{X_n\}$ be an irreducible Markov chain defined on $\mathcal{N} = \{0, 1, \cdots, N\}$. When considering the first passage time to state j, it is convenient to make the state j absorbing. Throughout this section, we choose $j = 0$ after renumbering the states if necessary. The transition matrix of the absorbing Markov chain is given by

$$\mathbf{P} = \begin{pmatrix} 1 & \mathbf{0}^\top \\ \mathbf{r} & \mathbf{T} \end{pmatrix}; \tag{3.42}$$

see (2.76). Note that $\mathbf{r} = \mathbf{1} - \mathbf{T1} \geq \mathbf{0}$ and $\mathbf{r} \neq \mathbf{0}$ under the irreducibility assumption. Let $\widetilde{\boldsymbol{\alpha}}^\top = (\alpha_0, \boldsymbol{\alpha}^\top)$ be the initial distribution of $\{X_n\}$ and let T be the first passage time of $\{X_n\}$ to state 0. Write

$$f_n = P_\alpha[T = n], \quad n = 0, 1, \cdots,$$

where we define $f_0 = \alpha_0$. It is assumed that $\alpha_0 < 1$. The vector $\boldsymbol{\alpha}$ is defined on $\mathcal{N} \setminus \{0\}$. Note that $\boldsymbol{\alpha}^\top \mathbf{1} = 1 - \alpha_0$. It is easy to see that

$$P_\alpha[T > n] = \boldsymbol{\alpha}^\top \mathbf{T}^n \mathbf{1}, \quad n = 0, 1, \cdots, \tag{3.43}$$

and

$$f_n = \boldsymbol{\alpha}^\top \mathbf{T}^{n-1}\mathbf{1} - \boldsymbol{\alpha}^\top \mathbf{T}^n \mathbf{1} = \boldsymbol{\alpha}^\top \mathbf{T}^{n-1}\mathbf{r}, \quad n = 1, 2, \cdots; \tag{3.44}$$

see (2.80). As in Section 2.7, we call the probability vector $\mathbf{f} = (f_n)$ the phase-type distribution generated by $(\boldsymbol{\alpha}, \mathbf{T})$.

In this section, we consider the distribution properties of T. The next result is a special case from Brown and Chaganty (1983). See Marshall and Shaked (1983) for related results. Li and Shaked (1995) give a survey of such distribution properties.

Theorem 3.20 *Suppose that* $\mathbf{P} \in \mathcal{M}_{\text{st}}$. *If* $\boldsymbol{\alpha} = \boldsymbol{\delta}_1$ $(\boldsymbol{\alpha} = \boldsymbol{\delta}_N$, *respectively) then* $T \in$ NWU ($T \in$ NBU).

Proof. Let $\boldsymbol{\beta}_n = \mathbf{T}^n \mathbf{1} = (\beta_{ni})$ and define

$$\tau_n \equiv \boldsymbol{\alpha}^\top \boldsymbol{\beta}_n = \boldsymbol{\alpha}^\top \mathbf{T}^n \mathbf{1} = P_\alpha[T > n], \quad n = 1, 2, \cdots.$$

Note that $\mathbf{P} \in \mathcal{M}_{\text{st}}$ implies that $\mathbf{U}^{-1}\mathbf{T}\mathbf{U} \geq \mathbf{O}$. It follows that

$$\mathbf{U}^{-1}\boldsymbol{\beta}_n = \mathbf{U}^{-1}\mathbf{T}^n\mathbf{U}\mathbf{U}^{-1}\mathbf{1} = (\mathbf{U}^{-1}\mathbf{T}\mathbf{U})^n \boldsymbol{\delta}_1 \geq \mathbf{0},$$

so that β_{ni} is increasing in i for every n. For $\boldsymbol{\alpha} = \boldsymbol{\delta}_1$, we have $\tau_n = \beta_{n1}$ and hence $\boldsymbol{\beta}_n \geq \tau_n \mathbf{1}$. It follows that

$$\tau_{m+n} = \boldsymbol{\alpha}^\top \mathbf{T}^{m+n}\mathbf{1} = \boldsymbol{\alpha}^\top \mathbf{T}^m \boldsymbol{\beta}_n \geq \tau_n \boldsymbol{\alpha}^\top \mathbf{T}^m \mathbf{1} = \tau_m \tau_n$$

for any $m, n = 1, 2, \cdots$, whence T is NWU (see Definition 3.9(iii)). The other case follows similarly. □

Suppose that $\alpha_0 = 0$. For the phase-type distribution $\mathbf{f} = (f_n)$ generated by $(\boldsymbol{\alpha}, \mathbf{T})$, let h_n be its hazard rate function. From (3.43) and (3.44), we have

$$h_n = \frac{\boldsymbol{\alpha}^\top \mathbf{T}^{n-1}\mathbf{r}}{\boldsymbol{\alpha}^\top \mathbf{T}^{n-1}\mathbf{1}} = 1 - \frac{\boldsymbol{\alpha}^\top \mathbf{T}^n \mathbf{1}}{\boldsymbol{\alpha}^\top \mathbf{T}^{n-1}\mathbf{1}}, \quad n = 1, 2, \cdots. \tag{3.45}$$

Recall that h_n is the hazard rate function of the first passage time T. The next theorem is proved by Shanthikumar (1988).

Theorem 3.21 *Suppose that* $\mathbf{P} \in \mathcal{M}_{\text{hr}}$, *i.e.,* $\mathbf{TU} \in \text{TP}_2$. *If* $\boldsymbol{\alpha} = \boldsymbol{\delta}_1$ $(\boldsymbol{\alpha} = \boldsymbol{\delta}_N$, *respectively) then* $T \in$ DHR ($T \in$ IHR).

Proof. Let $\mathbf{u}_0 = \boldsymbol{\alpha}$ and define $\mathbf{u}_n$ successively by $\mathbf{u}_n^\top = \mathbf{u}_{n-1}^\top \mathbf{T}$, $n = 1, 2, \cdots$. Then, from (3.45), one obtains

$$h_n = \frac{\mathbf{u}_{n-1}^\top \mathbf{r}}{\mathbf{u}_{n-1}^\top \mathbf{1}}, \quad n = 1, 2, \cdots, \tag{3.46}$$

where $\mathbf{r} = \mathbf{1} - \mathbf{T}\mathbf{1} = (r_i)$. Note that, under the condition, r_i is decreasing in i, whence $\mathbf{V}^{-1}\mathbf{r} \geq \mathbf{0}$. Also, since $\mathbf{u}_0 = \boldsymbol{\delta}_1$, $\begin{pmatrix} \mathbf{u}_0^\top \\ \mathbf{u}_1^\top \end{pmatrix} \mathbf{U}$ is TP_2, from which we can show that $\begin{pmatrix} \mathbf{u}_{n-1}^\top \\ \mathbf{u}_n^\top \end{pmatrix} \mathbf{U}$ is TP_2 for all n by induction using the composition law (Theorem 3.2). It follows that $\mathbf{u}_n / \mathbf{u}_n^\top \mathbf{1} \geq_{\text{hr}} \mathbf{u}_{n-1} / \mathbf{u}_{n-1}^\top \mathbf{1}$ and so $\mathbf{u}_n / \mathbf{u}_n^\top \mathbf{1} \geq_{\text{st}} \mathbf{u}_{n-1} / \mathbf{u}_{n-1}^\top \mathbf{1}$ for all $n = 1, 2, \cdots$. The latter result is equivalent to

$$\frac{\mathbf{u}_n^\top \mathbf{V}}{\mathbf{u}_n^\top \mathbf{1}} \leq \frac{\mathbf{u}_{n-1}^\top \mathbf{V}}{\mathbf{u}_{n-1}^\top \mathbf{1}}, \quad n = 1, 2, \cdots.$$

Since $\mathbf{V}^{-1}\mathbf{r} \geq \mathbf{0}$, it follows from (3.46) that $h_n \leq h_{n-1}$ for all $n = 1, 2, \cdots$. If $\boldsymbol{\alpha} = \boldsymbol{\delta}_N$, we consider $\begin{pmatrix} \mathbf{u}_1^\top \\ \mathbf{u}_0^\top \end{pmatrix} \mathbf{U}$ which is TP_2. The same argument then goes through to conclude that h_n is increasing in n. □

It is easily seen from the proof of Theorem 3.21 that the assumption $\boldsymbol{\alpha} = \boldsymbol{\delta}_1$ is replaced by the assumption $\begin{pmatrix} \boldsymbol{\alpha}^\top \\ \boldsymbol{\alpha}^\top \mathbf{T} \end{pmatrix} \mathbf{U} \in \text{TP}_2$ for the DHR case. Also, we can prove similar results under the assumptions that $\mathbf{TV} \in \text{TP}_2$ and $\mathbf{V}^{-1}\mathbf{TV} \geq \mathbf{O}$ (see Exercise 3.20).

Example 3.15 (Generalized Erlang distribution) A discrete *generalized Erlang distribution* is a phase-type distribution generated by

$$\mathbf{T} = \begin{pmatrix} 1-p_1 & 0 & 0 & \cdots & 0 \\ p_2 & 1-p_2 & 0 & \cdots & 0 \\ 0 & p_3 & 1-p_3 & \cdots & 0 \\ \vdots & & \ddots & \ddots & \vdots \\ 0 & \cdots & 0 & p_N & 1-p_N \end{pmatrix}; \quad \boldsymbol{\alpha} = \boldsymbol{\delta}_N = \begin{pmatrix} 0 \\ 0 \\ 0 \\ \vdots \\ 1 \end{pmatrix},$$

where $0 < p_i < 1$ for all i. It is readily seen that $\mathbf{TU}$ is TP_2 for any such p_is. Hence, from Theorem 3.21, any discrete generalized Erlang distribution is IHR.

So far, we have treated the case of finite state spaces. However, it is easily seen that the same arguments apply for the denumerable case. The result given in the next example is taken from Shanthikumar (1988).

Example 3.16 As in Example 3.14, we consider an M/G/1 queue with arrival rate λ and service time distribution $G(t)$. Let $\{X_n\}$ be a Markov chain describing the number of customers in the system just after the nth departure. Consider, then, the first passage time of $\{X_n\}$ to state 0, i.e.,

$$T = \inf\{n \geq 1 : X_n = 0 | X_0 = 0\}.$$

Note that T represents the number served during a busy period. In this example, we prove that if the service time distribution is IHR then T is DHR. According to (3.42) and (1.29), we define $\mathbf{r} = b_0 \boldsymbol{\delta}_1$ and

$$\mathbf{T} = \begin{pmatrix} b_1 & b_2 & b_3 & b_4 & \cdots \\ b_0 & b_1 & b_2 & b_3 & \cdots \\ 0 & b_0 & b_1 & b_2 & \cdots \\ 0 & 0 & b_0 & b_1 & \cdots \\ \vdots & \vdots & \vdots & \ddots & \ddots \end{pmatrix}.$$

Also, $\boldsymbol{\alpha}^\top = (b_1, b_2, \cdots)$. Let $f_n = P_0[T = n]$ for $n = 1, 2, \cdots$. Then

$$f_1 = 1 - \boldsymbol{\alpha}^\top \mathbf{1} = b_0$$

and

$$f_{n+1} = \boldsymbol{\alpha}^\top \mathbf{T}^{n-1} \mathbf{r}, \quad n = 1, 2, \cdots.$$

From (3.45), the hazard rate function of T is given by

$$h_1 = b_0; \quad h_{n+1} = \frac{\boldsymbol{\alpha}^\top \mathbf{T}^{n-1} \mathbf{r}}{\boldsymbol{\alpha}^\top \mathbf{T}^{n-1} \mathbf{1}}, \quad n = 1, 2, \cdots,$$

provided that the queue is stable. We have

$$h_2 = \frac{\boldsymbol{\alpha}^\top \mathbf{r}}{\boldsymbol{\alpha}^\top \mathbf{1}} = \frac{b_0 \, b_1}{1 - b_0} < b_0 = h_1.$$

As was noted following Theorem 3.21, it suffices to show that $\begin{pmatrix} \boldsymbol{\alpha}^\top \\ \boldsymbol{\alpha}^\top \mathbf{T} \end{pmatrix} \mathbf{U}$ is TP_2. To this end, note that $\boldsymbol{\alpha}^\top = \boldsymbol{\delta}_1^\top \mathbf{T}$. Hence

$$\begin{pmatrix} \boldsymbol{\alpha}^\top \\ \boldsymbol{\alpha}^\top \mathbf{T} \end{pmatrix} \mathbf{U} = \begin{pmatrix} \boldsymbol{\delta}_1^\top \\ \boldsymbol{\alpha}^\top \end{pmatrix} \mathbf{TU}.$$

Since $\mathbf{TU} \in \mathrm{TP}_2$ under the assumption (see Example 3.14 for the result) and $\begin{pmatrix} \boldsymbol{\delta}_1^\top \\ \boldsymbol{\alpha}^\top \end{pmatrix} \mathbf{U} \in \mathrm{TP}_2$, the composition law (Theorem 3.2) ensures the claim. Thus, if the service time distribution is IHR then the number served during a busy period is DHR.

When the transition matrix is lower triangular (as in Example 3.15) or upper triangular, the associated Markov chain is said to have a *decreasing* or *increasing path*, respectively, and in either case it is said to have a *monotone path*. Recall that $\mathbf{P} \in \mathcal{M}_{\mathrm{hr}}$ implies that $\mathbf{P} \in \mathcal{M}_{\mathrm{st}}$. Also, the former implies that $T \in$ IHR while the latter implies that $T \in$ NBU. An intermediate result $T \in$ IHRA can be derived under the monotone path assumption (see Exercise 3.21). The next result is a special case from Brown and Chaganty (1983). See Shaked and Shanthikumar (1987) for more general results.

Theorem 3.22 *Suppose that* $\mathbf{P} \in \mathcal{M}_{st}$ *and that* $\mathbf{P}$ *is lower triangular, i.e., the associated Markov chain has a decreasing path. If* $\boldsymbol{\alpha} = \boldsymbol{\delta}_N$ *then* $T \in$ IHRA.

Proof. We shall prove that

$$P_i^{n+1}[T > n] \geq P_i^n[T > n+1], \quad n = 1, 2, \cdots,$$

for each $i = 1, \cdots, N$. Denoting by $\mathbf{T}_i$ the $i \times i$ north-west corner truncation of $\mathbf{T}$ in (3.42), the above inequality is equivalent to

$$(\boldsymbol{\delta}_i^\top \mathbf{T}_i^n \mathbf{1})^{n+1} \geq (\boldsymbol{\delta}_i^\top \mathbf{T}_i^{n+1} \mathbf{1})^n, \quad n = 1, 2, \cdots, \tag{3.47}$$

because $\mathbf{T}$ is lower triangular. We prove (3.47) by induction on n. For $n = 1$, the result follows from Theorem 3.20 since $\mathbf{P}_i \in \mathcal{M}_{st}$ for every i, where $\mathbf{P}_i$ denotes the $(i+1) \times (i+1)$ north-west corner truncation of $\mathbf{P}$. Now suppose that (3.47) holds up until n. For $n+1$, note that

$$\boldsymbol{\delta}_i^\top \mathbf{T}_i^{n+1} \mathbf{1} = \sum_{j \leq i} \beta_j \boldsymbol{\delta}_j^\top \mathbf{T}_i^n \mathbf{1},$$

where $\boldsymbol{\delta}_i^\top \mathbf{T}_i = (\beta_j)$ with $\beta_j \geq 0$. But, by the monotone path assumption, we have

$$\boldsymbol{\delta}_j^\top \mathbf{T}_i^n \mathbf{1} = \boldsymbol{\delta}_j^\top \mathbf{T}_j^n \mathbf{1}, \quad j \leq i,$$

which, together with the induction hypothesis, yields

$$\boldsymbol{\delta}_i^\top \mathbf{T}_i^{n+1} \mathbf{1} = \sum_{j \leq i} \beta_j \boldsymbol{\delta}_j^\top \mathbf{T}_j^n \mathbf{1} \geq \sum_{j \leq i} \beta_j (\boldsymbol{\delta}_j^\top \mathbf{T}_j^{n+1} \mathbf{1})^{n/n+1}.$$

It is obvious that $\boldsymbol{\delta}_i^\top \mathbf{T}_i^n \mathbf{1} \geq \boldsymbol{\delta}_j^\top \mathbf{T}_j^n \mathbf{1}$ for all n whenever $j \leq i$. It follows that

$$\begin{aligned}(\boldsymbol{\delta}_i^\top \mathbf{T}_i^{n+1} \mathbf{1})^{n+2} &\geq \left[(\boldsymbol{\delta}_i^\top \mathbf{T}_i^{n+1} \mathbf{1})^{1/n+1} \left(\sum_{j \leq i} \beta_j (\boldsymbol{\delta}_j^\top \mathbf{T}_j^{n+1} \mathbf{1})^{n/n+1} \right) \right]^{n+1} \\ &\geq \left[\sum_{j \leq i} \beta_j (\boldsymbol{\delta}_j^\top \mathbf{T}_j^{n+1} \mathbf{1}) \right]^{n+1},\end{aligned}$$

the last term being equal to $(\boldsymbol{\delta}_i^\top \mathbf{T}_i^{n+2} \mathbf{1})^{n+1}$. This completes the proof. □

Suppose that state 0 can be reached from state 1 only, as in the random walk case. Let $\mathbf{T}$ be the substochastic matrix given in (3.42), i.e., $\mathbf{T}$ is obtained by deleting the row and column corresponding to state 0. Consider then the taboo transition probabilities ${}_0p_{ij}(n)$. The first-passage-time distribution $\mathbf{f}_i = (f_i(n))$ to state 0 starting from state i for the original Markov chain is given by

$$f_i(n) = {}_0p_{i1}(n-1)\, r_1, \quad n = 1, 2, \cdots,$$

where $\mathbf{r} = \mathbf{1} - \mathbf{T1} = (r_i) = r_1 \boldsymbol{\delta}_1$. It should be noted that all the results in the preceding section hold true even when the transition matrix is strictly substochastic. Therefore, the monotonicity and unimodality of the first-passage-time distribution $\mathbf{f}_i$ can be studied using those of the taboo transition probabilities ${}_0p_{i1}(n)$.

3.7 Bounds for quasi-stationary distributions

Let $\mathbf{T}$ be a strictly substochastic matrix which is finite, irreducible and aperiodic. We have seen in Section 2.8 that the quasi-stationary distribution and the doubly limiting conditional distribution of $\mathbf{T}$ can be obtained from the eigenvectors, positive componentwise, associated with the Perron–Frobenius (PF) eigenvalue of $\mathbf{T}$. However, it is in general nontrivial to compute the PF eigenvalue and the associated eigenvectors. In this section, we obtain some easily computed bounds on the quasi-stationary distribution of a monotone Markov chain. Recall that the stationary distribution is typically much simpler to obtain than its quasi-stationary counterpart. Therefore, a natural candidate to bound the quasi-stationary distribution of a Markov chain $\{X_n\}$ is the stationary distribution of an ergodic Markov chain which is naturally constructed from $\{X_n\}$. Throughout this section, we assume that the state space is restricted to $\mathcal{N} = \{1, 2, \cdots, N\}$ and that the matrix $\mathbf{T}$ defined on $\mathcal{N}$ is strictly substochastic, irreducible and aperiodic.

Let $\boldsymbol{\delta}_i$ denote the ith unit vector with the ith component equal to 1 and 0s elsewhere. For a strictly substochastic matrix $\mathbf{T}$, let $\mathbf{r} = \mathbf{1} - \mathbf{T1}$. It is easy to see that the matrix defined by

$$\mathbf{P}_i = \mathbf{T} + \mathbf{r}\,\boldsymbol{\delta}_i^\top, \quad i \in \mathcal{N}, \tag{3.48}$$

is stochastic. Note that the irreducibility and aperiodicity of $\mathbf{P}_i$ are inherited from $\mathbf{T}$. Hence, assuming that $\mathbf{T}$ is primitive (see Definition A.1), there exists a stationary distribution $\boldsymbol{\pi}_i = (\pi_j^i)$ of $\mathbf{P}_i$ satisfying

$$\boldsymbol{\pi}_i^\top = \boldsymbol{\pi}_i^\top \mathbf{P}_i, \quad \boldsymbol{\pi}_i^\top \mathbf{1} = 1,$$

for every $i \in \mathcal{N}$. The following results are taken from Kijima (1995).

First, we define two classes of nonnegative matrices. Let

$$\mathcal{M}_U = \{\mathbf{Q} \geq \mathbf{O} : \mathbf{U}^{-1}\mathbf{QU} \geq \mathbf{O}\}; \quad \mathcal{M}_V = \{\mathbf{Q} \geq \mathbf{O} : \mathbf{V}^{-1}\mathbf{QV} \geq \mathbf{O}\}.$$

Recall that if $\mathbf{Q}$ is stochastic then $\mathbf{Q} \in \mathcal{M}_U$ is equivalent to $\mathbf{Q} \in \mathcal{M}_V$.

However, for strictly substochastic matrices, the two classes are, in general, distinct.

Lemma 3.10 *For a strictly substochastic matrix* $\mathbf{T} = (p_{ij})$, *let* $\mathbf{r} = (r_i) = \mathbf{1} - \mathbf{T1}$. *Then,*

(i) $\mathbf{T} \in \mathcal{M}_U$ *if and only if* $\mathbf{P}_1 \in \mathcal{M}_{\text{st}}$ *and* r_i *is decreasing in* i;

(ii) $\mathbf{T} \in \mathcal{M}_V$ *if and only if* $\mathbf{P}_N \in \mathcal{M}_{\text{st}}$ *and* r_i *is increasing in* i,

where $\mathbf{P}_i$ *is defined in* (3.48).

Proof. To prove part (i), one has from (3.48) that

$$\mathbf{U}^{-1}\mathbf{P}_1\mathbf{U} = \mathbf{U}^{-1}\mathbf{T}\mathbf{U} + \mathbf{U}^{-1}\mathbf{r}\,\boldsymbol{\delta}_1^{\mathsf{T}}\mathbf{U} = \mathbf{U}^{-1}\mathbf{T}\mathbf{U} + \mathbf{U}^{-1}\mathbf{r}\,\boldsymbol{\delta}_1^{\mathsf{T}}.$$

Denoting the jth column vector of matrix $\mathbf{P}$ by $[\mathbf{P}]_j$, it follows that

$$[\mathbf{U}^{-1}\mathbf{P}_1\mathbf{U}]_j = [\mathbf{U}^{-1}\mathbf{T}\mathbf{U}]_j, \quad j \geq 2, \tag{3.49}$$

and

$$[\mathbf{U}^{-1}\mathbf{P}_1\mathbf{U}]_1 = [\mathbf{U}^{-1}\mathbf{T}\mathbf{U}]_1 + \mathbf{U}^{-1}\mathbf{r} = \boldsymbol{\delta}_1, \tag{3.50}$$

where the last equality follows since $\mathbf{P}_1$ is stochastic. Suppose $\mathbf{T} \in \mathcal{M}_U$. Then, from (3.49) and (3.50), one has $\mathbf{P}_1 \in \mathcal{M}_{\text{st}}$. Also, since the component $[\mathbf{U}^{-1}\mathbf{T}\mathbf{U}]_{i1}$ is given by

$$\sum_{j=1}^{N} p_{ij} - \sum_{j=1}^{N} p_{i-1,j} = 1 - r_i - (1 - r_{i-1}) = r_{i-1} - r_i \geq 0,$$

where $p_{0j} = 0$ and $r_0 = 1$, we observe that r_i is decreasing in i. Conversely, if r_i is decreasing in i, then the ith component of $\mathbf{U}^{-1}\mathbf{r}$ is $r_i - r_{i-1} \leq 0$, $i \geq 2$, and the first component is $r_1 \leq 1$. Hence, if in addition $\mathbf{P}_1 \in \mathcal{M}_{\text{st}}$, it follows from (3.49) and (3.50) that $\mathbf{T} \in \mathcal{M}_U$. Assertion (ii) follows similarly. □

In the next theorem, we compare the quasi-stationary distribution $\mathbf{q}$ with the stationary distribution $\boldsymbol{\pi}_i$. Recall that $\mathbf{q}^{\mathsf{T}}\mathbf{T} = \gamma\,\mathbf{q}^{\mathsf{T}}$, where γ is the PF eigenvalue of $\mathbf{T}$.

Theorem 3.23 *Let* $\mathbf{T}$ *be strictly substochastic and primitive. Let* $\mathbf{q}$ *be the quasi-stationary distribution of* $\mathbf{T}$ *and let* $\boldsymbol{\pi}_i$ *be the stationary distribution of* $\mathbf{P}_i$. *Then,*

(i) *If* $\mathbf{P}_1 \in \mathcal{M}_{\text{st}}$ *then* $\mathbf{q} \geq_{\text{st}} \boldsymbol{\pi}_1$.

(ii) *If* $\mathbf{P}_N \in \mathcal{M}_{\text{st}}$ *then* $\mathbf{q} \leq_{\text{st}} \boldsymbol{\pi}_N$.

Proof. To prove assertion (i), let $\mathbf{u}_0 = \boldsymbol{\delta}_1$ and define the sequence of probability vectors $\mathbf{u}_n$ successively by

$$\mathbf{u}_{n+1}^{\mathsf{T}} = \mathbf{u}_n^{\mathsf{T}}\mathbf{P}_1, \quad n = 0, 1, \cdots.$$

If $\mathbf{P}_1 \in \mathcal{M}_{\text{st}}$, it is readily shown that $\mathbf{u}_{n+1} \geq_{\text{st}} \mathbf{u}_n$ and $\mathbf{u}_n$ converges to $\boldsymbol{\pi}_1$ as $n \to \infty$. We shall prove that $\mathbf{q} \geq_{\text{st}} \mathbf{u}_n$ for all $n \geq 0$, from which assertion (i) follows. Evidently, $\mathbf{q} \geq_{\text{st}} \mathbf{u}_0 = \boldsymbol{\delta}_1$. So, assuming that $\mathbf{q} \geq_{\text{st}} \mathbf{u}_n$

for some n, we need to prove that $\mathbf{q} \geq_{\text{st}} \mathbf{u}_{n+1}$. Since $\mathbf{U}^{-1}\mathbf{P}_1\mathbf{U} \geq \mathbf{O}$ and $\mathbf{q}^\mathsf{T}\mathbf{U} \geq \mathbf{u}_n^\mathsf{T}\mathbf{U}$, it follows that

$$\mathbf{u}_{n+1}^\mathsf{T}\mathbf{U} = \mathbf{u}_n^\mathsf{T}\mathbf{U}(\mathbf{U}^{-1}\mathbf{P}_1\mathbf{U}) \leq \mathbf{q}^\mathsf{T}\mathbf{U}(\mathbf{U}^{-1}\mathbf{P}_1\mathbf{U}) = \mathbf{q}^\mathsf{T}\mathbf{P}_1\mathbf{U}. \tag{3.51}$$

Also, using $\mathbf{q}^\mathsf{T}\mathbf{T} = \gamma\,\mathbf{q}^\mathsf{T}$ and $\boldsymbol{\delta}_1^\mathsf{T}\mathbf{U} = \boldsymbol{\delta}_1^\mathsf{T}$, we have

$$\mathbf{q}^\mathsf{T}\mathbf{P}_1\mathbf{U} = \mathbf{q}^\mathsf{T}(\mathbf{T} + \mathbf{r}\,\boldsymbol{\delta}_1^\mathsf{T})\mathbf{U} = \gamma\,\mathbf{q}^\mathsf{T}\mathbf{U} + (\mathbf{q}^\mathsf{T}\mathbf{r})\boldsymbol{\delta}_1^\mathsf{T},$$

from which, since $\gamma < 1$,

$$[\mathbf{q}^\mathsf{T}\mathbf{P}_1\mathbf{U}]_i = \gamma\,[\mathbf{q}^\mathsf{T}\mathbf{U}]_i \leq [\mathbf{q}^\mathsf{T}\mathbf{U}]_i, \quad i \geq 2,$$

and $[\mathbf{q}^\mathsf{T}\mathbf{P}_1\mathbf{U}]_1 = [\mathbf{q}^\mathsf{T}\mathbf{U}]_1 = 1$. Hence $\mathbf{q}^\mathsf{T}\mathbf{P}_1\mathbf{U} \leq \mathbf{q}^\mathsf{T}\mathbf{U}$ so that, from (3.51), one concludes that $\mathbf{q} \geq_{\text{st}} \mathbf{u}_{n+1}$, as claimed. Assertion (ii) can be proved similarly. □

The next example illustrates Theorem 3.23.

Example 3.17 Consider the finite random walk given in Example 2.9. Starting from state $i \neq 0$, suppose that we are interested in a quasi-statistical equilibrium given that there is no visit to state 0. Then we need to calculate the quasi-stationary distribution of the strictly substochastic matrix

$$\mathbf{T} = \begin{pmatrix} r_1 & p_1 & 0 & 0 & \cdots & 0 \\ q_2 & r_2 & p_2 & 0 & \cdots & 0 \\ 0 & q_3 & r_3 & p_2 & \cdots & 0 \\ \vdots & & \ddots & \ddots & \ddots & \vdots \\ 0 & \cdots & 0 & q_{N-1} & r_{N-1} & p_{N-1} \\ 0 & \cdots & 0 & 0 & q_N & r_N \end{pmatrix},$$

where $r_1 + p_1 < 1$ and the other row sums are equal to 1. It is easily seen that $\mathbf{T} \in \mathcal{M}_U$ whenever $p_i + q_{i+1} \leq 1$ (see Example 3.12 for this result). The stationary distribution $\boldsymbol{\pi}_1$ of $\mathbf{P}_1 = \mathbf{T} + \mathbf{r}\,\boldsymbol{\delta}_1^\mathsf{T}$ is easily calculated as in Example 2.9. Theorem 3.23, together with Lemma 3.10, then ensures that $\mathbf{q} \geq_{\text{st}} \boldsymbol{\pi}_1$. Note that $\mathbf{T} \notin \mathcal{M}_V$ so that the relation between $\mathbf{q}$ and $\boldsymbol{\pi}_N$ is undetermined.

Let $\mathbf{q} = (q_i)$ be the quasi-stationary distribution and let $\mathbf{d} = (d_i)$ be the doubly limiting conditional distribution of $\mathbf{T}$. From Theorem 2.25, we know that $\mathbf{d} = \mathbf{q}_\mathrm{D}\mathbf{v}$, where $\mathbf{q}_\mathrm{D}$ is the diagonal matrix with diagonal components q_i and $\mathbf{v}$ denotes the right PF eigenvector, positive componentwise, such that $\mathbf{q}^\mathsf{T}\mathbf{v} = 1$. The next theorem compares the doubly limiting conditional distribution $\mathbf{d}$ with the quasi-stationary distribution $\mathbf{q}$.

Theorem 3.24 *Let* $\mathbf{T}$ *be strictly substochastic and primitive.*

(i) *If* $\mathbf{T} \in \mathcal{M}_U$ *then* $\mathbf{d} \geq_{\text{lr}} \mathbf{q}$.

(ii) *If* $\mathbf{T} \in \mathcal{M}_V$ *then* $\mathbf{d} \leq_{\text{lr}} \mathbf{q}$.

Proof. Since $\mathbf{Tv} = \gamma\,\mathbf{v}$, we have

$$\gamma\,\mathbf{U}^{-1}\mathbf{v} = (\mathbf{U}^{-1}\mathbf{T}\mathbf{U})\mathbf{U}^{-1}\mathbf{v}.$$

Note that, since $\mathbf{U}^{-1}\mathbf{T}\mathbf{U}$ is a similarity transform of $\mathbf{T}$, the eigenvalues of $\mathbf{U}^{-1}\mathbf{T}\mathbf{U}$ are the same as those of $\mathbf{T}$. Under the assumption of assertion (i), we have $\mathbf{U}^{-1}\mathbf{T}\mathbf{U} \geq \mathbf{O}$. Hence, the eigenvector $\mathbf{U}^{-1}\mathbf{v}$ associated with the PF eigenvalue γ of $\mathbf{U}^{-1}\mathbf{T}\mathbf{U}$ must be nonnegative componentwise (it cannot be nonpositive). But, $\mathbf{v} = \mathbf{q}_{\mathrm{D}}^{-1}\mathbf{d}$. It follows that $\mathbf{U}^{-1}\mathbf{q}_{\mathrm{D}}^{-1}\mathbf{d} \geq \mathbf{0}$. Hence, $\mathbf{q}_{\mathrm{D}}^{-1}\mathbf{d}$ is increasing, from Theorem 3.4, meaning that $\mathbf{d} \geq_{\mathrm{lr}} \mathbf{q}$ (see Definition 3.10). Assertion (ii) follows similarly. □

It is of interest to note that, from Lemma 3.10 and Theorems 3.23 and 3.24, if $\mathbf{T} \in \mathcal{M}_U$ ($\mathbf{T} \in \mathcal{M}_V$, respectively) then $\mathbf{d} \geq_{\mathrm{st}} \boldsymbol{\pi}_1$ ($\mathbf{d} \leq_{\mathrm{st}} \boldsymbol{\pi}_N$).

3.8 Renewal processes in discrete time

Let $T_1, T_2, \cdots$ be a sequence of IID (independent, identically distributed) nonnegative, integer-valued random variables with a common probability vector $\mathbf{a} = (a_i)$. Define $S_n = \sum_{i=1}^{n} T_i$, $n = 1, 2, \cdots$, with $S_0 \equiv 0$, and let

$$N_k = \max\{n : S_n \leq k\}, \quad k = 0, 1, \cdots.$$

To eliminate the trivial case, we assume throughout that $a_0 < 1$. Then it is not difficult to show that $P[N_k < \infty] = 1$ for every k (see, e.g., Ross, page 57, 1983). If T_i represents the lifetime of the ith (successively replaced) system, then N_k counts the number of failed systems up to time k. The process $\{N_k\}$ is called a *renewal process* in discrete time (see Figure 3.2). For a general treatment of such processes, we refer to Feller (Chapter 13, 1957).

The partial-sum process $\{S_n\}$ is a temporally and spatially homogeneous Markov chain with an increasing monotone path. The random quantity $N_k + 1$ coincides with the first passage time of the Markov chain $\{S_n\}$, starting from state 0, into the set $\{k+1, k+2, \cdots\}$, called an *upper set.* More precisely, $\{S_n\}$ is a Markov chain with transition matrix

$$\mathbf{A} = \begin{pmatrix} a_0 & a_1 & a_2 & \cdots \\ 0 & a_0 & a_1 & \cdots \\ 0 & 0 & a_0 & \cdots \\ \vdots & \vdots & \vdots & \ddots \end{pmatrix} \tag{3.52}$$

and the first passage time $N_k + 1$ for each $k \geq 0$ follows the phase-type distribution generated by $(\boldsymbol{\delta}_0, \mathbf{T}_k)$, where

$$\mathbf{T}_k = \begin{pmatrix} a_0 & a_1 & \cdots & a_k \\ 0 & a_0 & \cdots & a_{k-1} \\ \vdots & \vdots & \ddots & \vdots \\ 0 & 0 & \cdots & a_0 \end{pmatrix}.$$

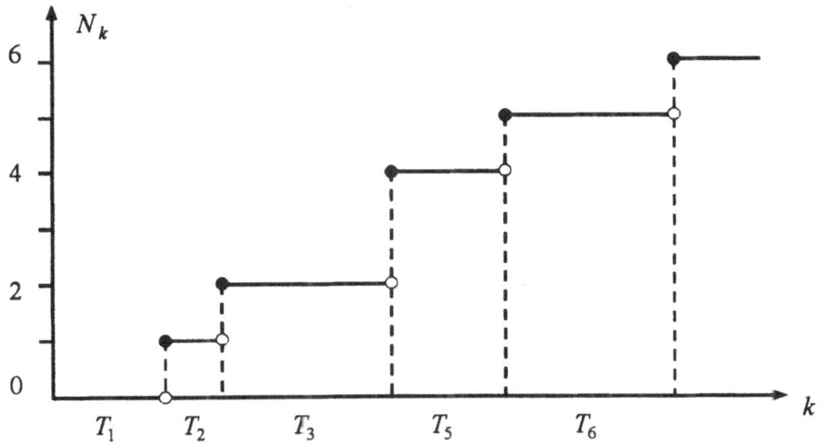

Figure 3.2 *A sample path of a renewal process.*

The next results are immediate consequences of Theorems 3.22 and 3.21 respectively. It should be noted that the order of states is reversed here.

Theorem 3.25 *Let* $\{N_k\}$ *be a renewal process with lifetime distribution* $\mathbf{a}$. *For each* $k \geq 0$,

(i) $N_k \in \text{IHRA}$.

(ii) *If* $\mathbf{a} \in \text{DRHR}$ *then* $N_k \in \text{IHR}$.

A fundamental identity in studying renewal processes is

$$\{N_k \geq n\} = \{S_n \leq k\}, \quad k,\ n = 0, 1, \cdots. \tag{3.53}$$

Since $\{S_n \leq k\} \subset \{S_n \leq k+1\}$, we have

$$P[N_k \geq n] \leq P[N_{k+1} \geq n], \quad n = 0, 1, \cdots,$$

so that $N_k \leq_{\text{st}} N_{k+1}$, $k = 0, 1, \cdots$. Hence, the renewal process $\{N_k\}$ is stochastically increasing for any lifetime distribution. Stronger results can be derived under hypotheses on the lifetime distribution. Let $\boldsymbol{\pi}(n)$ be the state distribution of the Markov chain $\{S_n\}$, i.e.,

$$\boldsymbol{\pi}^{\top}(n) = \boldsymbol{\delta}_0^{\top} \mathbf{A}^n, \quad n = 0, 1, \cdots.$$

Let $f_n(k) = P[N_k = n]$, $k = 0, 1, \cdots$, and define $\mathbf{f}_n = (f_n(k))$ for $n = 0, 1, \cdots$. From (3.53), we have

$$\begin{aligned} P[N_k = n] &= P[N_k \geq n] - P[N_k \geq n+1] \\ &= P[S_n \leq k] - P[S_{n+1} \leq k]. \end{aligned}$$

It follows that

$$\mathbf{f}_n^\top = \boldsymbol{\delta}_0^\top \mathbf{A}^n \mathbf{V} - \boldsymbol{\delta}_0^\top \mathbf{A}^{n+1} \mathbf{V}, \quad n = 0, 1, \cdots. \tag{3.54}$$

Exercise 3.24 asks the reader to prove that the vector $\mathbf{f}_n$ is PF_2 under the conditions of the next theorem. The next result is the discrete-time counterpart of Theorem 3 in Karlin and Proschan (1960).

Theorem 3.26 *Let $\{N_k\}$ be a renewal process with lifetime distribution $\mathbf{a} = (a_i)$. Suppose that $\mathbf{a} \in \mathrm{DLR}$. Then:*

(i) *$N_k \le_{\mathrm{lr}} N_{k+1}$ for every $k \ge 0$, i.e., $\{N_k\}$ is increasing in the sense of likelihood ratio ordering;*

(ii) *$N_k \in \mathrm{DLR}$.*

Proof. Note that, due to the structure of $\mathbf{A}$ given in (3.52), Lemma 3.3(ii) shows that

$$\boldsymbol{\delta}_0^\top \mathbf{A}^n \mathbf{V} = (\boldsymbol{\delta}_0^\top \mathbf{V})(\mathbf{V}^{-1}\mathbf{A}^n \mathbf{V}) = (\boldsymbol{\delta}_0^\top \mathbf{V})(\mathbf{V}^{-1}\mathbf{A}\mathbf{V})^n.$$

But, since $\mathbf{V}^{-1}\mathbf{A}\mathbf{V} = \mathbf{A}$, it follows that

$$\boldsymbol{\delta}_0^\top \mathbf{A}^n \mathbf{V} = (\boldsymbol{\delta}_0^\top \mathbf{V})\mathbf{A}^n.$$

(i) We have from (3.54) that

$$\begin{pmatrix} \mathbf{f}_n^\top \\ \mathbf{f}_{n+1}^\top \end{pmatrix} = \begin{pmatrix} \boldsymbol{\delta}_0^\top \mathbf{V} - \mathbf{a}^\top \mathbf{V} \\ \mathbf{a}^\top \mathbf{V} - \mathbf{a}^\top \mathbf{A}\mathbf{V} \end{pmatrix} \mathbf{A}^n, \quad n = 0, 1, \cdots.$$

Note that $\begin{pmatrix} \boldsymbol{\delta}_0^\top - \mathbf{a}^\top \\ \mathbf{a}^\top - \mathbf{a}^\top \mathbf{A} \end{pmatrix} \mathbf{V} \in \mathrm{TP}_2$ if and only if

$$\overline{A}_i(2) \ge \overline{A}_i \quad \text{and} \quad \overline{A}_i\, \overline{A}_{i+1}(2) \ge \overline{A}_{i+1}\, \overline{A}_i(2), \quad i = 1, 2, \cdots,$$

where $\overline{A}_i = \sum_{k=i}^\infty a_k = [\boldsymbol{\delta}_0^\top \mathbf{A}\mathbf{U}]_i$ and $\overline{A}_i(2) = [\mathbf{a}^\top \mathbf{A}\mathbf{U}]_i$. It is easy to see that $\mathbf{a}^\top \mathbf{A}\mathbf{U} \ge \boldsymbol{\delta}_0^\top \mathbf{A}\mathbf{U}$ and that if $\mathbf{a} \in \mathrm{DLR}$ then $\begin{pmatrix} \boldsymbol{\delta}_0^\top \\ \mathbf{a}^\top \end{pmatrix} \mathbf{A}\mathbf{U}$ is TP_2. Hence, the basic composition formula (Theorem 3.1) is applied to conclude that $\begin{pmatrix} \mathbf{f}_n^\top \\ \mathbf{f}_{n+1}^\top \end{pmatrix} \in \mathrm{TP}_2$. Therefore $N_k \le_{\mathrm{lr}} N_{k+1}$.

(ii) To prove assertion (ii), Lemma 3.3(ii) and (3.54) again yield

$$\mathbf{f}_{n+1}^\top = \boldsymbol{\delta}_0^\top \mathbf{A}^n \mathbf{V}(\mathbf{V}^{-1}\mathbf{A}\mathbf{V}) - \boldsymbol{\delta}_0^\top \mathbf{A}^{n+1}\mathbf{V}(\mathbf{V}^{-1}\mathbf{A}\mathbf{V}) = \mathbf{f}_n^\top \mathbf{A}.$$

For $f_n(k) > 0$, we then have

$$\frac{f_{n+1}(k)}{f_n(k)} = \sum_{i=0}^k a_{k-i} \frac{f_n(i)}{f_n(k)}, \quad k = 0, 1, \cdots.$$

But, we have proved in assertion (i) that

$$f_n(i)f_{n+1}(k) \geq f_n(k)f_{n+1}(i), \quad i \leq k.$$

Hence, using the convention $0/0 = 0$, it follows that

$$\frac{f_{n+1}(k)}{f_n(k)} \geq \sum_{i=0}^{k} a_{k-i}\frac{f_{n+1}(i)}{f_{n+1}(k)} = \frac{f_{n+2}(k)}{f_{n+1}(k)}, \quad k = 0, 1, \cdots,$$

proving the theorem. □

Other stochastic monotonicity results are the following.

Theorem 3.27 *Let* $\{N_k\}$ *be a renewal process with lifetime distribution* $\mathbf{a} = (a_i)$*:*

(i) *If* $\mathbf{a} \in$ IHR *then* $\{N_k\}$ *is increasing in the sense of reversed hazard rate ordering;*

(ii) *If* $\mathbf{a} \in$ DRHR *then* $\{N_k\}$ *is increasing in the sense of hazard rate ordering.*

Proof. Note that $\mathbf{a} \in$ IHR implies $\mathbf{AU} \in \mathrm{TP}_2$ while $\mathbf{a} \in$ DRHR implies $\mathbf{AV} \in \mathrm{TP}_2$. Hence, if $\mathbf{a} \in$ IHR then $\{S_n\}$ is increasing in the sense of hazard rate ordering (see Theorem 3.13(ii)). That is, we have

$$P[S_n > k]\, P[S_{n-1} > k+1] \geq P[S_n > k+1]\, P[S_{n+1} > k],$$

which, using (3.53), is equivalent to

$$P[N_k < n]\, P[N_{k+1} < n+1] \geq P[N_{k+1} < n]\, P[N_k < n+1],$$

so that $\{N_k\}$ is monotonically increasing in the sense of reversed hazard rate ordering. Assertion (ii) can be proved similarly. □

Of interest in the theory of renewal processes are random variables defined by

$$A_k \equiv k - S_{N_k}, \quad Z_k \equiv S_{N_k+1} - k; \quad k = 0, 1, \cdots.$$

Clearly, A_k is the elapsed time at time k since the last renewal, while Z_k represents the time to the next failure (see Figure 3.3). The process $\{A_k\}$ is called the *age process* (see Example 3.13) and $\{Z_k\}$ the *residual-lifetime process* of the renewal process $\{N_k\}$. For the lifetime distribution $\mathbf{a} = (a_i)$, let $h_i = a_i/\overline{A}_i$, $i = 0, 1, \cdots$, be the hazard rate function whenever defined. In order to study the processes $\{A_k\}$ and $\{Z_k\}$, we consider a Markov chain $\{X_n\}$ on the state space $\{1, 2, \cdots\}$, starting from state 1, with transition matrix

$$\mathbf{P} = \begin{pmatrix} h_1 & 1-h_1 & 0 & 0 & \cdots \\ h_2 & 0 & 1-h_2 & 0 & \cdots \\ h_3 & 0 & 0 & 1-h_3 & \cdots \\ \vdots & \vdots & \vdots & \ddots & \ddots \end{pmatrix}.$$

Then, a renewal having a positive lifetime in $\{N_k\}$ can be considered as an

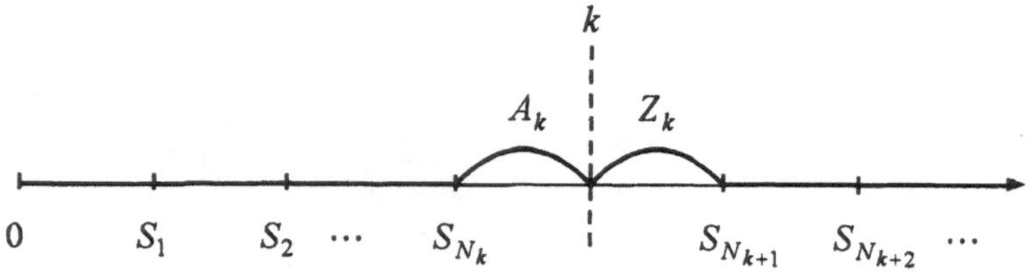

Figure 3.3 *The age and residual lifetime of a renewal process.*

entrance of $\{X_n\}$ to state 1. If $a_0 > 0$, however, we may have a renewal of zero lifetime, called a *degenerate renewal.* Although degenerate renewals do not appear in $\{X_n\}$, since they are irrelevant to the age A_k as well as to the residual lifetime Z_k, the stochastic behavior of the Markov chain $\{X_n\}$ suffices to determine that of the two processes. In fact $A_n = X_n - 1$. The next result is due to Kijima (1992c).

Theorem 3.28 *Suppose that the lifetime distribution* $\mathbf{a}$ *is* DHR. *Then both the processes* $\{A_k\}$ *and* $\{Z_k\}$ *are monotonically increasing in the sense of hazard rate ordering.*

Proof. Let $\boldsymbol{\pi}(n) = (\pi_i(n))$ denote the state distribution of $\{X_n\}$ at time n. Then

$$\begin{pmatrix} \boldsymbol{\pi}^{\top}(n) \\ \boldsymbol{\pi}^{\top}(n+1) \end{pmatrix} \mathbf{U} = \begin{pmatrix} \boldsymbol{\pi}^{\top}(n-1) \\ \boldsymbol{\pi}^{\top}(n) \end{pmatrix} \mathbf{PU}.$$

Since $X_0 = 1$ so that $X_0 \leq_{\text{hr}} X_1$, a simple induction argument using the composition law (Theorem 3.2) shows that $\{X_n\}$ is increasing in the sense of hazard rate ordering. Since $A_n = X_n - 1$ is the age at time n, the process $\{A_n\}$ is also increasing in the sense of hazard rate ordering.

For the residual-lifetime process $\{Z_k\}$, we note that

$$P[Z_n = j | X_n = i] = h_{i+j-1} \prod_{k=i}^{i+j-2} (1 - h_k), \tag{3.55}$$

which we denote by c_{ij} for $i,\ j = 1, 2, \cdots$, where the empty product is understood to be unity. It follows that

$$\sum_{m=j}^{\infty} c_{im} = \prod_{k=0}^{j-2} (1 - h_{i+k}), \quad i,\ j = 1, 2, \cdots.$$

Let $\mathbf{C} = (c_{ij})$. It is easy to show that if h_i is decreasing in i then $\mathbf{CU}$ is

TP_2 and $\mathbf{U}^{-1}\mathbf{C}\mathbf{U} \geq \mathbf{O}$. From (3.55), we have

$$\pi_j^Z(n) = \sum_{i=1}^{\infty} \pi_i(n)\, c_{ij}, \quad j = 1, 2, \cdots,$$

where $\pi_j^Z(n) = P[Z_n = j]$. Hence, defining $\boldsymbol{\pi}_Z(n) = (\pi_i^Z(n))$, it follows that

$$\begin{pmatrix} \boldsymbol{\pi}_Z^{\mathsf{T}}(n) \\ \boldsymbol{\pi}_Z^{\mathsf{T}}(n+1) \end{pmatrix} \mathbf{U} = \begin{pmatrix} \boldsymbol{\pi}^{\mathsf{T}}(n) \\ \boldsymbol{\pi}^{\mathsf{T}}(n+1) \end{pmatrix} \mathbf{C}\mathbf{U}.$$

The desired monotonicity of $\{Z_k\}$ now follows from the hazard rate monotonicity of $\boldsymbol{\pi}(k)$ and the composition law. □

For each $m = 1, 2, \cdots$, let $\Delta_k^m = N_{k+m} - N_k$, the number of renewals (including degenerate renewals) in the interval $(k, k+m]$. The next theorem provides the stochastic monotonicity of $\{\Delta_k^m\}$ for any DHR renewal process. Brown (1980) used an elegant coupling argument to prove the result for the continuous-time case.

Theorem 3.29 *Suppose that the lifetime distribution* $\mathbf{a}$ *is* DHR. *Then, for each* $m = 1, 2, \cdots$, *the process* $\{\Delta_k^m\}$ *is stochastically decreasing.*

Proof. Fix m and let $d_i(k) = P[\Delta_k^m = i]$. It can be shown that $d_0(k) = P[Z_k \geq m+1]$ and

$$d_i(k) = \sum_{j=0}^{m-1} P[Z_k = m-j]\, P[N_j = i-1], \quad i = 1, 2, \cdots,$$

where Z_k is the residual lifetime of the renewal process $\{N_k\}$ at time k. Let $\overline{D}_i(k) = \sum_{\ell=i}^{\infty} d_\ell(k) = P[\Delta_k^m \geq i]$. It follows that

$$\begin{aligned} \overline{D}_i(k) &= \sum_{j=0}^{m-1} P[Z_k = m-j]\ P[N_j \geq i-1] \\ &= \sum_{j=0}^{m-1} P[Z_k \leq m-j]\{P[N_j \geq i-1] - P[N_{j-1} \geq i-1]\}, \end{aligned}$$

where the convention $P[N_{-1} \geq i] = 0$ is used. If $\mathbf{a} \in$ DHR, we have $Z_k \leq_{\mathrm{hr}} Z_{k+1}$, from Theorem 3.28, so that

$$P[Z_k \leq i] \geq P[Z_{k+1} \leq i], \quad i = 1, 2, \cdots.$$

Also, as noted earlier, $N_{j-1} \leq_{\mathrm{st}} N_j$ so that $P[N_j \geq i-1] \geq P[N_{j-1} \geq i-1]$. Combining these results, the theorem is proved. □

Let $H_k = E[N_k]$, the *renewal function* of the renewal process $\{N_k\}$. If $\mathbf{a}$ is DHR, taking $m = 1$ in Theorem 3.29 reveals that the renewal function H_k is increasing and concave in k. This result is useful in various applications. See, e.g., Hirayama and Kijima (1992) for a single-machine scheduling problem. In particular, for the case of no degenerate renewals, we have

$$p_k \equiv P[N_k - N_{k-1} = 1] = H_k - H_{k-1},$$

so that the ***renewal probability*** p_k is decreasing in k if H_k is concave in k.

In the ordinary renewal processes, the lifetimes T_n are nonnegative and IID. Accordingly, the partial-sum process $\{S_n\}$ is a spatially and temporally homogeneous Markov chain with the transition matrix $\mathbf{A}$ given in (3.52). A natural generalization of the ordinary renewal processes is to assume that the distribution of T_{n+1} depends on the value of S_n. Such a renewal process is called a *g-renewal process*, first introduced by Kijima and Sumita (1986a). This generalization has its apparent use in studying the stochastic behavior of a repairable system with general repair in the reliability literature (see, e.g., Kijima, Morimura and Suzuki, 1988).

Associated with this generalization is the Markov chain with transition matrix

$$\mathbf{A} = \begin{pmatrix} a_{00} & a_{01} & a_{02} & \cdots \\ 0 & a_{10} & a_{11} & \cdots \\ 0 & 0 & a_{20} & \cdots \\ \vdots & \vdots & \vdots & \ddots \end{pmatrix}. \tag{3.56}$$

The distribution property of the g-renewal process $\{N_k\}$ can be analyzed by the phase-type distribution generated by $(\boldsymbol{\delta}_0, \mathbf{T}_k)$, where

$$\mathbf{T}_k = \begin{pmatrix} a_{00} & a_{01} & \cdots & a_{0k} \\ 0 & a_{10} & \cdots & a_{1,k-1} \\ \vdots & \vdots & \ddots & \vdots \\ 0 & 0 & \cdots & a_{k0} \end{pmatrix}.$$

Let $T(i)$ represent the generic lifetime when $S_n = i$, and denote the probability vector of $T(i) + i$ by $\mathbf{a}_i = (a_{ij})$. Note that $\mathbf{a}_i^{\top}$ corresponds to the ith row of $\mathbf{A}$ given in (3.56). The proof of the next result is parallel to that of Theorem 3.25 and is omitted.

Corollary 3.7 *Let $\{N_k\}$ be a g-renewal process governed by $\mathbf{A}$ given in* (3.56). *For each $k \geq 0$:*

(i) *If $T(i) + i$ is stochastically increasing then $N_k \in$ IHRA;*

(ii) *If $T(i)+i$ is increasing in the sense of reversed hazard rate ordering, then $N_k \in$ IHR.*

The conditions in Corollary 3.7 can be slightly weakened. Another generalization allowing time nonhomogeneity will be also possible, but we do not pursue it here. Analogous results to Theorem 3.27 in the g-renewal setting can also be derived without difficulty.

Let $(\boldsymbol{\alpha}, \mathbf{T})$ be a phase-type distribution and consider a Markov chain on the state space $\{1, \cdots, N\}$, where N denotes the number of phases, with transition matrix $\mathbf{T} + \mathbf{r}\boldsymbol{\alpha}^\top$ and initial distribution $\boldsymbol{\alpha}$. Here $\mathbf{r} = \mathbf{1} - \mathbf{T1} = (r_i)$. To this Markov chain, we adjoin an instantaneous state 0 such that the probability of visits to 0, given that the chain is in state i, is r_i. By choosing the return state to the set $\{1, \cdots, N\}$ according to multinomial trials, the successive visits to the instantaneous state form a discrete-time renewal process with underlying lifetimes distributed by the phase-type distribution $(\boldsymbol{\alpha}, \mathbf{T})$. Such a renewal process is called a *phase-type renewal process* (see Neuts, 1981). In the next theorem, we prove a somewhat stronger result than Theorem 3.29. For this purpose, we consider an absorbing Markov chain $\{Y_n\}$ with state space $\{0, 1, \cdots, N\}$, where state 0 is absorbing, and the transition matrix

$$\mathbf{P} = \begin{pmatrix} 1 & \mathbf{0}^\top \\ \mathbf{r} & \mathbf{T} \end{pmatrix}.$$

Let T be the first passage time of $\{Y_n\}$ to state 0 and define

$$u_{in} = P[T \le n | Y_0 = i], \quad i \in \{0, 1, \cdots, N\}; \quad n = 0, 1, \cdots.$$

Note that $u_{0n} = 1$ for all $n = 0, 1, \cdots$ and $u_{i0} = 0$ for $i = 1, \cdots, N$. Then one has

$$u_{in} = \sum_{k=0}^{N} p_{ik} u_{k,n-1}, \quad i \in \{0, 1, \cdots, N\}; \quad n = 1, 2, \cdots,$$

where $\mathbf{P} = (p_{ik})$. Hence, writing $\mathbf{u}_n = (u_{in})$, it follows that

$$\mathbf{u}_n = \mathbf{P}\,\mathbf{u}_{n-1}, \quad n = 1, 2, \cdots. \tag{3.57}$$

Recall that $\mathbf{T} \in \mathcal{M}_U$ if and only if $\mathbf{T} \ge \mathbf{O}$ and $\mathbf{U}^{-1}\mathbf{TU} \ge \mathbf{O}$.

Theorem 3.30 *Let* $\boldsymbol{\alpha} = \boldsymbol{\delta}_1$ *and suppose that* $\mathbf{T} \in \mathcal{M}_U$. *Then* $\{\Delta_k^m\}$ *is stochastically decreasing for each* $m = 1, 2, \cdots$.

Proof. Under the condition, we have $\mathbf{P} \in \mathcal{M}_{st}$. Note that $\mathbf{V}^{-1}\mathbf{u}_0 \ge \mathbf{0}$ since $\mathbf{u}_0 = \boldsymbol{\delta}_0$. Also, from (3.57), we have

$$\mathbf{V}^{-1}\mathbf{u}_n = (\mathbf{V}^{-1}\mathbf{PV})\mathbf{V}^{-1}\mathbf{u}_{n-1}, \quad n = 1, 2, \cdots.$$

It follows that $\mathbf{V}^{-1}\mathbf{u}_n \ge \mathbf{0}$ for all $n = 0, 1, \cdots$, i.e., the vector $\mathbf{u}_n$ is decreasing from Theorem 3.4(i). Now, define $\{X_n\}$ to be a Markov chain with initial distribution $\boldsymbol{\alpha} = \boldsymbol{\delta}_1$ and transition matrix

$$\mathbf{P}_1 = \mathbf{T} + \mathbf{r}\,\boldsymbol{\delta}_1^\top;$$

see (3.48). Under the condition, we have $\mathbf{P}_1 \in \mathcal{M}_{st}$ from Lemma 3.10, so that $\{X_n\}$ is stochastically increasing. Let $\{Z_n\}$ be the residual-lifetime process of the phase-type renewal process. Then

$$P[Z_k \le n] \;=\; \sum_{i=1}^{N} P[X_k = i]\, u_{in}$$

$$= u_{Nn} + \sum_{i=1}^{N-1} P[X_k \leq i](u_{in} - u_{i+1,n}).$$

Since $\{X_n\}$ is stochastically increasing and $\mathbf{u}_n$ is decreasing, it follows that $\{Z_n\}$ is stochastically increasing. The theorem now follows by the same argument as in the proof of Theorem 3.29. □

In a phase-type renewal process with lifetime distribution $(\boldsymbol{\alpha}, \mathbf{T})$, if $\boldsymbol{\alpha}^\top \mathbf{1} = 1$ then the renewal probability p_n that a renewal occurs at time n is given by

$$p_n = \boldsymbol{\alpha}^\top (\mathbf{T} + \mathbf{r}\,\boldsymbol{\alpha}^\top)^{n-1}\mathbf{r}, \quad n = 1, 2, \cdots.$$

Under the conditions of Theorem 3.30, we know that p_n is decreasing in n and, from Theorem 3.20, the lifetime of such a phase-type renewal process is NWU. Hence, in this setting, DHR lifetimes may not be necessary for a decreasing renewal probability (see Brown's conjecture, 1981, in a general setting).

3.9 Comparability of Markov chains

In this section, we consider two Markov chains $\{X_n^i\}$, $i = 1, 2$, defined on the state space $\mathcal{N} = \{0, 1, \cdots\}$ and compare them in the sense of stochastic ordering. For more general settings, see, e.g., Stoyan (1983) or Szekli (1995).

Definition 3.12 For two Markov chains $\{X_n^i\}$, $i = 1, 2$, let $\mathbf{P}_i$ be the transition matrix of $\{X_n^i\}$. The Markov chain $\{X_n^1\}$ is said to *stochastically dominate* $\{X_n^2\}$ if $\mathbf{P}_1\mathbf{U} \geq \mathbf{P}_2\mathbf{U}$ or, equivalently, $\mathbf{P}_1\mathbf{V} \leq \mathbf{P}_2\mathbf{V}$.

The next result justifies the use of the phrase 'stochastic dominance'.

Theorem 3.31 *For two Markov chains $\{X_n^i\}$, $i = 1, 2$, each with transition matrix $\mathbf{P}_i$ and initial distribution $\boldsymbol{\alpha}_i$, suppose that $\boldsymbol{\alpha}_1 \geq_{\text{st}} \boldsymbol{\alpha}_2$ and that $\{X_n^1\}$ dominates $\{X_n^2\}$. If either $\{X_n^1\}$ or $\{X_n^2\}$ is stochastically monotone, then $X_n^1 \geq_{\text{st}} X_n^2$ for every $n = 1, 2, \cdots$.*

Proof. Let $\boldsymbol{\pi}_i(n)$ denote the state distribution of $\{X_n^i\}$ at time n. Suppose that $\{X_n^1\}$ is stochastically monotone, i.e., $\mathbf{U}^{-1}\mathbf{P}_1\mathbf{U} \geq \mathbf{O}$, and that $X_n^1 \geq_{\text{st}} X_n^2$ for some n. Then

$$\begin{aligned}
&\boldsymbol{\pi}_1^\top(n+1)\mathbf{U} - \boldsymbol{\pi}_2^\top(n+1)\mathbf{U} \\
=\ &\boldsymbol{\pi}_1^\top(n)\mathbf{P}_1\mathbf{U} - \boldsymbol{\pi}_2^\top(n)\mathbf{P}_2\mathbf{U} \\
=\ &(\boldsymbol{\pi}_1^\top(n)\mathbf{U} - \boldsymbol{\pi}_2^\top(n)\mathbf{U})(\mathbf{U}^{-1}\mathbf{P}_1\mathbf{U}) + \boldsymbol{\pi}_2^\top(n)(\mathbf{P}_1\mathbf{U} - \mathbf{P}_2\mathbf{U}),
\end{aligned}$$

which is nonnegative under the assumptions. Hence $X_{n+1}^1 \geq_{\text{st}} X_{n+1}^2$. If $\{X_n^2\}$ is stochastically monotone, the roles of $\mathbf{P}_1$ and $\mathbf{P}_2$ are interchanged and the proof is similar. The theorem now follows by an induction argument. □

Example 3.18 Consider two M/G/1 queues with the same arrival rate $\lambda = 1$. Let $G_i(x)$, $i = 1, 2$, be the service time distribution of the ith queue, and suppose that $\overline{G}_1(x) \geq \overline{G}_2(x)$ for all $x \geq 0$. This means that the service time of the first queue is stochastically longer than that of the second queue. For each queue, let

$$b_n^i = \int_0^\infty \frac{x^n}{n!} e^{-x} dG_i(x), \quad n = 0, 1, \cdots,$$

and define

$$\mathbf{P}_i = \begin{pmatrix} b_0^i & b_1^i & b_2^i & b_3^i & \cdots \\ b_0^i & b_1^i & b_2^i & b_3^i & \cdots \\ 0 & b_0^i & b_1^i & b_2^i & \cdots \\ 0 & 0 & b_0^i & b_1^i & \cdots \\ \vdots & \vdots & \vdots & \ddots & \ddots \end{pmatrix};$$

see (1.29). Let X_n^i be the number of customers just after the nth departure in the ith queue. The process $\{X_n^i\}$ is a Markov chain with the transition matrix $\mathbf{P}_i$. As in Example 3.14, we have

$$\overline{B}_n^1 = \int_0^\infty \frac{x^{n-1}}{(n-1)!} e^{-x} \overline{G}_1(x) dx \geq \int_0^\infty \frac{x^{n-1}}{(n-1)!} e^{-x} \overline{G}_2(x) dx = \overline{B}_n^2,$$

where $\overline{B}_n^i = \sum_{k=n}^\infty b_k^i$. Hence $\mathbf{P}_1$ dominates $\mathbf{P}_2$ under the assumption, since

$$\mathbf{P}_i\mathrm{U} = \begin{pmatrix} 1 & \overline{B}_1^i & \overline{B}_2^i & \cdots \\ 1 & \overline{B}_1^i & \overline{B}_2^i & \cdots \\ 1 & 1 & \overline{B}_1^i & \cdots \\ \vdots & \vdots & \vdots & \ddots \end{pmatrix}.$$

As we saw in Example 3.14, both $\{X_n^1\}$ and $\{X_n^2\}$ are stochastically monotone. It follows from Theorem 3.31 that if $X_0^1 \geq_{\rm st} X_0^2$ then $X_n^1 \geq_{\rm st} X_n^2$ for every $n = 1, 2, \cdots$. Hence, if X_n^1 and X_n^2 converge in distribution to respective random variables X_∞^1 and X_∞^2, say, as $n \to \infty$, then $X_\infty^1 \geq_{\rm st} X_\infty^2$.

The next two examples answer the questions raised by Examples 3.1 and 3.2 in the introduction of this chapter.

Example 3.19 Consider an ergodic Markov chain $\{X_n\}$ with state space $\mathcal{N} = \{0, 1, \cdots\}$ and transition matrix $\mathbf{P}$. Let $\boldsymbol{\pi}$ be the stationary distribution of $\mathbf{P}$. We want to approximate $\boldsymbol{\pi}$ through finite north-west corner truncations of $\mathbf{P}$. We shall write

$$\mathbf{P} = \begin{pmatrix} \mathbf{T}_n & \mathbf{S}_n \\ \mathbf{R}_n & \mathbf{Q}_n \end{pmatrix},$$

where $\mathbf{T}_n$ denotes the truncation of size n. Let

$$\mathbf{P}_n = \mathbf{T}_n + \mathbf{r}_n \boldsymbol{\delta}_{n-1}^\top,$$

where $\mathbf{r}_n = \mathbf{1} - \mathbf{T}_n\mathbf{1}$; see (3.48). It is assumed that the stochastic matrices $\mathbf{P}_n$ are ergodic for all sufficiently large n. Then, there exists a unique stationary distribution of $\mathbf{P}_n$, which we denote by $\boldsymbol{\pi}_n$.

Suppose that $\mathbf{P}$ is stochastically monotone. Define the infinite matrix

$$\widetilde{\mathbf{P}}_n = \begin{pmatrix} \mathbf{P}_n & \mathbf{O} \\ \mathbf{R}_n & \mathbf{Q}_n \end{pmatrix}$$

and the infinite vector $\widetilde{\boldsymbol{\pi}}_n^\top = (\boldsymbol{\pi}_n^\top, \mathbf{0}^\top)$. It is easy to see that the $\widetilde{\mathbf{P}}_n$ are all stochastically monotone. Let $\mathbf{u}_0(n) = \boldsymbol{\delta}_0$ and define $\mathbf{u}_k(n)$ successively by

$$\mathbf{u}_{k+1}^\top(n) = \mathbf{u}_k^\top(n)\widetilde{\mathbf{P}}_n, \quad k = 0, 1, \cdots.$$

It is readily seen that $\mathbf{u}_k(n)$ is stochastically increasing in k and converges to $\widetilde{\boldsymbol{\pi}}_n$ as $k \to \infty$. Now compare $\widetilde{\boldsymbol{\pi}}_n$ and $\widetilde{\boldsymbol{\pi}}_{n+1}$. By construction, $\widetilde{\mathbf{P}}_{n+1}$ stochastically dominates $\widetilde{\mathbf{P}}_n$ for all n. Since $\mathbf{u}_0(n) = \mathbf{u}_0(n+1) = \boldsymbol{\delta}_0$, Theorem 3.31 shows that $\mathbf{u}_k(n+1) \geq_{\mathrm{st}} \mathbf{u}_k(n)$ for all k and hence $\widetilde{\boldsymbol{\pi}}_{n+1} \geq_{\mathrm{st}} \widetilde{\boldsymbol{\pi}}_n$. Similarly, it can be shown that $\boldsymbol{\pi} \geq_{\mathrm{st}} \widetilde{\boldsymbol{\pi}}_n$ for all $n = 1, 2, \cdots$. Since the denumerable Markov chain $\{X_n\}$ is ergodic, it follows that $\widetilde{\boldsymbol{\pi}}_n$ is stochastically increasing in n and converges to $\boldsymbol{\pi}$ as $n \to \infty$. This is part of the results obtained by Gibson and Seneta (1987a). See Gibson and Seneta (1987b) for other truncation results.

For each $i = 1, 2$, let $\{X_n^i\}$ be a sequence of IID positive, integer-valued random variables, and let N be a discrete random variable independent of $\{X_n^i\}$. Of interest is a comparison between $\sum_{j=1}^N X_j^1$ and $\sum_{j=1}^N X_j^2$. The quantity $\sum_{j=1}^N X_j$ represents a random sum of random variables X_j, where the empty sum is interpreted as zero. Such a random quantity arises in various fields of applied probability. The interested reader should consult Stoyan (1983) for further results (see also Exercise 3.27).

Theorem 3.32 *For the random variables N and $\{X_n^i\}$, $i = 1, 2$, given above, suppose that N is geometrically distributed and that $X_j^1 \geq_{\mathrm{hr}} X_j^2$ for all $j = 1, 2, \cdots$. If either X_j^1 or X_j^2 are* DHR, *then*

$$\sum_{j=1}^N X_j^1 \geq_{\mathrm{st}} \sum_{j=1}^N X_j^2.$$

Proof. Let h_n^i, $n = 1, 2, \cdots$ be the hazard rate function of X^i, and suppose that the random variable N is distributed by

$$P[N = k] = (1-\rho)\rho^k, \quad k = 0, 1, \cdots,$$

where $0 < \rho < 1$. Let $\{Z_n^i\}$ be an absorbing Markov chain with state space $\mathcal{N} = \{0, 1, \cdots\}$, where state 0 is absorbing, and transition matrix

$$\mathbf{P}_i = \begin{pmatrix} 1 & 0 & 0 & 0 & \cdots \\ (1-\rho)h_1^i & \rho h_1^i & 1-h_1^i & 0 & \cdots \\ (1-\rho)h_2^i & \rho h_2^i & 0 & 1-h_2^i & \cdots \\ (1-\rho)h_3^i & \rho h_3^i & 0 & 0 & \cdots \\ \vdots & \vdots & \vdots & \vdots & \ddots \end{pmatrix}. \tag{3.58}$$

It is not difficult to show that the first passage time of $\{Z_n^i\}$ from state 1 to the absorbing state is equal in distribution to $T_i \equiv \sum_{j=1}^N X_j^i$ given that $N \geq 1$ (see Shanthikumar, 1988). Since

$$\mathbf{P}_i\mathbf{U} = \begin{pmatrix} 1 & 0 & 0 & 0 & \cdots \\ 1 & 1-(1-\rho)h_1^i & 1-h_1^i & 0 & \cdots \\ 1 & 1-(1-\rho)h_2^i & 1-h_2^i & 1-h_2^i & \cdots \\ 1 & 1-(1-\rho)h_3^i & 1-h_3^i & 1-h_3^i & \cdots \\ \vdots & \vdots & \vdots & \vdots & \ddots \end{pmatrix},$$

if $X^1 \geq_{\text{hr}} X^2$, i.e., $h_n^1 \leq h_n^2$ for all $n = 1, 2, \cdots$, then $\{Z_n^1\}$ stochastically dominates $\{Z_n^2\}$. Also, if X^i is DHR, then $\{Z_n^i\}$ is stochastically monotone. It follows from Theorem 3.31 that $Z_n^1 \geq_{\text{st}} Z_n^2$ for all $n = 1, 2, \cdots$, provided that $Z_0^1 = Z_0^2 = 1$. Hence

$$P[Z_n^1 = 0 | Z_0^1 = 1] \leq P[Z_n^2 = 0 | Z_0^2 = 1].$$

But, since

$$P[Z_n^i = 0 | Z_0^i = 1] = P[T_i \leq n | N \geq 1],$$

the proof is complete. □

The next result is due to Kijima (1989b).

Corollary 3.8 *In Theorem 3.32, suppose that either X_j^1 or X_j^2 are geometrically distributed. If $X_j^1 \geq_{\text{hr}} X_j^2$ then*

$$\sum_{j=1}^N X_j^1 \geq_{\text{hr}} \sum_{j=1}^N X_j^2.$$

Proof. Suppose that X^1 is geometrically distributed, i.e.,

$$P[X^1 = k] = h(1-h)^{k-1}, \quad k = 1, 2, \cdots,$$

for some $0 < h < 1$. Then $h_n^1 = h$, $n = 1, 2, \cdots$, and

$$P[T_1 = k] = \begin{cases} 1-\rho, & k = 0, \\ (1-\rho)h(1-h+\rho h)^{k-1}, & k = 1, 2, \cdots, \end{cases}$$

where $T_1 = \sum_{j=1}^{N} X_j^1$, so that its hazard rate function H_n^1 is given by

$$H_n^1 = \begin{cases} 1-\rho, & n = 0, \\ (1-\rho)h, & n = 1, 2, \cdots. \end{cases}$$

Now consider the first passage time of $\{Z_n^2\}$ from state 1 to state 0, where $\{Z_n^2\}$ is an absorbing Markov chain with transition matrix (3.58). Then, denoting the hazard rate function of the first passage time by H_n^2, we have

$$H_n^2 = \sum_{k=1}^{\infty}(1-\rho)h_k^2 P_1[Z_{n-1}^2 = k | Z_{n-1}^2 \geq 1], \quad n = 1, 2, \cdots.$$

But, since $h_n^2 \geq h_n^1 = h$ by assumption, it follows that

$$H_n^2 \geq (1-\rho)h \sum_{k=1}^{\infty} P_1[Z_{n-1}^2 = k | Z_{n-1}^2 \geq 1] = (1-\rho)h = H_n^1,$$

whence the result. The other case can be proved similarly. □

Example 3.20 In the M/G/1 queue considered in Example 3.2, let X be the generic random variable distributed by

$$G_R(t) = \mu \int_0^t \overline{G}(x)dx, \quad t \geq 0,$$

where $\overline{G}(x) = 1 - G(x)$. The hazard rate function of X is given by

$$h_R(t) = \frac{\overline{G}(t)}{\int_t^\infty \overline{G}(x)dx}, \quad t \geq 0.$$

We note that Theorem 3.32 and Corollary 3.8 hold even when the X_js are continuous random variables. This fact can be verified by a routine limiting argument. Hence, denoting the stationary waiting time of the M/G/1 queue by W, if $h_R(t) \geq \mu'$ for some μ' with $\rho' = \lambda/\mu' < 1$, then we have from Theorem 3.32 that

$$P[W \leq t] \geq 1 - \rho' e^{-\mu'(1-\rho')t}, \quad t \geq 0, \tag{3.59}$$

while from Corollary 3.8 the hazard rate function of W is bounded by $\mu'(1-\rho')$ from below. The reversed inequality holds if $h_R(t) \leq \mu'$.

We note that $\int_t^\infty \overline{G}(x)dx/\overline{G}(t)$ represents the conditional mean remaining service time when the elapsed service time is t. If this is shorter than the (unconditional) mean service time then the service time distribution is called *new better than used in expectation* (NBUE). If this is the case, we have $h_R(t) \geq \mu$ and so the lower bound (3.59) on the distribution function of W holds true. If the conditional mean remaining time is longer than the mean service time, then the service time distribution is called *new worse than used in expectation* (NWUE) and we have an upper bound on the distribution function of W with $\mu' = \mu$. See Stoyan (page 82, 1983) for another derivation of this result. Corollary 3.8 reveals more. That is, if the service time distribution is NBUE (NWUE, respectively) in an M/G/1 queue, then the hazard rate function of the stationary waiting time is bounded by $(\mu - \lambda)$ from below (above), where μ and λ are the service and arrival rates respectively.

3.10 Exercises

Exercise 3.1 Complete the proof of Theorem 3.3.

Exercise 3.2 Prove Theorem 3.4.

Exercise 3.3 Let X be a discrete random variable with probability vector $\mathbf{a} = (a_i)$. Prove that if $X \in \text{DHR}$ then $a_i > 0$ for all i. (Hence a_i is strictly decreasing in i.)

Exercise 3.4 Let X be a discrete random variable with probability vector $\mathbf{a} = (a_i)$ defined on $\{0, 1, \cdots, n\}$. We call $X \in \text{IRHR}$ (increasing reversed hazard rate) if

$$A_{i+1}^2 \le A_i A_{i+1}, \quad i = 0, 1, \cdots, n-2,$$

where $A_i = \sum_{k=0}^{i} a_k$. Show that $X \in \text{IRHR}$ if and only if $r_i = a_i/A_i$ is increasing in $i = 1, 2, \cdots, n$. Also prove that $X \in \text{IRHR}$ implies $X \in \text{IHR}$.

Exercise 3.5 Prove CM $\subset$ ILR by using the Schwarz inequality. Also, prove ILR $\subset$ DHR by using the matrix

$$\mathbf{A} = \begin{pmatrix} a_0 & a_1 & a_2 & \cdots \\ a_1 & a_2 & a_3 & \cdots \end{pmatrix}.$$

Exercise 3.6 Obtain the hazard rate and reversed hazard rate functions of the binomial, negative binomial, and Poisson distributions.

Exercise 3.7 Let X and Y be discrete random variables on $\mathcal{Z}_+$ with probability vectors $\mathbf{a} = (a_i)$ and $\mathbf{b} = (b_i)$ respectively. Suppose that $a_i,\ b_i > 0$ for all $i \in \mathcal{Z}_+$. Show that $X \ge_{\text{lr}} Y$ ($X \ge_{\text{hr}} Y$ or $X \ge_{\text{rh}} Y$, respectively) if and only if their respective likelihood ratio functions (hazard rate functions or reversed hazard rate functions) are ordered as $\ell_i^X \ge \ell_i^Y$ ($h_i^X \le h_i^Y$ or $r_i^X \ge r_i^Y$) for all $i \in \mathcal{Z}_+$.

Exercise 3.8 Let $\succ$ represent each of $\ge_{\text{lr}}$, $\ge_{\text{hr}}$, $\ge_{\text{rh}}$, or $\ge_{\text{st}}$. Show that if $X \succ Y$ then $\max\{X, c\} \succ \max\{Y, c\}$ and $\min\{X, c\} \succ \min\{Y, c\}$ for any positive number c.

Exercise 3.9 For discrete random variables X and Y defined on $\mathcal{Z}$, prove that $X \ge_{\text{hr}} Y$ if and only if $-Y \ge_{\text{rh}} -X$.

Exercise 3.10 Show that the following families of discrete distributions are ordered in the sense of likelihood ratio ordering: geometric with α, binomial with p, negative binomial with p, and Poisson with λ. (See (3.19) and Examples 3.5–3.7.)

Exercise 3.11 Mimicking the proof of Theorem 3.10, show that the ordering $\geq_{st}$ satisfies the convolution property (C).

Exercise 3.12 For a sequence of discrete IID random variables $\{X_n\}$, define $M_k = \max\{X_1, \cdots, X_k\}$ and $m_k = \min\{X_1, \cdots, X_k\}$. Compare M_k with M_{k+1}, m_k with m_{k+1}, and M_k with m_k in some appropriate stochastic ordering relations.

Exercise 3.13 Complete the proof of Theorem 3.11(iii).

Exercise 3.14 Prove that a transition matrix $\mathbf{P}$ is stochastically monotone if and only if the vector $\mathbf{Pf}$ is increasing for any increasing real vector $\mathbf{f}$.

Exercise 3.15 Let τ_n denote the arrival time of the nth customer to a single-server queue, and define $U_n = \tau_n - \tau_{n-1}$, $n = 1, 2, \cdots$, where $\tau_0 \equiv 0$. Let S_n be the service time of the nth customer. Write $\xi_n = S_n - U_{n+1}$, and let W_n be the waiting time of the nth customer. If the service is first-come-first-served, we have the elementary relationship

$$W_{n+1} = [W_n + \xi_n]_+, \quad n = 0, 1, \cdots,$$

where $[x]_+ = \max\{0, x\}$ (see, e.g., Kleinrock, page 277, 1975). Suppose that ξ_n are IID integer-valued random variables. Prove that $\{W_n\}$ is a Markov chain with state space $\mathcal{Z}_+$ and determine its transition matrix. Also show that the waiting-time process $\{W_n\}$ is stochastically monotone (Keilson and Kester, 1977).

Exercise 3.16 Let $p(n,t)$ be defined as in Example 3.14 for $n = 1, 2, \cdots$. For $\varepsilon > 0$ sufficiently small, let $p(0,t) = \varepsilon e^{-\varepsilon t}$ so that $p(0,t) \to \delta(t)$ as $\varepsilon \to 0$. Prove that $p(n,t)$ is TP_2 in $n = 0, 1, \cdots$ and $t \geq 0$.

Exercise 3.17 For $\mathbf{P} \in \mathcal{S}$, prove that if $\mathbf{P} \in TP_3$ then both $\mathbf{U}^{-1}\mathbf{P}\mathbf{U}$ and $\mathbf{V}^{-1}\mathbf{P}\mathbf{V}$ are TP_2.

Exercise 3.18 Prove Lemma 3.9.

Exercise 3.19 Let $s_\alpha(n)$ denote the separation of an ergodic Markov chain $\{X_n\}$ with initial distribution $\boldsymbol{\alpha}$. Let $\boldsymbol{\pi}$ be the stationary distribution of $\{X_n\}$ and suppose that $\boldsymbol{\alpha}^\top \boldsymbol{\pi}_D^{-1}$ is increasing. Identify an absorbing Markov chain by which $s_\alpha(n)$ can be written as in (3.40).

Exercise 3.20 Let T be a random variable generated by a phase-type distribution $(\boldsymbol{\alpha}, \mathbf{T})$. Suppose that $\mathbf{TV} \in TP_2$ and $\mathbf{V}^{-1}\mathbf{TV} \geq \mathbf{O}$. Prove that if $\boldsymbol{\alpha} = \boldsymbol{\delta}_1$ then $T \in$ IHR, while if $\boldsymbol{\alpha} = \boldsymbol{\delta}_N$ then $T \in$ DHR.

Exercise 3.21 For a Markov chain $\{X_n\}$ with state space $\mathcal{Z}_+$ and transition matrix $\mathbf{P}$, let $T_N = \inf\{n \geq 1 : X_n \geq N\}$, i.e., T_N is the first passage time to the upper set $\{N, N+1, \cdots\}$. Prove that if $\mathbf{P} \in \mathcal{M}_{st}$ and $\{X_n\}$ has an increasing path then T_N is IHRA for all N (Brown and Chaganty, 1983).

Exercise 3.22 Under the conditions of Theorem 3.23, suppose that the dual of $\mathbf{P}_1$ is stochastically monotone. Prove that $\mathbf{q} \geq_{\text{lr}} \boldsymbol{\pi}_1$ (Kijima, 1995).

Exercise 3.23 For nonnegative, discrete random variables X and Y with probability vectors $\mathbf{a}$ and $\mathbf{b}$, respectively, suppose that

$$E[X - t|X > t] \geq E[Y - t|Y > t], \quad t = 0, 1, \cdots.$$

Represent this inequality in terms of $\mathbf{a}$, $\mathbf{b}$ and the matrix $\mathbf{U}$. What if $E[t - X|X \leq t] \geq E[t - Y|Y \leq t]$ for each t?

Exercise 3.24 Prove that if $\mathbf{a} \in$ DLR then $\mathbf{f}_n$ defined in (3.54) is PF_2.

Exercise 3.25 In a phase-type renewal process generated by $(\boldsymbol{\delta}_N, \mathbf{T})$ where N is the number of phases, suppose that $\mathbf{V}^{-1}\mathbf{r} \geq \mathbf{0}$ and that $\mathbf{T} + \mathbf{r}\boldsymbol{\delta}_N^{\mathsf{T}} \in \mathcal{M}_{\text{st}}$. Here $\mathbf{r} = \mathbf{1} - \mathbf{T1}$ as before. Prove that the renewal probability p_n is increasing in n.

Exercise 3.26 Let M and N be nonnegative, discrete random variables, and let $\{X_n\}$ be a sequence of IID random variables independent of M and N. Prove that if $N \geq_{\text{hr}} M$ ($N \geq_{\text{rh}} M$, respectively) and $X_i \in$ IHR ($X_i \in$ DRHR), then $\sum_{i=1}^N X_i \geq_{\text{hr}} \sum_{i=1}^M X_i$ ($\sum_{i=1}^N X_i \geq_{\text{rh}} \sum_{i=1}^M X_i$).

Exercise 3.27 Let N (M, respectively) be a nonnegative, discrete random variable, and let $\{X_n\}$ ($\{Y_n\}$) be a sequence of IID random variables. Suppose that N and $\{X_n\}$ (M and $\{Y_n\}$) are mutually independent and consider $\sum_{i=1}^N X_i$ ($\sum_{i=1}^M Y_i$). Prove that if $N \geq_{\text{st}} M$ and $X_i \geq_{\text{st}} Y_i$ then $\sum_{i=1}^N X_i \geq_{\text{st}} \sum_{i=1}^M Y_i$.

Exercise 3.28 A *portfolio selection problem* can be formulated as follows. For random variables X and Y representing future returns of financial assets, consider the maximization problem

$$\max_{k \in R} E[u(kX + (1 - k)Y)],$$

where u stands for the utility function of an investor and k is the fraction invested in the asset X. In the economics literature, one is often interested in the conditions under which the demand for X exceeds the demand for Y, i.e., when the optimal fraction, denoted by k^*, exceeds 1/2. Suppose that the utility function u is strictly concave and let

$$\phi(k) = E[u(kX + (1 - k)Y)], \quad k \in R.$$

Assume that interchange of differentiation and expectation is permissible.

(i) Prove that $k^* \geq 1/2$ if and only if $\phi'(1/2) \geq 0$.

(ii) Suppose that X and Y are mutually independent and $X \geq_{\text{st}} Y$. Prove that, using Theorem 3.11, if $x\,u'(x + y)$ is increasing in x for any y then $k^* \geq 1/2$ (Kijima and Ohnishi, 1996).

4

Continuous-time Markov chains

In this chapter, we consider the continuous-time analogs of discrete-time Markov chains. As in the discrete-time case, they are characterized by the Markov property that, given the present state, the future of the process is stochastically independent of the past.

4.1 Transition probability functions

Let $\mathcal{T} = [0, \infty)$ be the index set and consider a continuous-time stochastic process $\{X(t),\ t \in \mathcal{T}\}$ taking values on $\mathcal{N} = \{0, 1, 2, \cdots\}$. We say that the process $\{X(t)\}$ is a Markov chain in continuous time if, for each $s \geq 0,\ t > 0$ and each set A,

$$P[X(t+s) \in A | X(u),\, 0 \leq u \leq s] = P[X(t+s) \in A | X(s)];$$

see (1.5). More precisely, if, for each $s \geq 0,\ t > 0$, each $i,\ j \in \mathcal{N}$, and every history $x(u),\ 0 \leq u < s$,

$$\begin{aligned} & P[X(t+s) = j | X(s) = i,\ X(u) = x(u),\ 0 \leq u < s] \\ = \ & P[X(t+s) = j | X(s) = i], \end{aligned} \tag{4.1}$$

then the process $\{X(t)\}$ is called a *Markov chain in continuous time* or a *continuous-time Markov chain*. In other words, a continuous-time Markov chain is a stochastic process having the *Markov property* (4.1) that the conditional distribution of the future state, given the present state and all past states, depends only on the present state and is independent of the past.

Let $\{X(t)\}$ be a continuous-time Markov chain and define

$$p_{ij}(s,t) = P[X(t) = j | X(s) = i], \quad 0 \leq s < t.$$

The conditional probability $p_{ij}(s,t)$ is called the *transition probability function* from state i to state j and the matrix $\mathbf{P}(s,t) = (p_{ij}(s,t))$ is called the *transition matrix function*. For any fixed $0 \le s < t$, we assume that the matrix $\mathbf{P}(s,t)$ is stochastic, i.e., $\mathbf{P}(s,t) \ge \mathbf{O}$ and $\mathbf{P}(s,t)\mathbf{1} = \mathbf{1}$.

For $0 \le s < u < t$, by virtue of the equation

$$\begin{aligned} P[X(t) = j|X(s) = i] \\ &= \sum_k P[X(u) = k,\ X(t) = j|X(s) = i] \\ &= \sum_k P[X(u) = k|X(s) = i]\, P[X(t) = j|X(s) = i,\ X(u) = k] \end{aligned}$$

and the fact that, due to the Markov property

$$P[X(t) = j|X(s) = i,\ X(u) = k] = P[X(t) = j|X(u) = k],$$

we must have

$$p_{ij}(s,t) = \sum_k p_{ik}(s,u)\, p_{kj}(u,t), \quad 0 \le s < u < t. \tag{4.2}$$

In matrix notation, this is written as

$$\mathbf{P}(s,t) = \mathbf{P}(s,u)\,\mathbf{P}(u,t), \quad 0 \le s < u < t.$$

The above equation should be compared with its discrete-time counterpart (1.16) or (1.17). Equation (4.2) is called the *Chapman–Kolmogorov equation*.

When the transition probability functions $p_{ij}(s,t)$ depend only on the difference $t - s$, i.e.,

$$p_{ij}(t-s) = P[X(t) = j|X(s) = i], \quad 0 \le s < t,$$

for all $i,\ j \in \mathcal{N}$, the continuous-time Markov chain $\{X(t)\}$ is said to be *homogeneous* (*nonhomogeneous*, otherwise). For any homogeneous Markov chain, the Chapman–Kolmogorov equation (4.2) is expressed as

$$p_{ij}(s+t) = \sum_k p_{ik}(s)\, p_{kj}(t), \quad s,\ t > 0, \tag{4.3}$$

or, in matrix form,

$$\mathbf{P}(s+t) = \mathbf{P}(s)\,\mathbf{P}(t), \quad s,\ t > 0, \tag{4.4}$$

where $\mathbf{P}(t) = (p_{ij}(t))$ which satisfies $\mathbf{P}(t-s) = \mathbf{P}(s,t)$ for $t > s \ge 0$. From (4.3) and the fact that $\sum_j p_{kj}(t) \le 1$, we have

$$\sum_j p_{ij}(s+t) = \sum_{j,k} p_{ik}(s)\, p_{kj}(t) = \sum_k p_{ik}(s) \sum_j p_{kj}(t) \le \sum_k p_{ik}(s),$$

so that $\sum_j p_{ij}(t)$ is nonincreasing in t. Hence, if $\mathbf{P}(t)$ is stochastic then so is $\mathbf{P}(s)$ for all $s \le t$. For $s > t$, choose n so that $t > s/n$. Then, since $\mathbf{P}(s) = \mathbf{P}^n(s/n)$ from (4.4) and since $\mathbf{P}(t)\mathbf{1}$ is nonincreasing in t, we conclude that $\mathbf{P}(s)$ is also stochastic. Hence, if $\mathbf{P}(t) \in \mathcal{S}$ for some $t > 0$

then $\mathbf{P}(t) \in \mathcal{S}$ for all $t \geq 0$. In what follows, we consider the homogeneous case only (Exercise 4.2 treats the nonhomogeneous case) and assume that the transition matrix function $\mathbf{P}(t)$ is stochastic for all $t \geq 0$. Also, it is assumed throughout that every transition probability function $p_{ij}(t)$ is continuous in $t > 0$.*

Every transition probability function of a homogeneous Markov chain has the following property.

Theorem 4.1 *Let* $\mathbf{P}(t) = (p_{ij}(t))$ *be the transition matrix function of a continuous-time Markov chain. Then, for each* $h > 0$,

$$\sum_j |p_{ij}(t+h) - p_{ij}(t)|$$

is nonincreasing in t.

Proof. Let $0 < s < t$. From (4.3), we have

$$p_{ij}(t+h) - p_{ij}(t) = \sum_k \{p_{ik}(s+h) - p_{ik}(s)\} p_{kj}(t-s)$$

so that

$$\begin{aligned}\sum_j |p_{ij}(t+h) - p_{ij}(t)| &\leq \sum_k |p_{ik}(s+h) - p_{ik}(s)| \sum_j p_{kj}(t-s) \\ &= \sum_k |p_{ik}(s+h) - p_{ik}(s)|,\end{aligned}$$

where the equality follows since $\mathbf{P}(t)$ is stochastic. □

Definition 4.1 A transition matrix function $\mathbf{P}(t) = (p_{ij}(t))$ is called *standard* if, for every i and j, $\lim_{t \to 0+} p_{ij}(t) = \delta_{ij}$ or, equivalently,

$$\lim_{t \to 0+} \mathbf{P}(t) = \mathbf{I}.$$

Henceforth, we assume that the transition matrix function under consideration is standard unless stated otherwise.

Lemma 4.1 *For every* i, *the transition probability function* $p_{ii}(t)$ *is positive for all* $t \geq 0$. *If* $p_{ij}(s) > 0$ *then* $p_{ij}(t) > 0$ *for all* $t \geq s$.

Proof. From the Chapman–Kolmogorov equation (4.3), we have

$$p_{ii}(t) = \sum_k p_{ik}(t/2)\, p_{ki}(t/2) \geq [p_{ii}(t/2)]^2, \quad t > 0,$$

and, hence, in general

$$p_{ii}(t) \geq [p_{ii}(t/n)]^n, \quad n = 1, 2, \cdots.$$

* The continuity of $p_{ij}(t)$ is ensured by the measurability of $p_{ij}(t)$ in the sense of Lebesgue. See, e.g., Chung (page 120, 1967) for details.

Since $\lim_{t\to 0+} p_{ii}(t) = 1$, it follows that $p_{ii}(t) > 0$ for all $t \geq 0$. Further, again from (4.3),

$$p_{ij}(t) \geq p_{ij}(s)\,p_{jj}(t-s), \quad 0 < s < t.$$

Since $p_{jj}(t-s) > 0$, we conclude that if $p_{ij}(s) > 0$ then so is $p_{ij}(t)$ for any $t \geq s$. □

In fact, Lemma 4.1 can be sharpened into the following fundamental result, called Lévy's theorem. We state it without proof. The interested reader may consult, e.g., Freedman (Section 2.3, 1972) for the proof.

Theorem 4.2 *For every i and j, the transition probability function $p_{ij}(t)$ is either identically zero or positive for all $t > 0$.*

An important consequence of Theorem 4.2 is that there is no periodicity in continuous-time Markov chains. Another important result regarding the transition probability function is the following.

Theorem 4.3 *For every i and j, we have*

$$|p_{ij}(t+h) - p_{ij}(t)| \leq 1 - p_{ii}(h), \quad h > 0, \ t \geq 0.$$

Proof. From (4.3), we have

$$\begin{aligned} p_{ij}(t+h) - p_{ij}(t) &= \sum_k p_{ik}(h)\,p_{kj}(t) - p_{ij}(t) \\ &= -\{1 - p_{ii}(h)\}p_{ij}(t) + \sum_{k\neq i} p_{ik}(h)\,p_{kj}(t). \end{aligned}$$

Since $\mathbf{P}(t) = (p_{ij}(t))$ is stochastic,

$$0 \leq \sum_{k\neq i} p_{ik}(h)\,p_{kj}(t) \leq \sum_{k\neq i} p_{ik}(h) = 1 - p_{ii}(h).$$

If $p_{ij}(t+h) - p_{ij}(t) \geq 0$ then

$$0 \leq p_{ij}(t+h) - p_{ij}(t) \leq 1 - p_{ii}(h).$$

If it is nonpositive then

$$0 \geq p_{ij}(t+h) - p_{ij}(t) \geq -\{1 - p_{ii}(h)\}p_{ij}(t) \geq -\{1 - p_{ii}(h)\}.$$

Combining the two inequalities, the proof is complete. □

A consequence of Theorem 4.3 is the following.

Corollary 4.1 *For every i and j, $p_{ij}(t)$ is uniformly continuous in $t \geq 0$.*

We close this section with two illustrative examples. Other interesting examples can be found in, e.g., Çinlar (1975), Karlin and Taylor (1975, 1981) and Ross (1983).

Example 4.1 (Poisson process) Let $\{N(t)\}$ be a stochastic process with the properties that its sample paths are nondecreasing in t, increase by jumps of unit magnitude, are right-continuous, $N(0) = 0$ with probability one, and:

(i) for every t, $s \geq 0$, the distribution of the *increment* $N(t+s) - N(t)$ is independent of the past $\{N(u),\ u \leq t\}$;

(ii) for every t, $s \geq 0$, the distribution of the increment $N(t+s) - N(t)$ is independent of t;

(iii) $P[N(h) \geq 2] = o(h)$,

where $\lim_{h \to 0+} o(h)/h = 0$. Such a process is called a *Poisson process*. The random quantity $N(t)$ counts the number of events occurring in time $[0, t]$ (see Section 3.8 for renewal processes). A point in time at which a unit jump occurs is called an *arrival time*. It is readily seen that

$$\begin{aligned} & P[N(t+s) - N(t) = j \mid N(u),\ u \leq t] \\ = \ & P[N(t+s) - N(t) = j] \\ = \ & P[N(s) = j], \end{aligned}$$

where the first equality follows from property (i) and the second from (ii). It follows that the process $\{N(t)\}$ is a homogeneous Markov chain in continuous time with transition probability functions

$$p_{0j}(t) = P[N(t) = j], \quad j = 0, 1, \cdots.$$

Let $0 < z < 1$ and define

$$g(t, z) = E[z^{N(t)}] = \sum_{n=0}^{\infty} z^n P[N(t) = n];$$

see Appendix B.1. Writing

$$N(t+s) = N(t) + \{N(t+s) - N(t)\}, \quad t,\ s > 0,$$

we then have

$$g(t+s, z) = E[z^{N(t)}]\, E[z^{N(t+s)-N(t)}] = g(t, z)\, g(s, z), \tag{4.5}$$

where the first equality follows from property (i) and the second from (ii). It is well known that the solution to (4.5) is either identically zero or $g(t, z)$ must have the form

$$g(t, z) = e^{t\gamma(z)}, \quad t \geq 0; \quad 0 < z < 1. \tag{4.6}$$

Meantime, assume that $g(t, z) > 0$ for all $t \geq 0$. Note that $\gamma(z)$ is then given by the derivative of $g(t, z)$ at $t = 0$, i.e.,

$$\gamma(z) = \lim_{t \to 0+} \frac{g(t, z) - 1}{t}.$$

Now, since the identity

$$\{N(t+s)=0\}=\{N(t)=0,\ N(t+s)-N(t)=0\}$$

holds, we have

$$\begin{aligned} P[N(t+s)=0] &= P[N(t)=0]\,P[N(t+s)-N(t)=0] \\ &= P[N(t)=0]\,P[N(s)=0]. \end{aligned}$$

Hence the function $P[N(t)=0]$ satisfies the same equation as (4.5). This means that $P[N(t)=0]$ is either identically zero or has the form (4.6). Suppose $P[N(t)=0]=0$ for all $t>0$. Then, for any $t>0$, we must have $N(t+s)-N(t)\geq 1$ for all $s>0$. This in turn implies that, for any n,

$$N(t)=N(t_1)+\{N(t_2)-N(t_1)\}+\cdots+\{N(t)-N(t_{n-1})\}\geq n,$$

where $t_1<t_2<\cdots<t_{n-1}<t$. Hence, if $P[N(t)=0]$ is identically zero, then $N(t)=\infty$ for all $t>0$ with probability one. Therefore, we must have $P[N(t)=0]>0$ for some $t>0$ so that, as in (4.6),

$$P[N(t)=0]=e^{-\lambda t},\quad t\geq 0, \tag{4.7}$$

for some constant $\lambda>0$. ($\lambda=0$ corresponds to the degenerate case.) The parameter λ is called the *intensity* of the Poisson process $\{N(t)\}$. Observe that

$$g(t,z)\geq P[N(t)=0]>0,\quad t\geq 0;\quad 0<z<1,$$

and that

$$\lim_{t\to 0+}\frac{P[N(t)=0]-1}{t}=-\lambda.$$

Further, property (iii) implies that

$$\lim_{t\to 0+}\frac{P[N(t)\geq 2]}{t}=0,$$

so that

$$\lim_{t\to 0+}\frac{P[N(t)=1]}{t}=\lim_{t\to 0+}\frac{1-P[N(t)=0]}{t}-\lim_{t\to 0+}\frac{P[N(t)\geq 2]}{t}=\lambda.$$

It follows that, for $0<z<1$,

$$\begin{aligned} \gamma(z) &= \lim_{t\to 0+}\frac{P[N(t)=0]-1}{t}+\lim_{t\to 0+}\frac{z\,P[N(t)=1]}{t} \\ &\quad +\lim_{t\to 0+}\frac{\sum_{n=2}^{\infty}z^n P[N(t)=n]}{t} \\ &= -\lambda+\lambda z. \end{aligned}$$

Thus, from (4.6),

$$g(t,z)=e^{-\lambda t+\lambda tz}=e^{-\lambda t}\sum_{n=0}^{\infty}\frac{(\lambda tz)^n}{n!}=\sum_{n=0}^{\infty}\frac{(\lambda t)^n}{n!}e^{-\lambda t}z^n,\quad 0<z<1,$$

which, together with Corollary B.2, implies that

$$P[N(t)=n]=\frac{(\lambda t)^n}{n!}\,e^{-\lambda t},\quad n=0,1,\cdots.$$

Therefore, the random variable $N(t)$ follows a Poisson distribution with parameter λt. Note that, from (4.7), the interarrival times of successive events are exponentially distributed with parameter λ. Recall that this fact was used in Example 1.5. See Exercises 4.3 and 4.4 for related results.

Example 4.2 (Subordinated Markov chain) Let $\{X_n\}$ be a discrete-time Markov chain with state space $\mathcal{N}$ and transition matrix $\mathbf{S}=(s_{ij})$, and let $\{N(t)\}$ be a Poisson process with intensity λ independent of $\{X_n\}$. Consider a system that moves from one state in $\mathcal{N}$ to another in such a way that the successive states visited form a Markov chain $\{X_n\}$ and the times at which the system changes its state form a Poisson process $\{N(t)\}$. Let $Y(t)$ represent the state of the system at time t. Then

$$Y(t)=X_{N(t)},\quad t\ge 0. \tag{4.8}$$

The process $\{Y(t)\}$ is called a *subordinated Markov chain*, which is indeed a Markov chain in continuous time. To see this, we have from (4.8) that

$$\begin{aligned}
&P[Y(t+s)=j|Y(s)=i,\,Y(u)=y(u),\,0\le u<s]\\
&=\sum_{n=0}^{\infty}\frac{(\lambda t)^n}{n!}\,e^{-\lambda t}P[Y(t+s)=j|Y(s)=i,\,Y(u)=y(u),\,0\le u<s;\\
&\qquad\qquad N(t+s)-N(s)=n]\\
&=\sum_{n=0}^{\infty}\frac{(\lambda t)^n}{n!}\,e^{-\lambda t}P[X_{N(s)+n}=j|X_{N(s)}=i,\,N(t+s)-N(s)=n].
\end{aligned}$$

But, under the assumptions, the conditional probability of the last term in the above equation is just equal to $P[X_n=j|X_0=i]$. It follows that the subordinated process $\{Y(t)\}$ is a continuous-time Markov chain with transition probability function

$$p_{ij}(t)=\sum_{n=0}^{\infty}\frac{(\lambda t)^n}{n!}\,e^{-\lambda t}s_{ij}(n),\quad t\ge 0, \tag{4.9}$$

where $\mathbf{S}^n=(s_{ij}(n))$.

4.2 Finite Markov chains in continuous time

In this section, we consider a continuous-time Markov chain $\{X(t)\}$ on a finite state space, $\mathcal{N}=\{0,1,\cdots,N\}$, say. Let $\mathbf{P}(t)$ be the transition matrix function of $\{X(t)\}$. From Corollary 4.1, each $p_{ij}(t)$ is uniformly continuous in $t\ge 0$. Moreover, $p_{ij}(t)$ is differentiable with respect to $t\ge 0$, as we shall see below. (The derivative $p'_{ij}(0+)$ is the right-hand derivative at $t=0$.)

We say that the matrix $\mathbf{P}(t) = (p_{ij}(t))$ is differentiable with respect to t if each component $p_{ij}(t)$ is differentiable with respect to t. Also, $\int \mathbf{P}(t)dt$ means componentwise integral. The proof of the next theorem is taken from Iosifescu (1980).

Lemma 4.2 *There exists some $h > 0$ such that $\int_0^h \mathbf{P}(t)dt$ is nonsingular.*

Proof. Since $\lim_{h\to 0} p_{ij}(h) = \delta_{ij}$, Corollary 4.1 ensures the existence of $h > 0$ such that the matrix $\widehat{\mathbf{P}} = (p_{ij}(u_{ij}))$ is nonsingular whenever $0 \le u_{ij} \le h$. On the other hand, by the mean value theorem (see, e.g., Bartle, page 230, 1976), we can find u_{ij}s such that $0 \le u_{ij} \le h$ and

$$\int_0^h \mathbf{P}(u)du = h\widehat{\mathbf{P}},$$

which is nonsingular. □

Theorem 4.4 *The transition matrix function $\mathbf{P}(t)$ is differentiable with respect to $t > 0$.*

Proof. From the Chapman–Kolmogorov equation (4.4), we have

$$\begin{aligned}\int_t^{t+h} \mathbf{P}(u)du &= \int_0^h \mathbf{P}(t+u)du \\ &= \mathbf{P}(t)\left(\int_0^h \mathbf{P}(u)du\right) \\ &= \left(\int_0^h \mathbf{P}(u)du\right)\mathbf{P}(t).\end{aligned}$$

Fix $h > 0$ so that $\int_0^h \mathbf{P}(u)du$ is nonsingular. Then

$$\begin{aligned}\mathbf{P}(t) &= \int_t^{t+h} \mathbf{P}(u)du \left(\int_0^h \mathbf{P}(u)du\right)^{-1} \\ &= \left(\int_0^h \mathbf{P}(u)du\right)^{-1} \int_t^{t+h} \mathbf{P}(u)du.\end{aligned}$$

Since $\mathbf{P}(t)$ is continuous in t, $\int_t^{t+h} \mathbf{P}(u)du$ is differentiable with respect to $t > 0$, whence the theorem. □

From the proof of Theorem 4.4, the derivative of $\mathbf{P}(t)$ is given by

$$\begin{aligned}\mathbf{P}'(t) &= \{\mathbf{P}(t+h) - \mathbf{P}(t)\}\left(\int_0^h \mathbf{P}(u)du\right)^{-1} \\ &= \left(\int_0^h \mathbf{P}(u)du\right)^{-1} \{\mathbf{P}(t+h) - \mathbf{P}(t)\}, \qquad (4.10)\end{aligned}$$

which, in fact, shows that $\mathbf{P}(t)$ is infinitely differentiable with respect to $t > 0$.

Let us define

$$\mathbf{Q} \equiv \mathbf{P}'(0+).$$

The matrix $\mathbf{Q} = (q_{ij})$ is called the *infinitesimal generator*, or *generator** for short, which is of fundamental importance in the theory of continuous-time Markov chains. Since $\mathbf{P}(0) = \mathbf{I}$, we have

$$q_{ij} = \begin{cases} \lim_{h \to 0+} \dfrac{p_{ij}(h)}{h} \geq 0, & i \neq j, \\ \lim_{h \to 0+} \dfrac{p_{ii}(h) - 1}{h} \leq 0, & i = j, \end{cases} \tag{4.11}$$

or, in matrix form,

$$\mathbf{Q} = \lim_{h \to 0+} \frac{\mathbf{P}(h) - \mathbf{I}}{h}. \tag{4.12}$$

Also, since for any $h \geq 0$

$$1 - p_{ii}(h) = \sum_{j \neq i} p_{ij}(h),$$

dividing both sides by h and letting h decrease to zero yields the relation

$$-q_{ii} = \sum_{j \neq i} q_{ij}, \quad i \in \mathcal{N}. \tag{4.13}$$

Hence we shall define $q_i \equiv \sum_{j \neq i} q_{ij} = -q_{ii}$.

Interpretations of q_i and q_{ij}, $i \neq j$, are as follows. From (4.11), we have

$$q_i = \lim_{h \to 0+} \frac{1 - p_{ii}(h)}{h} = \lim_{h \to 0+} \frac{P[X(t+h) \neq i | X(t) = i]}{h},$$

so that

$$P[X(t+h) \neq i | X(t) = i] = q_i h + o(h).$$

Hence, $q_i h + o(h)$ is the conditional probability of no longer being in state i at time $t + h$ given that the process was in state i at time t. That is, the quantity q_i is the *intensity* of passage from state i. On the other hand, for $i \neq j$, (4.11) implies

$$q_{ij} = \lim_{h \to 0+} \frac{p_{ij}(h)}{h} = \lim_{h \to 0+} \frac{P[X(t+h) = j | X(t) = i]}{h},$$

so that

$$P[X(t+h) = j | X(t) = i] = q_{ij} h + o(h).$$

* Some authors, such as Anderson (1991), call the generator $\mathbf{Q}$ the Q-matrix.

It follows that q_{ij} is the intensity of transition from state i to state j. Note that if $q_i > 0$, then

$$\frac{q_{ij}}{q_i} = \lim_{h\to 0+} \frac{P[X(t+h)=j|X(t)=i]}{P[X(t+h)\neq i|X(t)=i]}. \tag{4.14}$$

Therefore, q_{ij}/q_i can be interpreted as the conditional probability of a transition from state i to state j given that a transition from state i has taken place. We shall return to this characterization shortly.

From (4.11) or (4.12), the infinitesimal generator $\mathbf{Q}$ has the following properties.

Theorem 4.5 *Let* $\mathbf{Q}$ *be the infinitesimal generator of a finite Markov chain. Then* $\mathbf{Q}$ *is finite componentwise, the diagonal elements of* $\mathbf{Q}$ *are nonpositive, the off-diagonal elements are nonnegative, and the row sums of* $\mathbf{Q}$ *are all zero, i.e.,* $\mathbf{Q1} = \mathbf{0}$.

Referring to the Chapman–Kolmogorov equation (4.4), we have

$$\frac{\mathbf{P}(t+h)-\mathbf{P}(t)}{h} = \frac{\mathbf{P}(h)-\mathbf{I}}{h}\mathbf{P}(t) = \mathbf{P}(t)\frac{\mathbf{P}(h)-\mathbf{I}}{h}, \quad h>0.$$

Since $\mathbf{P}(t)$ is differentiable, it follows from (4.12) that

$$\mathbf{P}'(t) = \mathbf{Q}\,\mathbf{P}(t); \quad \mathbf{P}'(t) = \mathbf{P}(t)\mathbf{Q}, \quad t \geq 0, \tag{4.15}$$

which are the systems of ordinary linear differential equations. The former is known as the *backward Kolmogorov equation* and the latter as the *forward Kolmogorov equation*. The unique solution to (4.15) under the initial condition $\mathbf{P}(0) = \mathbf{I}$ is given by

$$\mathbf{P}(t) = \exp\{\mathbf{Q}\,t\} = \sum_{n=0}^{\infty} \frac{\mathbf{Q}^n t^n}{n!}, \quad t \geq 0, \tag{4.16}$$

the proof of which is left to the reader (see Exercise 4.8).

For $\mathrm{Re}\,(s) > 0$, let $\pi_{ij}(s) = \int_0^\infty e^{-st} p_{ij}(t)dt$, i.e. the Laplace transform of $p_{ij}(t)$, and define the matrix $\boldsymbol{\Pi}(s) = (\pi_{ij}(s))$. From Theorem B.11(iv), the Laplace transform of $p'_{ij}(t)$ is given by $s\,\pi_{ij}(s) - \delta_{ij}$. Hence taking the Laplace transform of both sides of (4.15) yields

$$s\,\boldsymbol{\Pi}(s) - \mathbf{I} = \mathbf{Q}\boldsymbol{\Pi}(s) = \boldsymbol{\Pi}(s)\mathbf{Q}, \quad \mathrm{Re}\,(s) > 0.$$

From Theorem 4.5, the matrix $(s\mathbf{I} - \mathbf{Q})$ is invertible for $\mathrm{Re}\,(s) > 0$. It follows that

$$\boldsymbol{\Pi}(s) = (s\,\mathbf{I} - \mathbf{Q})^{-1}, \quad \mathrm{Re}\,(s) > 0. \tag{4.17}$$

Of course, this is equivalent to (4.16).

From (4.16) or (4.17), we can say that there is a one-to-one correspondence between the infinitesimal generator $\mathbf{Q}$ and the transition matrix function $\mathbf{P}(t)$. Thus, it is to be expected that the generator $\mathbf{Q}$ plays a role similar to that of one-step transition matrices in the theory of discrete-time Markov chains. In fact, as in the discrete-time case (see Theorem 1.2), suppose that we want to evaluate the joint probability

$$P[X(t_0) = i_0,\ X(t_1) = i_1, \cdots,\ X(t_n) = i_n],$$

where $0 = t_0 < t_1 < \cdots < t_n$. Let $\boldsymbol{\alpha} = (\alpha_i)$ be the initial distribution of $\{X(t)\}$. Then the chain rule of conditional probabilities in conjunction with repeated application of the Markov property (4.1) and homogeneity yields

$$\begin{aligned} P[X(t_0) = i_0,\ X(t_1) = i_1, \cdots,\ X(t_n) = i_n] \\ = \quad \alpha_{i_0}\, p_{i_0,i_1}(t_1) \cdots p_{i_{n-1},i_n}(t_n - t_{n-1}). \end{aligned} \tag{4.18}$$

Hence, the initial distribution $\boldsymbol{\alpha}$ and the transition matrix function $\mathbf{P}(t)$ together suffice to determine every joint distribution of the continuous-time Markov chain $\{X(t)\}$. Note the natural interpretation of the right-hand side of (4.18). Since the infinitesimal generator $\mathbf{Q}$ determines the transition matrix function $\mathbf{P}(t)$, by (4.16), the initial distribution $\boldsymbol{\alpha}$, together with the generator $\mathbf{Q}$, suffices to determine the stochastic behavior of the continuous-time Markov chain $\{X(t)\}$. Because of this property, any finite Markov chain in continuous time can be described by giving its infinitesimal generator and its initial distribution.

Example 4.3 Consider a system that can be in one of two states, *on* or *off*. Suppose initially it is on and it remains on for a time Y_1; it then goes off and remains off for a time Z_1; it then goes on for a time Y_2; then off for a time Z_2; then on, and so forth. Suppose that the random vectors $\{(Y_n, Z_n)\}$ are IID. Such a process is called an *alternating renewal process*. Now suppose that Y and Z are independent and exponentially distributed with parameters λ and μ respectively. Let $X(t)$ denote the state of the system at time t. Recalling the interpretation of the transition rate q_{ij}, it can be readily verified that $\{X(t)\}$ is a continuous-time Markov chain with the two states $\{0, 1\}$, where 0 means *on* and 1 means *off*, and an infinitesimal generator

$$\mathbf{Q} = \begin{pmatrix} -\lambda & \lambda \\ \mu & -\mu \end{pmatrix}.$$

The eigenvalues of $\mathbf{Q}$ are 0 and $-(\lambda + \mu)$. It is easily seen that $\mathbf{Q}$ can be decomposed as

$$\mathbf{Q} = \begin{pmatrix} 1 & \frac{\lambda}{\lambda+\mu} \\ 1 & \frac{-\mu}{\lambda+\mu} \end{pmatrix} \begin{pmatrix} 0 & 0 \\ 0 & -(\lambda+\mu) \end{pmatrix} \begin{pmatrix} \frac{\mu}{\lambda+\mu} & \frac{\lambda}{\lambda+\mu} \\ 1 & -1 \end{pmatrix}.$$

Therefore, from (4.16), we have

$$\mathbf{P}(t) = \begin{pmatrix} 1 & \frac{\lambda}{\lambda-\mu} \\ 1 & \frac{-\mu}{\lambda-\mu} \end{pmatrix} \begin{pmatrix} 1 & 0 \\ 0 & e^{-(\lambda+\mu)t} \end{pmatrix} \begin{pmatrix} \frac{\mu}{\lambda+\mu} & \frac{\lambda}{\lambda+\mu} \\ 1 & -1 \end{pmatrix}.$$

For example,

$$p_{00}(t) = \frac{\mu}{\lambda+\mu} + \frac{\lambda}{\lambda+\mu} e^{-(\lambda+\mu)t}, \quad t \geq 0.$$

Other transition probability functions can easily be obtained as well.

Example 4.4 Let $\{X(t)\}$ be a continuous-time Markov chain with state space $\{0, 1, \cdots, N\}$ and infinitesimal generator

$$\mathbf{Q} = \begin{pmatrix} -\lambda_0 & \lambda_0 & 0 & 0 & \cdots & 0 \\ \mu_1 & -\lambda_1 - \mu_1 & \lambda_1 & 0 & \cdots & 0 \\ 0 & \mu_2 & -\lambda_2 - \mu_2 & \lambda_2 & \cdots & 0 \\ \vdots & & \ddots & \ddots & \ddots & \vdots \\ 0 & \cdots & 0 & \mu_{N-1} & -\lambda_{N-1} - \mu_{N-1} & \lambda_{N-1} \\ 0 & \cdots & 0 & 0 & \mu_N & -\mu_N \end{pmatrix},$$

where λ_i, $\mu_{i+1} > 0$ for all i. If one interprets $X(t)$ as the size of a randomly varying population, the transition intensity λ_i is viewed as the *birth rate* while the intensity μ_i is the *death rate* when the population size is i. The process $\{X(t)\}$ is called a (finite) *birth–death process.* Note that $\{X(t)\}$ can jump only to adjacent states. Hence, birth–death processes are a continuous-time analog of the one-dimensional random walks given in Example 2.9. Birth–death processes will be considered in detail in the next chapter.

Consider a finite Markov chain $\{X(t)\}$ in continuous time with infinitesimal generator $\mathbf{Q} = (q_{ij})$. Recall that $q_i = -q_{ii} = \sum_{j \neq i} q_{ij}$.

Definition 4.2 (i) State i is called *absorbing* if $q_i = 0$.

(ii) State i is called *instantaneous* if $q_i = \infty$.

(iii) State i is called *stable* if $0 < q_i < \infty$.

From (4.11), we have $q_i = 0$ if and only if $p_{ii}(t) = 1$ for all $t \geq 0$. If state i is absorbing and if $\{X(t)\}$ enters state i at some time, then it stays there forever. A state for which $q_i = \infty$ is called instantaneous since, whenever entered, the chain $\{X(t)\}$ leaves it instantaneously. Note from Theorem 4.5 that no finite Markov chain has instantaneous states.

In what follows, we assume that the sample paths of $\{X(t)\}$ are right-continuous step functions. If not, we can consider a *modification* $\{\widehat{X}(t)\}$ of $\{X(t)\}$, i.e., $P[\widehat{X}(t) = X(t)] = 1$ for every $t \geq 0$, such that $\{\widehat{X}(t)\}$ has right-continuous sample paths. Now, define

$$T_i = \inf\{t \geq 0 : X(t) \neq i\}$$

if the set is not empty and $T_i = \infty$ otherwise. When $X(0) = i$, the random variable T_i is called the *holding time* in state i. If $X(0) = i$ and $q_i = 0$, i.e., i is an absorbing state, then $T_i = \infty$. As before, we write $P_i[A]$ for $P[A|X(0) = i]$.

Theorem 4.6 *For a finite Markov chain $\{X(t)\}$ in continuous time with infinitesimal generator $\mathbf{Q} = (q_{ij})$, suppose that $q_i > 0$. Then:*

(i) $P_i[T_i > t] = e^{-q_i t}$ *for* $t \geq 0$.

(ii) $P_i[X(T_i) = j] = q_{ij}/q_i$ *for* $i \neq j$.

Proof. (i) Write $\phi(t) = P_i[T_i > t]$. Then

$$\begin{aligned}
\phi(t+s) &= P_i[X(u) = i,\ 0 \leq u \leq t+s] \\
&= P_i[X(u) = i,\ 0 \leq u \leq t+s | X(u) = i,\ 0 \leq u \leq t]\, \phi(t) \\
&= P_i[X(u) = i,\ t \leq u \leq t+s | X(t) = i]\, \phi(t) \\
&= P_i[X(u) = i,\ 0 \leq u \leq s]\, \phi(t) \\
&= \phi(s)\, \phi(t),
\end{aligned}$$

where we have made use of the Markov property and homogeneity. On the other hand, the same argument as given in Example 4.1 shows that $\phi(t) > 0$ for some $t > 0$. Hence, $\phi(t)$ must be of the form $\phi(t) = e^{-ct}$ for some constant $c > 0$. But, for sufficiently small $h > 0$, we have

$$\phi(h) = P_i[X(u) = i,\ 0 \leq u \leq h] = p_{ii}(h) + o(h),$$

whence

$$c = \lim_{h \to 0+} \frac{1 - \phi(h)}{h} = \lim_{h \to 0+} \left[\frac{1 - p_{ii}(h)}{h} + \frac{o(h)}{h} \right] = q_i.$$

(ii) Since the sample paths are right-continuous, $X(T_i)$ represents the state that the Markov chain visits just after leaving state i. For $i \neq j$, let

$$R_{ij}(h) = P[X(t+h) = j | X(t) = i,\ X(t+h) \neq i], \quad h > 0.$$

Then $P_i[X(T_i) = j] = \lim_{h \to 0+} R_{ij}(h)$. Also, by homogeneity,

$$R_{ij}(h) = P_i[X(h) = j | X(h) \neq i] = \frac{P_i[X(h) = j]}{P_i[X(h) \neq i]}.$$

The result now follows from (4.14). □

Let $\{X(t)\}$ be a finite Markov chain in continuous time with infinitesimal generator $\mathbf{Q} = (q_{ij})$. From Theorem 4.6, we observe the following. Suppose that $X(0) = i$. If state i is absorbing, i.e., $q_i = 0$, then the chain remains there forever. If $q_i > 0$, then the chain stays in i for a finite but strictly positive amount of time T_i which is exponentially distributed with parameter q_i. At the end of this holding time, the chain moves to state j according to the transition law (q_{ij}/q_i). Now, because of homogeneity and the strong Markov property, the future behavior of the chain after T_i is independent of the past behavior and the chain repeats this procedure as if the initial state were j. Let J_n be the time epoch of the nth transition of the Markov chain $\{X(t)\}$, where $J_0 \equiv 0$. Since J_n is a sum of n independent, exponentially distributed random variables, it is obvious that $\lim_{n \to \infty} J_n = \infty$

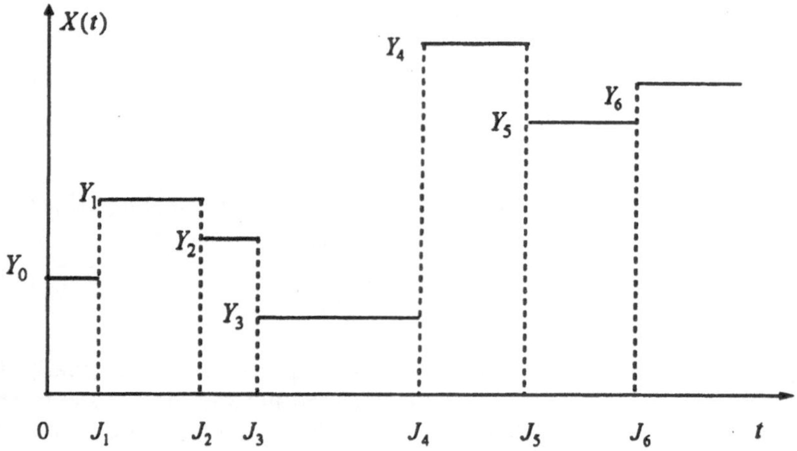

Figure 4.1 *A sample path of a continuous-time Markov chain.*

almost surely. Hence, repeating the above argument, we can describe the complete motion of $\{X(t)\}$ in terms of the infinitesimal generator $\mathbf{Q}$. A typical sample path of the Markov chain $\{X(t)\}$ is depicted in Figure 4.1.

Let $Y_n = X(J_n)$, the state of the chain at the nth transition, and define

$$p_{ij} = \begin{cases} \delta_{ij}, & q_i = 0, \\ 0, & q_i > 0 \text{ and } j = i, \\ q_{ij}/q_i, & q_i > 0 \text{ and } j \neq i. \end{cases} \tag{4.19}$$

It is clear that $\{Y_n\}$ is a discrete-time Markov chain with transition matrix (p_{ij}). The Markov chain $\{Y_n\}$ is called the *embedded Markov chain* associated with $\{X(t)\}$. The next theorem follows.

Theorem 4.7 *For any i, $j \in \mathcal{N}$ and $t > 0$, we have*

$$P[Y_{n+1} = j,\, J_{n+1} - J_n > t | Y_0, \cdots, Y_n = i;\, J_0, \cdots, J_n] = p_{ij} e^{-q_i t}.$$

The structure of continuous-time Markov chains suggests the construction of more general stochastic processes. For example, suppose that the holding time in each state follows an arbitrary distribution that is not necessarily exponential. Such a process is called a *semi-Markov process* and will be discussed later in this chapter.

Let T be an exponentially distributed random variable with parameter λ. Then, since $P[T > y] = e^{-\lambda y}$, we have

$$P[T - x > y | T > x] = \frac{P[T > x + y]}{P[T > x]} = \frac{e^{-\lambda(x+y)}}{e^{-\lambda x}} = e^{-\lambda y},$$

whence

$$P[T - x > y | T > x] = P[T > y], \quad x,\ y \geq 0. \tag{4.20}$$

If we think of T as the lifetime of a system, then (4.20) states that, given that the system has survived for x hours, the probability that the system will survive for more than $x + y$ hours in total is the same as the initial probability that it will survive for at least y hours. In other words, if the system is still operating at time x, then the distribution of its remaining life coincides with the original lifetime distribution. Since the history of the survival until time x is ignored, the property stated in (4.20) is called memorylessness. The *memoryless* property is equivalent to the condition

$$P[T > x + y] = P[T > x]\, P[T > y], \quad x,\ y \geq 0.$$

Therefore, as in the proof of Theorem 4.6(i), the memoryless property characterizes exponential distributions.

In many applications, the data for a continuous-time Markov chain are provided in terms of the parameters q_i and the transition probabilities p_{ij} given in (4.19), as the next example illustrates.

Example 4.5 Consider a single-server queueing system with a waiting space of size $N - 1$ so that the total number of customers in the system does not exceed N. Suppose that the arrival process is Poisson with intensity λ (see Example 4.1), that the service times are independent and exponentially distributed with parameter μ, and that the arrival and service processes are mutually independent. Such a queueing system is denoted by M/M/1/N. When considering the queue-size process, it is obvious that the transition epochs are the instants at which the queue size changes. If the change is due to an arrival, then the queue size increases by 1, and if the change is due to a departure, then the queue size decreases by 1. Changes of size more than 1 happen with probability zero.

Let $\{Y_n\}$ and $\{J_n\}$ be defined as above. If the event $\{Y_n = 0\}$ occurs, then the next change must be due to an arrival, and $J_{n+1} - J_n$ has the same distribution as the interarrival times, which is exponentially distributed with parameter λ. Hence, $q_0 = \lambda$, $p_{01} = 1$ and $p_{0i} = 0$ for $i \neq 1$. Similarly, if the event $\{Y_n = N\}$ occurs, then the next change must be due to a departure, and $J_{n+1} - J_n$ has the same distribution as the service times, which is exponential with parameter μ. Hence, $q_N = \mu$, $p_{N,N-1} = 1$ and $p_{Ni} = 0$ for $i \neq N - 1$.

Suppose that the event $\{Y_n = i\}$, $i \neq 0, N$, occurs. Let A be the time from J_n to the next arrival, and let S be the time from J_n to the completion of service being carried on. Because of the memoryless property of exponential distributions, A and S are exponentially distributed with parameters λ and μ respectively. Furthermore, $J_{n+1} - J_n = \min\{A, S\}$. Hence,

$$P[J_{n+1} - J_n > t | Y_n = i] = P[A > t,\ S > t] = e^{-(\lambda+\mu)t},$$

so that $q_i = \lambda + \mu$. Consider the queue size. If $S > A$, i.e., an arrival occurs before a service completion, then the queue size increases by 1, while if $S < A$ then the queue size decreases by 1. It follows that

$$\begin{aligned} p_{i,i+1} &= P[S > A] \\ &= \int_0^\infty P[S > x | A = x] \lambda e^{-\lambda x} dx \\ &= \int_0^\infty \lambda e^{-\lambda x} e^{-\mu x} dx \\ &= \frac{\lambda}{\lambda + \mu} \end{aligned}$$

and, similarly,

$$p_{i,i-1} = P[S < A] = \frac{\mu}{\lambda + \mu}.$$

Therefore, for $i = 1, \cdots, N-1$, we have

$$p_{ij} = \begin{cases} \dfrac{\lambda}{\lambda+\mu}, & j = i+1, \\ \dfrac{\mu}{\lambda+\mu}, & j = i-1, \\ 0, & \text{otherwise.} \end{cases}$$

It follows from (4.19) that the queue-size process is a continuous-time Markov chain with infinitesimal generator

$$\mathbf{Q} = \begin{pmatrix} -\lambda & \lambda & 0 & 0 & \cdots & 0 \\ \mu & -\lambda-\mu & \lambda & 0 & \cdots & 0 \\ 0 & \mu & -\lambda-\mu & \lambda & \cdots & 0 \\ \vdots & & \ddots & \ddots & \ddots & \vdots \\ 0 & \cdots & 0 & \mu & -\lambda-\mu & \lambda \\ 0 & \cdots & 0 & 0 & \mu & -\mu \end{pmatrix}. \tag{4.21}$$

This is a finite birth–death process with birth rate λ and death rate μ for every state.

The next theorem relates the parameters q_i and p_{ij} to the transition probability functions $p_{ij}(t)$.

Theorem 4.8 *For any i, $j \in \mathcal{N}$ and $t > 0$,*

$$p_{ij}(t) = \delta_{ij} e^{-q_i t} + \int_0^t q_i e^{-q_i s} \sum_{k \neq i} p_{ik} p_{kj}(t-s) ds.$$

Proof. Taking the Laplace transform of the given equation yields

$$\pi_{ij}(s) = \frac{\delta_{ij}}{s + q_i} + \frac{q_i}{s + q_i} \sum_{k \neq i} p_{ik} \pi_{kj}(s), \quad \operatorname{Re}(s) > 0,$$

where $\pi_{ij}(s) = \int_0^\infty e^{-st} p_{ij}(t)dt$; see Theorem B.11(viii). Using (4.19), the above transform can be written as

$$(s + q_i)\pi_{ij}(s) = \delta_{ij} + \sum_{k \neq i} q_{ik} \pi_{kj}(s),$$

so that

$$s\,\pi_{ij}(s) - \delta_{ij} = \sum_k q_{ik} \pi_{kj}(s), \quad \mathrm{Re}\,(s) > 0.$$

Hence $\boldsymbol{\Pi}(s) = (\pi_{ij}(s))$ satisfies (4.17) and the result follows due to the uniqueness of the Laplace transform. □

An interpretation of Theorem 4.8 is the following. Starting from state i, the Markov chain is in state j at time t if it either remains in state i, in which case $i = j$, or moves at some time $s < t$ to some state k and then it is in state j at time t. It should be noted that, due to the Markov property and homogeneity, the probability of the latter conditional event is equal to the conditional probability that, starting from state k, the Markov chain is in state j at time $t - s$.

4.3 Denumerable Markov chains in continuous time

In this section, we consider the case where the state space is denumerably infinite. Most of results for the finite case can be generalized, as they stand, to the denumerable case under mild conditions. However, the generalizations are not straightforward and require careful analytical considerations. We will not pursue the details here because they are beyond the scope of this book. The interested reader should consult, e.g., Anderson (1991). There are slight differences between the finite case and the denumerable case. As examples, the passage intensity q_i may be infinite (see Theorem 4.9 below). Also, the infinitesimal generator may no longer determine the transition probability functions uniquely even if all the states are stable, i.e., $q_i < \infty$ for all i.

Let $\{X(t)\}$ be a continuous-time Markov chain with state space $\mathcal{N} = \{0, 1, 2, \cdots\}$ and transition matrix function $\mathbf{P}(t) = (p_{ij}(t))$. As in the finite case, the initial distribution $\boldsymbol{\alpha} = (\alpha_i)$ and the transition matrix function $\mathbf{P}(t)$ together determine the joint distributions of $\{X(t)\}$, by (4.18). We begin with the right-hand derivative of $p_{ij}(t)$ at $t = 0$.

Theorem 4.9 *For each $i \in \mathcal{N}$,*

$$-p'_{ii}(0+) = \lim_{h \to 0+} \frac{1 - p_{ii}(h)}{h}$$

exists but may be infinite.

Proof. Let $\phi(t) = -\log p_{ii}(t)$, which is well defined since $p_{ii}(t) > 0$, from

Lemma 4.1. Since

$$p_{ii}(s+t) \geq p_{ii}(s)\,p_{ii}(t), \quad s,\ t \geq 0,$$

by the Chapman–Kolmogorov equation (4.3), the subadditivity property

$$\phi(s+t) \leq \phi(s) + \phi(t), \quad s,\ t \geq 0,$$

follows. Define

$$q_i \equiv \sup_{t>0} \frac{\phi(t)}{t}.$$

Suppose $q_i < \infty$. Then, for any $\varepsilon > 0$, there is some $t_0 > 0$ such that $\phi(t_0)/t_0 \geq q_i - \varepsilon$. Also, for any $t < t_0$, there are some integer n and real number c such that $t_0 = nt + c$, where $0 \leq c < t$. It then follows that

$$q_i - \varepsilon \leq \frac{\phi(nt+c)}{t_0} \leq \frac{n\phi(t) + \phi(c)}{t_0} = \frac{t_0 - c}{t_0}\frac{\phi(t)}{t} + \frac{\phi(c)}{t_0},$$

where the second inequality is due to the subadditivity of $\phi(t)$. Note that $\phi(c) \to 0$ as $t \to 0+$ since $p_{ii}(0+) = 1$. Hence

$$q_i - \varepsilon \leq \liminf_{t\to 0+} \frac{\phi(t)}{t} \leq \limsup_{t\to 0+} \frac{\phi(t)}{t} \leq q_i.$$

Since $\varepsilon > 0$ is arbitrary, we conclude that

$$q_i = \lim_{t\to 0+} \frac{\phi(t)}{t}.$$

If $q_i = \infty$, we replace $q_i - \varepsilon$ by a sufficiently large number, and the same conclusion holds. Since

$$\lim_{t\to 0+} \frac{\phi(t)}{t} = \lim_{t\to 0+} \frac{-\log[1 - \{1 - p_{ii}(t)\}]}{t} = \lim_{t\to 0+} \frac{1 - p_{ii}(t)}{t},$$

the theorem follows. □

Let us define $q_{ii} = p'_{ii}(0+)$ and $q_i = -q_{ii}$. From the above proof, we know that

$$q_i = \sup_{t>0} \frac{-\log p_{ii}(t)}{t} = \lim_{t\to 0+} \frac{1 - p_{ii}(t)}{t}, \quad i \in \mathcal{N}.$$

For $i \neq j$, the right-hand derivative $p'_{ij}(0+)$ also exists and is finite as the following theorem asserts. We shall write $q_{ij} = p'_{ij}(0+)$ as in the finite case. The matrix $\mathbf{Q} = (q_{ij})$ is called the *infinitesimal generator*, or *generator* for short, of $\{X(t)\}$. The proof of the next theorem is omitted since it is complicated and lengthy (see Anderson, Section 1.2, 1991).

Theorem 4.10 *For every pair i and j such that $i \neq j$,*

$$p'_{ij}(0+) = \lim_{h\to 0+} \frac{p_{ij}(h)}{h}$$

exists and is finite.

Since

$$1 - p_{ii}(h) = \sum_{j \neq i} p_{ij}(h)$$

for any $h > 0$, we have

$$q_i = \lim_{h \to 0+} \sum_{j \neq i} \frac{p_{ij}(h)}{h} \geq \sum_{j \leq N, j \neq i} \lim_{h \to 0+} \frac{p_{ij}(h)}{h} = \sum_{j \leq N, j \neq i} q_{ij}$$

for any N sufficiently large. Therefore, in general,

$$0 \leq \sum_{j \neq i} q_{ij} \leq q_i \leq \infty, \quad i \in \mathcal{N}.$$

Note the difference from the finite case (4.13).

Definition 4.3 Let $\{X(t)\}$ be a continuous-time Markov chain with state space $\mathcal{N} = \{0, 1, 2, \cdots\}$ and infinitesimal generator $\mathbf{Q} = (q_{ij})$:

(i) $\{X(t)\}$ or $\mathbf{Q}$ is said to be *stable* if $q_i < \infty$ for all $i \in \mathcal{N}$.

(ii) $\{X(t)\}$ or $\mathbf{Q}$ is said to be *conservative* if $q_i = \sum_{j \neq i} q_{ij}$ for all $i \in \mathcal{N}$.

Examples where $q_i > \sum_{j \neq i} q_{ij}$ can be found in, e.g., Chung (Examples 20.2–20.4, 1967). In what follows, we assume that the Markov chain under consideration is stable and conservative, unless stated otherwise.

Regarding the differentiability of the transition probability function $p_{ij}(t)$, we have the following result whose proof is beyond the scope of this book and is omitted. See, e.g., Anderson (Chapter 1, 1991) for the proof.

Theorem 4.11 *Every $p_{ij}(t)$ has a continuous derivative on $t \geq 0$. Moreover,*

$$p'_{ij}(s + t) = \sum_k p'_{ik}(s) \, p_{kj}(t), \quad t \geq 0, \ s > 0,$$

as well as

$$p'_{ij}(t + s) = \sum_k p_{ik}(t) \, p'_{kj}(s), \quad t \geq 0, \ s > 0,$$

hold, where the sums are absolutely convergent.

From Theorem 4.11 above, we have

$$p'_{ij}(t) = \lim_{s \to 0+} p'_{ij}(s + t) \geq \sum_{k \leq N} \lim_{s \to 0+} p'_{ik}(s) \, p_{kj}(t) = \sum_{k \leq N} q_{ik} p_{kj}(t)$$

for sufficiently large N, whence

$$p'_{ij}(t) \geq -q_i \, p_{ij}(t) + \sum_{k \neq i} q_{ik} p_{kj}(t), \quad t \geq 0. \tag{4.22}$$

Similarly, we have

$$p'_{ij}(t) \geq -p_{ij}(t) \, q_j + \sum_{k \neq j} p_{ik}(t) q_{kj}, \quad t \geq 0. \tag{4.23}$$

The next theorem shows that, in fact, equality holds in (4.22). For equality to hold in (4.23), however, an additional condition is needed. We shall address this problem in Theorem 4.15 below.

Theorem 4.12 *For every $i,\ j \in \mathcal{N}$, we have*

$$p'_{ij}(t) = -q_i\, p_{ij}(t) + \sum_{k \neq i} q_{ik} p_{kj}(t), \quad t \geq 0.$$

Proof. Suppose, first, that $q_i > 0$. (Recall that we have assumed that $\mathbf{Q}$ is stable.) From the Chapman–Kolmogorov equation (4.3), we have

$$\frac{p_{ij}(t+h) - p_{ij}(t)}{h} = -\frac{1 - p_{ii}(h)}{h} p_{ij}(t) + \sum_{k \neq i} \frac{p_{ik}(h)}{h} p_{kj}(t),$$

for $t \geq 0$ and $h > 0$. It is enough to justify the interchange of the limit and the summation on the right-hand side of the above equation. To this end, a similar argument to that given in the proof of (4.22) leads to

$$\liminf_{h \to 0+} \sum_{k \neq i} \frac{p_{ik}(h)}{h} p_{kj}(t) \geq \sum_{k \neq i} q_{ik} p_{kj}(t).$$

To prove the reverse inequality, note that, since $p_{kj}(t) \leq 1$, we have for arbitrary $N > i$,

$$\limsup_{h \to 0+} \sum_{k \neq i} \frac{p_{ik}(h)}{h} p_{kj}(t) \leq \limsup_{h \to 0+} \left[\sum_{k \leq N,\, k \neq i} \frac{p_{ik}(h)}{h} p_{kj}(t) + \sum_{k > N} \frac{p_{ik}(h)}{h} \right].$$

Also,

$$\sum_{k > N} \frac{p_{ik}(h)}{h} = \frac{1 - p_{ii}(h)}{h} - \sum_{k \leq N,\, k \neq i} \frac{p_{ik}(h)}{h}.$$

It follows that

$$\limsup_{h \to 0+} \sum_{k \neq i} \frac{p_{ik}(h)}{h} p_{kj}(t) \leq \sum_{k \leq N,\, k \neq i} q_{ik} p_{kj}(t) + q_i - \sum_{k \leq N,\, k \neq i} q_{ik}.$$

Since N is arbitrary, we then obtain

$$\limsup_{h \to 0+} \sum_{k \neq i} \frac{p_{ik}(h)}{h} p_{kj}(t) \leq \sum_{k \neq i} q_{ik} p_{kj}(t) + q_i - \sum_{k \neq i} q_{ik} = \sum_{k \neq i} q_{ik} p_{kj}(t).$$

The last equality holds since $\mathbf{Q}$ is conservative. Therefore the theorem is proved when $q_i > 0$. The case where $q_i = 0$ is trivial. □

As for the finite case, let J_n be the time just after the nth transition of $\{X(t)\}$. Then exactly the same proof as is given in Theorem 4.6 goes through to show that $J_{n+1} - J_n$ are independent and exponentially distributed. In contrast to the finite case, however, the property $\lim_{n \to \infty} J_n =$

∞ with probability one may no longer hold in general, and we need additional conditions to ensure that the process will never explode.

Definition 4.4 A continuous-time Markov chain $\{X(t)\}$ is called *regular** if $\lim_{n\to\infty} J_n = \infty$ with probability one.

Example 4.6 Consider a generalization of the Poisson process given in Example 4.1 where the time between the nth and the $(n+1)$th arrivals is exponentially distributed with parameter λ_n, which depends on n Such a process is called a *pure birth process* and is used to model, for example, the size of a bacterial colony. There are no deaths and the size increases without bound.

Let $X(t)$ denote the size of the population at time t and suppose that $X(0) = 1$. Since J_n is the sum of n independent and exponentially distributed random variables with means λ_i^{-1}, $i = 1, 2, \cdots, n$, we have

$$E[J_n] = \sum_{i=1}^{n} \frac{1}{\lambda_i},$$

so that, by the monotone convergence theorem,

$$E\left[\lim_{n\to\infty} J_n\right] = \sum_{n=1}^{\infty} \frac{1}{\lambda_n}.$$

If $\sum_{n=1}^{\infty} \lambda_n^{-1} < \infty$ then $\lim_{n\to\infty} J_n < \infty$ with probability one. If, on the other hand, $\sum_{n=1}^{\infty} \lambda_n^{-1} = \infty$ then $\lim_{n\to\infty} J_n = \infty$ with probability one and the process is regular. In particular, the *Yule process* where $\lambda_n = \beta n$ for some $\beta > 0$ is regular since $\sum_{n=1}^{\infty} n^{-1} = \infty$.

In general, it is not an easy task to identify whether or not a given continuous-time Markov chain is regular based on data such as $\mathbf{Q} = (q_{ij})$. Here we give two sets of sufficient conditions that ensure regularity.

Theorem 4.13 *Suppose there is some constant $c < \infty$ such that $q_i \leq c$ for all $i \in \mathcal{N}$. Then the Markov chain $\{X(t)\}$ is regular.*

Proof. Let $\{\widehat{J}_n\}$ be an almost surely increasing sequence of random variables such that $\widehat{J}_{n+1} - \widehat{J}_n$ are independent and exponentially distributed with common parameter c. Recall that $J_{n+1} - J_n$ are independent and exponentially distributed with parameter $q_i \leq c$ when $X(J_n) = i$. Hence we have

$$J_{n+1} - J_n \geq_{\text{st}} \widehat{J}_{n+1} - \widehat{J}_n, \quad n = 0, 1, \cdots,$$

* In the past ten years, it has become standard to call a Markov chain *regular* if its mean return time is finite. See, for instance, Nummelin (1984) and Meyn and Tweedie (1993). Recall that, in the discrete-time setting, a finite stochastic matrix is called 'regular' if it is primitive (see Section 2.4). Although we use the same term, the meanings are different.

and so $J_n \geq_{st} \widehat{J}_n$ for all n (see Lemma 3.7), where $J_0 = \widehat{J}_0 = 0$. Note that $\widehat{J}_n$ follows an Erlang distribution

$$P[\widehat{J}_n \leq t] = \sum_{k=n}^{\infty} \frac{(ct)^k}{k!} e^{-ct}, \quad t \geq 0,$$

whence

$$P[J_n \leq t] \leq \sum_{k=n}^{\infty} \frac{(ct)^k}{k!} e^{-ct}, \quad t \geq 0.$$

Since the event $\{J_n \leq t\}$ is decreasing in n, it follows that

$$P\left[\lim_{n\to\infty} J_n \leq t\right] = \lim_{n\to\infty} P[J_n \leq t] \leq \lim_{n\to\infty} \sum_{k=n}^{\infty} \frac{(ct)^k}{k!} e^{-ct} = 0, \quad t \geq 0.$$

Therefore, $\lim_{n\to\infty} J_n = \infty$ with probability one. □

If there are only finitely many states, we can always choose $c = \max_i q_i < \infty$ and, hence, any finite Markov chain is regular. In the next section, we will call a continuous-time Markov chain satisfying the condition in Theorem 4.13 ***uniformizable***.

Another sufficient condition to ensure regularity is the following. Let $Y_n = X(J_n)$, i.e. the embedded Markov chain. As in the finite case, the process $\{Y_n\}$ is a discrete-time Markov chain with transition probabilities given in (4.19).

Theorem 4.14 *Suppose that the embedded Markov chain $\{Y_n\}$ is recurrent. Then the continuous-time Markov chain $\{X(t)\}$ is regular.*

Proof. If the initial state i is recurrent in $\{Y_n\}$, then it is visited infinitely often and each visit lasts an exponentially distributed length of time with parameter q_i. The infinite sum of these exponentially distributed random variables is infinite with probability one, as was proved in Theorem 4.13. Hence, the total time spent in the recurrent state i alone is infinite, so that $\lim_{n\to\infty} J_n$ must be infinite with probability one. □

We saw in Theorem 4.8 that

$$p_{ij}(t) = \delta_{ij} e^{-q_i t} + \int_0^t e^{-q_i s} \sum_{k \neq i} q_{ik} p_{kj}(t-s) ds, \quad t \geq 0, \tag{4.24}$$

using the Laplace transform method for the finite state space case. Analogously to (4.24), we have

$$p_{ij}(t) = \delta_{ij} e^{-q_i t} + \int_0^t e^{-q_j s} \sum_{k \neq j} p_{ik}(t-s) q_{kj} ds, \quad t \geq 0. \tag{4.25}$$

Next we give a probabilistic derivation of these equations. The proof is taken from Anderson (1991).

Lemma 4.3 *For every $i,\ j \in \mathcal{N}$, (4.24) holds true. If, in addition, the Markov chain $\{X(t)\}$ is regular then (4.25) holds.*

Proof. Note that

$$p_{ij}(t) = P_i[X(t) = j,\ J_1 > t] + P_i[X(t) = j,\ J_1 \le t].$$

If $J_1 > t$ then $X(t) = X(0) = i$ so that

$$P_i[X(t) = j,\ J_1 > t] = \delta_{ij} P_i[J_1 > t] = \delta_{ij} e^{-q_i t}.$$

If $J_1 \le t$ then

$$\begin{aligned} p_{ij}(t) &= \sum_{k \ne i} \int_0^t P_i[X(t) = j,\ s < J_1 \le s + ds,\ X(J_1) = k] \\ &= \sum_{k \ne i} \int_0^t P_i[\text{no jump in } [0,s],\ X(s+ds) = k,\ X(t) = j] \\ &= \sum_{k \ne i} \int_0^t P_i[\text{no jump in } [0,s]]\, P[X(s+ds) = k | X(s) = i] \\ &\qquad\qquad \times P[X(t) = j | X(s+ds) = k] \\ &= \sum_{k \ne i} \int_0^t e^{-q_i s} q_{ik} p_{kj}(t-s) ds, \end{aligned}$$

where we have made use of the strong Markov property in the third equality and $p_{ik}(ds) = q_{ik} ds$ in the fourth. This proves the first assertion.

To prove the second assertion, we let T be the time of the last jump before time t. The regularity assumption ensures that there is such a last jump for any $t > 0$. The rest of the proof is similar (see Exercise 4.13). □

The next result shows that equality holds in (4.23) under the regularity condition.

Theorem 4.15 *Suppose that the continuous-time Markov chain $\{X(t)\}$ is regular. Then, for every pair i and j, we have*

$$p'_{ij}(t) = -p_{ij}(t) q_j + \sum_{k \ne j} p_{ik}(t) q_{kj}, \quad t \ge 0.$$

Proof. In (4.25), a change of variable $t - s = u$ yields

$$p_{ij}(t) = e^{-q_j t} \left[\delta_{ij} + \int_0^t e^{q_j u} \sum_{k \ne j} p_{ik}(u) q_{kj} du \right].$$

Differentiating with respect to t, we have

$$p'_{ij}(t) = -q_j p_{ij}(t) + \sum_{k \ne j} p_{ik}(t) q_{kj}, \quad t > 0,$$

as desired. □

At this point, we remark that the equation given in Theorem 4.12 is called the *backward Kolmogorov equation*, whereas the equation given in Theorem 4.15 is called the *forward Kolmogorov equation*. Recall that the backward Kolmogorov equation holds if the Markov chain is stable and conservative. For the forward Kolmogorov equation to hold, an additional regularity condition is needed.

Example 4.7 Consider a Poisson process $\{N(t)\}$ with intensity λ, where the transition probability functions are given by

$$p_{ij}(t) = \frac{(\lambda t)^{j-i}}{(j-i)!} e^{-\lambda t}, \quad j \geq i; \quad t \geq 0.$$

Observe that $p'_{ii}(t) = -\lambda e^{-\lambda t}$ and

$$p'_{ij}(t) = \lambda \left[\frac{(\lambda t)^{j-i-1}}{(j-i-1)!} - \frac{(\lambda t)^{j-i}}{(j-i)!} \right] e^{-\lambda t}, \quad j \geq i.$$

It follows that $q_i = -p'_{ii}(0) = \lambda$ and, for $i \neq j$,

$$q_{ij} = p'_{ij}(0) = \begin{cases} \lambda, & j = i+1, \\ 0, & \text{otherwise.} \end{cases}$$

Hence the infinitesimal generator of the Poisson process $\{N(t)\}$ is given by

$$\mathbf{Q} = \begin{pmatrix} -\lambda & \lambda & 0 & 0 & \cdots \\ 0 & -\lambda & \lambda & 0 & \cdots \\ 0 & 0 & -\lambda & \lambda & \cdots \\ \vdots & \vdots & \vdots & \ddots & \ddots \end{pmatrix}.$$

Since the Poisson process is regular (see Example 4.6), the backward and forward Kolmogorov equations are given by

$$p'_{ij}(t) = -\lambda p_{ij}(t) + \lambda p_{i+1,j}(t), \quad j > i,$$

and

$$p'_{ij}(t) = -\lambda p_{ij}(t) + \lambda p_{i,j-1}(t), \quad j > i$$

respectively.

In matrix form, the backward and forward Kolmogorov equations are written, respectively, as

$$\mathbf{P}'(t) = \mathbf{Q}\,\mathbf{P}(t); \quad \mathbf{P}'(t) = \mathbf{P}(t)\mathbf{Q}, \quad t \geq 0, \tag{4.26}$$

where $\mathbf{P}'(t) = (p'_{ij}(t))$; see (4.15). As for the finite case, a solution to these differential equations is

$$\mathbf{P}(t) = \exp\{\mathbf{Q}t\} = \mathbf{I} + \sum_{n=1}^{\infty} \frac{\mathbf{Q}^n t^n}{n!}, \quad t \geq 0, \tag{4.27}$$

provided the series converges. A formal proof of (4.27) for the uniformizable case will be given in the next section.

For each transition probability function $p_{ij}(t)$, let

$$\pi_{ij}(s) = \int_0^\infty e^{-st} p_{ij}(t)dt, \quad \text{Re}(s) > 0.$$

Since $p_{ij}(t)$ is continuous in t, the Laplace transform $\pi_{ij}(s)$ determines $p_{ij}(t)$ uniquely. From the backward Kolmogorov equation, we have

$$\begin{aligned}\int_0^\infty e^{-st} p'_{ij}(t)dt &= -q_i \int_0^\infty e^{-st} p_{ij}(t)dt + \int_0^\infty e^{-st} \sum_{k\neq i} q_{ik} p_{kj}(t)dt \\ &= -q_i \int_0^\infty e^{-st} p_{ij}(t)dt + \sum_{k\neq i} q_{ik} \int_0^\infty e^{-st} p_{kj}(t)dt,\end{aligned}$$

by Fubini's theorem. It follows from Theorem B.11(iv) that

$$s\,\pi_{ij}(s) - \delta_{ij} = \sum_k q_{ik}\pi_{kj}(s), \quad \text{Re}(s) > 0,$$

where $q_i = -q_{ii}$. Similarly, from the forward Kolmogorov equation, we have

$$s\,\pi_{ij}(s) - \delta_{ij} = \sum_k \pi_{ik}(s) q_{kj}, \quad \text{Re}(s) > 0.$$

In matrix form, these are equivalent to

$$(s\mathbf{I} - \mathbf{Q})\boldsymbol{\Pi}(s) = \boldsymbol{\Pi}(s)(s\mathbf{I} - \mathbf{Q}) = \mathbf{I}, \quad \text{Re}(s) > 0. \tag{4.28}$$

Hence, if the matrix $(s\mathbf{I} - \mathbf{Q})$ is invertible, we obtain

$$\boldsymbol{\Pi}(s) = (s\mathbf{I} - \mathbf{Q})^{-1}, \quad \text{Re}(s) > 0;$$

cf. (4.17).

Now consider the embedded Markov chain $\{Y_n\}$ of $\{X(t)\}$. If $\{Y_n\}$ is irreducible and recurrent, then, referring to the above construction, the continuous-time Markov chain $\{X(t)\}$ is also irreducible and recurrent (of course, it is regular, from Theorem 4.14). Recall that $\{X(t)\}$ cannot have a periodic structure because of the exponentiality of holding times. This is so even when the embedded Markov chain is periodic. Hence, it would be expected that a result similar to Theorem 2.8 regarding the stationary distribution holds for the continuous-time case, too. Suppose for simplicity that $\{Y_n\}$ is positive recurrent and aperiodic. Then, from Theorem 2.8, there is a strictly positive probability vector $\boldsymbol{\nu} = (\nu_i)$ such that

$$\nu_i = \sum_{j=0}^\infty \nu_j p_{ji}, \quad i \in \mathcal{N},$$

where the p_{ij} are given in (4.19). If $\sum_{j=0}^\infty \nu_j/q_j < \infty$, the probability vector

$\boldsymbol{\pi} = (\pi_i)$, defined by

$$\pi_i = \frac{\nu_i/q_i}{\sum_{j=0}^{\infty} \nu_j/q_j}, \quad i \in \mathcal{N},$$

satisfies

$$\sum_{j=0}^{\infty} \pi_j q_{ji} = 0, \quad i \in \mathcal{N}. \quad \text{(In matrix form, } \boldsymbol{\pi}^\top \mathbf{Q} = \mathbf{0}^\top.) \tag{4.29}$$

Furthermore, from (4.27), $\boldsymbol{\pi}^\top = \boldsymbol{\pi}^\top \mathbf{P}(t)$ and so

$$\pi_i = \sum_{j=0}^{\infty} \pi_j p_{ji}(t), \quad t \geq 0; \quad i \in \mathcal{N}. \tag{4.30}$$

Hence, the probability vector $\boldsymbol{\pi} = (\pi_i)$ satisfying (4.29) is the ***stationary distribution*** of $\{X(t)\}$. The system of equations in (4.29) in conjunction with the equation $\sum_{j=0}^{\infty} \pi_j = 1$ is called the ***stationary equation***; see (2.15) for the discrete-time case. It should be noted that, even if the embedded Markov chain $\{Y_n\}$ is positive recurrent, the continuous-time Markov chain $\{X(t)\}$ may not be so since $\sum_i \nu_i/q_i$ may diverge. Conversely, even if $\{X(t)\}$ is positive recurrent, $\{Y_n\}$ can be null recurrent. In fact, there are 2×2 possibilities of positive and null recurrence of $\{X(t)\}$ and $\{Y_n\}$.

Regarding the limit of the transition probability function $p_{ij}(t)$ as $t \to \infty$, we recall from Theorem 2.7 that ν_i/ν_j is the expected number of visits to state i during a return period of successive visits to state j. Suppose that $X(0) = j$ and consider the time to return to state j. The expected return time is

$$\frac{1}{q_j} + \sum_{i \neq j} \frac{\nu_i}{\nu_j} \frac{1}{q_i} = \frac{1}{\nu_j} \sum_{i=0}^{\infty} \frac{\nu_i}{q_i},$$

where q_j^{-1} equals the mean sojourn time at state j. The proportion of time that $\{X(t)\}$ spends in state j is

$$\frac{1/q_j}{\nu_j^{-1} \sum_{i=0}^{\infty} \nu_i/q_i} = \frac{\nu_j/q_j}{\sum_{i=0}^{\infty} \nu_i/q_i} = \pi_j,$$

which is equal to the long-run frequency of $\{X(t)\}$ being in state j. Hence the probability that $\{X(t)\}$ is in state j at time t for very large t is π_j, i.e.,

$$\pi_j = \lim_{t \to \infty} p_{ij}(t), \quad j \in \mathcal{N}.$$

Such a probability vector $\boldsymbol{\pi} = (\pi_i)$ is called the ***limiting distribution*** of the Markov chain $\{X(t)\}$, and if such a positive probability vector exists then the Markov chain is called ***ergodic***.

We summarize these results in a slightly more general form, which can be proved using Theorem 2.11.

Theorem 4.16 *Suppose that the embedded Markov chain $\{Y_n\}$ is recurrent. Then there exists a strictly positive vector $\boldsymbol{\pi} = (\pi_i)$ satisfying (4.29) and any other positive solution is a constant multiple of $\boldsymbol{\pi}$. If $\sum_{i=0}^{\infty} \pi_i < \infty$, then the vector normed to sum to unity is the stationary as well as the limiting distribution of $\{X(t)\}$, and the continuous-time Markov chain $\{X(t)\}$ is ergodic.*

Example 4.8 Consider a continuous-time Markov chain $\{X(t)\}$ with state space $\mathcal{Z}_+$ and infinitesimal generator

$$\mathbf{Q} = \begin{pmatrix} -\lambda_0 & \lambda_0 & 0 & 0 & \cdots \\ \mu_1 & -\lambda_1 - \mu_1 & \lambda_1 & 0 & \cdots \\ 0 & \mu_2 & -\lambda_2 - \mu_2 & \lambda_2 & \cdots \\ \vdots & \vdots & \ddots & \ddots & \ddots \end{pmatrix}.$$

The Markov chain $\{X(t)\}$ is a birth–death process defined on the state space $\mathcal{Z}_+$ (see Example 4.4). Write $\pi_0 = 1$ and define

$$\pi_n = \pi_{n-1} \frac{\lambda_{n-1}}{\mu_n} = \frac{\lambda_0 \cdots \lambda_{n-1}}{\mu_1 \cdots \mu_n}, \quad n = 1, 2, \cdots.$$

It is readily shown that the π_i satisfy (4.29) and if $\sum_{n=0}^{\infty} \pi_n < \infty$ then the birth–death process is ergodic. The transition matrix of the embedded Markov chain $\{Y_n\}$ is given by

$$\mathbf{P} = \begin{pmatrix} 0 & 1 & 0 & 0 & \cdots \\ \frac{\mu_1}{\lambda_1 - \mu_1} & 0 & \frac{\lambda_1}{\lambda_1 + \mu_1} & 0 & \cdots \\ 0 & \frac{\mu_2}{\lambda_2 + \mu_2} & 0 & \frac{\lambda_2}{\lambda_2 + \mu_2} & \cdots \\ \vdots & \vdots & \ddots & \ddots & \ddots \end{pmatrix}.$$

An invariant vector $\boldsymbol{\nu} = (\nu_n)$ of $\mathbf{P}$ is such that $\nu_n = \pi_n(\lambda_n + \mu_n)/\lambda_0$ for $n = 1, 2, \cdots$ and $\nu_0 = 1$.

For each $i,\ j \in \mathcal{N}$, let $f_{ij}^{(0)}(t) = \delta_{ij} e^{-q_i t}$ and define $f_{ij}^{(n)}(t)$ successively by

$$f_{ij}^{(n)}(t) = f_{ij}^{(0)}(t) + \int_0^t e^{-q_i s} \sum_{k \neq i} q_{ik} f_{kj}^{(n-1)}(t-s) ds, \quad n = 1, 2, \cdots; \tag{4.31}$$

see (4.24). Exercise 4.15 clarifies the meaning of $f_{ij}^{(n)}(t)$ by showing that

$$f_{ij}^{(n)}(t) = P_i[X(t) = j,\ J_{n+1} > t]. \tag{4.32}$$

It is obvious that $f_{ij}^{(n)}(t) \geq f_{ij}^{(n-1)}(t)$ for all $t \geq 0$ and every $i,\ j \in \mathcal{N}$. It follows that

$$f_{ij}(t) \equiv \lim_{n \to \infty} f_{ij}^{(n)}(t), \quad t \geq 0; \quad i,\ j \in \mathcal{N},$$

exists and, by the monotone convergence theorem, satisfies (4.24). The same proof as given in the proof of Theorem 4.15 applies to show that $f_{ij}(t)$ satisfies the backward Kolmogorov equation. Also, $\sum_{j=0}^{\infty} f_{ij}^{(0)}(t) = e^{-q_i t} \leq 1$ and

$$\sum_{j=0}^{\infty} f_{ij}^{(n)}(t) = e^{-q_i t} + \int_0^t e^{-q_i s} \sum_{k \neq i} q_{ik} \sum_{j=0}^{\infty} f_{kj}^{(n-1)}(t-s) ds, \quad n = 1, 2, \cdots.$$

Hence, by induction, one has $\sum_{j=0}^{\infty} f_{ij}^{(n)}(t) \leq 1$ for all $n = 1, 2, \cdots$. Since we have, for any N sufficiently large,

$$1 \geq \sum_{j=0}^{\infty} f_{ij}^{(n)}(t) \geq \sum_{j=0}^{N} f_{ij}^{(n)}(t),$$

it follows that $\sum_{j=0}^{\infty} f_{ij}(t) \leq 1$. Now let $p_{ij}(t)$ be any function satisfying (4.24), i.e., a solution to the backward Kolmogorov equation. Then $p_{ij}(t) \geq f_{ij}^{(0)}(t)$ and, by induction, $p_{ij}(t) \geq f_{ij}^{(n)}(t)$ for all n, from which we conclude that $p_{ij}(t) \geq f_{ij}(t)$ for all $t \geq 0$. Therefore, $f_{ij}(t)$ is the *minimal* solution to the backward Kolmogorov equation. Similarly, it can be shown (see Exercise 4.16) that $f_{ij}(t)$ also satisfies (4.25) and is minimal to the forward Kolmogorov equation. The reader is referred to Anderson (Chapter 2, 1991) for a complete discussion.

Theorem 4.17 *Let $f_{ij}(t)$ be generated as above. Then:*

(i) *$f_{ij}(t)$ is the minimal solution to both the backward and forward Kolmogorov equations.*

(ii) *If $\sum_{j=0}^{\infty} f_{ij}(t) = 1$ for all $t \geq 0$, then it is the unique solution to both the backward and forward Kolmogorov equations.*

Suppose that a continuous-time Markov chain $\{X(t)\}$ is regular. Since the event $\{J_n > t\}$ is increasing in n, we have from (4.32) that

$$f_{ij}(t) = \lim_{n \to \infty} f_{ij}^{(n)}(t) = P_i[X(t) = j],$$

so that $\sum_{j=0}^{\infty} f_{ij}(t) = 1$. Hence, if a Markov chain is regular, then its transition probability function $p_{ij}(t)$ is characterized as the unique solution to both the backward and forward Kolmogorov equations.

Example 4.9 Let $\{N(t)\}$ be a Poisson process with intensity λ, and let $p_n(t) = P[N(t) = n]$. In Example 4.7, we showed that

$$p_n'(t) = -\lambda p_n(t) + \lambda p_{n-1}(t), \quad n = 0, 1, \cdots; \quad t \geq 0,$$

where $p_{-1}(t) = 0$ for all $t \geq 0$. Under the initial condition $p_0(0) = 1$, we obtain $p_0(t) = e^{-\lambda t}$. For $n = 1$, one sees that

$$p_1'(t) = -\lambda p_1(t) + \lambda e^{-\lambda t}, \quad t \geq 0.$$

The solution to this under the initial condition $p_1(0) = 0$ is $p_1(t) = \lambda t e^{-\lambda t}$. In general, it can be shown that

$$p_n(t) = \frac{(\lambda t)^n}{n!} e^{-\lambda t}, \quad n = 0, 1, \cdots; \quad t \geq 0,$$

which coincides with the result obtained in Example 4.1.

4.4 Uniformization

We begin with the formal definition of uniformizable Markov chains in continuous time.

Definition 4.5 A continuous-time Markov chain is called *uniformizable* if its infinitesimal generator $\mathbf{Q} = (q_{ij})$ is stable and conservative and satisfies $\sup_i q_i < \infty$, where $q_i = -q_{ii}$.

Let $\{X(t)\}$ be a uniformizable Markov chain with infinitesimal generator $\mathbf{Q} = (q_{ij})$ and let $c = \sup_i q_i$. For any $\nu \geq c$, define

$$\mathbf{P}_\nu = \mathbf{I} + \frac{1}{\nu}\mathbf{Q}. \tag{4.33}$$

Note that the off-diagonal elements of $\mathbf{Q}$ are all nonnegative and that the diagonal elements of $\mathbf{Q}/\nu$ are not less than -1. Moreover $\mathbf{Q1} = \mathbf{0}$ since $\mathbf{Q}$ is conservative. It follows that $\mathbf{P}_\nu$ is a stochastic matrix for any $\nu \geq c$. Recall from Theorem 4.13 that $\{X(t)\}$ is regular. We formally prove that if $\{X(t)\}$ is uniformizable then the transition matrix function $\mathbf{P}(t)$ is given by (4.27), i.e.,

$$\mathbf{P}(t) = \exp\{\mathbf{Q}t\} = \sum_{n=0}^{\infty} \frac{\mathbf{Q}^n t^n}{n!}, \quad t \geq 0.$$

To prove the uniqueness of the transition matrix function, the next lemma is needed.

Lemma 4.4 (i) *Suppose that the backward Kolmogorov equation in Theorem 4.12 has a solution. If the system of equations*

$$\lambda x_i = \sum_j q_{ij} x_j, \quad |x_i| \leq 1,$$

has no nontrivial solution for some $\lambda > 0$, then the solution of the backward equation is unique.

(ii) *Suppose that the forward Kolmogorov equation in Theorem 4.15 has a solution. If the system of equations*

$$\lambda y_j = \sum_i y_i q_{ij}, \quad |y_i| \leq 1, \quad \sum_j |y_j| < \infty,$$

has no nontrivial solution for some $\lambda > 0$, then the solution of the forward equation is unique.

Proof. We prove assertion (ii) only. Assertion (i) can be proved in a similar manner (see Exercise 4.17). Let $p_{ij}^{\ell}(t)$, $\ell = 1, 2$, be distinct solutions of the forward equation, i.e.,

$$p_{ij}^{\ell}{}'(t) = -p_{ij}^{\ell}(t)q_j + \sum_{k \neq j} p_{ik}^{\ell}(t)q_{kj}, \quad t \geq 0; \quad \ell = 1, 2.$$

Fix i and let $s > 0$. Define $\pi_j^{\ell}(s) = \int_0^{\infty} e^{-st} p_{ij}^{\ell}(t)dt$, the Laplace transform of $p_{ij}^{\ell}(t)$. It follows from Theorem B.11(iv) and Fubini's theorem that

$$s\,\pi_j^{\ell}(s) - \delta_{ij} = -\pi_j^{\ell}(s)q_j + \sum_{k \neq j} \pi_k^{\ell}(s)q_{kj}, \quad s > 0; \quad \ell = 1, 2,$$

whence, writing $\pi_j(s) = \pi_j^1(s) - \pi_j^2(s)$, we obtain

$$s\,\pi_j(s) = \sum_k \pi_k(s)q_{kj}, \quad s > 0.$$

If $p_{ij}^1(t)$ and $p_{ij}^2(t)$ are distinct, then the $\pi_j(s)$ are nonzero. Since

$$|p_{ij}^1(t) - p_{ij}^2(t)| \leq 1,$$

we have

$$|\pi_j(s)| \leq \int_0^{\infty} e^{-st}|p_{ij}^1(t) - p_{ij}^2(t)|dt \leq \frac{1}{s}, \quad s > 0.$$

Also, since

$$|\pi_j(s)| \leq \int_0^{\infty} e^{-st} p_{ij}^1(t)dt + \int_0^{\infty} e^{-st} p_{ij}^2(t)dt,$$

it follows that

$$s \sum_j |\pi_j(s)| \leq s \int_0^{\infty} e^{-st} \sum_j p_{ij}^1(t)dt + s \int_0^{\infty} e^{-st} \sum_j p_{ij}^2(t)dt = 2$$

for $s > 0$. Hence, taking $y_j = s\,\pi_j(s)$, we conclude that

$$s\,y_j = \sum_k y_k q_{kj}, \quad |y_j| \leq 1, \quad \sum_j |y_j| \leq 2,$$

has a nontrivial solution for all $s > 0$. This proves assertion (ii). □

Theorem 4.18 *Suppose that a continuous-time Markov chain $\{X(t)\}$ is uniformizable. Then the transition matrix function is given by*

$$\mathbf{P}(t) = \exp\{\mathbf{Q}\,t\} = \sum_{n=0}^{\infty} \frac{\mathbf{Q}^n t^n}{n!}, \quad t \geq 0.$$

Proof. Let $\mathbf{P}_\nu$ be the stochastic matrix given by (4.33). Denote its nth power by $\mathbf{P}_\nu^n = (p_{ij}^\nu(n))$ and define

$$p_{ij}(t) = e^{-\nu t}\sum_{n=0}^{\infty}\frac{(\nu t)^n}{n!}\,p_{ij}^\nu(n), \quad t \geq 0. \tag{4.34}$$

If we regard $\sum_{n=0}^{\infty}(\nu t)^n p_{ij}^\nu(n)/n!$ as a generating function $g(t) = \sum_{n=0}^{\infty} a_n t^n$ where $a_n = \nu^n p_{ij}^\nu(n)/n!$, the radius of convergence is infinity and so term-by-term differentiation is allowed (see Corollary B.1). It follows that

$$p_{ij}'(t) = -\nu\, p_{ij}(t) + \nu\, e^{-\nu t}\sum_{n=0}^{\infty}\frac{(\nu t)^n}{n!}\,p_{ij}^\nu(n+1), \quad t \geq 0.$$

Now, by the Chapman–Kolmogorov equation (2.9), we have

$$p_{ij}^\nu(n+1) = \sum_k p_{ik}^\nu p_{kj}^\nu(n) = \sum_k p_{ik}^\nu(n) p_{kj}^\nu, \quad n = 0, 1, \cdots,$$

where $\mathbf{P}_\nu = (p_{ij}^\nu)$. Hence, applying Fubini's theorem,

$$\begin{aligned}
p_{ij}'(t) &= -\nu\, p_{ij}(t) + \nu\, e^{-\nu t}\sum_k p_{ik}^\nu \sum_{n=0}^{\infty}\frac{(\nu t)^n}{n!}\,p_{ij}^\nu(n) \\
&= -\nu\, p_{ij}(t) + \nu \sum_k p_{ik}^\nu p_{kj}(t) \\
&= -\nu\, p_{ij}(t) + \nu\left(1 - \frac{1}{\nu}q_i\right) p_{ij}(t) + \sum_{k\neq i} q_{ik} p_{kj}(t) \\
&= -q_i\, p_{ij}(t) + \sum_{k\neq i} q_{ik} p_{kj}(t),
\end{aligned}$$

whence $p_{ij}(t)$ defined in (4.34) satisfies the backward Kolmogorov equation. Similarly, it can be shown that $p_{ij}(t)$ satisfies the forward Kolmogorov equation. Now, in matrix notation, (4.34) can be written as

$$\mathbf{P}(t) = e^{-\nu t}\sum_{n=0}^{\infty}\frac{(\nu t)^n}{n!}\mathbf{P}_\nu^n = e^{-\nu t}\exp\{\nu t\,\mathbf{P}_\nu\} = \exp\{\mathbf{Q}\,t\},$$

where the last equality follows since $\mathbf{Q} = -\nu(\mathbf{I} - \mathbf{P}_\nu)$, from (4.33). It remains to show the uniqueness of the Kolmogorov equations. To this end, suppose that $\mathbf{y} = (y_j)$ is a solution of the equation given in Lemma 4.4(ii), i.e.,

$$(\lambda + q_j)y_j = \sum_{i\neq j} y_i q_{ij}, \quad |y_j| \leq 1, \quad \sum_j |y_j| < \infty,$$

for some $\lambda > 0$. It follows that

$$(\lambda + q_j)|y_j| \leq \sum_{i\neq j} |y_i| q_{ij}, \quad \sum_j |y_j| < \infty.$$

Summing over j yields

$$\lambda \sum_j |y_j| + \sum_j q_j |y_j| \le \sum_i |y_i| \sum_{j \neq i} q_{ij} = \sum_i q_i |y_i|,$$

since the chain is conservative. Note that $\lambda > 0$ and

$$\sum_i q_i |y_i| < c \sum_i |y_i| < \infty,$$

where $c = \sup_i q_i$. This means that the above inequality only holds if $y_j = 0$ for all j. Hence, the solution of the forward equation is unique, from Lemma 4.4(ii). The proof of the uniqueness of the solution of the backward equation is left to the reader. □

Suppose that $\{X(t)\}$ is uniformizable and let ν be any number such that $\nu \ge \sup_i q_i$. As in the proof of Theorem 4.18, we have

$$\mathbf{P}(t) = \sum_{n=0}^{\infty} \frac{(\nu t)^n}{n!} e^{-\nu t} \mathbf{P}_\nu^n, \quad t \ge 0. \tag{4.35}$$

Let $\{Z_n^\nu\}$ be a discrete-time Markov chain with transition matrix $\mathbf{P}_\nu$ given by (4.33). We will call $\{Z_n^\nu\}$ a *uniformized Markov chain* of $\{X(t)\}$. Note that the uniformized Markov chain $\{Z_n^\nu\}$ is different from the embedded Markov chain $\{Y_n\}$ considered in the preceding section. Namely, if $Z_0^\nu = i$, then the uniformized chain may stay in i for n steps with positive probability and, given that it is leaving state i, the probability that it moves to state j is

$$\frac{p_{ij}^\nu}{1 - p_{ii}^\nu} = \frac{q_{ij}}{q_i}.$$

In contrast, if $Y_0 = i$, then the embedded chain moves to state j with probability q_{ij}/q_i and *cannot* stay in state i. Let $\{N(t)\}$ be a Poisson process with intensity ν which is independent of the uniformized chain $\{Z_n^\nu\}$, and consider the subordinated Markov chain $\{Z_{N(t)}^\nu\}$ (see Example 4.2). A sojourn time in state i of the subordinated chain $\{Z_{N(t)}^\nu\}$ is a geometric sum of exponential interarrival times, which is exponentially distributed with parameter $\nu(1 - p_{ii}^\nu) = q_i$.

Uniformization plays the role of a bridge between discrete-time Markov chains and continuous-time Markov chains (see Keilson, 1979, for details).

Theorem 4.19 *For a uniformizable Markov chain $\{X(t)\}$, define the subordinated process $\{Z_{N(t)}^\nu\}$ as above. Then $\{X(t)\} \stackrel{\mathrm{d}}{=} \{Z_{N(t)}^\nu\}$ for any $\nu \ge \sup_i q_i$.*

Proof. Let $p_{ij}^\nu(t)$ be the transition probability function of $\{Z_{N(t)}^\nu\}$. It suffices to show that $p_{ij}(t) = p_{ij}^\nu(t)$ for every $i, j \in \mathcal{N}$ and all $t \ge 0$. But we have already shown in Example 4.2 that $p_{ij}^\nu(t)$ is given by (4.9), i.e. the right-hand side of (4.35). □

From Theorem 4.19, when investigating the continuous-time Markov chain $\{X(t)\}$, we may want to investigate the uniformized Markov chain $\{Z_n^\nu\}$ rather than $\{X(t)\}$ directly. For example, for *some* $\nu \geq c = \sup_i q_i$, suppose that $\{Z_n^\nu\}$ has a stationary distribution $\boldsymbol{\pi} = (\pi_i)$, i.e.,

$$\boldsymbol{\pi}^\top = \boldsymbol{\pi}^\top \mathbf{P}_\nu^n, \quad n = 0, 1, \cdots,$$

where $\mathbf{P}_\nu$ is given by (4.33). From (4.34), we have

$$\sum_i \pi_i p_{ij}(t) = e^{-\nu t} \sum_{n=0}^{\infty} \frac{(\nu t)^n}{n!} \sum_i \pi_i p_{ij}^\nu(n).$$

But, since $\sum_i \pi_i p_{ij}^\nu(n) = \pi_j$, it follows that

$$\sum_i \pi_i p_{ij}(t) = e^{-\nu t} \sum_{n=0}^{\infty} \frac{(\nu t)^n}{n!} \pi_j = \pi_j, \quad t \geq 0,$$

whence $\boldsymbol{\pi}$ is also a stationary distribution of $\{X(t)\}$; see (4.30). Moreover, from (4.33),

$$\boldsymbol{\pi}^\top \mathbf{Q} = -\nu \boldsymbol{\pi}^\top (\mathbf{I} - \mathbf{P}_\nu) = \mathbf{0}^\top,$$

so that the stationary distribution $\boldsymbol{\pi}$ satisfies the stationary equation; see (4.29). Conversely, if $\boldsymbol{\pi}$ satisfies the stationary equation, it also satisfies the stationary equation of the uniformized Markov chain. Hence $\boldsymbol{\pi}$ is also a stationary distribution of $\{Z_n^\nu\}$.

Recall that state j is said to be recurrent if, starting from state j, the probability of returning to j is unity. If the uniformized chain $\{Z_n^\nu\}$ is recurrent for *some* ν, then $\{X(t)\}$ is also recurrent. If $\{Z_n^\nu\}$ is positive recurrent for *some* ν, then $\{X(t)\}$ is positive recurrent. The irreducibility of $\{X(t)\}$ also follows from the irreducibility of $\{Z_n^\nu\}$ for *some* $\nu \geq c$. These facts can be readily seen from Theorem 4.19. Note that no uniformized Markov chain is periodic. This follows since $\mathbf{P}_\nu$ has positive diagonal components for sufficiently large ν.

From Theorem 4.16, we have the following.

Theorem 4.20 *Suppose that a continuous-time Markov chain* $\{X(t)\}$ *with infinitesimal generator* $\mathbf{Q}$ *is uniformizable and that the uniformized chain* $\{Z_n^\nu\}$ *is positive recurrent for some* $\nu \geq \sup_i q_i$. *Then there exists a unique probability vector* $\boldsymbol{\pi}$ *positive componentwise that satisfies the stationary equation* $\boldsymbol{\pi}^\top \mathbf{Q} = \mathbf{0}^\top$. *The probability vector is the stationary distribution and is also the limiting distribution of both* $\{X(t)\}$ *and* $\{Z_n^\nu\}$.

Note that the continuous-time Markov chain $\{X(t)\}$ is ergodic if there is some $\nu \geq \sup_i q_i$ such that the uniformized Markov chain $\{Z_n^\nu\}$ is ergodic (see Definition 2.6 for the discrete-time case).

4.5 More on finite Markov chains

In this section, we assume that the state space is given by $\mathcal{N} = \{0, 1, \cdots, N\}$, where $N < \infty$. Since we can always choose $c = \max_{0 \le i \le N} q_i < \infty$, any finite Markov chain in continuous time is uniformizable. This section concerns two topics, time reversibility and the rate of convergence to stationarity, of continuous-time Markov chains via uniformization. Keilson's book (1979) is the main source of this section.

For an ergodic Markov chain $\{X(t)\}$ with infinitesimal generator $\mathbf{Q} = (q_{ij})$, let $\nu \ge c \equiv \max_{0 \le i \le N} |q_{ii}|$ and define $\mathbf{P}_\nu = (p^\nu_{ij})$ as in (4.33). Suppose that the detailed balance equation in (2.43) holds for $\mathbf{P}_\nu$, i.e.,

$$\pi_i p^\nu_{ij} = \pi_j p^\nu_{ji}, \quad i,\ j \in \mathcal{N},$$

where $\boldsymbol{\pi} = (\pi_i)$ is the stationary distribution of $\mathbf{P}_\nu$. Since

$$p^\nu_{ij} = \delta_{ij} + \frac{1}{\nu} q_{ij}, \quad i,\ j \in \mathcal{N},$$

the detailed balance equation can be rewritten as

$$\pi_i q_{ij} = \pi_j q_{ji}, \quad i,\ j \in \mathcal{N}. \tag{4.36}$$

The left-hand side of (4.36) describes the transition intensity from state i to state j under equilibrium, while the right-hand side describes the transition intensity from state j to state i. Let $\boldsymbol{\pi}_{\mathrm{D}}$ denote the diagonal matrix with diagonal elements π_i and define

$$\mathbf{Q}_{\mathrm{R}} \equiv \boldsymbol{\pi}_{\mathrm{D}}^{-1} \mathbf{Q}^\top \boldsymbol{\pi}_{\mathrm{D}}. \tag{4.37}$$

The diagonal elements of $\mathbf{Q}_{\mathrm{R}}$ are negative, the off-diagonal elements of $\mathbf{Q}_{\mathrm{R}}$ are nonnegative, and since $\boldsymbol{\pi}^\top \mathbf{Q} = \mathbf{0}^\top$ we have

$$\mathbf{Q}_{\mathrm{R}} \mathbf{1} = \boldsymbol{\pi}_{\mathrm{D}}^{-1} \mathbf{Q}^\top \boldsymbol{\pi} = \boldsymbol{\pi}_{\mathrm{D}}^{-1} (\boldsymbol{\pi}^\top \mathbf{Q})^\top = \mathbf{0}.$$

Hence, the matrix $\mathbf{Q}_{\mathrm{R}}$ can be considered as an infinitesimal generator of a continuous-time Markov chain on $\mathcal{N}$. Moreover,

$$\boldsymbol{\pi}^\top \mathbf{Q}_{\mathrm{R}} = \mathbf{1}^\top \mathbf{Q}^\top \boldsymbol{\pi}_{\mathrm{D}} = (\mathbf{Q}\mathbf{1})^\top \boldsymbol{\pi}_{\mathrm{D}} = \mathbf{0}^\top,$$

whence $\boldsymbol{\pi}$ is the stationary distribution of $\mathbf{Q}_{\mathrm{R}}$, too; see (4.29). The generator $\mathbf{Q}_{\mathrm{R}}$ is called the *dual* of $\mathbf{Q}$, since $(\mathbf{Q}_{\mathrm{R}})_{\mathrm{R}} = \mathbf{Q}$. Now, in matrix notation, the detailed balance equation (4.36) says that $\boldsymbol{\pi}_{\mathrm{D}} \mathbf{Q}$ is symmetric, i.e.,

$$\boldsymbol{\pi}_{\mathrm{D}} \mathbf{Q} = (\boldsymbol{\pi}_{\mathrm{D}} \mathbf{Q})^\top = \mathbf{Q}^\top \boldsymbol{\pi}_{\mathrm{D}},$$

so that $\mathbf{Q} = \mathbf{Q}_{\mathrm{R}}$. An ergodic Markov chain or its infinitesimal generator $\mathbf{Q}$ is said to be *reversible in time* if $\mathbf{Q}$ and its dual $\mathbf{Q}_{\mathrm{R}}$ are identical. The above arguments show that if the uniformized Markov chain is reversible in time then so is the ergodic Markov chain $\{X(t)\}$. Conversely, if $\{X(t)\}$ is reversible in time, then there is some $\nu \ge c$ such that the uniformized chain is reversible in time. The proof for this statement is carried out by reversing the above arguments.

Suppose that the ergodic Markov chain $\{X(t)\}$ is reversible in time. Then, the matrix $\boldsymbol{\pi}_D^{1/2}\mathbf{Q}\boldsymbol{\pi}_D^{-1/2}$ is symmetric and, from Theorem 4.18,

$$\boldsymbol{\pi}_D^{1/2}\mathbf{P}(t)\boldsymbol{\pi}_D^{-1/2} = \sum_{n=0}^{\infty} \frac{t^n}{n!}\left(\boldsymbol{\pi}_D^{1/2}\mathbf{Q}\boldsymbol{\pi}_D^{-1/2}\right)^n \tag{4.38}$$

is symmetric, where $\boldsymbol{\pi}_D^{1/2}$ and $\boldsymbol{\pi}_D^{-1/2}$ are the diagonal matrices with diagonal elements $\pi_i^{1/2}$ and $\pi_i^{-1/2}$ respectively. It follows that $\boldsymbol{\pi}_D\mathbf{P}(t)$ is symmetric for all $t \geq 0$, i.e.,

$$\pi_i p_{ij}(t) = \pi_j p_{ji}(t), \quad i,\ j \in \mathcal{N}; \quad t \geq 0.$$

The term 'reversible in time' can be justified as in the discrete-time case (see Section 2.5).

Since $p_{ij}^{\nu} = q_{ij}/\nu$ for $i \neq j$, by definition, Theorem 2.13 leads to the *Kolmogorov criterion* for time reversibility. The proof is left to the reader.

Theorem 4.21 *An ergodic Markov chain with infinitesimal generator* $\mathbf{Q} = (q_{ij})$ *is reversible in time if and only if*

$$q_{i,i_1}q_{i_1,i_2}\cdots q_{i_k,i} = q_{i,i_k}q_{i_k,i_{k-1}}\cdots q_{i_1,i}$$

for all $k \geq 2$ *and all states* $i,\ i_1, \cdots, i_k$.

Our first example in this section is concerned with birth–death processes, a continuous-time analog of random walks (see Example 2.17).

Example 4.10 Let $\{X(t)\}$ be the finite birth–death process considered in Example 4.4, and define $\pi_0 = 1$ and

$$\pi_{i+1} = \pi_i \frac{\lambda_i}{\mu_{i+1}} = \frac{\lambda_0 \cdots \lambda_i}{\mu_1 \cdots \mu_{i+1}}, \quad i = 0, 1, \cdots, N-1;$$

see Example 4.8. With $\lambda_{-1} = \lambda_N = \mu_0 = \mu_{N+1} = 0$, we have

$$\pi_{i-1}\lambda_{i-1} - \pi_i(\lambda_i + \mu_i) + \pi_{i+1}\mu_{i+1} = 0, \quad i = 0, 1, \cdots, N.$$

Hence, the probability vector $\boldsymbol{\pi}/\boldsymbol{\pi}^{\mathrm{T}}\mathbf{1}$, where $\boldsymbol{\pi} = (\pi_i)$, is the stationary distribution of the finite birth–death process $\{X(t)\}$. Also,

$$\pi_i q_{i,i+1} = \pi_i\lambda_i = \pi_{i+1}\mu_{i+1} = \pi_{i+1}q_{i+1,i}, \quad i = 0, 1, \cdots, N.$$

It follows that the birth–death process is reversible in time.

Let λ_i^{ν}, $i = 0, 1, \cdots, N$, be the eigenvalues of $\mathbf{P}_{\nu}$, where $\lambda_0^{\nu} = 1$, and let $\mathbf{u}_i$ and $\mathbf{v}_i$ be, respectively, the left and right eigenvectors associated with λ_i^{ν}, i.e.,

$$\lambda_i^{\nu}\mathbf{u}_i^{\mathrm{T}} = \mathbf{u}_i^{\mathrm{T}}\mathbf{P}_{\nu}; \quad \lambda_i^{\nu}\mathbf{v}_i = \mathbf{P}_{\nu}\mathbf{v}_i, \quad i = 0, 1, \cdots, N.$$

From (4.33), it is readily seen that

$$\lambda_i = -\nu(1-\lambda_i^\nu), \quad i = 0, 1, \cdots, N,$$

are the eigenvalues of $\mathbf{Q}$, and $\mathbf{u}_i$ and $\mathbf{v}_i$ are the associated left and right eigenvectors respectively. In particular, $\lambda_0 = 0$, $\mathbf{u}_0 = \boldsymbol{\pi}$ and $\mathbf{v}_0 = \mathbf{1}$. (The fact that $\lambda_0 = 0$ also follows from $\mathbf{Q1} = \mathbf{0}$.)

Suppose that $\{X(t)\}$ is reversible in time. Then $\boldsymbol{\pi}_{\mathrm{D}}^{1/2}\mathbf{Q}\boldsymbol{\pi}_{\mathrm{D}}^{-1/2}$ is symmetric, so that the eigenvalues λ_i are all real. Since $|\lambda_i^\nu| \le 1$ and $\lambda_i = -\nu(1-\lambda_i^\nu)$, we can assume that

$$0 = \lambda_0 > \lambda_1 \ge \cdots \ge \lambda_N \ge -2\nu.$$

The spectral decomposition of the symmetric matrix $\boldsymbol{\pi}_{\mathrm{D}}^{1/2}\mathbf{Q}\boldsymbol{\pi}_{\mathrm{D}}^{-1/2}$ is given by

$$\boldsymbol{\pi}_{\mathrm{D}}^{1/2}\mathbf{Q}\boldsymbol{\pi}_{\mathrm{D}}^{-1/2} = \sum_{j=0}^{N} \lambda_j \, \mathbf{x}_j \mathbf{x}_j^\top,$$

where $\mathbf{x}_j$ is the eigenvector associated with λ_j and $\mathbf{x}_i^\top \mathbf{x}_j = \delta_{ij}$. It follows that

$$\boldsymbol{\pi}_{\mathrm{D}}^{1/2}\mathbf{Q}^n\boldsymbol{\pi}_{\mathrm{D}}^{-1/2} = \sum_{j=0}^{N} \lambda_j^n \, \mathbf{x}_j \mathbf{x}_j^\top, \quad n = 0, 1, \cdots. \tag{4.39}$$

Let $\mathbf{u}_j = \boldsymbol{\pi}_{\mathrm{D}}^{1/2}\mathbf{x}_j$ and $\mathbf{v}_j = \boldsymbol{\pi}_{\mathrm{D}}^{-1/2}\mathbf{x}_j$. Then, from Theorem 4.18, we have

$$\mathbf{P}(t) = \sum_{j=0}^{N} e^{\lambda_j t} \, \mathbf{v}_j \mathbf{u}_j^\top, \quad t \ge 0.$$

Writing $\xi_i = -\lambda_i > 0$ for $i = 1, \cdots, N$, we finally have

$$\mathbf{P}(t) = \mathbf{1}\boldsymbol{\pi}^\top + \sum_{j=1}^{N} e^{-\xi_j t} \, \mathbf{v}_j \mathbf{u}_j^\top, \quad t \ge 0, \tag{4.40}$$

which should be compared with the result given in Theorem A.7. The decomposition (4.40) is a continuous-time analog of (2.49). If the initial distribution is $\boldsymbol{\alpha}$, then the state distribution is given by

$$\boldsymbol{\pi}(t) = \boldsymbol{\pi} + \sum_{j=1}^{N} e^{-\xi_j t} (\boldsymbol{\alpha}^\top \mathbf{v}_j) \mathbf{u}_j, \quad t \ge 0, \tag{4.41}$$

which tends to $\boldsymbol{\pi}$ as $t \to \infty$ irrespective of the initial distribution $\boldsymbol{\alpha}$.

Theorem 4.22 *If a finite, ergodic Markov chain is reversible in time, then the transition probability function $p_{ii}(t)$ is completely monotone for every $i \in \mathcal{N}$.*

Proof. From (4.40), we have

$$p_{ii}(t) = \pi_i + \sum_{j=1}^{N} e^{-\xi_j t} x_{ji}^2, \quad t \geq 0,$$

where x_{ji} is the ith component of $\mathbf{x}_j$. Define

$$F(x) = \pi_i + \sum_{x_{ji} \leq x} x_{ji}^2, \quad x \geq 0.$$

The result follows from Theorem B.13. □

Based on the above theorem, an approximation of the transition probability function $p_{ii}(t)$ is given by

$$p_{ii}(t) \approx \pi_i + (1 - \pi_i) e^{-\eta_i t}, \quad t \geq 0, \tag{4.42}$$

where we choose η_i so that

$$\int_0^\infty (1 - \pi_i) e^{-\eta_i t} dt = \int_0^\infty \{p_{ii}(t) - \pi_i\} dt.$$

That is, $(1 - \pi_i)/\eta_i$ equals the (i, i)th component of

$$\mathbf{Z} \equiv \int_0^\infty \{\mathbf{P}(t) - \mathbf{1}\boldsymbol{\pi}^\top\} dt.$$

The matrix $\mathbf{Z}$ is called the *fundamental matrix* of the ergodic Markov chain $\{X(t)\}$ in continuous time. The fundamental matrix can be written in terms of the generator $\mathbf{Q}$ and the stationary distribution $\boldsymbol{\pi}$ (see Exercise 4.21). See Iosifescu (1980) and Keilson (1979) for discussions of the fundamental matrix in the context of ergodic Markov chains in continuous time.

Example 4.11 Let $\{X(t)\}$ be the queue-size process of the M/M/1/N queue considered in Example 4.5, and let $\mathbf{Q}$ be the infinitesimal generator given in (4.21). From Example 4.10, $\{X(t)\}$ is reversible in time. Also,

$$\mathbf{U}^{-1}\mathbf{Q}\mathbf{U} = \begin{pmatrix} 0 & \mathbf{a}^\top \\ \mathbf{0} & \mathbf{A} \end{pmatrix},$$

where $\mathbf{a}^\top = (\lambda, 0, \cdots, 0)$ and

$$\mathbf{A} = \begin{pmatrix} -\lambda-\mu & \lambda & 0 & 0 & \cdots & 0 \\ \mu & -\lambda-\mu & \lambda & 0 & \cdots & 0 \\ 0 & \mu & -\lambda-\mu & \lambda & \cdots & 0 \\ \vdots & & \ddots & \ddots & \ddots & \vdots \\ 0 & \cdots & 0 & \mu & -\lambda-\mu & \lambda \\ 0 & \cdots & 0 & 0 & \mu & -\lambda-\mu \end{pmatrix}.$$

Since $\mathbf{U}^{-1}\mathbf{Q}\mathbf{U}$ is a similarity transform, the eigenvalues of $\mathbf{Q}$ other than zero coincide with the eigenvalues of $\mathbf{A}$. Let π_i be defined as in Example 4.10, i.e.,

$$\pi_i = \rho^{i-1}, \quad i = 1, \cdots, N; \quad \rho = \frac{\lambda}{\mu}.$$

It is easily seen that

$$\boldsymbol{\pi}_{\mathrm{D}}^{1/2} \mathbf{A} \boldsymbol{\pi}_{\mathrm{D}}^{-1/2} = \sqrt{\lambda\mu} \begin{pmatrix} -k & 1 & 0 & 0 & \cdots & 0 \\ 1 & -k & 1 & 0 & \cdots & 0 \\ 0 & 1 & -k & 1 & \cdots & 0 \\ \vdots & & \ddots & \ddots & \ddots & \vdots \\ 0 & \cdots & 0 & 1 & -k & 1 \\ 0 & \cdots & 0 & 0 & 1 & -k \end{pmatrix}, \quad k = \frac{\lambda + \mu}{\sqrt{\lambda\mu}}.$$

The eigenvalues of $\boldsymbol{\pi}_{\mathrm{D}}^{1/2} \mathbf{A} \boldsymbol{\pi}_{\mathrm{D}}^{-1/2}$ are given by

$$\lambda_i = -(\lambda + \mu) + 2\sqrt{\lambda\mu} \cos \frac{i\pi}{N+1}, \quad i = 1, \cdots, N,$$

and the associated eigenvector $\mathbf{x}_i = (x_{ij})$ is

$$x_{ij} = \sin \frac{ij\pi}{N+1}, \quad j = 1, \cdots, N.$$

See, e.g., Noble and Daniel (page 296, 1977). The rest of the calculations leading to the spectral decomposition of the transition matrix $\mathbf{P}(t)$ of the M/M/1/N queue are left to the reader (see Exercise 4.22).

We now turn our attention to the rate of convergence to the stationarity of finite, ergodic Markov chains in continuous time. Let $\{X(t)\}$ be an ergodic Markov chain with infinitesimal generator $\mathbf{Q} = (q_{ij})$. After uniformization, suppose that $\mathbf{P}_\nu$ is regular for some $\nu \geq c = \max_i q_i$. Then all the eigenvalues of $\mathbf{P}_\nu$ other than unity are strictly less than unity in magnitude. If we denote the eigenvalues of $\mathbf{P}_\nu$ by λ_i^ν, then the $\lambda_i = -\nu(1 - \lambda_i^\nu)$ are the eigenvalues of $\mathbf{Q}$. Note that the real part of λ_i other than zero is strictly negative. Thus we assume that the λ_i are ordered as

$$0 > \mathrm{Re}\,(\lambda_1) \geq \mathrm{Re}\,(\lambda_2) \geq \cdots \geq \mathrm{Re}\,(\lambda_N),$$

where $\lambda_0 = 0$ and $\mathrm{Re}\,(\lambda)$ denotes the real part of the complex number λ. Let $\rho = -\mathrm{Re}\,(\lambda_1) > 0$. Then, from Theorem A.7, we have

$$\mathbf{P}(t) = \mathbf{1}\boldsymbol{\pi}^\top + O(e^{-\rho t}), \quad \text{as } t \to \infty.$$

The value ρ is called the ***decay parameter*** of $\{X(t)\}$ and the Markov chain is said to be ***exponentially ergodic***. The relaxation time for the continuous-time Markov chain is defined by

$$T_{\mathrm{REL}}(\mathbf{Q}) = \frac{1}{\rho}. \tag{4.43}$$

An easily computable upper bound for ρ is therefore of practical importance. See Friedland and Gurvits (1994) for an upper bound on the decay parameter.

For a stochastic matrix $\mathbf{P}$, let $\tau(\mathbf{P})$ be the coefficient of ergodicity as defined in (2.52). For each t, we define $\tau(t) = \tau(\mathbf{P}(t))$. After uniformization, we have

$$\|(\mathbf{x}^\top \mathbf{P}(t))^\top\|_1 \le \sum_{n=0}^{\infty} \frac{(\nu t)^n}{n!} e^{-\nu t} \|(\mathbf{x}^\top \mathbf{P}_\nu^n)^\top\|_1,$$

where $\|\cdot\|_1$ denotes the ℓ_1-norm and the above inequality follows from the triangle inequality. From (2.54) and property (c) of τ, we have

$$\begin{aligned}\tau(t) &\le \sum_{n=0}^{\infty} \frac{(\nu t)^n}{n!} e^{-\nu t} \tau(\mathbf{P}_\nu^n) \\ &\le \sum_{n=0}^{\infty} \frac{(\nu t)^n}{n!} e^{-\nu t} \tau^n(\mathbf{P}_\nu) \\ &= \exp\{-\nu(1-\tau(\mathbf{P}_\nu))t\}.\end{aligned}$$

Recall that $\tau(\mathbf{P}) = 0$ if and only if $\mathbf{P} = \mathbf{1}\boldsymbol{\pi}^\top$. Also, as in Theorem 2.15, it is easily shown that

$$\|\mathbf{P}(t) - \mathbf{1}\boldsymbol{\pi}^\top\|_\infty \le C\tau(t),$$

where C is a constant independent of t. Therefore, we conclude that if $\tau(\mathbf{P}_\nu) < 1$ for some ν, then the continuous-time Markov chain $\{X(t)\}$ is exponentially ergodic.

Corresponding to the ergodic Markov chain $\{X(t)\}$, let $\{\widehat{X}(t)\}$ be a stationary Markov chain with infinitesimal generator $\mathbf{Q}$ and the initial distribution $\boldsymbol{\pi}$. As for the discrete-time case (2.57), we define the *correlation coefficient* of $\mathbf{Q}$ by

$$d_t(\mathbf{Q}) = \sup_{f,g} \operatorname{Corr}[f(\widehat{X}(0)),\, g(\widehat{X}(t))], \quad t \ge 0.$$

It is a simple matter to show that

$$d_t(\mathbf{Q}) = \sup_{f,g} \frac{\mathbf{f}^\top \boldsymbol{\pi}_{\mathrm{D}}(\mathbf{P}(t) - \mathbf{1}\boldsymbol{\pi}^\top)\mathbf{g}}{\sqrt{\mathbf{f}^\top \boldsymbol{\pi}_{\mathrm{D}}\mathbf{f} - (\mathbf{f}^\top\boldsymbol{\pi})^2}\sqrt{\mathbf{g}^\top \boldsymbol{\pi}_{\mathrm{D}}\mathbf{g} - (\mathbf{g}^\top\boldsymbol{\pi})^2}};$$

cf. (2.58). After uniformization, as in (2.60), we write

$$\widetilde{\mathbf{P}}_\nu = \sqrt{\boldsymbol{\pi}}\sqrt{\boldsymbol{\pi}}^\top + \widetilde{\boldsymbol{\Delta}}_\nu,$$

where $\widetilde{\mathbf{P}}_\nu = \boldsymbol{\pi}_{\mathrm{D}}^{1/2}\mathbf{P}_\nu\boldsymbol{\pi}_{\mathrm{D}}^{-1/2}$. It is easily seen that

$$\widetilde{\mathbf{P}}(t) - \sqrt{\boldsymbol{\pi}}\sqrt{\boldsymbol{\pi}}^\top = \sum_{n=0}^{\infty} \frac{(\nu t)^n}{n!} e^{-\nu t} \widetilde{\boldsymbol{\Delta}}_\nu^n - e^{-\nu t}\sqrt{\boldsymbol{\pi}}\sqrt{\boldsymbol{\pi}}^\top, \quad t \ge 0.$$

Therefore, by mimicking the arguments given in Section 2.6, it follows that

$$d_t(\mathbf{Q}) = \sup_{\mathbf{x},\mathbf{y}\in W} \sum_{n=0}^{\infty} \frac{(\nu t)^n}{n!} e^{-\nu t} \frac{\mathbf{x}^\top \widetilde{\mathbf{P}}_\nu^n \mathbf{y}}{\|\mathbf{x}\|_2 \, \|\mathbf{y}\|_2}, \tag{4.44}$$

where $W = \{\mathbf{y} : \mathbf{y}^\top \sqrt{\boldsymbol{\pi}} = 0\} \subset R^{N+1}$; see (2.61). Obvious implications of (4.44) are the following.

Theorem 4.23 *Let $\{X(t)\}$ be a finite, ergodic Markov chain with infinitesimal generator $\mathbf{Q}$.*

(i) *If $\{X(t)\}$ is reversible in time, then $d_t(\mathbf{Q}) = e^{-|\lambda_1|t}$, where λ_1 is the largest negative eigenvalue of $\mathbf{Q}$.*

(ii) *If $\mathbf{Q}$ has a real eigenvalue λ different from zero, then*

$$d_t(\mathbf{Q}) \geq \frac{|\mathbf{u}^\top \mathbf{v}|}{\|\mathbf{u}\|_2 \, \|\mathbf{v}\|_2} e^{-|\lambda|t}, \quad t \geq 0,$$

where $\mathbf{u}$ and $\mathbf{v}$ are the left and right eigenvectors, respectively, associated with λ.

Proof. If we denote the eigenvalues of $\widetilde{\mathbf{P}}_\nu$ by λ_i^ν, then $\lambda_i = -\nu(1 - \lambda_i^\nu)$ are the eigenvalues of $\mathbf{Q}$. If $\{X(t)\}$ is reversible in time, the matrix $\widetilde{\boldsymbol{\Delta}}_\nu$ is symmetric and hence the λ_i^ν are all real. Moreover,

$$\widetilde{\boldsymbol{\Delta}}_\nu^n = \sum_{i=1}^{N} (\lambda_i^\nu)^n \, \mathbf{x}_i \mathbf{x}_i^\top, \quad n = 0, 1, \cdots,$$

where the $\mathbf{x}_i$ are the associated orthonormal eigenvectors. Note that, by choosing ν sufficiently large, we can assume that $1 > \lambda_1^\nu \geq \lambda_i^\nu \geq 0$ for all i. It follows that

$$\begin{aligned} d_t(\mathbf{Q}) &= \sup_{\mathbf{x},\mathbf{y}\in W} \sum_{n=0}^{\infty} \frac{(\nu t)^n}{n!} e^{-\nu t} \frac{\mathbf{x}^\top \widetilde{\mathbf{P}}_\nu^n \mathbf{y}}{\|\mathbf{x}\|_2 \, \|\mathbf{y}\|_2} \\ &= \sum_{n=0}^{\infty} \frac{(\nu t)^n}{n!} e^{-\nu t} \frac{\mathbf{x}_1^\top \widetilde{\boldsymbol{\Delta}}_\nu^n \mathbf{x}_1}{\|\mathbf{x}_1\|_2^2} \\ &= \sum_{n=0}^{\infty} \frac{(\nu t)^n}{n!} e^{-\nu t} (\lambda_1^\nu)^n \\ &= e^{-\nu t(1-\lambda_1^\nu)}, \end{aligned}$$

whence $d_t(\mathbf{Q}) = e^{\lambda_1 t}$. Part (ii) follows similarly. □

Let $\mathbf{Q}_\mathrm{R} = \boldsymbol{\pi}_\mathrm{D}^{-1} \mathbf{Q}^\top \boldsymbol{\pi}$ be the dual of $\mathbf{Q}$. Recall that, for the discrete-time case, the stochastic matrices defined in (2.62) provide an upper bound on the correlation coefficient (see Theorem 2.17). For the continuous-time case, this role is played by the infinitesimal generator

$$\mathbf{Q}^* = \frac{1}{2}(\mathbf{Q} + \mathbf{Q}_\mathrm{R}). \tag{4.45}$$

Note that, since $\mathbf{Q}_\mathrm{R}^* = \mathbf{Q}^*$, the Markov chain governed by the generator $\mathbf{Q}^*$ is reversible in time. Hence, from Theorem 4.23(i), its correlation coefficient is given by $d_t(\mathbf{Q}^*) = e^{-\rho t}$, where ρ is the largest negative eigenvalue of $\mathbf{Q}^*$. The next result is due to Kijima (1989a).

Theorem 4.24 *For any finite infinitesimal generator* $\mathbf{Q}$, *we have*

$$d_t(\mathbf{Q}) \le d_t(\mathbf{Q}^*), \quad t \ge 0.$$

Proof. From (4.44), one easily sees that

$$d_t(\mathbf{Q}) \le \sum_{n=0}^{\infty} \frac{(\nu t)^n}{n!} e^{-\nu t} d(\mathbf{P}_\nu^n) \le \exp\{-\nu(1 - \|\widetilde{\mathbf{\Delta}}_\nu\|_2)t\}, \tag{4.46}$$

where the first inequality follows from the definition of $d(\mathbf{P}_\nu^n)$ and the second follows from the proof of Theorem 2.18. Consider the symmetric matrix

$$\widetilde{\mathbf{P}}_\nu^\mathsf{T}\widetilde{\mathbf{P}}_\nu = \mathbf{I} + \frac{2}{\nu}\widetilde{\mathbf{Q}}^* + \frac{1}{\nu^2}\widetilde{\mathbf{Q}}^\mathsf{T}\widetilde{\mathbf{Q}},$$

where $\widetilde{\mathbf{Q}} = \boldsymbol{\pi}_\mathrm{D}^{1/2}\mathbf{Q}\boldsymbol{\pi}_\mathrm{D}^{-1/2}$. Denote the second largest eigenvalue of symmetric matrix $\mathbf{A}$ by $\lambda(\mathbf{A})$. Then, since

$$\widetilde{\mathbf{P}}_\nu^\mathsf{T}\widetilde{\mathbf{P}}_\nu = \sqrt{\boldsymbol{\pi}}\sqrt{\boldsymbol{\pi}}^\mathsf{T} + \widetilde{\mathbf{\Delta}}_\nu^\mathsf{T}\widetilde{\mathbf{\Delta}}_\nu,$$

we have

$$\begin{aligned} -\nu(1 - \|\widetilde{\mathbf{\Delta}}_\nu\|_2) &= -\nu\left(1 - \sqrt{\lambda\left(\mathbf{I} + \frac{2}{\nu}\widetilde{\mathbf{Q}}^* + \frac{1}{\nu^2}\widetilde{\mathbf{Q}}^\mathsf{T}\widetilde{\mathbf{Q}}\right)}\right) \\ &= \frac{\lambda\left(2\widetilde{\mathbf{Q}}^* + \widetilde{\mathbf{Q}}^\mathsf{T}\widetilde{\mathbf{Q}}/\nu\right)}{1 + \sqrt{\lambda\left(\mathbf{I} + \frac{2}{\nu}\widetilde{\mathbf{Q}}^* + \frac{1}{\nu^2}\widetilde{\mathbf{Q}}^\mathsf{T}\widetilde{\mathbf{Q}}\right)}}. \end{aligned}$$

But, since $\lambda(2\widetilde{\mathbf{Q}}^* + \widetilde{\mathbf{Q}}^\mathsf{T}\widetilde{\mathbf{Q}}/\nu) \le 0$ and

$$0 \le \lambda\left(\mathbf{I} + \frac{2}{\nu}\widetilde{\mathbf{Q}}^* + \frac{1}{\nu^2}\widetilde{\mathbf{Q}}^\mathsf{T}\widetilde{\mathbf{Q}}\right) \le 1,$$

it follows that

$$-\nu(1 - \|\widetilde{\mathbf{\Delta}}_\nu\|_2) \le \lambda\left(\widetilde{\mathbf{Q}}^* + \frac{1}{2\nu}\widetilde{\mathbf{Q}}^\mathsf{T}\widetilde{\mathbf{Q}}\right).$$

Write $\lambda(\nu) \equiv \lambda\left(2\widetilde{\mathbf{Q}}^* + \widetilde{\mathbf{Q}}^\mathsf{T}\widetilde{\mathbf{Q}}/\nu\right)$ and let $\mathbf{v}(\nu)$ be the right eigenvector associated with $\lambda(\nu)$, i.e.,

$$\left(2\widetilde{\mathbf{Q}}^* + \frac{1}{\nu}\widetilde{\mathbf{Q}}^\mathsf{T}\widetilde{\mathbf{Q}}\right)\mathbf{v}(\nu) = \lambda(\nu)\,\mathbf{v}(\nu).$$

Suppose $\|\mathbf{v}(\nu)\|_2 = 1$. Note that $\mathbf{v}^\top(\nu)$ is the left eigenvector. Differentiating with respect to ν componentwise in the above equation and then pre-multiplying by $\mathbf{v}^\top(\nu)$ on both sides yields

$$\lambda'(\nu) = -\frac{1}{\nu^2}\,\|\tilde{\mathbf{Q}}\,\mathbf{v}(\nu)\|_2^2 \le 0.$$

It follows that

$$\inf_{\nu \ge \max q_i} \lambda\left(\tilde{\mathbf{Q}}^* + \frac{1}{2\nu}\,\tilde{\mathbf{Q}}^\top\tilde{\mathbf{Q}}\right) = \lambda(\tilde{\mathbf{Q}}^*).$$

Therefore

$$\inf_{\nu \ge \max q_i} \{-\nu(1 - \|\tilde{\mathbf{\Delta}}_\nu\|_2)\} \le \lambda(\tilde{\mathbf{Q}}^*),$$

and the theorem follows from (4.46) and Theorem 4.23(i). □

When $\{X(t)\}$ is reversible in time, the infinitesimal generator $\mathbf{Q}^*$ defined in (4.45) coincides with $\mathbf{Q}$ itself, so that the inequality in Theorem 4.24 is tight. An extended notion of time reversibility may be formulated as $\mathbf{Q}\mathbf{Q}_{\mathrm{R}} = \mathbf{Q}_{\mathrm{R}}\mathbf{Q}$. Then

$$\frac{1}{t}\int_0^t \mathbf{P}(u)\mathbf{P}_{\mathrm{R}}(t-u)du = \frac{1}{t}\int_0^t \mathbf{P}_{\mathrm{R}}(u)\mathbf{P}(t-u)du, \quad t > 0,$$

where $\mathbf{P}_{\mathrm{R}}(t) = \exp\{\mathbf{Q}_{\mathrm{R}}t\}$. In this case, Theorem 4.24 also holds with strict equality (see Kijima, 1989a).

4.6 Absorbing Markov chains in continuous time

Consider an absorbing Markov chain $\{X(t)\}$ on the state space $\mathcal{N}$ with k absorbing states. Recall that state i is absorbing if and only if $q_i = 0$. Since $q_{ij} \ge 0$ and $\sum_{j \ne i} q_{ij} \le q_i$, the row corresponding to the absorbing state is a zero row vector. Therefore, by renumbering states, the infinitesimal generator can be written in canonical form as

$$\mathbf{Q} = \begin{pmatrix} \mathbf{O} & \mathbf{O} \\ \mathbf{R} & \mathbf{T} \end{pmatrix}. \tag{4.47}$$

The submatrix $\mathbf{T}$ is square and corresponds to the nonabsorbing states. The diagonal elements of $\mathbf{T}$ are strictly negative and the off-diagonal elements are nonnegative. It is assumed that all the nonabsorbing states communicate with each other and that they are transient. Hence $\mathbf{T1} \le \mathbf{0}$ with at least one strict inequality. Such a matrix is called a *lossy generator*. As before, we denote the set of absorbing states by $\mathcal{A}$ and the set of transient states by $\mathcal{A}^c$. The matrix $\mathbf{R}$ is nonnegative and nonzero, but need not be square. Throughout this section, we assume that the state space $\mathcal{N}$ is finite to eliminate technical difficulties.

From (4.47), a routine computation yields

$$\mathbf{Q}^n = \begin{pmatrix} \mathbf{O} & \mathbf{O} \\ \mathbf{T}^{n-1}\mathbf{R} & \mathbf{T}^n \end{pmatrix}, \quad n = 1, 2, \cdots;$$

cf. (2.67). Let $\mathbf{P}(t)$ denote the transition matrix function of the absorbing Markov chain $\{X(t)\}$. We obtain

$$\mathbf{P}(t) = \sum_{n=0}^{\infty} \frac{\mathbf{Q}^n t^n}{n!} = \begin{pmatrix} \mathbf{I} & \mathbf{O} \\ \mathbf{R}(t) & \mathbf{T}(t) \end{pmatrix}, \tag{4.48}$$

where

$$\mathbf{T}(t) \equiv \sum_{n=0}^{\infty} \frac{\mathbf{T}^n t^n}{n!} = \exp\{\mathbf{T}\, t\}$$

and

$$\mathbf{R}(t) \equiv \sum_{n=1}^{\infty} \frac{\mathbf{T}^{n-1} t^n}{n!} \mathbf{R} = \left(\int_0^t \mathbf{T}(u) du \right) \mathbf{R}.$$

Here, term-by-term integration is allowed by Corollary B.1. Under the assumptions mentioned above, we have $\lim_{t \to \infty} \mathbf{T}(t) = \mathbf{O}$ and

$$\mathbf{N} \equiv \int_0^{\infty} \mathbf{T}(t) dt \tag{4.49}$$

exists, as will be seen shortly. As in the discrete-time case, the matrix $\mathbf{N}$ is called the *fundamental matrix* of the absorbing Markov chain $\{X(t)\}$. It follows from (4.48) that

$$\lim_{t \to \infty} \mathbf{P}(t) = \begin{pmatrix} \mathbf{I} & \mathbf{O} \\ \mathbf{NR} & \mathbf{O} \end{pmatrix}; \tag{4.50}$$

see the discrete-time counterpart (2.70).

Theorem 4.25 *The matrix* $\mathbf{T}$ *is nonsingular and*

$$\mathbf{N} = -\mathbf{T}^{-1}.$$

Proof. We have

$$\mathbf{T} \int_0^t \mathbf{T}(s) ds = \mathbf{T} \sum_{n=1}^{\infty} \frac{\mathbf{T}^{n-1} t^n}{n!} = \sum_{n=1}^{\infty} \frac{\mathbf{T}^n t^n}{n!} = \mathbf{T}(t) - \mathbf{I}.$$

Hence, letting $t \to \infty$, it follows that

$$\mathbf{TN} = \lim_{t \to \infty} \mathbf{T}(t) - \mathbf{I} = -\mathbf{I}.$$

Here the fact that $\lim_{t \to \infty} \mathbf{T}(t) = \mathbf{O}$ follows since the corresponding states are transient. Similarly, $\mathbf{NT} = -\mathbf{I}$ and the theorem follows. □

Let T_j be the random variable representing a time at which absorption at state $j \in \mathcal{A}$ occurs. Let

$$a_{ij} \equiv P_i[T_j < \infty], \quad i \in \mathcal{A}^c, \; j \in \mathcal{A},$$

and define $\mathbf{A} = (a_{ij})$. From (4.50), it is readily seen that

$$\mathbf{A} = \mathbf{NR}. \tag{4.51}$$

Since $\mathbf{Q1} = \mathbf{0}$, we have $\mathbf{R1} = -\mathbf{T1}$. It follows from Theorem 4.25 that

$$\mathbf{A1} = \mathbf{NR1} = -\mathbf{NT1} = \mathbf{1},$$

whence the Markov chain is eventually absorbed at one of the absorbing states. Alternatively, (4.51) can be obtained as follows. Consider the embedded Markov chain $\{Y_n\}$ defined in Section 4.2. The transition matrix of $\{Y_n\}$ is given by

$$\mathbf{P}_e = \begin{pmatrix} \mathbf{I} & \mathbf{O} \\ \mathbf{R}_e & \mathbf{T}_e \end{pmatrix}; \quad \mathbf{T}_e = \mathbf{I} + \mathbf{T}_{\mathrm{D}}^{-1}\mathbf{T}, \quad \mathbf{R}_e = \mathbf{T}_{\mathrm{D}}^{-1}\mathbf{R},$$

where $\mathbf{T}_{\mathrm{D}}$ denotes the diagonal matrix with diagonal elements q_i and the suffix e indicates the embedded chain. For the discrete-time Markov chain $\{Y_n\}$, the fundamental matrix is given by

$$\mathbf{N}_e = (\mathbf{I} - \mathbf{T}_e)^{-1} = -\mathbf{T}^{-1}\mathbf{T}_{\mathrm{D}} = \mathbf{N}\mathbf{T}_{\mathrm{D}},$$

see (2.68). Note that the absorption probabilities of $\{Y_n\}$ are equal to those of $\{X(t)\}$. Hence

$$\mathbf{A} = \mathbf{A}_e = \mathbf{N}_e\mathbf{R}_e = \mathbf{NR},$$

as desired.

The fundamental matrix plays a central role in the theory of absorbing Markov chains. For example, let N_j represent the time spent in state $j \in \mathcal{A}^c$ until absorption, i.e.,

$$N_j = \int_0^\infty \mathrm{I}_{\{X(t)=j\}} dt, \quad j \in \mathcal{A}^c.$$

Note that

$$E_i[\mathrm{I}_{\{X(t)=j\}}] = P_i[X(t) = j] = p_{ij}(t).$$

Hence, for $i,\ j \in \mathcal{A}^c$, we have

$$E_i[N_j] = E_i\left[\int_0^\infty \mathrm{I}_{\{X(t)=j\}} dt\right] = \int_0^\infty p_{ij}(t)dt,$$

so that

$$(E_i[N_j]) = \int_0^\infty \mathbf{T}(t)dt = \mathbf{N}.$$

The mean time spent in the transient states, starting from state i, is given by $\sum_{j\in\mathcal{A}^c} E_i[N_j]$ or, in matrix form, $[\mathbf{N1}]_i$.

We now turn our attention to the first passage time of an ergodic Markov chain in continuous time. Let $\{X(t)\}$ be an ergodic Markov chain with state space $\mathcal{N} = \{0, 1, \cdots, N\}$ and let T_j be the first passage time to state $j \in \mathcal{N}$. For $i \neq j$, define

$$F_{ij}(t) = P_i[T_j \leq t], \quad t \geq 0.$$

In the study of the first-passage-time distribution $F_{ij}(t)$, we consider an absorbing Markov chain constructed from the original Markov chain $\{X(t)\}$ by making state j absorbing. As in the discrete-time case, it is assumed that $j = 0$ without loss of generality. Then, the absorbing Markov chain has the infinitesimal generator

$$\mathbf{Q} = \begin{pmatrix} 0 & \mathbf{0}^\top \\ \mathbf{r} & \mathbf{T} \end{pmatrix}, \tag{4.52}$$

if the original generator is

$$\begin{pmatrix} -q_0 & \mathbf{q}_0^\top \\ \mathbf{r} & \mathbf{T} \end{pmatrix},$$

with $\mathbf{q}_0^\top = (q_{0i})$. Note that $\mathbf{r} = (q_{j0}) = -\mathbf{T}\mathbf{1}$. If the original Markov chain is irreducible, the absorbing state 0 is reached from any state so that $\mathbf{r}$ is nonzero and $\mathbf{T}$ is a lossy generator. Let

$$\phi_i(s) = \int_0^\infty e^{-st} dF_{i0}(t), \quad \mathrm{Re}\,(s) > 0; \quad i = 1, \cdots, N,$$

i.e. the Laplace–Stieltjes transform of $F_{i0}(t)$, and denote $\boldsymbol{\phi}(s) = (\phi_i(s))$. The next theorem is analogous to Theorem 2.19.

Theorem 4.26 *For the ergodic Markov chain* $\{X(t)\}$, *we have*

$$\boldsymbol{\phi}(s) = (s\mathbf{I} - \mathbf{T})^{-1}\mathbf{r}, \quad \mathrm{Re}\,(s) > 0.$$

Proof. Since

$$P_i[t < T_0 \leq t + dt] = \sum_{j \neq 0} p_{ij}(t) q_{j0} dt + o(dt),$$

it is readily seen that

$$\boldsymbol{\phi}(s) = \boldsymbol{\Pi}(s)\,\mathbf{r}, \quad \mathrm{Re}\,(s) > 0,$$

where $\boldsymbol{\Pi}(s) = \int_0^\infty e^{-st}\mathbf{T}(t)dt$. But, the same argument leading to (4.17) holds equally well for any lossy generator, so that

$$\boldsymbol{\Pi}(s) = (s\mathbf{I} - \mathbf{T})^{-1}, \quad \mathrm{Re}\,(s) > 0,$$

whence the result. □

Since $\mathbf{r} = -\mathbf{T}\mathbf{1}$, we have from Theorem 4.26 that

$$\boldsymbol{\phi}(0) = (-\mathbf{T})^{-1}\mathbf{r} = \mathbf{1}.$$

Hence $\phi_i(0) = P_i[T_0 < \infty] = 1$, i.e., the first-passage-time distributions are nondefective. Also, differentiation of $\phi(s)$ with respect to s gives us

$$\phi'(s) = -(s\mathbf{I} - \mathbf{T})^{-2}\mathbf{r}, \quad \mathrm{Re}\,(s) > 0.$$

It follows that

$$-\phi'(0) = (-\mathbf{T})^{-2}\mathbf{r} = \mathbf{N1}.$$

A higher order moment of T_0 can be obtained by further differentiation of $\phi'(s)$ and its evaluation at $s = 0$.

Now, suppose that the original ergodic Markov chain is reversible in time, i.e., (4.36) holds for all $i,\ j \in \mathcal{N}$. In the absorbing Markov chain with infinitesimal generator (4.52), since $\mathbf{T}$ is a submatrix of the original generator, it is clear that the set of equations

$$\pi_i q_{ij} = \pi_j q_{ji}, \quad i,\ j = 1, \cdots, N,$$

still holds, where $\boldsymbol{\pi} = (\pi_i)$ is the stationary distribution of the original ergodic Markov chain $\{X(t)\}$. That is, $\boldsymbol{\pi}_{\mathrm{D}}\mathbf{T}$ is symmetric and hence so is $\boldsymbol{\pi}_{\mathrm{D}}^{1/2}\mathbf{T}\boldsymbol{\pi}_{\mathrm{D}}^{-1/2}$. Therefore, the matrix $\boldsymbol{\pi}_{\mathrm{D}}^{1/2}\mathbf{T}\boldsymbol{\pi}_{\mathrm{D}}^{-1/2}$ admits the spectral decomposition

$$\boldsymbol{\pi}_{\mathrm{D}}^{1/2}\mathbf{T}\boldsymbol{\pi}_{\mathrm{D}}^{-1/2} = \sum_{j=1}^{N} \lambda_j\, \mathbf{x}_j\mathbf{x}_j^{\mathsf{T}},$$

where the λ_j are the eigenvalues of $\mathbf{T}$, which are strictly negative, and the $\mathbf{x}_j$ are the associated orthonormal eigenvectors. It follows that

$$\mathbf{T}^n = \sum_{j=1}^{N} \lambda_j^n\, \mathbf{v}_j\mathbf{u}_j^{\mathsf{T}}, \quad n = 0, 1, \cdots,$$

where $\mathbf{u}_j = \boldsymbol{\pi}_{\mathrm{D}}^{1/2}\mathbf{x}_j$ and $\mathbf{v}_j = \boldsymbol{\pi}_{\mathrm{D}}^{-1/2}\mathbf{x}_j$. Since

$$\left(\mathbf{I} - \frac{\mathbf{T}}{s}\right)^{-1} = \sum_{n=0}^{\infty} \frac{\mathbf{T}^n}{s^n} = \sum_{j=1}^{N} \frac{s}{s - \lambda_j}\, \mathbf{v}_j\mathbf{u}_j^{\mathsf{T}},$$

Theorem 4.26 yields

$$\phi(s) = \sum_{j=1}^{N} \frac{1}{s + \xi_j}\, \mathbf{v}_j(\mathbf{u}_j^{\mathsf{T}}\mathbf{r}), \quad \mathrm{Re}\,(s) > 0,$$

where $\xi_j = -\lambda_j > 0$. Note that $\xi/(s + \xi)$ is the Laplace transform of the exponential density

$$e(t; \xi) = \xi\, e^{-\xi t}, \quad t \geq 0.$$

Hence the first-passage-time density $f_{i0}(t)$ is given by

$$f_{i0}(t) = \sum_{j=1}^{N} \beta_{ij}\, \xi_j\, e^{-\xi_j t}, \quad t \geq 0; \quad \beta_{ij} \equiv \frac{(\boldsymbol{\delta}_i^{\mathsf{T}}\mathbf{v}_j)(\mathbf{u}_j^{\mathsf{T}}\mathbf{r})}{\xi_j}.$$

If $\beta_{ij} > 0$, then $f_{i0}(t)$ is a mixture of exponential distributions, which is completely monotone (CM).

Example 4.12 Let $\{X(t)\}$ be the finite birth–death process considered in Example 4.4. Suppose that its first passage time to state 0 from state 1 is of interest. In this case, we define

$$\mathbf{T} = \begin{pmatrix} -\lambda_1 - \mu_1 & \lambda_1 & 0 & \cdots & 0 \\ \mu_2 & -\lambda_2 - \mu_2 & \lambda_2 & \cdots & 0 \\ \vdots & \ddots & \ddots & \ddots & \vdots \\ 0 & \cdots & \mu_{N-1} & -\lambda_{N-1} - \mu_{N-1} & \lambda_{N-1} \\ 0 & \cdots & 0 & \mu_N & -\mu_N \end{pmatrix}$$

and $\mathbf{r} = (\mu_1, 0, \cdots, 0)^\top$. Since $\mathbf{T}$ is tridiagonal and its off-diagonal elements are nonnegative, it is well known that the eigenvalues of $\mathbf{T}$, λ_j say, are simple and real. Moreover, since $\mathbf{T}$ is a lossy generator, the λ_j are strictly negative. Defining $\mathbf{x}_j = (x_{ij})$ and $\xi_j = -\lambda_j$, since $\mathbf{r} = \mu_1 \boldsymbol{\delta}_1$, we then have

$$\phi_1(s) = \sum_{j=1}^{N} \frac{1}{s + \xi_j} x_{1j}^2 \mu_1, \quad \text{Re}(s) > 0.$$

Therefore, the first-passage-time density is a mixture of exponential densities (see Keilson, 1971). This result should be compared with the result obtained in Example 2.24.

Let $\mathbf{T}$ be a lossy generator and let $\boldsymbol{\alpha}$ be a probability vector defined on the state space $\{1, 2, \cdots, N\}$. Let $\mathbf{r} = -\mathbf{T}\mathbf{1}$, which is nonnegative and nonzero. For the pair $(\boldsymbol{\alpha}, \mathbf{T})$, define

$$f(t) = \boldsymbol{\alpha}^\top \exp\{\mathbf{T}\, t\}\, \mathbf{r}, \quad t \geq 0. \tag{4.53}$$

The function $f(t)$ is nonnegative and

$$\int_0^\infty f(t) dt = \boldsymbol{\alpha}^\top \int_0^\infty \exp\{\mathbf{T}\, t\} dt\, \mathbf{r} = \boldsymbol{\alpha}^\top \mathbf{N} \mathbf{r} = \boldsymbol{\alpha}^\top \mathbf{1} = 1,$$

whence $f(t)$ is a probability density function. The density function $f(t)$ generated by the pair $(\boldsymbol{\alpha}, \mathbf{T})$ is called a (continuous) ***phase-type distribution*** (see Neuts, 1981, for details). Recalling (4.52), the phase-type density function $f(t)$ can be interpreted as the first-passage-time density to state 0 with the initial distribution $\boldsymbol{\alpha}$.

We now apply a uniformization technique to the infinitesimal generator $\mathbf{Q}$ given in (4.52). For $\nu \geq \max_{1 \leq i \leq N} q_i$, let

$$\mathbf{T}_\nu = \mathbf{I} + \frac{1}{\nu} \mathbf{T}, \quad \mathbf{r}_\nu = \frac{1}{\nu} \mathbf{r}.$$

From (4.35), we have

$$\exp\{\mathbf{T}\,t\} = \sum_{n=0}^{\infty} \frac{(\nu t)^n}{n!} e^{-\nu t}\,\mathbf{T}_\nu^n, \quad t \geq 0,$$

which together with (4.53) shows that the continuous phase-type density function generated from $(\boldsymbol{\alpha}, \mathbf{T})$ is given by

$$f(t) = \sum_{n=0}^{\infty} \frac{(\nu t)^n}{n!} e^{-\nu t}\,\boldsymbol{\alpha}^\top \mathbf{T}_\nu^n\,\mathbf{r}, \quad t \geq 0.$$

Define

$$g_n = \boldsymbol{\alpha}^\top \mathbf{T}_\nu^{n-1}\mathbf{r}_\nu, \quad n = 1, 2, \cdots,$$

and denote the density function of the Erlang distribution of order n by

$$f(n,t) = \frac{\nu^n t^{n-1}}{(n-1)!} e^{-\nu t}, \quad t \geq 0; \quad n = 1, 2, \cdots.$$

We then have

$$f(t) = \sum_{n=1}^{\infty} g_n f(n,t), \quad t \geq 0. \tag{4.54}$$

Note that (g_n) is a discrete phase-type distribution generated by $(\boldsymbol{\alpha}, \mathbf{T}_\nu)$; see (2.80). In fact,

$$\sum_{n=1}^{\infty} g_n = \boldsymbol{\alpha}^\top \left(\sum_{n=0}^{\infty} \mathbf{T}_\nu^n\right) \mathbf{r}_\nu = \boldsymbol{\alpha}^\top (\mathbf{I} - \mathbf{T}_\nu)^{-1}\mathbf{r}_\nu = \boldsymbol{\alpha}^\top(-\mathbf{T})^{-1}\mathbf{r} = 1.$$

Therefore, the continuous phase-type distribution generated by $(\boldsymbol{\alpha}, \mathbf{T})$ is a mixture of the Erlang distributions with weight (g_n).

The Laplace transform of $f(n,t)$ is given by $\{\nu/(s+\nu)\}^n$. Hence, the Laplace transform of $f(t)$ is

$$\phi(s) = \sum_{n=1}^{\infty} g_n \left(\frac{\nu}{s+\nu}\right)^n, \quad \mathrm{Re}\,(s) > 0.$$

In general, a continuous distribution with the Laplace transform

$$\phi(s) = \sum_{n=0}^{\infty} g_n \left(\frac{\nu}{s+\nu}\right)^n, \quad \mathrm{Re}\,(s) > 0,$$

is called a *generalized phase-type distribution* (see Shanthikumar, 1985). Note that the probability g_0 may be positive here.

Finally, we consider quasi-stationary distributions of continuous-time Markov chains. Let $\mathbf{T}$ be a lossy generator defined on $\mathcal{N} = \{1, 2, \cdots, N\}$. As in the discrete-time case (see Section 2.8), the Markov chain $\{X(t)\}$ governed by a lossy generator is called a *lossy Markov chain*. The (lossy) transition matrix function of $\{X(t)\}$ is given by

$$\mathbf{T}(t) = \exp\{\mathbf{T}\,t\} = \sum_{n=0}^{\infty} \frac{\mathbf{T}^n t^n}{n!}, \quad t \geq 0.$$

Suppose $\mathbf{T}$ is irreducible (see Definition A.4). Then, from Theorem A.6, there exists a unique probability vector $\mathbf{u}$ positive componentwise such that

$$r\,\mathbf{u}^\top = \mathbf{u}^\top \mathbf{T}, \quad \mathbf{u}^\top \mathbf{1} = 1, \tag{4.55}$$

where r is the PF eigenvalue of $\mathbf{T}$. The PF eigenvalue is strictly negative and largest in the real part. We define $\gamma = -r > 0$. It follows from (4.55) that

$$\mathbf{u}^\top \mathbf{T}(t) = \mathbf{u}^\top \sum_{n=0}^{\infty} \frac{(-\gamma t)^n}{n!} = e^{-\gamma t}\mathbf{u}^\top, \quad t \geq 0.$$

Now, choosing $\mathbf{u} = (u_i)$ as the initial distribution, consider the conditional probabilities

$$q_j(t) = P_u[X(t) = j | X(t) \in \mathcal{N}], \quad t \geq 0; \quad j \in \mathcal{N}.$$

Writing $\mathbf{T}(t) = (p_{ij}(t))$, we have

$$q_j(t) = \frac{\sum_i u_i p_{ij}(t)}{\sum_j \sum_i u_i p_{ij}(t)}.$$

Hence, the conditional distribution $\mathbf{q}(t) = (q_j(t))$ is given by

$$\mathbf{q}^\top(t) = \frac{\mathbf{u}^\top \mathbf{T}(t)}{\mathbf{u}^\top \mathbf{T}(t)\mathbf{1}} = \frac{e^{-\gamma t}\,\mathbf{u}^\top}{e^{-\gamma t}\,\mathbf{u}^\top \mathbf{1}} = \mathbf{u}^\top.$$

That is, $\mathbf{u}$ is the probability vector with the property that, starting with $\mathbf{u}$, the conditional distribution at any time given that the Markov chain is in $\mathcal{N}$ is equal to the initial distribution $\mathbf{u}$. In this respect, the PF left eigenvector $\mathbf{u}$ is called the ***quasi-stationary distribution*** and equation (4.55) is called the ***quasi-stationary equation*** (see Section 2.8 for the discrete-time case). When the state space $\mathcal{N}$ is denumerably infinite, however, verification of the existence of the quasi-stationary distribution becomes quite delicate. The reader interested in the theory should consult, e.g., Pollett (1988, 1989), Nair and Pollett (1993), and Ferrari, Kesten, Martínez and Picco (1995), and references therein.

Associated with the PF eigenvalue r, there also exists a unique positive vector such that

$$r\,\mathbf{v} = \mathbf{T}\mathbf{v}, \quad \mathbf{u}^\top \mathbf{v} = 1.$$

From Theorem A.7, it follows that

$$\mathbf{T}(t) = e^{-\gamma t}\,\mathbf{v}\mathbf{u}^\top + o(e^{-\gamma t}) \quad \text{as } t \to \infty.$$

The next result is due to Darroch and Seneta (1967). The proof is left to the reader (see Exercise 4.27).

Theorem 4.27 *For any initial distribution* $\boldsymbol{\alpha}$, *we have*

$$\lim_{t\to\infty} P_\alpha[X(t)=i|X(t)\in\mathcal{N}] = u_i, \quad i\in\mathcal{N},$$

where $\mathbf{u}=(u_i)$ *is the unique solution of* (4.55).

From Theorem 4.27, the conditional distribution converges to a non-defective distribution as $t\to\infty$. The limiting distribution $\mathbf{u}$ is positive componentwise, independent of the initial distribution, and equal to the quasi-stationary distribution. The limiting distribution is often called the ***quasi-limiting distribution*** of the lossy Markov chain $\{X(t)\}$. We shall consider the quasi-limiting distributions of birth–death processes in detail in the next chapter. The ***doubly limiting conditional distribution*** also exists in the finite case and is given by $\mathbf{u}_\mathrm{D}\mathbf{v}=(u_iv_i)$, where $\mathbf{u}_\mathrm{D}$ is the diagonal matrix with diagonal elements u_i. The proof of this fact is quite similar to the proof of Theorem 2.25 and is omitted.

4.7 Calculation of transition probability functions

Consider an irreducible Markov chain $\{X(t)\}$ in continuous time with state space $\mathcal{N}$ and infinitesimal generator $\mathbf{Q}=(q_{ij})$. The Markov chain $\{X(t)\}$ may be lossy, in which case $\mathbf{Q1}\le\mathbf{0}$ with at least one strict inequality. In any case, we assume throughout this section that the transition matrix function $\mathbf{P}(t)=(p_{ij}(t))$ is given by

$$\mathbf{P}(t)=\exp\{\mathbf{Q}t\}=\sum_{n=0}^{\infty}\frac{\mathbf{Q}^n t^n}{n!}, \quad t\ge 0. \tag{4.56}$$

Then, in principle, the transition probability functions $p_{ij}(t)$ can be computed by solving the backward Kolmogorov equation numerically:

$$p'_{ij}(t) = -q_i\,p_{ij}(t)+\sum_{k\ne i} q_{ik}p_{kj}(t), \quad t\ge 0,$$

or the forward Kolmogorov equation

$$p'_{ij}(t) = -p_{ij}(t)q_j+\sum_{k\ne j} p_{ik}(t)q_{kj}, \quad t\ge 0,$$

with the initial condition $p_{ij}(0)=\delta_{ij}$. Alternatively, from (4.28), the Laplace transform matrix $\boldsymbol{\Pi}(s)=(\pi_{ij}(s))$ satisfies

$$\boldsymbol{\Pi}(s)(s\mathbf{I}-\mathbf{Q})=\mathbf{I}, \quad \mathrm{Re}\,(s)>0,$$

where $\pi_{ij}(s)=\int_0^\infty e^{-st}p_{ij}(t)dt$. The transition matrix function $\mathbf{P}(t)=(p_{ij}(t))$ can then be evaluated by inverting $\boldsymbol{\Pi}(s)$ numerically. See Stewart (1994) for other information on the numerical solution of Markov chains.

4.7.1 Numerical methods for finite Markov chains

Suppose that the state space is finite, $\mathcal{N} = \{0, 1, \cdots, N\}$, say. The transition matrix function is given by (4.56). There are several methods for evaluating numerically transition probability functions of finite (possibly lossy) Markov chains in continuous time.

Let λ_j, $j = 0, 1, \cdots, N$, be the eigenvalues of $\mathbf{Q}$ and suppose that $\mathbf{Q}$ can be written as

$$\mathbf{Q} = \mathbf{M}\boldsymbol{\lambda}_{\mathrm{D}}\mathbf{M}^{-1} \tag{4.57}$$

by some nonsingular matrix $\mathbf{M}$, where $\boldsymbol{\lambda}_{\mathrm{D}}$ is the diagonal matrix with diagonal elements λ_j. If there exists such a matrix $\mathbf{M}$, the generator $\mathbf{Q}$ is said to be *diagonalizable*. Let $\mathbf{v}_j$ be the right eigenvector associated with λ_j, i.e.,

$$\mathbf{Q}\mathbf{v}_j = \lambda_j \mathbf{v}_j, \quad j = 0, 1, \cdots, N.$$

Writing $\mathbf{M} = (\mathbf{v}_0, \mathbf{v}_1, \cdots, \mathbf{v}_N)$, we have

$$\mathbf{Q}\mathbf{M} = \mathbf{M}\boldsymbol{\lambda}_{\mathrm{D}}.$$

Note that the set of eigenvectors $\mathbf{v}_i$, $i = 0, 1, \cdots, N$, is linearly independent if and only if $\mathbf{M}$ is nonsingular. Thus, $\mathbf{Q}$ is diagonalizable if and only if $\mathbf{Q}$ has a linearly independent set of $(N+1)$ eigenvectors. A sufficient condition for this is that the eigenvalues are all distinct. To see this, suppose, on the contrary, that the set is linearly dependent. Then the smallest integer k exists such that $\mathbf{v}_k$ can be written as a linear combination of $\mathbf{v}_0, \mathbf{v}_1, \cdots, \mathbf{v}_{k-1}$, say

$$\mathbf{v}_k = \alpha_0\mathbf{v}_0 + \alpha_1\mathbf{v}_1 + \cdots + \alpha_{k-1}\mathbf{v}_{k-1},$$

where not all the $\alpha_0, \alpha_1, \cdots, \alpha_{k-1}$ are zero. Then

$$\lambda_k\mathbf{v}_k = \mathbf{Q}\mathbf{v}_k = \alpha_0\lambda_0\mathbf{v}_0 + \alpha_1\lambda_1\mathbf{v}_1 + \cdots + \alpha_{k-1}\lambda_{k-1}\mathbf{v}_{k-1}.$$

Of course, $\lambda_k\mathbf{v}_k = \alpha_0\lambda_k\mathbf{v}_0 + \alpha_1\lambda_k\mathbf{v}_1 + \cdots + \alpha_{k-1}\lambda_k\mathbf{v}_{k-1}$, whence

$$0 = \alpha_0(\lambda_0 - \lambda_k)\mathbf{v}_0 + \alpha_1(\lambda_1 - \lambda_k)\mathbf{v}_1 + \cdots + \alpha_{k-1}(\lambda_{k-1} - \lambda_k)\mathbf{v}_{k-1}.$$

Since $\lambda_i \neq \lambda_k$ and not all the α_i, $i = 0, \cdots, k-1$, are zero, the set of vectors $\mathbf{v}_i$, $i = 0, \cdots, k-1$, is linearly dependent, so that $\mathbf{v}_j$ for some $j < k$ can be written as a linear combination of $\mathbf{v}_0, \mathbf{v}_1, \cdots, \mathbf{v}_{j-1}$. This contradicts the assumption that k is the smallest. Thus, the set of eigenvectors is linearly independent so that $\mathbf{Q}$ is diagonalizable.

Now suppose that (4.57) holds. Then

$$\mathbf{Q}^n = \mathbf{M}\boldsymbol{\lambda}_{\mathrm{D}}^n\mathbf{M}^{-1}, \quad n = 0, 1, \cdots,$$

so that

$$\mathbf{P}(t) = \sum_{n=0}^{\infty} \frac{\mathbf{Q}^n t^n}{n!} = \mathbf{M}\left(\sum_{n=0}^{\infty} \frac{\boldsymbol{\lambda}_{\mathrm{D}}^n t^n}{n!}\right)\mathbf{M}^{-1} = \mathbf{M}\boldsymbol{\lambda}_{\mathrm{D}}(t)\mathbf{M}^{-1}, \tag{4.58}$$

where $\boldsymbol{\lambda}_D(t)$ is the diagonal matrix with diagonal elements $e^{\lambda_j t}$. Recall that, if $\mathbf{Q}$ is symmetrizable, then the eigenvalues are all real and

$$\mathbf{P}(t) = \sum_{j=0}^{N} e^{\lambda_j t}\, \mathbf{v}_j \mathbf{u}_j^\top, \quad t \geq 0,$$

where

$$\mathbf{M}^{-1} = \begin{pmatrix} \mathbf{u}_0^\top \\ \mathbf{u}_1^\top \\ \vdots \\ \mathbf{u}_N^\top \end{pmatrix}.$$

In particular, if $\{X(t)\}$ is ergodic, then $\mathbf{Q1} = \mathbf{0}$ so that $\lambda_0 = 0$, $\mathbf{v}_0 = \mathbf{1}$ and $\mathbf{u}_0 = \boldsymbol{\pi}$, the stationary distribution of $\{X(t)\}$; see (4.40).

An easy way to compute the eigenvalues λ_j is to use the relation

$$\sum_{j=0}^{N} \lambda_j = -\sum_{j=0}^{N} q_j = \mathrm{tr}(\mathbf{Q}),$$

the *trace* of $\mathbf{Q}$, which is the sum of the diagonal elements of $\mathbf{Q}$. Since the eigenvalues of $\mathbf{Q}^n$ are the nth powers of the eigenvalues of $\mathbf{Q}$, it follows that

$$\mathrm{tr}(\mathbf{Q}^n) = \sum_{j=0}^{N} \lambda_j^n, \quad n = 1, 2, \cdots.$$

Therefore, the eigenvalues λ_j, $j = 0, \cdots, N$, are obtained as solutions of the set of equations

$$\lambda_0^n + \lambda_1^n + \cdots + \lambda_N^n = \mathrm{tr}(\mathbf{Q}^n), \quad n = 1, \cdots, N+1. \tag{4.59}$$

If the Markov chain $\{X(t)\}$ is ergodic, we know $\lambda_0 = 0$. Hence, in this case, N equations in (4.59) suffice to determine the eigenvalues λ_j, $j = 1, \cdots, N$.

Example 4.13 Let $\{X(t)\}$ be a continuous-time Markov chain on the state space $\{0, 1, 2\}$ with infinitesimal generator

$$\mathbf{Q} = \begin{pmatrix} -1 & 1 & 0 \\ 1 & -2 & 1 \\ 1 & 1 & -2 \end{pmatrix}.$$

Note that $\lambda_0 = 0$ since $\mathbf{Q1} = \mathbf{0}$. We have $\mathrm{tr}(\mathbf{Q}) = -5$ and $\mathrm{tr}(\mathbf{Q}^2) = 13$ so that, from (4.59), $\lambda_1 = -2$ and $\lambda_2 = -3$. Since the eigenvalues are distinct, the matrix $\mathbf{M}$ is nonsingular and

$$\mathbf{M} = \begin{pmatrix} 1 & 1 & 1 \\ 1 & -1 & -2 \\ 1 & -1 & 1 \end{pmatrix}; \quad \mathbf{M}^{-1} = \begin{pmatrix} 1/2 & 1/3 & 1/6 \\ 1/2 & 0 & -1/2 \\ 0 & -1/3 & 1/3 \end{pmatrix}.$$

Therefore, from (4.58), we have

$$\begin{aligned} p_{00}(t) &= \frac{1}{2} + \frac{1}{2} e^{-2t}, \\ p_{12}(t) &= \frac{1}{6} + \frac{1}{2} e^{-2t} - \frac{2}{3} e^{-3t}, \\ p_{21}(t) &= \frac{1}{3} - \frac{1}{3} e^{-3t}, \end{aligned}$$

etc. These analytical values will be compared with approximations derived in the following examples. Note that the relaxation time defined in (4.43) is given by

$$T_{\mathrm{REL}}(\mathbf{Q}) = \frac{1}{|\lambda_1|} = \frac{1}{2}.$$

However, it is observed (see Table 4.1 below) that the transition probability function $p_{12}(t)$ seems to settle down to the limiting probability 1/6 only after $t \geq 3$.

Since the transition matrix function $\mathbf{P}(t)$ of the Markov chain $\{X(t)\}$ is given by (4.56), $\mathbf{P}(t)$ can in principle be approximated arbitrarily well by truncation of the infinite sum in the right-hand side of (4.56). But appropriate choice of such truncation is usually very difficult. A technique commonly used in practice is to discretize the continuous time based on *uniformization* (4.35), i.e.,

$$\mathbf{P}(t) = \sum_{n=0}^{\infty} \frac{(\nu t)^n}{n} e^{-\nu t} \mathbf{P}_\nu^n; \quad \mathbf{P}_\nu = \mathbf{I} + \frac{1}{\nu}\mathbf{Q}, \quad \nu \geq \max_i q_i.$$

Then, since $\mathbf{P}_\nu^n \geq \mathbf{O}$ for all $n = 0, 1, \cdots$, any truncation of the infinite sum provides a lower bound of $\mathbf{P}(t)$. Several modifications and numerical experiments are reported in Sumita and Shanthikumar (1986).

Now suppose that the Markov chain $\{X(t)\}$ is ergodic. Then, we can write $\mathbf{P}_\nu = \mathbf{1}\boldsymbol{\pi}^\top + \boldsymbol{\Delta}_\nu$, where $\boldsymbol{\pi}$ is the stationary distribution of $\{X(t)\}$. Since $\mathbf{1}\boldsymbol{\pi}^\top\boldsymbol{\Delta}_\nu = \boldsymbol{\Delta}_\nu\mathbf{1}\boldsymbol{\pi}^\top = \mathbf{O}$, we have

$$\mathbf{P}_\nu^n = \mathbf{1}\boldsymbol{\pi}^\top + \boldsymbol{\Delta}_\nu^n, \quad n = 1, 2, \cdots.$$

It follows that

$$\mathbf{P}(t) = (1 - e^{-\nu t})\mathbf{1}\boldsymbol{\pi}^\top + \sum_{n=0}^{\infty} \frac{(\nu t)^n}{n!} e^{-\nu t} \boldsymbol{\Delta}_\nu^n, \quad t \geq 0. \tag{4.60}$$

Since the spectral radius of $\boldsymbol{\Delta}_\nu$ is strictly less than unity, we would expect that the infinite sum on the right-hand side converges geometrically fast. A numerical example is provided below.

Example 4.14 In Example 4.13 above, the stationary distribution is given

Table 4.1 *Uniformization: Approximation of* $p_{12}(t)$

		$\nu = 2$		$\nu = 4$	
t	Exact	Direct	Modification	Direct	Modification
0.1	.082153	.082153	.082153	.082152	.082153
0.2	.135952	.135952	.135952	.135920	.135951
0.3	.170026	.170020	.170026	.169766	.170016
0.4	.190535	.190506	.190537	.189488	.190494
0.5	.201853	.201759	.201858	.198985	.201746
0.6	.207065	.206826	.207076	.200897	.206843
0.7	.208328	.207818	.208352	.197088	.207940
0.8	.207136	.206173	.207180	.188970	.206538
0.9	.204512	.202855	.204584	.177690	.203670
1.0	.201143	.198492	.201252	.164230	.200042
1.5	.184154	.170599	.184585	.089712	.182098
2.0	.174172	.139199	.175010	.037438	.172232
2.5	.169667	.106772	.170778	.012903	.168389
3.0	.167824	.076593	.168980	.003867	.167144
3.5	.167104	.051574	.168123	.001043	.166788
4.0	.166830	.032831	.167625	.000259	.166696
4.5	.166727	.019907	.167292	.000060	.166673
5.0	.166689	.011577	.167063	.000013	.166668

by $\boldsymbol{\pi}^{\top} = (1/2,\ 1/3,\ 1/6)$. Hence, for $\nu \geq 2$, we have

$$\boldsymbol{\Delta}_\nu = \begin{pmatrix} \frac{1}{2} - \frac{1}{\nu} & \frac{1}{\nu} - \frac{1}{3} & -\frac{1}{6} \\ \frac{1}{\nu} - \frac{1}{2} & \frac{2}{3} - \frac{2}{\nu} & \frac{1}{\nu} - \frac{1}{6} \\ \frac{1}{\nu} - \frac{1}{2} & \frac{1}{\nu} - \frac{1}{3} & \frac{5}{6} - \frac{2}{\nu} \end{pmatrix}.$$

In Table 4.1, we compare the exact value $p_{12}(t)$ obtained in Example 4.13 with approximations based on a direct uniformization (4.35) and modification (4.60), where the infinite sum is truncated at $n = 5$. It is observed that our modification (4.60) provides a much better approximation. A choice of larger ν may be better for large t while a smaller ν seems appropriate for small t. In any case, however, taking the truncation point larger than $n = 20$ makes no difference in the choice of ν.

Let $\{S_n\}$ be a sequence of nonnegative random variables that converges in distribution to a constant $t \geq 0$ as $n \to \infty$. For a matrix function $\mathbf{F}(x)$ which is bounded and continuous in x componentwise, one then has

$$\lim_{n \to \infty} E[\mathbf{F}(S_n)] = \mathbf{F}(t),$$

see, e.g., Williams (1991). Hence, if $E[\mathbf{F}(S_n)]$ is easy to calculate, $E[\mathbf{F}(S_n)]$ at appropriate n will provide an approximation of $\mathbf{F}(t)$. Let $Y_1,\ Y_2, \cdots$ be nonnegative, independent and identically distributed (IID) random variables with mean t and define $S_n = \sum_{k=1}^{n} Y_k/n$. Then, by the strong law of large numbers, S_n converges to the mean t as $n \to \infty$ almost surely and, hence, converges to t in distribution. It remains to choose Y_n so that the $E[\mathbf{F}(S_n)]$ are easy to evaluate. For this purpose, the most appropriate choice would be exponential random variables, because one can then relate $E[\mathbf{F}(S_n)]$ to the matrix Laplace transform of $\mathbf{F}(x)$.

Now consider a continuous-time Markov chain with infinitesimal generator $\mathbf{Q}$. Let $\mathbf{P}(t)$ denote its transition matrix function. Let $Y_1, Y_2, \cdots$ be a sequence of independent and exponentially distributed random variables with the common mean t. Then S_n follows an Erlang distribution of order n with mean t whose density function is given by

$$\frac{(n/t)^n x^{n-1}}{(n-1)!} e^{-nx/t}, \quad x \geq 0; \quad n = 1, 2, \cdots.$$

Let $\boldsymbol{\Pi}(s) = \int_0^\infty e^{-sx}\mathbf{P}(x)dx$ be the matrix Laplace transform of $\mathbf{P}(t)$. From Theorem B.11(vi), it is easy to show that

$$E[\mathbf{P}(S_n)] = \frac{(n/t)^n(-1)^{n-1}}{(n-1)!} \boldsymbol{\Pi}^{(n-1)}(n/t), \quad t \geq 0, \tag{4.61}$$

where $\boldsymbol{\Pi}^{(n)}(s)$ denotes the nth derivative of $\boldsymbol{\Pi}(s)$ with respect to s. Recall from (4.17) that

$$\boldsymbol{\Pi}(s) = (s\mathbf{I} - \mathbf{Q})^{-1}, \quad \mathrm{Re}\,(s) > 0.$$

It follows from (4.61) that

$$E[\mathbf{P}(S_n)] = \left(\mathbf{I} - \frac{t}{n}\mathbf{Q}\right)^{-n}, \quad n = 1, 2, \cdots, \tag{4.62}$$

which converges to $\exp\{\mathbf{Q}t\}$ as $n \to \infty$. Observe the resemblance of the matrix in parentheses of the right-hand side of (4.62) to uniformization (4.33). In this regard, Ross (1987) called (4.62) *external uniformization*. The above derivation is taken from Kijima (1992a). For extensive numerical experiments on external uniformization, see Yoon and Shanthikumar (1989).

Example 4.15 For the infinitesimal generator given in Example 4.13, we have

$$\mathbf{I} - \frac{t}{n}\mathbf{Q} = \begin{pmatrix} 1+t/n & -t/n & 0 \\ -t/n & 1+2t/n & -t/n \\ -t/n & -t/n & 1+2t/n \end{pmatrix},$$

which is nonsingular for any $t \geq 0$ and $n \geq 1$, since the real part of any eigenvalue of $\mathbf{Q}$ is nonpositive. In Table 4.2, we examine the convergence speed of the external uniformization with respect to n of (4.62).

Motivated by (4.62), we also have

$$\left(\mathbf{I}+\frac{t}{n}\mathbf{Q}\right)^n \quad \rightarrow \quad \exp\{\mathbf{Q}\,t\} \quad \text{as } n\rightarrow\infty. \tag{4.63}$$

Table 4.3 lists the computational results for the same parameter set as is used for Table 4.2. It should be noted that if $n/t < \max_i q_i$ then the matrix $\mathbf{I}+\mathbf{Q}\,t/n$ is not nonnegative. In fact, for the case of $n=5$, the values for $t>2.5$ are rather peculiar. As n increases, the approximations based on external uniformization and (4.63) become close each other. Note, however, that external uniformization requires a matrix inversion. See also Ross (1989) for a discussion of these approximations.

4.7.2 Some inequalities

Let $\{X(t)\}$ ($\{Y(t)\}$, respectively) be a continuous-time Markov chain with state space $\mathcal{N}$ and (possibly lossy) generator $\mathbf{Q}_X=(q_{ij}^X)$ ($\mathbf{Q}_Y=(q_{ij}^Y)$). Let $p_{ij}^X(t)$ ($p_{ij}^Y(t)$) be the transition probability functions of $\{X(t)\}$ ($\{Y(t)\}$). The state space $\mathcal{N}$ may be denumerably infinite, in which case we assume that both $p_{ij}^X(t)$ and $p_{ij}^Y(t)$ are minimal solutions to the backward and forward Kolmogorov equations. As in (4.31), we define $f_{ij}^{(0)}(t)=\delta_{ij}e^{-q_i^X t}$ and generate $f_{ij}^{(n)}(t)$ successively by

$$f_{ij}^{(n)}(t)=f_{ij}^{(0)}(t)+\int_0^t e^{-q_i^X s}\sum_{k\neq i}q_{ik}^X f_{kj}^{(n-1)}(t-s)ds, \quad n=1,2,\cdots.$$

Similarly, for $\{Y(t)\}$, we define $g_{ij}^{(0)}(t)=\delta_{ij}e^{-q_i^Y t}$ and

$$g_{ij}^{(n)}(t)=g_{ij}^{(0)}(t)+\int_0^t e^{-q_i^Y s}\sum_{k\neq i}q_{ik}^Y g_{kj}^{(n-1)}(t-s)ds, \quad n=1,2,\cdots.$$

We saw in Theorem 4.17 that

$$p_{ij}^X(t)=\lim_{n\rightarrow\infty}f_{ij}^{(n)}(t) \quad \text{and} \quad p_{ij}^Y(t)=\lim_{n\rightarrow\infty}g_{ij}^{(n)}(t).$$

Lemma 4.5 *Suppose* $\mathbf{Q}_X\geq\mathbf{Q}_Y$, *i.e.,* $q_{ij}^X\geq q_{ij}^Y$ *for all* $i,\ j\in\mathcal{N}$. *Then*

$$p_{ij}^X(t)\geq p_{ij}^Y(t), \quad t\geq 0; \quad i,\ j\in\mathcal{N}.$$

Proof. Since $q_{ii}^X=-q_i^X\geq -q_i^Y=q_{ii}^Y$, we have

$$f_{ij}^{(0)}(t)=\delta_{ij}e^{-q_i^X t}\geq\delta_{ij}e^{-q_i^Y t}=g_{ij}^{(0)}(t).$$

By induction, it is readily shown that $f_{ij}^{(n)}(t)\geq g_{ij}^{(n)}(t)$ for all $n\geq 1$ and the result follows from Theorem 4.17. □

Table 4.2 *External uniformization: Approximation of* $p_{12}(t)$

t	Exact	$n=5$	$n=10$	$n=20$
0.1	.082153	.079458	.080778	.081459
0.2	.135952	.128674	.132186	.134035
0.3	.170026	.158974	.164257	.167076
0.4	.190535	.177318	.183615	.186990
0.5	.201853	.188053	.194649	.198169
0.6	.207065	.193931	.200277	.203615
0.7	.208328	.196716	.202443	.205373
0.8	.207136	.197548	.202439	.204830
0.9	.204512	.197167	.201123	.202917
1.0	.201143	.196056	.199061	.200255
1.5	.184154	.187426	.186709	.185704
2.0	.174172	.180192	.177889	.176201
2.5	.169667	.175465	.172863	.171289
3.0	.167824	.172495	.170127	.168903
3.5	.167104	.170617	.168638	.167759
4.0	.166830	.169408	.167816	.167209
4.5	.166727	.168611	.167352	.166941
5.0	.166689	.168073	.167085	.166808

Table 4.3 *Approximation of* $p_{12}(t)$ *based on* (4.63)

t	Exact	$n=5$	$n=10$	$n=20$
0.1	.082153	.085084	.083587	.082863
0.2	.135952	.144386	.140006	.137941
0.3	.170026	.183373	.176363	.173118
0.4	.190535	.206738	.198194	.194264
0.5	.201853	.218460	.209756	.205711
0.6	.207065	.221860	.214285	.210623
0.7	.208328	.219655	.214196	.211281
0.8	.207136	.214016	.211258	.209305
0.9	.204512	.206623	.206740	.205826
1.0	.201143	.198720	.201522	.201615
1.5	.184154	.171780	.179102	.181973
2.0	.174172	.167040	.169620	.171899
2.5	.169667	.187500	.167154	.168197
3.0	.167824	.384960	.166719	.167061
3.5	.167104	1.23522	.166670	.166757
4.0	.166830	3.71328	.166667	.166685
4.5	.166727	9.46854	.166648	.166670
5.0	.166689	21.0000	.166016	.166667

Consider, now, a continuous-time Markov chain $\{X(t)\}$ with state space $\{0,1,2,\cdots\}$ and infinitesimal generator $\mathbf{Q} = (q_{ij})$. Let $\mathbf{Q}_n$ be the $n \times n$ north-west corner truncation of $\mathbf{Q}$, and let $\mathbf{P}_n(t) = (p_{ij}^{(n)}(t))$ be the transition matrix function defined by

$$\mathbf{P}_n(t) = \exp\{\mathbf{Q}_n t\}, \quad t \geq 0.$$

With the aid of Lemma 4.5, we can prove the following important result.

Theorem 4.28 *For each $t \geq 0$ and all $i,\ j \in \mathcal{N}$, $p_{ij}^{(n)}(t)$ is monotonically nondecreasing in n and converges as $n \to \infty$ to the minimal transition probability function $f_{ij}(t)$ of $\{X(t)\}$.*

Proof. To prove that $p_{ij}^{(n)}(t) \geq p_{ij}^{(n-1)}(t)$, we define the $n \times n$ matrix $\widehat{\mathbf{Q}}_n$ whose $(n-1) \times (n-1)$ principal submatrix coincides with $\mathbf{Q}_{n-1}$ and whose other off-diagonal components are zero. Then $\mathbf{Q}_n \geq \widehat{\mathbf{Q}}_n$ so that, from Lemma 4.5, the claim holds. For each n, $p_{ij}^{(n)}(t)$ satisfies the backward Kolmogorov equations. That is,

$$p_{ij}^{(n)}(t) = \delta_{ij} e^{-q_i t} + \int_0^t e^{-q_i s} \sum_{k \neq i,\, k \leq n-1} q_{ik}\, p_{kj}^{(n)}(t-s) ds.$$

Let $n \to \infty$ and use the monotone convergence theorem. Then, writing $p_{ij}(t) = \lim_{n\to\infty} p_{ij}^{(n)}(t)$, it follows that

$$p_{ij}(t) = \delta_{ij} e^{-q_i t} + \int_0^t e^{-q_i s} \sum_{k \neq i} q_{ik}\, p_{kj}(t-s) ds,$$

i.e., $p_{ij}(t)$ satisfies the backward Kolmogorov equation. Similarly, we can prove that $p_{ij}(t)$ satisfies the forward Kolmogorov equation. Hence, by the minimality of $f_{ij}(t)$, we have $f_{ij}(t) \leq p_{ij}(t)$. But, from Lemma 4.5, it is readily seen that $p_{ij}^{(n)}(t) \leq f_{ij}(t)$ for any n, so that $p_{ij}(t) \leq f_{ij}(t)$. This completes the proof. □

4.8 Stochastic monotonicity

In this section, we shall discuss the stochastic monotonicity of a continuous-time Markov chain $\{X(t)\}$. Recall that the stochastic monotonicity of a discrete-time Markov chain is characterized in terms of its transition matrix $\mathbf{P}$. Namely, $\mathbf{P}$ is stochastically monotone, denoted by $\mathbf{P} \in \mathcal{M}_{st}$, if $\mathbf{U}^{-1}\mathbf{P}\mathbf{U} \geq \mathbf{O}$, where $\mathbf{U}$ is defined in (3.5). As a consequence (see Lemma 3.8), we have $\mathbf{P}^n \in \mathcal{M}_{st}$ for all $n = 1, 2, \cdots$. By analogy with this, we define stochastic monotonicity of continuous-time Markov chains as follows.

Definition 4.6 A continuous-time Markov chain $\{X(t)\}$ with transition matrix function $\mathbf{P}(t)$ is said to be *stochastically monotone* if $\mathbf{P}(t) \in \mathcal{M}_{st}$ for all $t \geq 0$.

An immediate consequence of the definition is the following, which may help to clarify the term 'stochastic monotonicity'.

Theorem 4.29 *Suppose that a continuous-time Markov chain $\{X(t)\}$ with state space $\mathcal{N} = \mathcal{Z}_+$ is stochastically monotone. If $X(0) = 0$, then $X(t) \geq_{\rm st} X(s)$ for all $t > s \geq 0$.*

Proof. Let $\boldsymbol{\pi}(t)$ be the state distribution of the Markov chain $\{X(t)\}$ at time t. From the Chapman–Kolmogorov equation (4.4), we have

$$\boldsymbol{\pi}^{\mathsf{T}}(t) = \boldsymbol{\pi}^{\mathsf{T}}(t-s)\mathbf{P}(s), \quad t > s \geq 0,$$

where $\boldsymbol{\pi}^{\mathsf{T}}(s) = \boldsymbol{\pi}^{\mathsf{T}}(0)\mathbf{P}(s)$. It follows that

$$\boldsymbol{\pi}^{\mathsf{T}}(t)\mathbf{U} - \boldsymbol{\pi}^{\mathsf{T}}(s)\mathbf{U} = \{\boldsymbol{\pi}^{\mathsf{T}}(t-s)\mathbf{U} - \boldsymbol{\pi}^{\mathsf{T}}(0)\mathbf{U}\}\mathbf{U}^{-1}\mathbf{P}(s)\mathbf{U}.$$

The fact that $X(0) = 0$ implies that $\boldsymbol{\pi}^{\mathsf{T}}(t-s)\mathbf{U} \geq \boldsymbol{\pi}^{\mathsf{T}}(0)\mathbf{U}$ for any $\boldsymbol{\pi}(t-s)$. Also, $\mathbf{P}(s) \in \mathcal{M}_{\rm st}$, by assumption. Therefore, we have

$$\boldsymbol{\pi}^{\mathsf{T}}(t)\mathbf{U} \geq \boldsymbol{\pi}^{\mathsf{T}}(s)\mathbf{U}, \quad t > s \geq 0,$$

whence $X(t) \geq_{\rm st} X(s)$, as desired. □

Stochastic monotonicity of a continuous-time Markov chain $\{X(t)\}$ can be characterized in terms of its infinitesimal generator $\mathbf{Q} = (q_{ij})$. In what follows, we assume that $\{X(t)\}$ is uniformizable. Then, as in (4.33), we define

$$\mathbf{P}_\nu = \mathbf{I} + \frac{1}{\nu}\mathbf{Q}; \quad \nu \geq c \equiv \sup_i q_i.$$

The transition matrix function $\mathbf{P}(t)$ of $\{X(t)\}$ is given by

$$\mathbf{P}(t) = \sum_{n=0}^{\infty} \frac{(\nu t)^n}{n!} e^{-\nu t}\, \mathbf{P}_\nu^n, \quad t \geq 0;$$

see (4.35).

Lemma 4.6 $\mathbf{P}_\nu \in \mathcal{M}_{\rm st}$ *for all $\nu \geq 2c$ if and only if*

$$\sum_{j \geq k} q_{ij} \leq \sum_{j \geq k} q_{i+1,j}, \quad k \neq i+1, \tag{4.64}$$

for all $i \in \mathcal{N}$.

Proof. Recall that $\mathbf{P}_\nu = (p_{ij}^\nu) \in \mathcal{M}_{\rm st}$ if and only if

$$\sum_{j \geq k} p_{ij}^\nu \leq \sum_{j \geq k} p_{i+1,j}^\nu, \quad k \in \mathcal{N}, \tag{4.65}$$

for all $i \in \mathcal{N}$. By the definition of $\mathbf{P}_\nu$, it is readily seen that (4.65) coincides with (4.64) for $k \geq i+2$. For $k \leq i$, we have from (4.65) that

$$1 + \frac{1}{\nu}\sum_{j \geq k} q_{ij} \leq 1 + \frac{1}{\nu}\sum_{j \geq k} q_{i+1,j}.$$

Hence $\mathbf{P}_\nu \in \mathcal{M}_{st}$ implies (4.64). Conversely, suppose that (4.64) holds. Then, obviously, (4.65) holds for all $k \neq i+1$. For $k = i+1$, since $\nu \geq 2c$ and $c \geq q_i$, we have

$$1 \geq \frac{2c}{\nu} \geq \frac{1}{\nu}(q_{i+1} + q_{i,i+1}), \quad i \in \mathcal{N}.$$

It follows that

$$\frac{1}{\nu} q_{i,i+1} + \frac{1}{\nu} \sum_{j \geq i+2} q_{ij} \leq 1 - \frac{1}{\nu} q_{i+1} + \frac{1}{\nu} \sum_{j \geq i+2} q_{i+1,j},$$

whence (4.65). □

Since $\mathbf{Q}$ is conservative by our early assumption, (4.64) can be rewritten as

$$\sum_{j \leq k} q_{ij} \geq \sum_{j \leq k} q_{i+1,j}, \quad k \neq i. \tag{4.66}$$

We are now in a position to state the main result of this section. For a more general result, see Anderson (Section 7.3, 1991).

Theorem 4.30 *Suppose that the Markov chain $\{X(t)\}$ is uniformizable. Then it is stochastically monotone if and only if (4.64) holds for all $i \in \mathcal{N}$.*

Proof. Suppose $\mathbf{P}(t) = (p_{ij}(t)) \in \mathcal{M}_{st}$ for all $t \geq 0$. For $k \leq i-1$, we have

$$\sum_{j \leq k} \frac{p_{ij}(t)}{t} \geq \sum_{j \leq k} \frac{p_{i+1,j}(t)}{t}, \quad t > 0,$$

so that, by letting $t \to 0$, (4.66) is obtained. On the other hand, if $k \geq i+1$, then

$$\begin{aligned} \sum_{j \leq k; j \neq i} \frac{p_{ij}(t)}{t} - \frac{1 - p_{ii}(t)}{t} &= \sum_{j \leq k} \frac{p_{ij}(t)}{t} - \frac{1}{t} \\ &\geq \sum_{j \leq k} \frac{p_{i+1,j}(t)}{t} - \frac{1}{t} \\ &= \sum_{j \leq k; j \neq i+1} \frac{p_{i+1,j}(t)}{t} - \frac{1 - p_{i+1,i+1}(t)}{t}. \end{aligned}$$

It follows that

$$\sum_{j \leq k; j \neq i} q_{ij} - q_i = \sum_{j \leq k} q_{ij} \geq \sum_{j \leq k} q_{i+1,j} = \sum_{j \leq k; j \neq i+1} q_{i+1,j} - q_{i+1},$$

so that (4.66) holds, which is equivalent to (4.64). Conversely, let $\nu \geq 2c$. If (4.64) holds, then we have, from Lemma 4.6, that $\mathbf{P}_\nu \in \mathcal{M}_{st}$, so that

$\mathbf{U}^{-1}\mathbf{P}_\nu^n\mathbf{U} \geq \mathbf{O}$ for all $n = 0, 1, \cdots$. Hence

$$\mathbf{U}^{-1}\mathbf{P}(t)\mathbf{U} = \sum_{n=0}^{\infty} \frac{(\nu t)^n}{n!} e^{-\nu t}\, \mathbf{U}^{-1}\mathbf{P}_\nu^n\mathbf{U} \geq \mathbf{O},$$

completing the proof. □

Example 4.16 Consider a birth–death process with birth rates λ_n and death rates μ_n. For $k \geq i+2$, we have

$$\sum_{j\geq k} q_{ij} = 0 \leq \lambda_{i+1} = \sum_{j\geq k} q_{i+1,j}.$$

Also, for $k = i$,

$$\sum_{j\geq k} q_{ij} = -\mu_i \leq 0 = \sum_{j\geq k} q_{i+1,j},$$

while, for $k \leq i-1$, $\sum_{j\geq k} q_{ij} = \sum_{j\geq k} q_{i+1,j} = 0$. Hence (4.64) holds, so that any birth–death process is stochastically monotone. It should be noted that

$$\sum_{j\geq i+1} q_{ij} = \lambda_i > -\mu_{i+1} = \sum_{j\geq i+1} q_{i+1,j},$$

but the case where $k = i+1$ is excluded in the condition (4.64).

In the context of continuous-time Markov chains, monotonicity in the sense of likelihood ratio ordering determines the zero structure of infinitesimal generators. Karlin and McGregor (1959) proved that if the transition matrix function $\mathbf{P}(t)$ of a continuous-time Markov chain is TP_2 for all $t \geq 0$, then the Markov chain must be a birth–death process (see Theorem C.2). See Kijima (1996) for results about monotonicities in the sense of hazard rate and reversed hazard rate orderings.

By analogy with Definition 3.12, we define stochastic comparability of continuous-time Markov chains as follows. For other definitions of comparability, see Stoyan (Section 4.2, 1983).

Definition 4.7 For two continuous-time Markov chains $\{X^k(t)\}$, $k = 1, 2$, let $\mathbf{P}_k(t)$ be the transition matrix function of $\{X^k(t)\}$. The continuous-time Markov chain $\{X^1(t)\}$ is said to *stochastically dominate* $\{X^2(t)\}$ if $\mathbf{P}_1(t)\mathbf{U} \geq \mathbf{P}_2(t)\mathbf{U}$ or, equivalently, $\mathbf{P}_1(t)\mathbf{V} \leq \mathbf{P}_2(t)\mathbf{V}$ for all $t \geq 0$.

The proof of the next theorem is left to the reader as an exercise (see Exercise 4.30).

Theorem 4.31 *Suppose that the two Markov chains $\{X^k(t)\}$, $k = 1, 2$, are uniformizable and one of them is stochastically monotone. Let $\mathbf{Q}_k = (q_{ij}^{(k)})$ denote the infinitesimal generator of $\{X^k(t)\}$. If*

$$\sum_{j\geq k} q_{ij}^{(1)} \leq \sum_{j\geq k} q_{ij}^{(2)}, \quad k \in \mathcal{N},$$

for all $i \in \mathcal{N}$, then $\{X^1(t)\}$ stochastically dominates $\{X^2(t)\}$.

Defining $\mathbf{P}_\nu^n = (p_{ij}^\nu(n))$, the transition probability functions of $\{X(t)\}$ are written as

$$p_{ij}(t) = \sum_{n=0}^{\infty} \frac{(\nu t)^n}{n!} e^{-\nu t} p_{ij}^\nu(n), \quad t \geq 0. \tag{4.67}$$

Since $p_{ij}(t)$ is differentiable and term-by-term differentiation of the right-hand side in (4.67) is possible, we have

$$\begin{aligned} p'_{ij}(t) &= -\nu e^{-\nu t}\delta_{ij} + \nu \sum_{n=1}^{\infty} \left\{ \frac{(\nu t)^{n-1}}{(n-1)!} - \frac{(\nu t)^n}{n!} \right\} e^{-\nu t} p_{ij}^\nu(n) \\ &= \nu \sum_{n=0}^{\infty} \frac{(\nu t)^n}{n!} e^{-\nu t} \{p_{ij}^\nu(n+1) - p_{ij}^\nu(n)\}, \quad t \geq 0. \end{aligned}$$

Therefore, if $p_{ij}^\nu(n)$ is nonincreasing (nondecreasing, respectively) in n then $p_{ij}(t)$ is nonincreasing (nondecreasing) in t. From Theorem 3.14, we know that $p_{00}^\nu(n)$ is nonincreasing in n if $\mathbf{P}_\nu \in \mathcal{M}_{\mathrm{st}}$. Therefore we have proved the following.

Theorem 4.32 *Suppose that $\{X(t)\}$ is stochastically monotone. Then the transition probability function $p_{00}(t)$ is nonincreasing in t. If the state space is given by $\mathcal{N} = \{0, 1, \cdots, N\}$ then $p_{0N}(t)$ is nondecreasing in t.*

Alternatively, Theorem 4.32 can be proved by using the variation diminishing property (VDP; see Theorem C.4) of the Poisson kernel

$$K(t, n) = \frac{(\nu t)^n}{n!} e^{-\nu t}, \quad t \geq 0; \quad n = 0, 1, \cdots.$$

Let $t_1 < t_2$ and let $n_2 = n_1 + \ell$ with $\ell > 0$. Then,

$$\frac{K(t_1, n_1)\, K(t_2, n_2)}{K(t_1, n_2)\, K(t_2, n_1)} = \frac{t_1^{n_1} t_2^{n_2}}{t_1^{n_2} t_2^{n_1}} = \left(\frac{t_2}{t_1}\right)^\ell > 1.$$

Hence, the Poisson kernel $K(t, n)$ is TP_2 in $t \geq 0$ and $n = 0, 1, \cdots$. In fact, it can be shown that $K(t, n)$ is totally positive of every order. From (4.67), we consider

$$p_{ij}(t) - \alpha = \sum_{n=0}^{\infty} \frac{(\nu t)^n}{n!} e^{-\nu t} \{p_{ij}^\nu(n) - \alpha\}, \quad t \geq 0, \tag{4.68}$$

where α is any real number. Suppose that $p_{ij}^\nu(n)$ is nonincreasing in n so that $p_{ij}^\nu(n) - \alpha$ changes sign at most once for any α and, if it does so once, it changes sign from + to −. Since $K(t, n)$ is TP_2, the VDP applied to (4.68) reveals that $p_{ij}(t) - \alpha$ changes sign at most once for any α and, if it does so once, it changes from + to −. This implies that $p_{ij}(t)$ is nonincreasing in t. Similarly, suppose that $p_{ij}^\nu(n)$ is strictly unimodal in n so that $p_{ij}^\nu(n) - \alpha$ changes sign at most twice for any α and, if it does so twice, it changes from $-$ to $+$ and then to $-$. Since $K(t,n)$ is TP_3, the VDP applied to (4.68) proves that $p_{ij}(t)$ is unimodal in t (though not necessarily strictly).

Before closing this section, we provide, without proof, a tool which enables us to transfer the distributional properties of discrete distributions to those of continuous distributions. For a discrete distribution $\mathbf{a} = (a_n)$ defined on $\mathcal{Z}_+$, consider the transformation

$$f(t) = \nu \sum_{n=1}^{\infty} \frac{(\nu t)^{n-1}}{(n-1)!} e^{-\nu t} a_{n-1}, \quad t \geq 0;$$

see (4.54). Let $F(t) = \int_0^t f(u)du$ and let $\overline{F}(t) = 1 - F(t)$. It is not difficult to show that

$$\overline{F}(t) = \sum_{n=0}^{\infty} \frac{(\nu t)^n}{n!} e^{-\nu t} \overline{A}_n, \quad t \geq 0,$$

where $\overline{A}_n = \sum_{k=n}^{\infty} a_k$. These transformations naturally arise from a *Poisson shock model*. Namely, consider a device subject to a sequence of shocks occurring randomly in time as events in a Poisson process with intensity ν. If $\overline{A}_n$ is interpreted as the probability of surviving the first n shocks, then $\overline{F}(t)$ is the probability that the device survives beyond time t. The next result is due to Esary, Marshall and Proschan (1973).

Theorem 4.33 (i) *$F(t)$ is* DLR *if* $\mathbf{a} = (a_n)$ *is* DLR.

(ii) *$F(t)$ is* IHR (DHR, *respectively*) *if* $\mathbf{a}$ *is* IHR (DHR).

(iii) *$F(t)$ is* IHRA (DHRA) *if* $\mathbf{a}$ *is* IHRA (DHRA).

(iv) *$F(t)$ is* NBU (NWU) *if* $\mathbf{a}$ *is* NBU (NWU).

4.9 Semi-Markov processes

Let $\{X(t)\}$ be a continuous-time stochastic process with state space $\mathcal{N}$. Suppose that no states are absorbing (the definition is given below) and that $X(0) = i$. In the case where $\{X(t)\}$ is a stable Markov chain, the process stays in state i for a finite, but strictly positive amount of time, called the *holding time*, which is exponentially distributed. A possible generalization of this construction is to allow the holding times to follow general distributions. That is, letting $F_i(t)$ be the holding-time distribution when the process is in state i, we can construct a stochastic process $\{X(t)\}$ as follows. If $X(0) = i$, then the process stays in state i for a time with distribution function $F_i(t)$. At the end of the holding time, the process moves to state j, which can be equal to i, according to the Markovian law $\mathbf{P} = (p_{ij})$. The process stays in state j for a time with distribution function $F_j(t)$ and then moves to some state according to $\mathbf{P}$. Under some regularity conditions, we can construct a stochastic process by repeating the above procedure.

We can introduce more dependent structure into the holding times.

Namely, when $X(0) = i$, we choose the next state j and the holding time simultaneously according to a joint distribution $F_{ij}(t)$. Given the next state j, the holding-time distribution is given by $F_{ij}(t)/F_{ij}(\infty)$. After the holding time, a transition to state j occurs and, at the same time, the next state k as well as the holding time is determined according to a joint distribution $F_{jk}(t)$. A stochastic process constructed in this way is called a *semi-Markov process*. Note that any discrete-time Markov chain is also a special case of semi-Markov processes, in which case we define $F_{ij}(t) = F_{ij}(\infty)$ for $t \geq 1$ and $F_{ij}(t) = 0$ for $t < 1$.

We first give a formal definition of semi-Markov processes. Let $\mathcal{N}$ denote the state space and let $\{Y_n\}$ be a sequence of random variables taking values on $\mathcal{N}$. Let $\{V_n\}$ be a sequence of random variables taking values on $R_+ = [0, \infty)$ and let $\tau_n = \sum_{k=0}^{n-1} V_k$, $n = 1, 2, \cdots$, with $\tau_0 \equiv 0$. We define

$$\tau(t) = \max\{n : \tau_n \leq t\}, \quad t \geq 0,$$

the renewal process associated with $\{V_n\}$.

Definition 4.8 With the above notation, suppose that

$$\begin{aligned} & P[Y_{n+1} = j, V_n \leq t | Y_0, \cdots, Y_n = i; V_0, \cdots, V_{n-1}] \\ = \; & P[Y_{n+1} = j, V_n \leq t | Y_n = i] \end{aligned} \tag{4.69}$$

for all $n = 0, 1, \cdots$; $i, j \in \mathcal{N}$, and $t \geq 0$. Then the stochastic process $\{X(t)\}$ defined by

$$X(t) = Y_{\tau(t)}, \quad t \geq 0,$$

is called a *semi-Markov process*.

Consider the two-dimensional process $\{Y_n, \tau_n\}$ in discrete time. It is obvious that the process is Markovian. Note that (4.69) is a generalization of the Markov property (1.1), but can be viewed as a special form of the two-dimensional Markov property. The process $\{Y_n, \tau_n\}$ is sometimes called a *Markov renewal process* (see, e.g., Nollau, 1980). The reader interested in more details of semi-Markov processes or Markov renewal processes should consult the excellent summary by Çinlar (1969).

Suppose that the right-hand side of (4.69) is independent of n. In this case, we define

$$F_{ij}(t) = P[Y_{n+1} = j, V_n \leq t | Y_n = i], \quad i, j \in \mathcal{N}; \quad t \geq 0. \tag{4.70}$$

A semi-Markov process with this property is said to be *homogeneous*. In what follows, we consider homogeneous semi-Markov processes only. The matrix $\mathbf{F}(t) = (F_{ij}(t))$ is called the *transition matrix* of the semi-Markov process $\{X(t)\}$. It is readily seen that $\mathbf{P} = (p_{ij})$ with

$$p_{ij} \equiv \lim_{t \to \infty} F_{ij}(t)$$

forms a transition matrix of the *embedded Markov chain* $\{Y_n\}$. Also,

$$P[V_n \le t | Y_n = i, Y_{n+1} = j] = \frac{F_{ij}(t)}{p_{ij}}, \quad t \ge 0,$$

which is the holding-time distribution given that the current state is i and the next state is j. The holding-time distribution in state i is given by

$$P[V_n \le t | Y_n = i] = \sum_{j \in \mathcal{N}} F_{ij}(t), \quad t \ge 0.$$

If $P[V_n < \infty | Y_n = i] = 0$, state i is called *absorbing*.

Let $\boldsymbol{\alpha}$ be the initial distribution of $\{X(t)\}$. Then the initial distribution $\boldsymbol{\alpha}$ and the transition matrix $\mathbf{F}(t)$ together determine the joint distribution

$$P[Y_0 = i_0, \cdots, Y_{n-1} = i_{n-1}, Y_n = i; V_0 \le t_0, \cdots, V_{n-1} \le t_{n-1}]$$

for any n and all i, $i_j \in \mathcal{N}$ and $t_j \ge 0$, $j = 0, \cdots, n-1$. The proof of this fact is left to the reader (see Exercise 4.32) since it is quite similar to the (ordinary) Markov chain case.

In what follows, we assume that the state space is finite and given by $\mathcal{N} = \{0, 1, \cdots, N\}$. The holding time in each state is assumed to be finite but nonzero with positive probability. That is,

$$P[V_n < \infty Y_n = i] = 1 \text{ and } P[V_n = 0 | Y_n = i] < 1$$

for all $i \in \mathcal{N}$. Let

$$T_j = \inf\{t > V_0 : X(t) = j\}, \quad j \in \mathcal{N},$$

and define

$$G_{ij}(t) = P[T_j \le t | X(0) = i], \quad i, j \in \mathcal{N}; \quad t > 0. \tag{4.71}$$

Two states i and j of the semi-Markov process $\{X(t)\}$ are said to *communicate* if either $i = j$ or $\lim_{t\to\infty} G_{ij}(t) > 0$ and $\lim_{t\to\infty} G_{ji}(t) > 0$. Note that under the regularity condition given above, the two states communicate if and only if they communicate in the embedded Markov chain $\{Y_n\}$ (see Section 2.2 for the definition). The notions of irreducibility, recurrence and transience of the semi-Markov process are also inherited from the embedded Markov chain.

For the finite semi-Markov process $\{X(t)\}$ with transition matrix $\mathbf{F}(t)$, define

$$p_{ij}(t) = P_i[X(t) = j], \quad i, j \in \mathcal{N}; \quad t \ge 0,$$

where $P_i[A] = P[A | X(0) = i]$. As in the continuous-time Markov chain case, the conditional probability $p_{ij}(t)$ is called the *transition probability function* of $\{X(t)\}$ and the matrix $\mathbf{P}(t) = (p_{ij}(t))$ is called the *transition matrix function*. For any fixed $t \ge 0$, the matrix $\mathbf{P}(t)$ is stochastic. Note that

$$p_{ij}(t) = P_i[X(t) = j, V_0 > t] + P_i[X(t) = j, V_0 \le t]$$

and that $P_i[X(t)=j,\, V_0>t]=\delta_{ij}\{1-F_i(t)\}$ where

$$F_i(t)\equiv\sum_{k\in\mathcal{N}}F_{ik}(t),\quad i\in\mathcal{N};\quad t\ge 0.$$

On the other hand,

$$\begin{aligned}&P_i[X(t)=j,\, V_0\le t]\\ =&\sum_{k\in\mathcal{N}}\int_0^t P_i[Y_1=k,\, u<V_0\le u+du]\,P[X(t)=j|Y_1=k,\, V_0=u]\\ =&\sum_{k\in\mathcal{N}}\int_0^t P_k[X(t-u)=j]dF_{ik}(u).\end{aligned}$$

Here we have made use of homogeneity for the second equality. It follows that

$$p_{ij}(t)=\delta_{ij}\{1-F_i(t)\}+\sum_{k\in\mathcal{N}}\int_0^t p_{kj}(t-u)dF_{ik}(u),\quad t\ge 0, \tag{4.72}$$

or, in matrix form,

$$\mathbf{P}(t)=\mathbf{I}-\mathbf{F}_{\mathrm{D}}(t)+\int_0^t d\mathbf{F}(u)\mathbf{P}(t-u),\quad t\ge 0, \tag{4.73}$$

where $\mathbf{F}_{\mathrm{D}}(t)$ denotes the diagonal matrix with diagonal elements $F_i(t)$. By taking the Laplace transform of (4.73), we have

$$\boldsymbol{\Pi}(s)=\frac{1}{s}\left[\mathbf{I}-\boldsymbol{\Phi}_{\mathrm{D}}(s)\right]+\boldsymbol{\Phi}(s)\boldsymbol{\Pi}(s),\quad \mathrm{Re}\,(s)>0.$$

Here and hereafter, we use the notation

$$\boldsymbol{\Pi}(s)=\int_0^\infty e^{-st}\mathbf{P}(t)dt,\quad \boldsymbol{\Phi}(s)=\int_0^\infty e^{-st}d\mathbf{F}(t)$$

and

$$\boldsymbol{\Phi}_{\mathrm{D}}(s)=\int_0^\infty e^{-st}d\mathbf{F}_{\mathrm{D}}(t).$$

It follows that

$$\boldsymbol{\Pi}(s)=\frac{1}{s}\left[\mathbf{I}-\boldsymbol{\Phi}(s)\right]^{-1}\left[\mathbf{I}-\boldsymbol{\Phi}_{\mathrm{D}}(s)\right],\quad \mathrm{Re}\,(s)>0. \tag{4.74}$$

The transition probability functions $p_{ij}(t)$ are then evaluated by inverting the matrix Laplace transform equation (4.74). The matrix Laguerre transform developed by Sumita (1981) may be a useful tool for inverting matrix Laplace transforms. The limiting distribution of the semi-Markov process $\{X(t)\}$ can be obtained by taking the limit

$$\lim_{s\to 0+} s\,\boldsymbol{\alpha}^{\top}\boldsymbol{\Pi}(s)=\lim_{s\to 0+}\boldsymbol{\alpha}^{\top}\left[\mathbf{I}-\boldsymbol{\Phi}(s)\right]^{-1}\left[\mathbf{I}-\boldsymbol{\Phi}_{\mathrm{D}}(s)\right],$$

if it exists (see Theorem B.12(ii) in Appendix B), where α denotes the initial distribution of $\{X(t)\}$.

Suppose that the transition matrix $\mathbf{P} = \mathbf{F}(\infty)$ of the embedded Markov chain $\{Y_n\}$ is irreducible and aperiodic. Further, suppose, for simplicity, that the $F_i(t)$ are all absolutely continuous. The probability density function of the $F_i(t)$ is denoted by $f_i(t)$. Then it can be shown that $s\boldsymbol{\Pi}(s)$ is analytic at $s = 0$ and the limit $\lim_{s\to 0+} s\boldsymbol{\Pi}(s)$ exists. The next lemma provides the asymptotic expansion of $[\mathbf{I} - \boldsymbol{\Phi}(s)]^{-1}$ (see Keilson, 1969).

Lemma 4.7 *Under the assumptions given above, we have*

$$[\mathbf{I} - \boldsymbol{\Phi}(s)]^{-1} = \frac{1}{s\,\boldsymbol{\pi}^\top \mathbf{m}} \mathbf{1}\boldsymbol{\pi}^\top + \mathbf{O}(1), \quad \text{as } s \to 0+,$$

where $\boldsymbol{\pi}$ *is the stationary distribution of* $\mathbf{P}$, *i.e.,* $\boldsymbol{\pi}^\top\mathbf{P} = \boldsymbol{\pi}^\top$, $\mathbf{m} = (m_i)$ *is the column vector with* $m_i = \int_0^\infty t\,dF_i(t)$, *and* $\mathbf{O}(1)$ *denotes a matrix such that* $\lim_{s\to 0+} s\,\mathbf{O}(1) = \mathbf{O}$.

Proof. For sufficiently small $\varepsilon > 0$, Keilson and Wishart (1964) showed that $\boldsymbol{\Phi}(s)$ on $0 < s < \varepsilon$ has a simple eigenvalue $\lambda(s)$ that is largest in magnitude among the eigenvalues of $\boldsymbol{\Phi}(s)$, which is continuous at $s = 0$, twice differentiable in $0 < s < \varepsilon$ and $|\lambda(s)| \le 1$. Let $\mathbf{J}(s)$ be a matrix satisfying

$$\lambda(s)\mathbf{J}(s) = \boldsymbol{\Phi}(s)\mathbf{J}(s) = \mathbf{J}(s)\boldsymbol{\Phi}(s) \tag{4.75}$$

and $\mathbf{J}^2(s) = \mathbf{J}(s)$. Such a matrix always exists on $0 < s < \varepsilon$, by standard matrix theory. The matrix $\mathbf{J}(s)$ is continuous at $s = 0$ and differentiable in $0 < s < \varepsilon$. Now define

$$\mathbf{L}(s) = \boldsymbol{\Phi}(s) - \lambda(s)\mathbf{J}(s).$$

Then, $\mathbf{J}(s)\mathbf{L}(s) = \mathbf{L}(s)\mathbf{J}(s) = \mathbf{O}$, so that

$$\boldsymbol{\Phi}^n(s) = \lambda^n(s)\mathbf{J}(s) + \mathbf{L}^n(s), \quad n = 1, 2, \cdots.$$

Summing both sides with respect to n yields

$$[\mathbf{I} - \boldsymbol{\Phi}(s)]^{-1} = \frac{\lambda(s)}{1 - \lambda(s)}\mathbf{J}(s) + [\mathbf{I} - \mathbf{L}(s)]^{-1}, \quad 0 < s < \varepsilon. \tag{4.76}$$

Since $\boldsymbol{\Phi}(s) \to \mathbf{P}$ as $s \to 0+$, we have $\lambda(s) \to 1$ and $\mathbf{J}(s) \to \mathbf{1}\boldsymbol{\pi}^\top$ as $s \to 0+$. It follows that

$$\lim_{s\to 0+} [\mathbf{I} - \mathbf{L}(s)]^{-1} = (\mathbf{I} - \mathbf{P} + \mathbf{1}\boldsymbol{\pi}^\top)^{-1},$$

i.e. the fundamental matrix of $\mathbf{P}$; see (2.63). On the other hand, differentiation of (4.75) with respect to s gives

$$\lambda'(s)\mathbf{J}(s) + \lambda(s)\mathbf{J}'(s) = \boldsymbol{\Phi}'(s)\mathbf{J}(s) + \boldsymbol{\Phi}(s)\mathbf{J}'(s).$$

Letting $s \to 0+$ and then pre-multiplying by $\boldsymbol{\pi}^\top$ and post-multiplying by $\mathbf{1}$, we obtain

$$\lambda'(0) = \boldsymbol{\pi}^\top\boldsymbol{\Phi}'(0)\mathbf{1} = -\boldsymbol{\pi}^\top\mathbf{m}.$$

It follows from (4.76) that

$$\lim_{s\to 0+} s\,[\mathbf{I}-\boldsymbol{\Phi}(s)]^{-1} = \lim_{s\to 0+} \frac{s\,\lambda(s)}{1-\lambda(s)}\,\mathbf{1}\boldsymbol{\pi}^{\top} = \frac{1}{-\lambda'(0)}\,\mathbf{1}\boldsymbol{\pi}^{\top}.$$

Hence the lemma is proved. □

We are ready to derive the limiting distribution of the semi-Markov process $\{X(t)\}$. An interpretation of the next result is immediate.

Theorem 4.34 *For a finite semi-Markov process, suppose that the transition matrix* $\mathbf{P} = \mathbf{F}(\infty)$ *is irreducible and aperiodic. Further, suppose that the holding-time distributions* $F_i(t)$ *are absolutely continuous. Then*

$$\lim_{t\to\infty} p_{ij}(t) = \frac{\pi_j m_j}{\sum_{k\in\mathcal{N}} \pi_k m_k}, \quad j \in \mathcal{N},$$

independently of the initial state i.

Proof. The asymptotic expansion of $\mathbf{I}-\boldsymbol{\Phi}_{\mathrm{D}}(s)$ is given by

$$\mathbf{I}-\boldsymbol{\Phi}_{\mathrm{D}}(s) = s\,\mathbf{m}_{\mathrm{D}} + s^2\,\mathbf{O}(1), \quad \text{as } s\to 0+,$$

where $\mathbf{m}_{\mathrm{D}}$ denotes the diagonal matrix with diagonal elements m_i. Under the assumptions, the limit of $s\,\boldsymbol{\Pi}(s)$ as $s\to 0+$ exists and so, from (4.74) and Lemma 4.7, we have

$$\begin{aligned}\lim_{t\to\infty} \mathbf{P}(t) &= \lim_{s\to 0+} s\,\boldsymbol{\Pi}(s)\\ &= \lim_{s\to 0+} s\,[\mathbf{I}-\boldsymbol{\Phi}(s)]^{-1}(\mathbf{m}_{\mathrm{D}} + s\,\mathbf{O}(1))\\ &= \frac{1}{\boldsymbol{\pi}^{\top}\mathbf{m}}\,\mathbf{1}\boldsymbol{\pi}^{\top}\mathbf{m}_{\mathrm{D}},\end{aligned}$$

as desired. □

For the finite semi-Markov process $\{X(t)\}$, define

$$\mathbf{Q} = \mathbf{m}_{\mathrm{D}}^{-1}(\mathbf{P}-\mathbf{I}).$$

Since $\mathbf{Q1} = \mathbf{0}$ and the off-diagonal elements of $\mathbf{Q}$ are nonnegative, $\mathbf{Q}$ can be considered as an infinitesimal generator of a continuous-time Markov chain $\{Z(t)\}$. Suppose that the transition matrix $\mathbf{P}$ is ergodic and let $\boldsymbol{\pi}$ be the stationary distribution of $\mathbf{P}$. Then the limiting distribution of $\{Z(t)\}$ is given by $\boldsymbol{\pi}^{\top}\mathbf{m}_{\mathrm{D}}/\boldsymbol{\pi}^{\top}\mathbf{m}$, the same as that of $\{X(t)\}$. A relation between $\{X(t)\}$ and $\{Z(t)\}$ for large t may be of interest. Some results concerning this problem are given by Kijima and Sumita (1986b).

Finally, in a finite semi-Markov process, suppose that the holding times are exponentially distributed and independent of the next state, i.e., say,

$$P[V_n \le t \mid Y_n = i,\, Y_{n+1} = j] = 1 - e^{-q_i t}, \quad q_i > 0; \quad i,\, j \in \mathcal{N}.$$

Further, suppose that the transition matrix $\mathbf{P} = (p_{ij})$ of the embedded

Markov chain has the property $p_{ii} = 0$. Then, for $i \neq j$, we have

$$F'_{ij}(t) = q_{ij}e^{-q_i t}; \quad q_{ij} \equiv q_i p_{ij},$$

so that

$$\boldsymbol{\Phi}(s) = \boldsymbol{\Phi}_{\mathrm{D}}(s)\,\mathbf{P}, \quad \mathrm{Re}\,(s) > 0,$$

where $\boldsymbol{\Phi}_{\mathrm{D}}(s)$ is the diagonal matrix with diagonal elements $q_i/(s+q_i)$. It follows from (4.74) that

$$\boldsymbol{\Pi}(s) = \frac{1}{s}\,[\mathbf{I} - \boldsymbol{\Phi}_{\mathrm{D}}(s)\mathbf{P}]^{-1}\,[\mathbf{I} - \boldsymbol{\Phi}_{\mathrm{D}}(s)], \quad \mathrm{Re}\,(s) > 0.$$

Write $\mathbf{Q} = (q_{ij})$ with $q_{ii} = -q_i$ and let $\mathbf{q}_{\mathrm{D}}$ be the diagonal matrix with diagonal elements q_i. Note that $\mathbf{P} = \mathbf{q}_{\mathrm{D}}^{-1}\mathbf{Q} + \mathbf{I}$ and that

$$[\mathbf{I} - \boldsymbol{\Phi}_{\mathrm{D}}(s)]^{-1}\boldsymbol{\Phi}_{\mathrm{D}}(s) = \frac{1}{s}\,\mathbf{q}_{\mathrm{D}}.$$

Using these relations, it is readily seen that

$$\boldsymbol{\Pi}(s) = (s\,\mathbf{I} - \mathbf{Q})^{-1}, \quad \mathrm{Re}\,(s) > 0,$$

which coincides with (4.17), i.e., the Laplace transform of the transition matrix function of a continuous-time Markov chain with infinitesimal generator $\mathbf{Q}$. Further relationships between Markov chains and semi-Markov processes via uniformization are discussed in Kijima (1987a).

4.10 Exercises

Exercise 4.1 For a regular Markov chain $\{X(t)\}$ with state space $\mathcal{N}$ and infinitesimal generator $\mathbf{Q} = (q_{ij})$, let T_j be the first passage time to state j and define

$$F_{ij}(t) = P_i[T_j \leq t], \quad t \geq 0; \quad i,\ j \in \mathcal{N}.$$

Let $p_{ij}(t)$ be the transition probability functions of $\{X(t)\}$. Show that

$$p_{ij}(t) = \delta_{ij}e^{-q_i t} + \int_0^t p_{jj}(t-s)\,dF_{ij}(s),$$

i.e. a continuous-time analog of Theorem 2.1. Moreover, taking the Laplace transform, prove that state j is recurrent if and only if $\int_0^\infty p_{jj}(t)dt = \infty$.

Exercise 4.2 Let $\{X(t)\}$ be a *nonhomogeneous* Markov chain with a finite state space and denote its transition matrix function by $\mathbf{P}(s,t)$. Assuming differentiability, prove the *backward Kolmogorov equation*

$$\frac{\partial}{\partial s}\mathbf{P}(s,t) = -\mathbf{Q}(s)\mathbf{P}(s,t), \quad 0 \leq s < t,$$

and the *forward Kolmogorov equation*

$$\frac{\partial}{\partial t}\mathbf{P}(s,t) = -\mathbf{P}(s,t)\mathbf{Q}(t), \quad 0 \leq s < t,$$

where

$$\mathbf{Q}(t) = \lim_{h\to 0} \frac{\mathbf{P}(t,t+h)-\mathbf{I}}{h} = \lim_{h\to 0} \frac{\mathbf{P}(t-h,t)-\mathbf{I}}{h}.$$

Also, prove that

$$\mathbf{P}(s,t) = \mathbf{I} + \int_s^t \mathbf{Q}(u)\mathbf{P}(u,t)du, \quad 0 \le s < t,$$

and

$$\mathbf{P}(s,t) = \mathbf{I} + \int_s^t \mathbf{P}(s,u)\mathbf{Q}(u)du, \quad 0 \le s < t,$$

satisfy both the backward and forward Kolmogorov equations. In particular, verify that

$$\mathbf{P}(s,t) = \mathbf{I} + \int_s^t \mathbf{Q}(u)du + \int_s^t ds_1 \int_{s_1}^t ds_2 \mathbf{Q}(s_1)\mathbf{Q}(s_2) + \cdots$$

(Iosifescu, 1980).

Exercise 4.3 Let $\{N(t)\}$ be a Poisson process with intensity λ (see Example 4.1).

(i) Calculate $P[N(t_1) = k_1, N(t_2) = k_2, N(t_3) = k_3]$ where $0 \le t_1 < t_2 < t_3$ and $0 \le k_1 \le k_2 \le k_3$.

(ii) Show that $P[N(t) = k | N(t+s) = n]$, $k = 0, 1, \cdots, n$, follows a binomial distribution.

Exercise 4.4 Let $\{N(t)\}$ be a Poisson process with intensity λ and let S_n denote the nth arrival time.

(i) Prove that the conditional joint density of $(S_1, \cdots, S_n)$ given that $N(t) = n$ exists and is given by

$$f(t_1, \cdots, t_n) = \frac{n!}{t^n}, \quad 0 < t_1 < \cdots < t_n < t.$$

That is, given that $N(t) = n$, the n arrival times $S_1, \cdots, S_n$ have the same distribution as the *order statistics* corresponding to n independent random variables uniformly distributed on the interval $(0, t)$.

(ii) Using the result stated in (i), show that $E\left[\sum_{i=1}^{N(t)} S_i\right] = \lambda t^2/2$.

Exercise 4.5 A stochastic process $\{X(t)\}$ is said to be a *compound Poisson process* if it is represented by

$$X(t) = \sum_{i=1}^{N(t)} Y_i, \quad t \ge 0,$$

where $\{N(t)\}$ is a Poisson process, $\{Y_n\}$ is a sequence of IID random variables, and $\{N(t)\}$ and $\{Y_n\}$ are mutually independent. Obtain the moment generating function $E[e^{tX(t)}]$ in terms of the intensity λ of $\{N(t)\}$ and the moment generating function $\phi_Y(s) = E[e^{sY}]$.

Exercise 4.6 In the postulates given in Example 4.1, suppose instead of (ii) that

(ii′) $P[N(t+h) - N(t) = 1] = \lambda(t)h + o(h)$

for some function $\lambda(t)$. Accordingly, postulate (iii) is replaced by

(iii′) $P[N(t+h) - N(t) \geq 2] = o(h)$.

Such a counting process is called a *nonhomogeneous Poisson process*. Let

$$m(t,s) = \int_t^s \lambda(u)du, \quad s > t \geq 0.$$

Fix t and define $p_n(s) = P[N(t+s) - N(t) = n], \ s \geq 0$.

(i) Derive the forward differential equation

$$p_0'(s) = -\lambda(t+s)\, p_0(s), \quad s > 0,$$

so that $p_0(s) = e^{-m(t,t+s)}$.

(ii) By a similar argument, derive

$$p_n'(s) = -\lambda(t+s)\, p_n(s) + \lambda(t+s)\, p_{n-1}(s), \quad s > 0,$$

and show that

$$p_n(s) = \frac{\{m(t,t+s)\}^n}{n!} e^{-m(t,t+s)}, \quad t, \ s \geq 0; \quad n = 0, 1, \cdots.$$

Exercise 4.7 Let $\{X(t)\}$ be a Markov chain in continuous time with infinitesimal generator $\mathbf{Q} = (q_{ij})$, where $q_i = q_{i,i-1} = i\mu, \ i = 1, 2, \cdots,$ for some $\mu > 0$. Such a Markov chain is called a *pure death process*. Find $P_i[X(t) = n], \ n = 0, 1, \cdots, i$.

Exercise 4.8 Show that $\mathbf{P}(t) = \exp\{\mathbf{Q}\,t\}$ satisfies the Kolmogorov equation (4.15) by a direct substitution. Also show that the solution is unique.

Exercise 4.9 Consider two machines that are maintained by a single repairman. Machine i functions for an exponential time with parameter μ_i before breaking down. The repair times for either machine are exponentially distributed with parameter λ. Formulate this system by a continuous-time Markov chain (see Example 5.1) and obtain the mean number of functioning machines in equilibrium.

Exercise 4.10 Consider a graph with vertex set $\mathcal{N} = \{1, 2, \cdots, n\}$ and edge set $E = \{(i,j) : i \neq j, \ i, \ j \in \mathcal{N}\}$. Suppose that a particle moves along this graph as follows. Events occur along the edges (i,j) according to independent Poisson processes with intensities λ_{ij}. An event along edge (i,j) causes that edge to become *excited*. If the particle is at vertex i at the moment that (i,j) becomes excited, it instantaneously moves to vertex j.

Let $X(t)$ denote the vertex at which the particle is located at time t. Formulate $\{X(t)\}$ by a continuous-time Markov chain and find the stationary distribution (Ross, 1989).

Exercise 4.11 Consider the two-state Markov chain $\{X(t)\}$ given in Example 4.3, where the state space is $\{0, 1\}$ and the infinitesimal generator is

$$\mathbf{Q} = \begin{pmatrix} -\lambda & \lambda \\ \mu & -\mu \end{pmatrix}.$$

Of interest is the total time up to time t that the chain has been in a given state. That is, for $t > 0$, let

$$S_1(t) = \int_0^t X(s)ds, \quad S_0(t) = \int_0^t \{1 - X(s)\}ds.$$

Compute the mean and the variance of $S_0(t)$ given that $X(0) = 0$.

Exercise 4.12 Let $\mathbf{P}(t)$ denote the transition matrix function of a finite Markov chain in continuous time. Show that $\det(\mathbf{P}(t)) > 0$ for all $t > 0$ (Karlin and Taylor, 1981).

Exercise 4.13 Mimicking the proof of Lemma 4.3, prove (4.25) under the regularity assumption.

Exercise 4.14 Let $\{X(t)\}$ be a continuous-time Markov chain with infinitesimal generator $\mathbf{Q}$. For a nonnegative vector $\mathbf{g} = (g(i))$, let

$$\gamma_i(s) = \int_0^\infty e^{-st} E_i[g(X(t))]dt, \quad \text{Re}\,(s) > 0.$$

Show that $\boldsymbol{\gamma}(s) = (\gamma_i(s))$ is a solution of the linear equation

$$(s\mathbf{I} - \mathbf{Q})\boldsymbol{\gamma}(s) = \mathbf{g}.$$

Exercise 4.15 Let $f_{ij}^{(0)}(t) = \delta_{ij}e^{-q_i t}$ and let $f_{ij}^{(n)}(t)$, $n = 1, 2, \cdots$, be generated by (4.31). Prove by induction on n that

$$f_{ij}^{(n)}(t) = P_i[X(t) = j,\ J_{n+1} > t],$$

where J_n denotes the time just after the nth transition of $\{X(t)\}$.

Exercise 4.16 Let $g_{ij}^{(0)}(t) = \delta_{ij}e^{-q_i t}$ and define $g_{ij}^{(n)}(t)$ successively by

$$g_{ij}^{(n)}(t) = g_{ij}^{(0)}(t) + \int_0^t e^{-q_j s} \sum_{k \neq j} g_{ik}^{(n-1)}(t-s) q_{kj} ds, \quad n = 1, 2, \cdots.$$

Prove that $g_{ij}^{(n)}(t) = f_{ij}^{(n)}(t)$ for all n, where the $f_{ij}^{(n)}(t)$ are generated by (4.31).

Exercise 4.17 Prove Lemma 4.4(i). Using this, prove the uniqueness of the backward Kolmogorov equation in Theorem 4.18.

Exercise 4.18 Let T_j be the first passage time to state j in a uniformizable continuous-time Markov chain. After uniformization, let $T_j(\nu)$ be the first passage time to state j of the uniformized Markov chain. Prove that

$$\pi_j = \frac{1}{E_j[T_j(\nu)]} = \frac{1/q_i}{E_j[T_j]},$$

independently of the choice of ν.

Exercise 4.19 In Exercise 4.11, we wanted to determine the conditional distribution of $S_0(t)$ given that $X(0) = 0$. To this end, let $\{N(t)\}$ be a Poisson process with intensity $\nu = \lambda + \mu$ and consider uniformization. Then

$$P_0[S_0(t) \le s] = \sum_{n=1}^{\infty} \frac{(\nu t)^n}{n!} e^{-\nu t} P_0[S_0(t) \le s | N(t) = n], \quad s < t.$$

Show that

$$\begin{aligned} & P_0[S_0(t) \le s | N(t) = n] \\ = & \sum_{k=1}^{n} {}_nC_{k-1} \left(\frac{\mu}{\nu}\right)^{k-1} \left(\frac{\lambda}{\nu}\right)^{n-k+1} \sum_{i=k}^{n} {}_nC_i \left(\frac{s}{t}\right)^i \left(1 - \frac{s}{t}\right)^{n-i} \end{aligned}$$

(Ross, 1983).

Exercise 4.20 Let $\{X(t)\}$ be a uniformizable, ergodic Markov chain with state space $\mathcal{N}$ and infinitesimal generator $\mathbf{Q}$. Let $\boldsymbol{\pi}$ be the stationary distribution of $\{X(t)\}$.

(i) For any function f on $\mathcal{N}$ for which $\boldsymbol{\pi}^\top \mathbf{f}$ converges absolutely, show that

$$\lim_{t\to\infty} \frac{1}{t} E_i \left[\int_0^t f(X(s))ds \right] = \boldsymbol{\pi}^\top \mathbf{f}.$$

(ii) For any functions f and g on $\mathcal{N}$ for which $\boldsymbol{\pi}^\top \mathbf{f}$ and $\boldsymbol{\pi}^\top \mathbf{g}$ converge absolutely and $\boldsymbol{\pi}^\top \mathbf{g} \neq 0$, show that

$$\lim_{t\to\infty} \frac{E_i[\int_0^t f(X(s))ds]}{E_j[\int_0^t g(X(s))ds]} = \frac{\boldsymbol{\pi}^\top \mathbf{f}}{\boldsymbol{\pi}^\top \mathbf{g}}$$

independently of i, $j \in \mathcal{N}$, and

$$\lim_{t\to\infty} \frac{\int_0^t f(X(s))ds}{\int_0^t g(X(s))ds} = \frac{\boldsymbol{\pi}^\top \mathbf{f}}{\boldsymbol{\pi}^\top \mathbf{g}}$$

almost surely.

Exercise 4.21 For a finite, continuous-time ergodic Markov chain with infinitesimal generator $\mathbf{Q}$, let $\mathbf{Z}$ be the fundamental matrix defined by

$$\mathbf{Z} = \int_0^\infty \{\mathbf{P}(t) - \mathbf{1}\boldsymbol{\pi}^\top\} dt,$$

where $\boldsymbol{\pi}$ is the stationary distribution. Prove that

$$\mathbf{Z} = (\mathbf{1}\boldsymbol{\pi}^\top - \mathbf{Q})^{-1} - \mathbf{1}\boldsymbol{\pi}^\top.$$

Exercise 4.22 By continuing the analysis given in Example 4.11, obtain the transition probability functions of the M/M/1/N queue explicitly.

Exercise 4.23 For an absorbing Markov chain, let ξ be the time spent in the nonabsorbing states until absorption. Prove that $(E_i[\xi^k]) = k!\,\mathbf{N}^k\mathbf{1}$, where $i \in \mathcal{A}^c$ and $\mathbf{N}$ is the fundamental matrix of the absorbing Markov chain (Iosifescu, 1980).

Exercise 4.24 Let $\{X(t)\}$ be an absorbing Markov chain on the state space $\{0, 1, \cdots, N\}$ with absorbing states 0 and N. Let N_j be the time spent in state j until absorption and define $m_{ij} = E_i[N_j]$ for $i,\ j \neq 0,\ N$. Show that

$$P_i[N_j = 0] = 1 - \frac{m_{ij}}{m_{jj}}$$

and

$$P_i[0 < N_j \le t] = \frac{m_{ij}}{m_{jj}}[1 - e^{-t/m_{jj}}], \quad t > 0.$$

Exercise 4.25 In Exercise 4.24, let ξ be the total time until absorption in states 0 or N, i.e., $\xi = \sum_{j=1}^{N-1} N_j$. Show that, conditional on $X(0) = i$, ξ has density function $f_i(t) = -\boldsymbol{\delta}_i^\top \mathbf{T}\exp\{\mathbf{T}\,t\}\mathbf{1}$, where $\mathbf{T}$ is the lossy generator corresponding to the nonabsorbing states.

Exercise 4.26 For the phase-type density $f(t)$ generated by $(\boldsymbol{\alpha}, \mathbf{T})$, let $\overline{F}(t) = \int_t^\infty f(u)du$. Prove that

$$\overline{F}(t) = \boldsymbol{\alpha}^\top \exp\{\mathbf{T}\,t\}\mathbf{1}, \quad t \ge 0.$$

Exercise 4.27 Prove Theorem 4.27.

Exercise 4.28 Let $\{X(t)\}$ be an irreducible Markov chain with finite state space $\mathcal{N} = \{0, 1, \cdots, N\}$ and infinitesimal generator $\mathbf{Q} = (q_{ij})$. The *entropy* of $\{X(t)\}$ given that $X(0) = i$ is defined by

$$e_i(t) = -\sum_{j=0}^{N} p_{ij}(t)\log p_{ij}(t),$$

where the $p_{ij}(t)$ denote the transition probability functions of $\{X(t)\}$. Suppose $q_{ij} = q_{ji}$. Prove that $e_i(t)$ is nondecreasing in t for all $i \in \mathcal{N}$ (Karlin and Taylor, 1981).

Exercise 4.29 Let $\{X(t)\}$ be an ergodic Markov chain with state space $\{0, 1, \cdots, N\}$ and infinitesimal generator $\mathbf{Q}$. Let $\mathbf{T}$ be a submatrix of $\mathbf{Q}$ corresponding to the states $\{1, 2, \cdots, N\}$. Let $\mathbf{r} = -\mathbf{T}\mathbf{1}$ and define $\mathbf{Q}_1 = \mathbf{T} + \mathbf{r}\boldsymbol{\delta}_1^\top$. The matrix $\mathbf{Q}_1$ is a generator of an ergodic Markov chain. Let $\mathbf{q}$ denote the quasi-stationary distribution of the lossy generator $\mathbf{T}$, and let

$\boldsymbol{\pi}_1$ be the stationary distribution of the generator $\mathbf{Q}_1$. Prove that if the original Markov chain $\{X(t)\}$ is stochastically monotone then $\mathbf{q} \geq_{\rm st} \boldsymbol{\pi}_1$.

Exercise 4.30 Prove Theorem 4.31.

Exercise 4.31 Let $\{X^k(t)\}$, $k = 1, 2$, be birth–death processes with birth rates λ_n^k and death rates μ_n^k respectively. Specify a condition in terms of the birth and death rates so that $\{X^1(t)\}$ stochastically dominates $\{X^2(t)\}$.

Exercise 4.32 Let $\boldsymbol{\alpha} = (\alpha_i)$ be the initial distribution of a semi-Markov process $\{X(t)\}$ and let $\mathbf{F}(t) = (F_{ij}(t))$ be its transition matrix. Prove that the joint distribution

$$P[Y_0 = i_0, \cdots, Y_{n-1} = i_{n-1}, Y_n = i;\ V_0 \leq t_0, \cdots, V_{n-1} \leq t_{n-1}]$$

can be expressed in terms of $\boldsymbol{\alpha}$ and $\mathbf{F}(t)$.

Exercise 4.33 In the notation given in Section 4.9, let $\tau_n = \sum_{k=0}^{n-1} V_k$, $n = 1, 2, \cdots$. Prove that

$$P_i[\tau_n \leq t] = \sum_{j \in \mathcal{N}} F_{ij}^{(n)}(t), \quad t \geq 0,$$

where $F_{ij}^{(1)}(t) = F_{ij}(t)$ and

$$F_{ij}^{(n)}(t) = \sum_{k \in \mathcal{N}} \int_0^t F_{kj}(t-s) dF_{ik}^{(n-1)}(s), \quad n = 1, 2, \cdots.$$

Exercise 4.34 Let $\{X(t)\}$ be a finite Markov chain with state space $\mathcal{N}$ and transition matrix function $\mathbf{P}(t)$. For a nonnegative function f on $\mathcal{N}$, define

$$v_i = \sup_T E_i[f(X(T))],$$

where the supremum is taken over all stopping times T. Write $\mathbf{v} = (v_i)$.

(i) Show that $\mathbf{v} \geq \mathbf{0}$ is a subinvariant vector of $\mathbf{P}(t)$, i.e., $\mathbf{v} \geq \mathbf{P}(t)\mathbf{v}$, for all $t > 0$, and $\mathbf{v} \geq \mathbf{f}$.

(ii) Show that if $\mathbf{h} \geq \mathbf{0}$ is another subinvariant vector of $\mathbf{P}(t)$ for all $t > 0$ with $\mathbf{h} \geq \mathbf{f}$ then $\mathbf{h} \geq \mathbf{v}$.

(iii) Show that

$$T_0 = \inf\{t : f(X(t)) = v(X(t))\}$$

is an optimal *stopping time* in the sense that

$$E_i[f(X(T_0))] = v_i, \quad i \in \mathcal{N}$$

(Çinlar, 1975).

5

Birth–death processes

In the preceding chapter, we saw birth–death processes as a special class of continuous-time Markov chains. Let $\{X(t)\}$ denote a birth–death process. In Example 4.4, $X(t)$ represents the size of a population at time t. A 'birth' increases the size by 1 and a 'death' decreases it by 1. However, this is indeed a rich and important class in modeling a variety of phenomena not only in biology but also in, e.g., operations research, demography, economics and engineering. Typical examples of problems that can be formulated as birth–death processes are the following.

Example 5.1 A system is composed of N identical machines which are served by one repairman. Each machine operates independently for a random length of time until failure. When it fails, it stands idle until the repairman can repair it. Once repaired, the machine becomes as good as new. Suppose that the failure-time distribution is exponential with parameter λ and that the repair-time distribution is also exponential with parameter μ. Let $X(t)$ denote the number of failed machines at time t. If $X(t) = i$, then $N - i$ machines are working, and the time until the next failure is exponentially distributed with parameter $\lambda(N - i)$ if no machines are repaired in the meantime. Hence $\{X(t)\}$ is a birth–death process with state space $\{0, 1, \cdots, N\}$, birth rates $\lambda_i = \lambda(N - i)$ and death rates $\mu_i = \mu$. This sort of model is called a *machine repair problem*.

Example 5.2 Consider the following queueing system. Customers arrive according to a Poisson process with intensity λ; service times are exponentially distributed with parameter μ; there are s servers working independently of each other; and the waiting room is of infinite size. Such a queueing system is called an *M/M/s queue*. Let $X(t)$ denote the number of customers in the system at time t. If there are n customers in the system, then $\min\{n, s\}$ servers are busy, and thus the time until a service completion is exponentially distributed with parameter $\mu \min\{n, s\}$ if no arrivals occur in the meantime. Hence $\{X(t)\}$ is a birth–death process with state space $\mathcal{Z}_+$, birth rates $\lambda_i = \lambda$ and death rates $\mu_i = \mu \min\{i, s\}$.

Example 5.3 Suppose we consider a population whose size $X(t)$ ranges between two integers N_1 and N_2, $N_1 < N_2$, for all $t \geq 0$. Suppose that the birth and death rates per individual given that $X(t) = n$ are $\alpha(N_2 - n)$ and $\beta(n - N_1)$, respectively, and that the individuals of the population act independently of each other. Then $\{X(t)\}$ is a birth–death process with state space $\{N_1, N_1 + 1, \cdots, N_2 - 1, N_2\}$, birth rates $\lambda_n = \alpha n(N_2 - n)$ and death rates $\mu_n = \beta n(n - N_1)$. This birth–death process is often called a *logistic process.*

What makes birth–death processes so useful is that standard methods of analysis are available for determining numerous important quantities such as stationary distributions and mean first passage times. In this chapter, we focus on birth–death processes and investigate their properties in some detail.

5.1 Boundary classification

Consider a birth–death process $\{X(t)\}$ defined on the nonnegative integers $\mathcal{N} = \{0, 1, 2, \cdots\}$ with birth rates λ_n and death rates μ_n, all of which are finite and positive except $\mu_0 \geq 0$. In the case where $\mu_0 > 0$, we imagine that there is an ignored absorbing state, -1 say, which can be reached via state 0. The infinitesimal generator of $\{X(t)\}$ is given by

$$\mathbf{Q} = \begin{pmatrix} -\lambda_0 - \mu_0 & \lambda_0 & 0 & 0 & \cdots \\ \mu_1 & -\lambda_1 - \mu_1 & \lambda_1 & 0 & \cdots \\ 0 & \mu_2 & -\lambda_2 - \mu_2 & \lambda_2 & \cdots \\ \vdots & \vdots & \ddots & \ddots & \ddots \end{pmatrix}. \tag{5.1}$$

Throughout this chapter, we write $\pi_0 = 1$ and

$$\pi_n = \pi_{n-1}\frac{\lambda_{n-1}}{\mu_n} = \frac{\lambda_0 \cdots \lambda_{n-1}}{\mu_1 \cdots \mu_n}, \quad n = 1, 2, \cdots,$$

regardless of whether $\mu_0 > 0$ or $\mu_0 = 0$. The quantities π_n are called the *potential coefficients* of the birth–death process $\{X(t)\}$ (see Keilson, 1979). Note that

$$\lambda_n \pi_n = \mu_{n+1}\pi_{n+1}, \quad n = 0, 1, \cdots. \tag{5.2}$$

Let $\boldsymbol{\pi}_D$, $\boldsymbol{\pi}_D^{1/2}$ and $\boldsymbol{\pi}_D^{-1/2}$ be the diagonal matrices with diagonal elements π_i, $\pi_i^{1/2}$ and $\pi_i^{-1/2}$ respectively. Then, using (5.2), it is easily seen that

$$\boldsymbol{\pi}_D^{1/2}\mathbf{Q}\boldsymbol{\pi}_D^{-1/2} = \begin{pmatrix} -\lambda_0 - \mu_0 & \sqrt{\lambda_0\mu_1} & 0 & 0 & \cdots \\ \sqrt{\lambda_0\mu_1} & -\lambda_1 - \mu_1 & \sqrt{\lambda_1\mu_2} & 0 & \cdots \\ 0 & \sqrt{\lambda_1\mu_2} & -\lambda_2 - \mu_2 & \sqrt{\lambda_2\mu_3} & \cdots \\ \vdots & \vdots & \ddots & \ddots & \ddots \end{pmatrix},$$

whence the generator $\mathbf{Q}$ is symmetrizable in terms of the potential coefficients π_n.

We introduce the notation

$$A = \sum_{n=0}^{\infty} \frac{1}{\lambda_n \pi_n}, \qquad B = \sum_{n=0}^{\infty} \pi_n,$$
$$C = \sum_{n=0}^{\infty} \frac{1}{\lambda_n \pi_n} \sum_{i=0}^{n} \pi_i, \quad D = \sum_{n=0}^{\infty} \frac{1}{\lambda_n \pi_n} \sum_{i=n+1}^{\infty} \pi_i. \tag{5.3}$$

It is readily seen that

$$AB = C + D.$$

Hence $A + B = \infty$ if and only if at least one of C and D diverges. Also, from (5.3), we have

$$A = \infty \quad \Rightarrow \quad C = \infty,$$
$$B = \infty \quad \Rightarrow \quad D = \infty. \tag{5.4}$$

We shall give an interpretation of the values C and D below. The next result is due to Karlin and McGregor (1957b). The result follows immediately from the interpretations of the quantities A and B given in Theorem 5.9 below and Example 4.8 respectively.

Lemma 5.1 *Suppose that $\mu_0 = 0$. The birth–death process $\{X(t)\}$ is recurrent (transient, respectively) if $A = \infty$ ($A < \infty$) and, when $\{X(t)\}$ is recurrent, it is positive recurrent (null recurrent) if $B < \infty$ ($B = \infty$) in addition.*

Let us denote the transition matrix function of the birth–death process $\{X(t)\}$ by $\mathbf{P}(t) = (p_{ij}(t))$. Throughout this chapter, it is assumed that the $\mathbf{P}(t)$ satisfies the forward and backward Kolmogorov equations

$$\mathbf{P}'(t) = \mathbf{P}(t)\,\mathbf{Q}, \quad \mathbf{P}'(t) = \mathbf{Q}\,\mathbf{P}(t), \quad t \geq 0 \tag{5.5}$$

respectively. The Chapman–Kolmogorov equation can be written as

$$\mathbf{P}(t + s) = \mathbf{P}(t)\,\mathbf{P}(s), \quad t,\ s \geq 0. \tag{5.6}$$

In order to establish the uniqueness of a solution of the Kolmogorov equations, we need to introduce a classification of boundaries at infinity of birth–death processes. We denote the boundary by ∞. The next boundary classification is due to Feller (1959). See Callaert and Keilson (1973a, b) for a further classification.

Definition 5.1 The boundary at infinity of a birth–death process is said to be *regular* if C, $D < \infty$, *exit* if $C < \infty$ and $D = \infty$, *entrance* if $C = \infty$ and $D < \infty$, and *natural* if $C = D = \infty$.

Regardless of whether $\mu_0 > 0$ or $\mu_0 = 0$, the transition probability functions are uniquely determined by the birth and death rates if and only

Table 5.1 *Implications of boundary classification*

Boundary classification	Condition		State classification
exit	$C < \infty,\ D = \infty$	$\Rightarrow$	transient
entrance	$C = \infty,\ D < \infty$	$\Rightarrow$	positive recurrent
natural	$C = \infty,\ D = \infty$	$\Rightarrow$	transient or recurrent

Table 5.2 *Implications of state classification*

State classification	Condition		Boundary classification
transient	$A < \infty$	$\Rightarrow$	exit or natural
positive recurrent	$A = \infty,\ B < \infty$	$\Rightarrow$	entrance or natural
null recurrent	$A = \infty,\ B = \infty$	$\Rightarrow$	natural

if $A+B = \infty$, i.e., at least one of C and D diverges. If the series C diverges, then the birth–death process $\{X(t)\}$ is *nonexplosive*, which means that the process makes at most finitely many jumps in a finite time with probability one.* If $C < \infty$ while the series D diverges (∞ is an exit boundary), then the process is *explosive* and the boundary ∞ is absorbing. In the case of a regular boundary, the transition probabilities are not uniquely determined by the birth and death rates. That is, both the backward and the forward Kolmogorov equations have infinitely many solutions. In this case, we say that the rate problem associated with $\{X(t)\}$ is *indeterminate* (see van Doorn, 1987a, for details).

The boundary classification given in Definition 5.1 and the state classification are not independent of each other. For example, an exit boundary implies that the process is transient since $C < \infty$ implies $A < \infty$, from (5.4). Table 5.1 summarizes such implications (see Callaert and Keilson, 1973a). Conversely, suppose that the boundary ∞ is not regular. A transient process implies that ∞ is either exit or natural since if $A < \infty$ then $B = \infty$ (the boundary ∞ is not regular) so that $D = \infty$, from (5.4). Table 5.2 summarizes these implications.

The next result, which is due to Callaert and Keilson (1973a), provides a sufficient condition for a natural boundary at infinity.

Theorem 5.1 *Let λ_n and μ_n be the birth and death rates respectively. If*

$$\liminf_{n\to\infty}(\lambda_n + \mu_n) < \infty$$

then the boundary at infinity is natural.

* This implies that the process is *regular* (see Definition 4.4 in Section 4.3). If the birth–death process has a regular boundary at infinity then the process is *not* regular.

Proof. The condition implies that

$$\sum_{n=0}^{\infty} \lambda_n^{-1} = \sum_{n=1}^{\infty} \mu_n^{-1} = \infty.$$

Since $\pi_n < \sum_{i=0}^{n} \pi_i$ so that

$$\frac{1}{\lambda_n \pi_n} \sum_{i=1}^{n} \pi_i > \frac{1}{\lambda_n},$$

it follows that

$$C = \sum_{n=0}^{\infty} \frac{1}{\lambda_n \pi_n} \sum_{i=0}^{n} \pi_i > \sum_{n=0}^{\infty} \frac{1}{\lambda_n} = \infty.$$

Similarly, the fact $D = \infty$ follows since

$$\frac{1}{\lambda_n \pi_n} \sum_{i=n+1}^{\infty} \pi_i = \frac{1}{\mu_{n+1} \pi_{n+1}} \sum_{i=n+1}^{\infty} \pi_i > \frac{1}{\mu_{n+1}},$$

where the equality follows from (5.2). □

From the proof of Theorem 5.1, the following result is immediate.

Corollary 5.1 *An exit boundary implies* $\lim_{n\to\infty} \lambda_n = \infty$, *while an entrance boundary implies* $\lim_{n\to\infty} \mu_n = \infty$. *A regular boundary implies both* $\lim_{n\to\infty} \lambda_n = \infty$ *and* $\lim_{n\to\infty} \mu_n = \infty$.

Example 5.4 Consider an M/M/1 queue with arrival rate λ and service rate μ. Writing $\rho = \lambda/\mu$, one obtains

$$\pi_n = \rho^n, \quad n = 0, 1, \cdots.$$

If $\rho \le 1$, then $A = \infty$ so that the queue-size process is recurrent. If $\rho < 1$, it is positive recurrent since $B < \infty$. It should be noted from Theorem 5.1 that the boundary at infinity is natural for any $\rho > 0$.

An interpretation of the values C and D in (5.3) is given by Callaert and Keilson (1973a). Let τ_n^+ be the mean first passage time of $\{X(t)\}$ from state n to state $n+1$. According to the skip-free nature of birth–death processes, after an exponential sojourn in state $n \ge 1$, it moves up to state $n+1$ with probability $\lambda_n/(\lambda_n+\mu_n)$, or down to state $n-1$ with probability $\mu_n/(\lambda_n+\mu_n)$. In the latter case, the process $\{X(t)\}$ reaches state $n+1$ in a random time with mean $\tau_{n-1}^+ + \tau_n^+$. Thus,

$$\begin{aligned} \tau_n^+ &= \frac{\lambda_n}{\lambda_n+\mu_n} \frac{1}{\lambda_n+\mu_n} + \frac{\mu_n}{\lambda_n+\mu_n} \left(\frac{1}{\lambda_n+\mu_n} + \tau_{n-1}^+ + \tau_n^+ \right) \\ &= \frac{1}{\lambda_n+\mu_n} + \frac{\mu_n}{\lambda_n+\mu_n} (\tau_{n-1}^+ + \tau_n^+), \end{aligned}$$

so that

$$\tau_n^+ = \frac{1}{\lambda_n} + \frac{\mu_n}{\lambda_n}\tau_{n-1}^+, \quad n = 1, 2, \cdots.$$

For $n = 0$, we assume that $\mu_0 = 0$ since, otherwise, the first passage time to state 1 is infinity with positive probability. Then, of course, $\tau_0^+ = \lambda_0^{-1}$. It is easily seen that

$$\tau_n^+ = \frac{1}{\lambda_n \pi_n} \sum_{i=0}^{n} \pi_i, \quad n = 0, 1, \cdots, \tag{5.7}$$

from which, denoting the mean first passage time to the boundary ∞ at infinity from state 0 by $\tau_{0\infty} = \sum_{n=0}^{\infty} \tau_n^+$, we have

$$\tau_{0\infty} = \sum_{n=0}^{\infty} \frac{1}{\lambda_n \pi_n} \sum_{i=0}^{n} \pi_i = C.$$

To describe the value D, we consider a finite birth–death process defined on the state space $\{0, 1, \cdots, N\}$ with the same birth and death rates as the original denumerable birth–death process except that $\lambda_N = 0$. Let τ_n^- be the mean first passage time of the finite birth–death process from state n to state $n - 1$, where $n = 1, \cdots, N$. The same considerations as are given above lead to

$$\tau_n^- = \frac{1}{\lambda_n + \mu_n} + \frac{\lambda_n}{\lambda_n + \mu_n}(\tau_{n+1}^- + \tau_n^-), \quad n = 1, \cdots, N - 1,$$

so that

$$\tau_n^- = \frac{1}{\mu_n} + \frac{\lambda_n}{\mu_n}\tau_{n+1}^-, \quad n = 1, \cdots, N - 1.$$

For $n = N$, we have $\tau_N^- = \mu_N^{-1}$, by assumption. It is easily seen that

$$\tau_n^- = \frac{1}{\mu_n \pi_n} \sum_{i=n}^{N} \pi_i, \quad n = 1, \cdots, N,$$

so that, defining $\tau_{\infty 0} = \lim_{N\to\infty} \sum_{n=1}^{N} \tau_n^-$, the mean first passage time to state 0 from the boundary ∞ at infinity, we have

$$\tau_{\infty 0} = \sum_{n=1}^{\infty} \frac{1}{\mu_n \pi_n} \sum_{i=n}^{\infty} \pi_i = \sum_{n=0}^{\infty} \frac{1}{\lambda_n \pi_n} \sum_{i=n+1}^{\infty} \pi_i = D,$$

where we have used (5.2) in the second equality.

5.2 Birth–death polynomials

In the analysis of birth–death processes, a prominent role is played by a sequence of polynomials $\{Q_n(x)\}$, called ***birth–death polynomials***, satisfying the recurrence relation

$$-x\,Q_n(x) = \mu_n Q_{n-1}(x) - (\lambda_n + \mu_n)Q_n(x) + \lambda_n Q_{n+1}(x) \tag{5.8}$$

for $n = 0, 1, \cdots$, where $Q_{-1}(x) = 0$ and $Q_0(x) = 1$. Denote the column vector with components $Q_i(x)$ by $\mathbf{q}(x) = (Q_i(x))$. Then (5.8) can be written in matrix notation as

$$-x\,\mathbf{q}(x) = \mathbf{Q}\,\mathbf{q}(x), \tag{5.9}$$

where $\mathbf{Q}$ is the infinitesimal generator given by (5.1). Let $p_{ij}(t)$ be the (minimal) transition probability functions corresponding to $\mathbf{Q}$ and write $\mathbf{P}(t) = (p_{ij}(t))$. As we will show later, $p_{ij}(t)$ can be represented as

$$p_{ij}(t) = \pi_j \int_0^\infty e^{-xt} Q_i(x) Q_j(x) d\psi(x), \quad i,\, j \in \mathcal{N}; \quad t \geq 0, \tag{5.10}$$

called the *Karlin–McGregor representation*, where ψ is a positive Borel measure of total mass 1 with support in $[0, \infty)$. The probability distribution function ψ is called the *spectral measure* of the transition probability function $p_{ij}(t)$. The sequence of polynomials $\{Q_n(x)\}$ constitutes an *orthogonal polynomial* sequence with respect to ψ, since taking $t = 0$ in (5.10) yields

$$\pi_j \int_0^\infty Q_i(x) Q_j(x) d\psi(x) = \delta_{ij}, \quad i,\, j \in \mathcal{N}.$$

An immediate consequence of (5.10) is the following.

Theorem 5.2 *The transition probability function $p_{ii}(t)$ is completely monotone for every $i \in \mathcal{N}$.*

Proof. For each i, define

$$\mu(x) = \pi_i \int_0^x Q_i^2(y) d\psi(y), \quad x \geq 0.$$

Then $\mu(x)$ is nondecreasing in x and $\mu(\infty) = 1$. Hence μ is a probability distribution function on $[0, \infty)$ and

$$p_{ii}(t) = \int_0^\infty e^{-xt} d\mu(x), \quad t \geq 0.$$

The theorem follows from Theorem B.13. □

Suppose that $\mu_0 > 0$ and let T_0 be the first passage time from state 0 to the absorbing state -1. Then, since -1 is absorbing, we have

$$f_0(t)dt = P_0[X(t) = 0,\, X(t + dt) = -1] = P_0[X(t) = 0]\,\mu_0 dt,$$

where $f_0(t)$ denotes the density function of T_0. It follows that the density function is given by $f_0(t) = \mu_0\, p_{00}(t)$. The next result is then immediate (see Example 4.12 for the finite case).

Corollary 5.2 *Suppose $\mu_0 > 0$ and let T_0 be the first passage time from state 0 to the absorbing state -1. Then the density function of T_0 is completely monotone.*

The next lemma is well known. See, e.g., Szegö (Section 3.3, 1959) for the proof and related results concerning orthogonal polynomials.

Lemma 5.2 *The birth–death polynomial $Q_n(x)$ has n positive, simple zeros x_{ni}, $i = 1, \cdots, n$, which satisfy the interlacing property*

$$0 < x_{n+1,i} < x_{ni} < x_{n+1,i+1}, \quad i = 1, \cdots, n; \quad n = 1, 2, \cdots.$$

As an immediate consequence, the limits

$$\xi_i \equiv \lim_{n\to\infty} x_{ni}, \quad i = 1, 2, \cdots,$$

exist and $0 \le \xi_i \le \xi_{i+1} < \infty$.

Following Karlin and McGregor (1957b), we introduce a dual generator of the infinitesimal generator $\mathbf{Q}$ (see also van Doorn, 1985). Suppose, firstly, that $\mu_0 > 0$ and define

$$\lambda_n^d = \mu_n, \quad \mu_0^d = 0, \quad \mu_{n+1}^d = \lambda_n, \quad n = 0, 1, \cdots. \tag{5.11}$$

Accordingly, we define $\pi_0^d = 1$ and

$$\pi_n^d = \pi_{n-1}^d \frac{\lambda_{n-1}^d}{\mu_n^d} = \frac{\mu_0}{\lambda_{n-1}\pi_{n-1}} = \frac{\mu_0}{\mu_n \pi_n}, \quad n = 1, 2, \cdots.$$

Secondly, when $\mu_0 = 0$, we define

$$\lambda_n^d = \mu_{n+1}, \quad \mu_n^d = \lambda_n, \quad n = 0, 1, \cdots, \tag{5.12}$$

so that $\pi_0^d = 1$ and

$$\pi_n^d = \pi_{n-1}^d \frac{\lambda_{n-1}^d}{\mu_n^d} = \frac{\lambda_0}{\lambda_n \pi_n} = \frac{\lambda_0}{\mu_{n+1}\pi_{n+1}}, \quad n = 1, 2, \cdots.$$

The infinitesimal generator $\mathbf{Q}^d$ with birth rates λ_n^d and death rates μ_n^d is called the *dual generator* of $\mathbf{Q}$, and a birth–death process $\{X^d(t)\}$ with the dual generator is called a *dual process* of $\{X(t)\}$. When $\mu_0 > 0$, the dual generator is given by

$$\mathbf{Q}^d = \begin{pmatrix} -\mu_0 & \mu_0 & 0 & 0 & \cdots \\ \lambda_0 & -\mu_1 - \lambda_0 & \mu_1 & 0 & \cdots \\ 0 & \lambda_1 & -\mu_2 - \lambda_1 & \mu_2 & \cdots \\ \vdots & \vdots & \ddots & \ddots & \ddots \end{pmatrix}, \tag{5.13}$$

while, when $\mu_0 = 0$, it is given by

$$\mathbf{Q}^d = \begin{pmatrix} -\mu_1 - \lambda_0 & \mu_1 & 0 & 0 & \cdots \\ \lambda_1 & -\mu_2 - \lambda_1 & \mu_2 & 0 & \cdots \\ 0 & \lambda_2 & -\mu_3 - \lambda_2 & \mu_3 & \cdots \\ \vdots & \vdots & \ddots & \ddots & \ddots \end{pmatrix}. \tag{5.14}$$

It should be noted that the dual of the dual generator $\mathbf{Q}^d$ is just the original infinitesimal generator **Q** since, from (5.11) and (5.12), we have, when $\mu_0 > 0$,

$$(\lambda_n^d)^d = \mu_{n+1}^d = \lambda_n, \quad (\mu_n^d)^d = \lambda_n^d = \mu_n,$$

while, when $\mu_0 = 0$,

$$(\lambda_n^d)^d = \mu_n^d = \lambda_n, \quad (\mu_{n+1}^d)^d = \lambda_n^d = \mu_{n+1}.$$

Exercise 5.8 asks the reader to determine the boundary classification of the dual birth–death process.

Let $\{Q_n^d(x)\}$ be the sequence of birth–death polynomials, called the *dual polynomials*, of the dual generator $\mathbf{Q}^d$, i.e.,

$$-x\,Q_n^d(x) = \mu_n^d Q_{n-1}^d(x) - (\lambda_n^d + \mu_n^d)Q_n^d(x) + \lambda_n^d Q_{n+1}^d(x) \qquad (5.15)$$

for $n = 0, 1, \cdots$, where $Q_{-1}^d(x) = 0$ and $Q_0^d(x) = 1$. The next lemma links the birth–death polynomials $Q_n(x)$ and the dual polynomials $Q_n^d(x)$.

Lemma 5.3 (i) *Suppose that* $\mu_0 > 0$. *Then*

$$Q_{n+1}^d(x) = \frac{\lambda_n \pi_n}{\mu_0}\{Q_{n+1}(x) - Q_n(x)\}, \quad n = 0, 1, \cdots.$$

(ii) *Suppose that* $\mu_0 = 0$. *Then*

$$Q_n^d(x) = \frac{\lambda_n \pi_n}{-x}\{Q_{n+1}(x) - Q_n(x)\}, \quad n = 0, 1, \cdots.$$

Proof. Suppose that $\mu_0 > 0$. From (5.8), we have

$$\begin{aligned}
&-x\,\lambda_{n-1}\pi_{n-1}Q_n(x) \\
=\;& \mu_n\lambda_{n-1}\pi_{n-1}Q_{n-1}(x) + \mu_n\lambda_n\pi_n Q_{n+1}(x) \\
&\qquad -(\mu_n\lambda_n\pi_n + \mu_n\lambda_{n-1}\pi_{n-1})Q_n(x) \\
=\;& -\lambda_n^d\lambda_{n-1}\pi_{n-1}\{Q_n(x) - Q_{n-1}(x)\} + \lambda_n^d\lambda_n\pi_n\{Q_{n+1}(x) - Q_n(x)\}
\end{aligned}$$

and

$$\begin{aligned}
&-x\,\lambda_{n-1}\pi_{n-1}Q_{n-1}(x) \\
=\;& \lambda_{n-1}\lambda_{n-2}\pi_{n-2}Q_{n-2}(x) + \lambda_{n-1}^2\pi_{n-1}Q_n(x) \\
&\qquad -(\lambda_{n-1}^2\pi_{n-1} + \lambda_{n-1}\lambda_{n-2}\pi_{n-2})Q_{n-1}(x) \\
=\;& -\mu_n^d\lambda_{n-2}\pi_{n-2}\{Q_{n-1}(x) - Q_{n-2}(x)\} \\
&\qquad +\mu_n^d\lambda_{n-1}\pi_{n-1}\{Q_n(x) - Q_{n-1}(x)\},
\end{aligned}$$

where we have used (5.2) and (5.11). Subtracting the above second identity from the first and then dividing the result by μ_0 yields (5.15). Assertion (ii) follows similarly. □

Interpreting the empty sum $\sum_{k=0}^{-1}$ as zero, we have the following result.

Theorem 5.3 (i) *Suppose that* $\mu_0 > 0$. *Then*

$$Q_n(x) = \sum_{k=0}^{n} \pi_k^d Q_k^d(x) = 1 + \sum_{k=0}^{n-1} Q_{k+1}^d(x)\frac{\mu_0}{\lambda_k \pi_k}, \quad n = 0, 1, \cdots,$$

and

$$Q_n^d(x) = 1 - \frac{x}{\lambda_0^d}\sum_{k=0}^{n-1} \pi_k Q_k(x) = 1 - \frac{x}{\mu_0}\sum_{k=0}^{n-1} \pi_k Q_k(x), \quad n = 0, 1, \cdots.$$

(ii) *Suppose that* $\mu_0 = 0$. *Then*

$$Q_n(x) = 1 - \frac{x}{\lambda_0}\sum_{k=0}^{n-1} \pi_k^d Q_k^d(x) = 1 - x\sum_{k=0}^{n-1} Q_k^d(x)\frac{1}{\lambda_k \pi_k}, \quad n = 0, 1, \cdots,$$

and

$$Q_n^d(x) = \sum_{k=0}^{n} \pi_k Q_k(x), \quad n = 0, 1, \cdots.$$

Proof. Suppose that $\mu_0 > 0$. From Lemma 5.3(i), we have

$$Q_{n+1}(x) - Q_n(x) = \frac{\mu_0}{\lambda_n \pi_n} Q_{n+1}^d(x), \quad n = 0, 1, \cdots.$$

Summing both sides, it follows that

$$Q_n(x) = 1 + \sum_{k=0}^{n-1} Q_{k+1}^d(x)\frac{\mu_0}{\lambda_k \pi_k} = \sum_{k=0}^{n} \pi_k^d Q_k^d(x).$$

Next, suppose $\mu_0 = 0$ so that $\mu_0^d > 0$. Since the dual of the dual generator is the original generator, we then have

$$Q_n^d(x) = \sum_{k=0}^{n} (\pi_k^d)^d (Q_k^d(x))^d = \sum_{k=0}^{n} \pi_k Q_k(x).$$

The other identities can be proved similarly. □

Recall that the sequence of polynomials $\{Q_n^d(x)\}$ constitutes a sequence of birth–death polynomials. Hence, from Lemma 5.2, the polynomial $Q_n^d(x)$ has n positive, simple zeros x_{ni}^d, $i = 1, \cdots, n$ which satisfy the interlacing property

$$0 < x_{n+1,i}^d < x_{ni}^d < x_{n+1,i+1}^d, \quad i = 1, \cdots, n; \quad n = 1, 2, \cdots,$$

whence the limits

$$\xi_i^d \equiv \lim_{n\to\infty} x_{ni}^d, \quad i = 1, 2, \cdots,$$

exist and $0 \le \xi_i^d \le \xi_{i+1}^d < \infty$. The next theorem is a consequence of a theorem from Chihara (Theorem I.7.2, 1978). The proof is omitted.

Theorem 5.4 (i) *Suppose that* $\mu_0 > 0$. *Then*

$$0 < x_{ni}^d < x_{ni} < x_{n,i+1}^d < x_{n,i+1}, \quad i = 1, \cdots, n-1,$$

and

$$0 \leq \xi_i^d \leq \xi_i \leq \xi_{i+1}^d < \infty, \quad i = 1, 2, \cdots.$$

(ii) *Suppose that* $\mu_0 = 0$. *Then*

$$0 < x_{ni} < x_{ni}^d < x_{n,i+1} < x_{n,i+1}^d, \quad i = 1, \cdots, n-1,$$

and

$$0 \leq \xi_i \leq \xi_i^d \leq \xi_{i+1} < \infty, \quad i = 1, 2, \cdots.$$

5.3 Finite birth–death processes

In this section, we consider a finite birth–death process $\{X_N(t)\}$ defined on the state space $\{0, 1, \cdots, N-1\}$ with birth rates λ_n and death rates μ_n all positive except μ_0, $\lambda_{N-1} \geq 0$. The (possibly lossy) infinitesimal generator is denoted by $\mathbf{Q}_N$. When $\lambda_{N-1} > 0$, we imagine that there is an absorbing state, N say, which can be reached via state $N-1$ only. In this case, the lossy generator $\mathbf{Q}_N$ is the $N \times N$ north-west corner truncation of $\mathbf{Q}$ given by (5.1). If $\mu_0 = \lambda_{N-1} = 0$, i.e., the generator $\mathbf{Q}_N$ is conservative, then the birth–death process $\{X_N(t)\}$ is ergodic.

Let $Q_n(x)$ be the birth–death polynomials defined in (5.8) and let $\mathbf{q}_N(x)$ be the column vector with components $Q_n(x)$, $n = 0, 1, \cdots, N-1$. Recall that all the eigenvalues of $\mathbf{Q}_N$ are real and nonpositive, since $\mathbf{Q}_N$ is tridiagonal, its off-diagonal elements are nonnegative, and $\mathbf{Q}_N \mathbf{1} \leq \mathbf{0}$.

Lemma 5.4 *Let x be a nonnegative number. Then, $-x$ is an eigenvalue of $\mathbf{Q}_N$ if and only if*

$$-x\, Q_{N-1}(x) = \mu_{N-1} Q_{N-2}(x) - (\lambda_{N-1} + \mu_{N-1}) Q_{N-1}(x)$$

holds, regardless of whether $\lambda_{N-1} > 0$ or $\lambda_{N-1} = 0$.

Proof. The 'if' part holds since, then,

$$-\, x\, \mathbf{q}_N(x) = \mathbf{Q}_N\, \mathbf{q}_N(x). \tag{5.16}$$

Conversely, suppose that $-x$ is an eigenvalue of $\mathbf{Q}_N$. Let (q_i) be a nonzero right eigenvector associated with $-x$. Then

$$\begin{cases} -xq_0 &= -(\lambda_0 + \mu_0)q_0 + \lambda_0 q_1, \\ -xq_n &= \mu_n q_{n-1} - (\lambda_n + \mu_n)q_n + \lambda_n q_{n+1}, \quad n = 1, \cdots, N-2, \\ -xq_{N-1} &= \mu_{N-1} q_{N-2} - (\lambda_{N-1} + \mu_{N-1}) q_{N-1}. \end{cases}$$

If $q_0 = 0$ then $q_n = 0$ for all $n = 1, 2, \cdots, N-1$. Hence $q_0 \neq 0$ so that, we have $Q_n(x) = q_n / q_0$ and the 'only if' part follows. □

Suppose, firstly, that $\lambda_{N-1} > 0$. Let x_{Nk}, $k = 1, \cdots, N$, be the zeros of $Q_N(x)$. Since $Q_N(x_{Nk}) = 0$, Lemma 5.4 shows that the $-x_{Nk}$ are the eigenvalues of $\mathbf{Q}_N$ and the $\mathbf{q}_N(x_{Nk})$ are the right eigenvectors each associated with $-x_{Nk}$. Let $\boldsymbol{\pi}_{\mathrm{D}}$ denote the diagonal matrix with diagonal elements π_k, $k = 0, 1, \cdots, N-1$, where the π_k are the potential coefficients. It is easily seen that the matrix $\boldsymbol{\pi}_{\mathrm{D}}^{1/2}\mathbf{Q}_N\boldsymbol{\pi}_{\mathrm{D}}^{-1/2}$ is symmetric. Hence, letting

$$\mathbf{u}_k = \alpha_k \boldsymbol{\pi}_{\mathrm{D}}^{1/2}\mathbf{q}_N(x_{Nk}); \quad \alpha_k = \left\{ \sum_{j=0}^{N-1} \pi_j Q_j^2(x_{Nk}) \right\}^{-1/2}$$

for $k = 1, \cdots, N$, we obtain the spectral decomposition

$$\boldsymbol{\pi}_{\mathrm{D}}^{1/2}\mathbf{Q}_N\boldsymbol{\pi}_{\mathrm{D}}^{-1/2} = -\sum_{k=1}^{N} x_{Nk}\, \mathbf{u}_k \mathbf{u}_k^{\mathsf{T}}.$$

Let $\mathbf{P}_N(t) = (p_{ij}^N(t))$ be the transition matrix function of the finite (lossy) birth–death process $\{X_N(t)\}$. Since $\mathbf{P}_N(t) = \exp\{\mathbf{Q}_N t\}$, we have

$$\boldsymbol{\pi}_{\mathrm{D}}^{1/2}\mathbf{P}_N(t)\boldsymbol{\pi}_{\mathrm{D}}^{-1/2} = \sum_{k=1}^{N} e^{-x_{Nk}t}\, \mathbf{u}_k \mathbf{u}_k^{\mathsf{T}}, \quad t \geq 0,$$

from which it follows that

$$\mathbf{P}_N(t) = \sum_{k=1}^{N} \alpha_k^2\, e^{-x_{Nk}t}\mathbf{q}_N(x_{Nk})\mathbf{q}_N^{\mathsf{T}}(x_{Nk})\boldsymbol{\pi}_{\mathrm{D}}, \quad t \geq 0, \tag{5.17}$$

or, componentwise,

$$p_{ij}^N(t) = \pi_j \sum_{k=1}^{N} \alpha_k^2\, e^{-x_{Nk}t} Q_i(x_{Nk})Q_j(x_{Nk}), \quad t \geq 0. \tag{5.18}$$

Recall that $\sum_{k=1}^{N} \mathbf{u}_k \mathbf{u}_k^{\mathsf{T}} = \mathbf{I}$. Hence, considering the $(0,0)$th component of $\sum_{k=1}^{N} \mathbf{u}_k \mathbf{u}_k^{\mathsf{T}}$, we have $\sum_{k=1}^{N} \alpha_k^2 = 1$ since $\pi_0 = Q_0(x) = 1$. Also, $\alpha_k^2 > 0$. Thus, we can define a probability distribution function

$$\psi_N(x) = \begin{cases} 0, & x < x_{N1}, \\ \sum_{i=1}^{k} \alpha_i^2, & x_{Nk} \leq x < x_{N,k+1}, \\ 1, & x \geq x_{NN}. \end{cases} \tag{5.19}$$

Using ψ_N, (5.18) can then be rewritten as

$$p_{ij}^N(t) = \pi_j \int_0^\infty e^{-xt} Q_i(x)Q_j(x)d\psi_N(x), \quad t \geq 0. \tag{5.20}$$

In particular, since $p_{ij}^N(0) = \delta_{ij}$, we have

$$\pi_j \int_0^\infty Q_i(x) Q_j(x) d\psi_N(x) = \delta_{ij} \tag{5.21}$$

for $i,\ j = 0, 1, \cdots, N-1$. See (5.10) for the infinite case.

Suppose, secondly, that $\lambda_{N-1} = 0$ and consider $\mathbf{U}^{-1}\mathbf{Q}_N\mathbf{U}$, where $\mathbf{U}$ is given by (3.1). It is not difficult to see that $(\mathbf{U}^{-1}\mathbf{Q}_N\mathbf{U})^\top$ is equal to the matrix

$$\mathbf{Q}_N^d = \begin{pmatrix} -\mu_0 & \mu_0 & 0 & \cdots & 0 \\ \lambda_0 & -\mu_1 - \lambda_0 & \mu_1 & \cdots & 0 \\ \vdots & \ddots & \ddots & \ddots & \vdots \\ 0 & \cdots & \lambda_{N-3} & -\mu_{N-2} - \lambda_{N-3} & \mu_{N-2} \\ 0 & \cdots & 0 & \lambda_{N-2} & -\mu_{N-1} - \lambda_{N-2} \end{pmatrix}.$$

If $\mu_0 > 0$ then the matrix $\mathbf{Q}_N^d$ is just the $N \times N$ north-west corner truncation of the dual generator $\mathbf{Q}^d$ given by (5.13). Since $\mathbf{U}^{-1}\mathbf{Q}_N\mathbf{U}$ is a similarity transform, the values $-x_{Nk}^d$, where the x_{Nk}^d are the zeros of the dual polynomial $Q_N^d(x)$, are the eigenvalues of $\mathbf{Q}_N$. This follows from Lemma 5.4 since $\mu_{N-1} > 0$. If $\mu_0 = 0$ then the state 0 associated with $\mathbf{Q}_N^d$ is absorbing. Since the submatrix of $\mathbf{Q}_N^d$ corresponding to the transient states $\{1, \cdots, N-1\}$ is just the $(N-1) \times (N-1)$ north-west corner truncation of the dual generator $\mathbf{Q}^d$ given by (5.14), the eigenvalues of $\mathbf{Q}_N$ other than 0 are given by $-x_{N-1,k}^d$ (see Example 5.5 below). Define $y_{Nk} = x_{Nk}^d$ for $k = 1, \cdots, N$ if $\mu_0 > 0$, and $y_{N1} = 0$ and $y_{Nk} = x_{N-1,k-1}^d$ for $k = 2, \cdots, N$ if $\mu_0 = 0$. Then, in either case, we have (5.16) with $x = y_{Nk},\ k = 1, \cdots, N$. Hence if $\lambda_{N-1} = 0$ then, defining

$$\psi_N^d(x) = \begin{cases} 0, & x < y_{N1}, \\ \sum_{i=1}^k \beta_i^2, & y_{Nk} \le x < y_{N,k+1}, \\ 1, & x \ge y_{NN}, \end{cases}$$

where $\beta_k = \left\{\sum_{j=0}^{N-1} \pi_j Q_j^2(y_{Nk})\right\}^{-1/2}$, the transition probability functions are given by

$$p_{ij}^N(t) = \pi_j \int_0^\infty e^{-xt} Q_i(x) Q_j(x) d\psi_N^d(x), \quad t \ge 0.$$

The orthogonality condition (5.21) holds for this case as well.

Let $\pi_{ij}^N(s) = \int_0^\infty e^{-st} p_{ij}^N(t) dt$ be the Laplace transform of $p_{ij}^N(t)$. From (4.17), we have

$$\pi_{ij}^N(s) = [(s\mathbf{I} - \mathbf{Q}_N)^{-1}]_{ij}, \quad \mathrm{Re}(s) > -x_{N1},$$

where $[\mathbf{A}]_{ij}$ means the (i,j)th component of matrix $\mathbf{A}$. Let γ_k denote the eigenvalues of $-\mathbf{Q}_N$. Recall that $\gamma_k = x_{Nk}$ if $\lambda_{N-1} > 0$ and $\gamma_k = y_{Nk}$ if $\lambda_{N-1} = 0$. Let $b_{ij}(s)$ be the (i,j)th cofactor of $(s\,\mathbf{I} - \mathbf{Q}_N)$. Then it is well known (see, e.g., Noble and Daniel, page 208, 1977) that

$$\pi_{ij}^N(s) = \frac{b_{ji}(s)}{\det(s\,\mathbf{I} - \mathbf{Q}_N)}, \quad \mathrm{Re}\,(s) > -\gamma_1,$$

where $\det(\mathbf{A})$ denotes the determinant of matrix $\mathbf{A}$. On the other hand, by the tridiagonality of $(s\,\mathbf{I} - \mathbf{Q}_N)$, we have

$$b_{0,N-1}(s) = \prod_{k=1}^{N-1} \mu_k.$$

The characteristic polynomial of $\mathbf{Q}_N$ is given by

$$\det(s\,\mathbf{I} - \mathbf{Q}_N) = \prod_{k=1}^{N} (s + \gamma_k).$$

It follows that

$$\mu_0\, \pi_{N-1,0}^N(s) = \prod_{k=1}^{N} \frac{\mu_{k-1}}{s + \gamma_k}, \quad \mathrm{Re}\,(s) > -\gamma_1, \tag{5.22}$$

provided that $\mu_0 > 0$. In particular, when $\lambda_{N-1} = 0$, the identity

$$\prod_{k=1}^{N} \gamma_k = \prod_{k=1}^{N} x_{Nk}^d = \prod_{k=1}^{N} \mu_{k-1} \tag{5.23}$$

holds, from which we obtain

$$\mu_0\, \pi_{N-1,0}^N(s) = \prod_{k=1}^{N} \frac{\gamma_k}{s + \gamma_k}, \quad \mathrm{Re}\,(s) > -\gamma_1. \tag{5.24}$$

Note that $\mu_0\, \pi_{N-1,0}^N(s)$ is the Laplace transform of the first passage time T_{N-1} from state $N-1$ to the absorbing state -1 of the finite birth–death process $\{X_N(t)\}$. Equation (5.24) implies that the first passage time T_{N-1} is a sum of N independent exponential random variables with distinct parameters γ_k. This result should be compared with the (discrete-time) random walk case given in Example 2.24. To prove (5.23), let $\mathbf{A}_n$ be the $n \times n$ south-east corner truncation of $-\mathbf{Q}_N$, $n = 1, \cdots, N$. Expanding the determinant yields

$$\det(\mathbf{A}_n) = (\lambda_{N-n} + \mu_{N-n}) \det(\mathbf{A}_{n-1}) - \lambda_{N-n}\, \mu_{N-n+1} \det(\mathbf{A}_{n-2}),$$

where $\det(\mathbf{A}_0) = 1$ and $\det(\mathbf{A}_1) = \mu_{N-1}$ since $\lambda_{N-1} = 0$. The solution to this equation is $\det(\mathbf{A}_n) = \prod_{k=1}^n \mu_{N-k}$, which can be readily verified by a direct induction argument. But, since $\det(\mathbf{A}_N) = \det(-\mathbf{Q}_N)$ and $\det(-\mathbf{Q}_N) = \prod_{k=1}^N \gamma_k$, (5.23) follows. The next theorem is due to Keilson (1971, 1979).

Theorem 5.5 *Suppose that $\lambda_{N-1} = 0$ and $\mu_0 > 0$. Let T_{N-1} be the first passage time of $\{X_N(t)\}$ from state $N-1$ to the absorbing state -1. Then T_{N-1} is a sum of N independent and exponentially distributed random variables with distinct parameters.*

The Laplace transform in (5.24) can be inverted and the density function of T_{N-1} is given by

$$f_{N-1}(t) = \sum_{k=1}^{N} \left(\prod_{j \neq k} \frac{\gamma_j}{\gamma_j - \gamma_k} \right) \gamma_k \, e^{-\gamma_k t}, \quad t \geq 0,$$

where $\lambda_{N-1} = 0$ so that $\gamma_k = x_{Nk}^d$. The proof of this result is left to the reader (see Exercise 5.9).

Analogously to Definition 4.6, we define the following.

Definition 5.2 A continuous-time Markov chain $\{X(t)\}$ with transition matrix function $\mathbf{P}(t)$ is said to be *monotone in the sense of likelihood ratio ordering* if $\mathbf{P}(t) \in \mathrm{TP}_2$ for all $t \geq 0$.

A very special property that birth–death processes possess is the following, which is a special case of Karlin and McGregor (1959). See Exercises 5.10 and 5.11 for other interesting properties.

Theorem 5.6 (i) *The transition matrix function of a finite birth–death process $\{X_N(t)\}$ is strictly totally positive for every $t > 0$. As a result, $\{X_N(t)\}$ is monotone in the sense of likelihood ratio ordering.*

(ii) *If a finite Markov chain in continuous time is monotone in the sense of likelihood ratio ordering, then it must be a birth–death process.*

Let $\gamma_1, \cdots, \gamma_N$ denote the eigenvalues of $-\mathbf{Q}_N$, where

$$0 \leq \gamma_1 < \gamma_2 < \cdots < \gamma_N.$$

Suppose that $\mu_0 = \lambda_{N-1} = 0$ so that $\mathbf{Q}_N$ is conservative and the finite birth–death process $\{X_N(t)\}$ is ergodic. Then we have $\gamma_1 = 0$. Also, from Theorem 5.3(ii), $Q_n(0) = 1$ for all n. Hence $\mathbf{q}_N(0) = \mathbf{1}$ and, from (5.17), we have

$$\mathbf{P}_N(t) = \mathbf{1}\mathbf{m}^{\mathsf{T}} + \sum_{k=2}^{N} \alpha_k^2 \, e^{-\gamma_k t} \mathbf{q}_N(\gamma_k) \mathbf{q}_N^{\mathsf{T}}(\gamma_k) \boldsymbol{\pi}_{\mathrm{D}}, \quad t \geq 0,$$

where $\mathbf{m} = \left(\pi_i / \sum_{i=0}^{N-1} \pi_i\right)$. Since $\gamma_k > 0$ for $k = 2, \cdots, N$, the decay parameter of the ergodic birth–death process $\{X_N(t)\}$ is given by γ_2. The decay parameter can be calculated based on the following representation result, due to van Doorn (1987b), of the smallest eigenvalue of a sign-symmetric, tridiagonal matrix. The proof is left to the reader (see Exercise 5.12).

Lemma 5.5 *Let*

$$\mathbf{T}_n = \begin{pmatrix} c_1 & b_2 & 0 & 0 & \cdots & 0 \\ a_2 & c_2 & b_3 & 0 & \cdots & 0 \\ 0 & a_3 & c_3 & b_4 & \cdots & 0 \\ \vdots & & \ddots & \ddots & \ddots & \vdots \\ 0 & \cdots & 0 & a_{n-1} & c_{n-1} & b_n \\ 0 & \cdots & 0 & 0 & a_n & c_n \end{pmatrix},$$

and assume that $\beta_i = a_i b_i > 0$ for all i. Then the smallest eigenvalue γ_1 of $\mathbf{T}_n$ is given by

$$\gamma_1 = \max_{\boldsymbol{\chi}} \left\{ \min_{1 \le i \le n} \left\{ c_i - \frac{\beta_i}{\chi_i} - \chi_{i+1} \right\} \right\},$$

where $\boldsymbol{\chi} = \{\chi_1, \chi_2, \cdots, \chi_{n+1}\}$ ranges over all sequences such that $\chi_1 = \infty$, $\chi_{n+1} = 0$ and $\chi_i > 0$ for $i = 2, \cdots, n$.

Corollary 5.3 *Let $\beta_1 = \beta_{n+1} = 0$ and $\beta_i = a_i b_i$ for $i = 2, \cdots, n$. Then,*

$$\min_{1 \le i \le n} \left\{ c_i - \sqrt{\beta_i} - \sqrt{\beta_{i+1}} \right\} \le \gamma_1 \le \min_{1 \le i \le n} c_i.$$

Proof. The lower bound is obtained by choosing $\chi_i = \sqrt{\beta_i}$ in Lemma 5.5. The upper bound follows since $\chi_i,\ \beta_i > 0$. □

From Lemma 5.5, we have

$$\chi_{i+1} = c_i - \frac{\beta_i}{\chi_i} - \gamma_1, \quad i = 1, \cdots, n, \tag{5.25}$$

where $\chi_1 = \infty$, $\chi_{n+1} = 0$ and $\chi_i > 0$ for $i = 2, \cdots, n$. Choose r arbitrarily and generate $\chi_i(r)$ successively by

$$\chi_{i+1}(r) = c_i - \frac{\beta_i}{\chi_i(r)} - r, \quad i = 1, \cdots, n, \tag{5.26}$$

where $\chi_1(r) = \infty$. First, $\chi_2(r) = c_1 - r$ so that $\chi_2(r)$ is strictly decreasing in r and is positive in $r < c_1 \equiv S_2$. Next, $\chi_3(r)$ is strictly decreasing in r and both $\chi_2(r)$ and $\chi_3(r)$ are positive in $r < S_3$ for some $S_3 < S_2$. Repeating the argument, we can prove the following facts. There exist some S_i such that

(P1) $\chi_i(r)$ is strictly decreasing in $r < S_i$,

(P2) $\chi_j(r),\ j = 2, \cdots, i$, are positive in $r < S_i$,

(P3) S_i is strictly decreasing in i.

From Lemma 5.5 and (5.25), if we find r such that $\chi_i(r) > 0$ for $i = 2, \cdots, n$ and $\chi_{n+1}(r) = 0$, then this r must be equal to the smallest eigenvalue

γ_1. Based on the above properties, we can develop an algorithm using a bisection search for finding γ_1. The algorithm is very similar to the one given in Example 2.23, where ruin probabilities in a random walk were calculated.

Algorithm Let $\varepsilon > 0$ be a prespecified error.

Step 1 $L \leftarrow$ any lower bound and $R \leftarrow$ any upper bound.

Step 2 If $R - L < \varepsilon$ holds, then $\gamma_1 \leftarrow L$ and terminate. Otherwise, $r \leftarrow (L+R)/2$ and $i \leftarrow 1$.

Step 3 Calculate $\chi_{i+1}(r)$ by (5.26). If $\chi_{i+1}(r) \leq 0$ then $R \leftarrow r$ and go to Step 2.

Step 4 $i \leftarrow i+1$. If $i < N-1$ then go to Step 3.

Step 5 Calculate $\chi_N(r)$ by (5.26). If $\chi_N(r) > 0$ (< 0, respectively) then $L \leftarrow r$ ($R \leftarrow r$) and go to Step 2. If $\chi_N(r) = 0$ then $\gamma_1 \leftarrow r$ and terminate.

Note that the algorithm provides a lower bound γ_L of γ_1 such that $\gamma_1 - \gamma_L < \varepsilon$. If an upper bound is preferred, we replace the statement $\gamma_1 \leftarrow L$ in Step 2 by $\gamma_1 \leftarrow R$.

Example 5.5 We apply the above result to the finite birth–death process $\{X_N(t)\}$. Suppose that $\mu_0 = \lambda_{N-1} = 0$ so that the process is ergodic. We saw that the decay parameter γ_2 of $\{X_N(t)\}$ is equal to the smallest eigenvalue of $\mathbf{A}$, where

$$\mathbf{A} = \begin{pmatrix} \mu_1 + \lambda_0 & -\mu_1 & 0 & \cdots & 0 \\ -\lambda_1 & \mu_2 - \lambda_1 & -\mu_2 & \cdots & 0 \\ \vdots & \ddots & \ddots & \ddots & \vdots \\ 0 & \cdots & -\lambda_{N-3} & \mu_{N-2} + \lambda_{N-3} & -\mu_{N-2} \\ 0 & \cdots & 0 & -\lambda_{N-2} & \mu_{N-1} + \lambda_{N-2} \end{pmatrix},$$

which is the $(N-1) \times (N-1)$ north-west corner truncation of $-\mathbf{Q}^d$, see (5.14). The above algorithm is then applied to the tridiagonal matrix $\mathbf{A}$ to evaluate the smallest eigenvalue γ_2.

Another application of the algorithm is to calculate the quasi-stationary distribution of a lossy birth–death process. Suppose that either μ_0 or λ_{N-1} is positive so that $\mathbf{Q}_N$ is a lossy generator. Let γ_1 be the smallest eigenvalue of $-\mathbf{Q}_N$, i.e., $\gamma_1 = x_{N1}$ if $\lambda_{N-1} > 0$, while $\gamma_1 = x_{N1}^d$ if $\lambda_{N-1} = 0$ and $\mu_0 > 0$. Then $\gamma_1 > 0$ and the quasi-stationary distribution of the lossy birth–death process $\{X_N(t)\}$ is given by the left eigenvector, positive componentwise and normed to sum to unity, of $\mathbf{Q}_N$ associated with the largest negative eigenvalue $-\gamma_1$. That is, denoting the quasi-stationary distribution by $\mathbf{q} = (q_i)$, we have

$$-\gamma_1 \mathbf{q}^\top = \mathbf{q}^\top \mathbf{Q}_N, \quad \mathbf{q}^\top \mathbf{1} = 1.$$

Note from (5.16) with $x = \gamma_1$ that, since $\boldsymbol{\pi}_{\mathrm{D}}\mathbf{Q}_N$ is symmetric,

$$-\gamma_1 \mathbf{q}_N^\top(\gamma_1)\boldsymbol{\pi}_{\mathrm{D}} = \mathbf{q}_N^\top(\gamma_1)\boldsymbol{\pi}_{\mathrm{D}}\mathbf{Q}_N.$$

Hence, the quasi-stationary distribution is given by

$$\mathbf{q}^\top = \frac{\mathbf{q}_N^\top(\gamma_1)\boldsymbol{\pi}_{\mathrm{D}}}{\mathbf{q}_N^\top(\gamma_1)\boldsymbol{\pi}_{\mathrm{D}}\mathbf{1}}.$$

For $0 \le r \le x_{N-1,1}$, let

$$q_i(r) = \frac{\pi_i Q_i(r)}{\sum_{i=0}^{N-1} \pi_i Q_i(r)}, \quad i = 0, 1, \cdots, N-1, \tag{5.27}$$

and define $\mathbf{q}(r) = (q_i(r))$. The quasi-stationary distribution $\mathbf{q}$ is then given by $\mathbf{q}(\gamma_1)$. The smallest eigenvalue γ_1 of $-\mathbf{Q}_N$ can be calculated by the above algorithm. If $\mu_0 = \lambda_{N-1} = 0$ so that $\mathbf{Q}_N$ is conservative, then $\gamma_1 = 0$, $Q_i(0) = 1$ and the quasi-stationary distribution $\mathbf{q}(0)$ coincides with the ordinary stationary distribution.

Recall that the algorithm provides either an upper bound or a lower bound on γ_1. The next result is therefore important when the above algorithm is applied to finding γ_1. Before proceeding, we define the following stochastic ordering relation; see Definition 3.10(i).

Definition 5.3 For two probability vectors $\mathbf{a} = (a_i)$ and $\mathbf{b} = (b_i)$, $\mathbf{a}$ is said to be *strictly* greater than $\mathbf{b}$ in the sense of likelihood ratio ordering, denoted by $\mathbf{a} \ge_{\mathrm{slr}} \mathbf{b}$, if

$$a_i b_j > a_j b_i \quad \text{for all } i > j.$$

Of course, $\mathbf{a} \ge_{\mathrm{slr}} \mathbf{b}$ implies $\mathbf{a} \ge_{\mathrm{lr}} \mathbf{b}$. Recall that $\gamma_1 = x_{N1}$ if $\lambda_{N-1} > 0$ and $\gamma_1 = x_{N1}^d$ if $\lambda_{N-1} = 0$. Hence, from Lemma 5.2 and Theorem 5.4(i), we have $\gamma_1 < x_{N-1,1}$ for either case.

Theorem 5.7 *Suppose that either μ_0 or λ_{N-1} is positive. Then, for $0 \le r_1 < r_2 < x_{N-1,1}$, we have $\mathbf{q}(r_1) \ge_{\mathrm{slr}} \mathbf{q}(r_2)$.*

Proof. As in (5.26), define

$$\chi_{i+1}(r) = \lambda_i + \mu_i - \frac{\lambda_{i-1}\mu_i}{\chi_i(r)} - r, \quad i = 0, 1, \cdots, N-1,$$

where $\chi_0(r) = \infty$. It follows that

$$\chi_i(r) = \lambda_{i-1}\frac{Q_i(r)}{Q_{i-1}(r)}, \quad i = 1, \cdots, N-1, \tag{5.28}$$

since, then,

$$\lambda_i Q_{i+1}(r) = (\lambda_i + \mu_i - r)Q_i(r) - \mu_i Q_{i-1}(r).$$

Since $\chi_i(r)$ is strictly decreasing in $r < x_{N-1,1}$, we have

$$\lambda_{i-1}\frac{Q_i(r_1)}{Q_{i-1}(r_1)} > \lambda_{i-1}\frac{Q_i(r_2)}{Q_{i-1}(r_2)}, \qquad 0 \le r_1 < r_2 < x_{N-1,1},$$

so that

$$\pi_i Q_i(r_1)\,\pi_{i-1}Q_{i-1}(r_2) > \pi_{i-1}Q_{i-1}(r_1)\,\pi_i Q_i(r_2).$$

The result follows at once from (5.27). □

5.4 The Karlin–McGregor representation theorem

In this section, we prove the Karlin–McGregor representation (5.10) for minimal birth–death processes (see Section 4.3 for the definition of minimal processes). Throughout this section, we denote the probability distribution function defined by (5.19) by ψ_n. The proof of the next lemma is taken from Ledermann and Reuter (1954).

Lemma 5.6 *Let m be a nonnegative integer. Then, for a given $\varepsilon > 0$, there exists $K > 0$ such that*

$$\int_K^\infty x^m\, d\psi_n(x) < \varepsilon$$

for all $n > m + 1$.

Proof. Note that

$$0 \le \int_K^\infty x^m\, d\psi_n(x) \le \frac{1}{K}\int_K^\infty x^{m+1} d\psi_n(x) \le \frac{1}{K}\int_0^\infty x^{m+1} d\psi_n(x).$$

But, since x^{m+1} can be expressed uniquely as a linear combination of $Q_i(x)$ for $i = 0, 1, \cdots, m+1$, i.e.,

$$x^{m+1} = \sum_{i=0}^{m+1} c_i\, Q_i(x)$$

for some c_i, we have

$$\int_0^\infty x^{m+1} d\psi_n(x) = \sum_{i=0}^{m+1} c_i \int_0^\infty Q_i(x) d\psi_n(x) = c_0,$$

where we have used the orthogonality condition (5.21) with $j = 0$ for $i < n$. Thus, choosing $K > c_0/\varepsilon$, it follows that

$$\int_K^\infty x^m\, d\psi_n(x) \le \frac{c_0}{K} < \varepsilon.$$

This proves the lemma. □

Lemma 5.6 shows that the sequence of probability distribution functions $\{\psi_n\}$ is *tight*. Hence, from the Helly–Bray theorem (see, e.g., Williams, Chapter 17, 1991), there exist a probability distribution function ψ defined on $[0,\infty)$ and a subsequence $\{N_k\}$ such that ψ_{N_k} converges weakly to ψ and

$$\lim_{N_k\to\infty}\int_0^\infty e^{-xt}x^m d\psi_{N_k}(x) = \int_0^\infty e^{-xt}x^m d\psi(x), \quad t \ge 0, \tag{5.29}$$

for each $m = 0, 1, \cdots$. We are now ready to prove the main result of this section.

Theorem 5.8 *For a minimal birth-death process with transition probability functions $p_{ij}(t)$, there exists a probability distribution function ψ defined on $[0,\infty)$ such that*

$$p_{ij}(t) = \pi_j \int_0^\infty e^{-xt} Q_i(x)Q_j(x)d\psi(x), \quad i,\ j = 0, 1, \cdots; \quad t \ge 0.$$

Proof. From Theorem 4.28, we know that each $p_{ij}^N(t)$ in (5.20) converges monotonically from below, as $N \to \infty$, to the minimal transition probability function $p_{ij}(t)$. On the other hand, $Q_i(x)Q_j(x)$ is a polynomial in x of degree $i+j$. Hence,

$$\begin{aligned} p_{ij}(t) &= \lim_{N\to\infty} p_{ij}^N(t) \\ &= \lim_{N\to\infty} \pi_j \int_0^\infty e^{-xt}Q_i(x)Q_j(x)d\psi_N(x) \\ &= \pi_j \int_0^\infty e^{-xt}Q_i(x)Q_j(x)d\psi(x), \end{aligned}$$

where the last equality is due to (5.29). □

Suppose that the Karlin–McGregor representation (5.10) holds for (not necessarily minimal) transition probability functions $p_{ij}(t)$. Then, differentiation of $p_{ij}(t)$ with respect to t is permissible and the derivative is given by

$$p_{ij}'(t) = -\pi_j \int_0^\infty x e^{-xt} Q_i(x)Q_j(x)d\psi(x).$$

This is so since, from the proof of Lemma 5.6, the differentiated integral is uniformly convergent for $t \ge 0$. It is an easy exercise to check (see Exercise 5.13) that the transition probability functions $p_{ij}(t)$ satisfy both the forward and backward Kolmogorov equations (5.5). Also,

$$\lim_{t\to\infty} p_{00}(t) = \lim_{t\to\infty}\int_0^\infty e^{-xt}d\psi(x) = \psi(\{0\}),$$

so that ψ has an atom at 0 if and only if the birth–death process is positive recurrent. Hence, if the process is positive recurrent, we have

$$\lim_{t\to\infty} p_{ij}(t) = \psi(\{0\})\pi_j.$$

The atom at 0 must be equal to $\psi(\{0\}) = \left(\sum_{j=0}^\infty \pi_j\right)^{-1} = B^{-1}$. If $\mu_0 > 0$, ψ has no atom at 0.

When $\mu_0 > 0$, we have assumed that there is an absorbing state -1 which can be reached through state 0. Let T denote the (possibly defective) random variable representing the time at which absorption at -1 occurs. Let $a_i = P_i[T < \infty]$ be the absorption probability at -1 with the initial state i. We shall show shortly that

$$a_i = \mu_0 \int_0^\infty \frac{Q_i(x)}{x} d\psi(x), \quad i = 0, 1, \cdots. \tag{5.30}$$

Then, from (5.8), we obtain

$$\begin{aligned} a_i &= \frac{\mu_0}{\lambda_{i-1}} \int_0^\infty \frac{1}{x}\{(\lambda_{i-1} + \mu_{i-1} - x)Q_{i-1}(x) - \mu_{i-1}Q_{i-2}(x)\}d\psi(x) \\ &= \frac{\lambda_{i-1} + \mu_{i-1}}{\lambda_{i-1}} a_{i-1} - \frac{\mu_{i-1}}{\lambda_{i-1}} a_{i-2} - \frac{\mu_0}{\lambda_{i-1}} \delta_{i-1,0}, \end{aligned}$$

where $\delta_{ij} = 1$ for $i = j$ and $\delta_{ij} = 0$ for $i \neq j$. It follows that

$$\begin{aligned} a_{i-1} - a_i &= \frac{\mu_{i-1}}{\lambda_{i-1}}(a_{i-2} - a_{i-1}) \\ &= \frac{\mu_{i-1} \cdots \mu_0}{\lambda_{i-1} \cdots \lambda_0}(1 - a_0) \\ &= \frac{\mu_0}{\lambda_{i-1}\pi_{i-1}}(1 - a_0), \end{aligned}$$

so that

$$a_i = a_0 - \mu_0(1 - a_0) \sum_{n=0}^{i-1} \frac{1}{\lambda_n \pi_n}, \quad i = 1, 2, \cdots, \tag{5.31}$$

provided that $a_0 = \mu_0 \int_0^\infty x^{-1} d\psi(x)$ is well defined. On the other hand, we have

$$P_i[T \le t] = \mu_0 \int_0^t p_{i0}(u)du, \tag{5.32}$$

and

$$P_i[t < T < \infty] = \mu_0 \int_t^\infty p_{i0}(u)du.$$

It follows from (5.10) that

$$\begin{aligned} P_i[t < T < \infty] &= \mu_0 \int_t^\infty \int_0^\infty e^{-xu} Q_i(x) d\psi(x) du \\ &= \mu_0 \int_0^\infty \frac{e^{-xt}}{x} Q_i(x) d\psi(x), \end{aligned} \tag{5.33}$$

where the interchange of the integrals is ensured by Fubini's theorem if $\int_0^\infty d\psi(x)/x < \infty$. Equation (5.30) now follows by taking $t = 0$. It remains to prove that $\int_0^\infty d\psi(x)/x$ is well defined. To this end, Fubini's theorem yields

$$\mu_0 \int_0^\infty \frac{d\psi(x)}{x} = \mu_0 \int_0^\infty \int_0^\infty e^{-xt} d\psi(x) dt = P_0[T < \infty] \le 1.$$

The following result is due to Karlin and McGregor (1957b). See Kijima, Nair, Pollett and van Doorn (1997, to appear) for a more general result.

Lemma 5.7 *For a minimal birth–death process with $\mu_0 > 0$, we have*

$$a_0 = 1 - \lim_{n \to \infty} \frac{1}{Q_n(0)}.$$

Proof. Let $\{X_N(t)\}$ be the finite birth–death process defined in Section 5.3, where μ_0 and λ_{N-1} are positive. Let a_i^N, $i = 0, \cdots, N-1$ be the absorption probability of $\{X_N(t)\}$ at -1 with the initial state i. It is readily seen that

$$a_i^N = \frac{\mu_i}{\lambda_i + \mu_i} a_{i-1}^N + \frac{\lambda_i}{\lambda_i + \mu_i} a_{i+1}^N, \quad i = 0, 1, \cdots, N-1,$$

where $a_{-1}^N = 1$ and $a_N^N = 0$. Hence, defining $b_i = 1 - a_i^N$, it follows that

$$0 = \mu_i b_{i-1} - (\lambda_i + \mu_i) b_i + \lambda_i b_{i+1}, \quad i = 0, 1, \cdots, N-1,$$

where $b_{-1} = 0$ and $b_N = 1$. Comparing these equations with (5.8), we obtain $b_i = Q_i(0)/Q_N(0)$ since $Q_N(0) > 0$, so that

$$a_0^N = 1 - b_0 = 1 - \frac{1}{Q_N(0)}.$$

On the other hand, as in (5.32), we have

$$a_0^N = \mu_0 \int_0^\infty p_{00}^N(t) dt,$$

where $p_{ij}^N(t)$ denotes the transition probability functions of $\{X_N(t)\}$. Since $p_{00}^N(t)$ converges monotonically from below to $p_{00}(t)$ as $N \to \infty$ (see Theorem 4.28), the monotone convergence theorem yields

$$a_0 = \mu_0 \int_0^\infty p_{00}(t) dt = \lim_{N \to \infty} \mu_0 \int_0^\infty p_{00}^N(t) dt = \lim_{N \to \infty} a_0^N,$$

proving the lemma. □

Theorem 5.9 *For a minimal birth–death process with $\mu_0 > 0$ and initial state i, the absorption probability at -1 is given by*

$$a_i = \frac{\mu_0 \sum_{n=i}^\infty \frac{1}{\lambda_n \pi_n}}{1 + \mu_0 \sum_{n=0}^\infty \frac{1}{\lambda_n \pi_n}} = \frac{\sum_{n=i+1}^\infty \pi_n^d}{\sum_{n=0}^\infty \pi_n^d}, \quad i = 0, 1, \cdots, \tag{5.34}$$

where a_i should be interpreted as unity if $A = \sum_{n=0}^\infty (\lambda_n \pi_n)^{-1} = \infty$. Hence, absorption at state -1 is certain if and only if $A = \infty$.

Proof. From Theorem 5.3(i), we have $Q_n^d(0) = 1$ and

$$Q_n(0) = 1 + \mu_0 \sum_{k=0}^{n-1} \frac{1}{\lambda_k \pi_k}, \quad n = 0, 1, \cdots.$$

It follows from Lemma 5.7 that

$$a_0 = \lim_{n\to\infty} \frac{\mu_0 \sum_{k=0}^{n-1} \frac{1}{\lambda_k \pi_k}}{1 + \mu_0 \sum_{k=0}^{n-1} \frac{1}{\lambda_k \pi_k}} = \frac{\mu_0 \sum_{n=0}^{\infty} \frac{1}{\lambda_n \pi_n}}{1 + \mu_0 \sum_{n=0}^{\infty} \frac{1}{\lambda_n \pi_n}}.$$

The result follows from (5.31). □

The next example illustrates the results obtained so far.

Example 5.6 Consider an M/M/1 queue with arrival rate λ and service rate μ. Write $\rho = \lambda/\mu$. From Example 5.4, if $\rho < 1$ then the queue-size process $\{X(t)\}$ is ergodic so that the spectral measure ψ has an atom at 0. According to Karlin and McGregor (1958), if $\rho < 1$, then the atom at 0 is $1 - \rho$ and the continuous part of the spectral measure ψ is given by

$$\psi'(x) = \frac{1}{2\pi} \frac{\sqrt{4\rho - (1 + \rho - x)^2}}{x}, \quad (1 - \sqrt{\rho})^2 \le x \le (1 + \sqrt{\rho})^2.$$

It follows from (5.10) that the transition probability functions are given by

$$\begin{aligned} & p_{ij}(t) - (1 - \rho)\rho^j \\ = \; & \frac{\rho^j}{2\pi} \int_{(1-\sqrt{\rho})^2}^{(1+\sqrt{\rho})^2} e^{-xt} Q_i(x) Q_j(x) \frac{\sqrt{4\rho - (1 + \rho - x)^2}}{x} \, dx. \end{aligned} \quad (5.35)$$

If $\rho = 1$ then the process is null recurrent, while if $\rho > 1$ then it is transient. The transition probability functions for $\rho \ge 1$ are given by

$$p_{ij}(t) = \frac{\rho^j}{2\pi} \int_{(1-\sqrt{\rho})^2}^{(1+\sqrt{\rho})^2} e^{-xt} Q_i(x) Q_j(x) \frac{\sqrt{4\rho - (1 + \rho - x)^2}}{x} \, dx.$$

The numerical integration involved in (5.35) is stable and easy to calculate (see Abate and Whitt, 1989).

Since the process is spatially homogeneous, the quantity a_i given in (5.34) can be considered as the probability that the first busy period terminates when the process starts with i customers, $i \ge 1$. Of course, $a_i = 1$ for all $i \ge 1$ if the process is recurrent. If $\rho > 1$, we have from (5.34) that

$$a_i = \frac{\sum_{n=i-1}^{\infty} \rho^{-n-1}}{1 + \sum_{n=0}^{\infty} \rho^{-n-1}} = \frac{1}{\rho^i}, \quad i = 1, 2, \cdots.$$

5.5 Asymptotics of birth–death polynomials

Throughout this section, we assume that $\mu_0 > 0$ and consider the asymptotics of the birth–death polynomials $Q_n(x)$. We shall not be concerned with results for the case where $\mu_0 = 0$, since the reader should be able to produce such results with no difficulty by using reasoning corresponding to the case where $\mu_0 = 0$. For example, the reader should use Theorem 5.3(ii) to obtain a counterpart for the case $\mu_0 = 0$ wherever we derive a result by using Theorem 5.3(i) for the case where $\mu_0 > 0$.

From Theorem 5.3(i), substitution of $Q_n^d(x)$ into $Q_n(x)$ gives us

$$Q_n(x) = \sum_{k=0}^{n} \pi_k^d - \frac{x}{\mu_0} \sum_{k=1}^{n} \pi_k^d \sum_{i=0}^{k-1} \pi_i Q_i(x), \quad n = 0, 1, \cdots, \tag{5.36}$$

which can be written as

$$Q_n(x) = \sum_{k=0}^{n} \pi_k^d - \frac{x}{\mu_0} \sum_{i=0}^{n-1} \pi_i Q_i(x) \sum_{k=i+1}^{n} \pi_k^d,$$

whence

$$\frac{Q_n(x)}{\sum_{k=0}^{n} \pi_k^d} = 1 - \frac{x}{\mu_0} \sum_{i=0}^{n-1} \frac{\sum_{k=i+1}^{n} \pi_k^d}{\sum_{k=0}^{n} \pi_k^d} \pi_i Q_i(x), \quad n = 0, 1, \cdots. \tag{5.37}$$

Similarly, substitution of $Q_n(x)$ into $Q_n^d(x)$ yields

$$Q_n^d(x) = 1 - \frac{x}{\mu_0} \sum_{k=0}^{n-1} \pi_k \sum_{i=0}^{k} \pi_i^d Q_i^d(x), \quad n = 0, 1, \cdots. \tag{5.38}$$

The next result is taken from Karlin and McGregor (1957a).*

Theorem 5.10 (i) *$Q_n(x)$ is bounded as $n \to \infty$ for at least one $x < 0$ if and only if $C < \infty$.*

(ii) *$Q_n^d(x)$ is bounded as $n \to \infty$ for at least one $x < 0$ if and only if $D < \infty$.*

Proof. Let x_{ni}, $i = 1, \cdots, n$, be the zeros of $Q_n(x)$. Then

$$Q_n(x) = Q_n(0) \prod_{i=1}^{n} \left(1 - \frac{x}{x_{ni}}\right). \tag{5.39}$$

It follows that

$$|Q_n(x)| \le Q_n(0) \prod_{i=1}^{n} \left(1 + \frac{|x|}{x_{ni}}\right) \le Q_n(0) \left(1 + \frac{|x|}{n} \sum_{i=1}^{n} \frac{1}{x_{ni}}\right)^n,$$

* Actually, it goes back to Stieltjes (1894).

where the second inequality holds since $\log x$ is concave in x so that

$$\frac{1}{n}\sum_{i=1}^{n}\log c_i \le \log\frac{\sum_{i=1}^{n} c_i}{n}, \quad c_i > 0.$$

From (5.39), differentiation with respect to x yields

$$Q_n'(x) = Q_n(0)\sum_{i=1}^{n}\frac{-1}{x_{ni}}\prod_{j\neq i}\left(1-\frac{x}{x_{nj}}\right),$$

from which we have

$$\sigma_n \equiv \sum_{i=1}^{n}\frac{1}{x_{ni}} = -\frac{Q_n'(0)}{Q_n(0)}.$$

It follows that

$$|Q_n(x)| \le Q_n(0)\left(1+\frac{\sigma_n|x|}{n}\right)^n.$$

Since $(1+x/n)^n$ converges monotonically from below to e^x as $n\to\infty$, we finally have

$$|Q_n(x)| \le Q_n(0)\, e^{\sigma_n|x|}.$$

Hence, in order that $Q_n(x)$ be bounded for some $x<0$, it is sufficient that $Q_n(0)$ and σ_n are both bounded. Note from (5.4) that $C<\infty$ implies $A<\infty$. Also, from Theorem 5.3(i), we observe that $Q_n^d(0)=1$ and so

$$Q_n(0) = 1+\sum_{k=0}^{n-1}\frac{\mu_0}{\lambda_k\pi_k}, \quad n=0,1,\cdots,$$

which is monotonically increasing in n and is bounded by $1+\mu_0 A$ from above. On the other hand, from (5.36), we have

$$-Q_n'(0) = \frac{1}{\mu_0}\sum_{k=1}^{n}\pi_k^d\sum_{i=0}^{k-1}\pi_i Q_i(0),$$

which, together with the fact that $Q_n(0)$ is increasing in n, yields

$$\sigma_n = -\frac{Q_n'(0)}{Q_n(0)} \le \sum_{k=0}^{n-1}\frac{1}{\lambda_k\pi_k}\sum_{i=0}^{k}\pi_i < C.$$

Hence, if $C<\infty$ then $Q_n(0)$ as well as σ_n is bounded.

Conversely, from (5.39), we have

$$Q_n(-|x|) \ge Q_n(0)(1+\sigma_n|x|).$$

Hence, if $Q_n(x)$ is bounded as $n\to\infty$ for some $x<0$, then both $Q_n(0)$ and σ_n are bounded. Since

$$\sigma_n = -\frac{Q_n'(0)}{Q_n(0)} = \sum_{k=0}^{n-1}\frac{1}{\lambda_k\pi_k}\sum_{i=0}^{k}\pi_i\frac{Q_i(0)}{Q_n(0)} \ge \frac{1}{K}\sum_{k=0}^{n-1}\frac{1}{\lambda_k\pi_k}\sum_{i=0}^{k}\pi_i,$$

where $K = \sup_n Q_n(0) < \infty$, assertion (i) is proved. Part (ii) can be proved similarly. □

If $Q_n(x)$ is bounded for some $x < 0$, then $Q_n(x)$ is uniformly bounded in every circle $|x| \le R$ and, since $\{Q_n(x)\}$ is a monotone sequence for each $x < 0$, $Q_n(x)$ converges for each $x < 0$. It can be shown in fact that $Q_n(x)$ converges uniformly on bounded sets to an *entire function* whose zeros are simple and are precisely the points ξ_i, $i = 1, 2, \cdots$ given in Lemma 5.2 (see van Doorn, 1986).

To investigate the asymptotics of birth–death polynomials, it is helpful to consider the following four cases based on the type of boundaries at infinity. The following results are restatements of more general results obtained in Kijima and van Doorn (1995).

Case 1: Regular boundary $(C,\ D < \infty)$

In this case, we have $A,\ B < \infty$ since $AB = C + D$. As noted above, both $Q_n(x)$ and $Q_n^d(x)$ converge uniformly on bounded sets to respective entire functions $Q_\infty(x)$ and $Q_\infty^d(x)$, say, whose zeros are ξ_i and ξ_i^d. Moreover,

$$0 < \xi_i^d < \xi_i < \xi_{i+1}^d, \quad i = 1, 2, \cdots.$$

Hence, letting $n \to \infty$ in Theorem 5.3(i), we have

$$\frac{x}{\mu_0} \sum_{n=0}^{\infty} \pi_n Q_n(x) = 1 - Q_\infty^d(x), \tag{5.40}$$

while (5.37), together with Theorem 5.9, yields

$$\frac{x}{\mu_0} \sum_{n=0}^{\infty} a_n \pi_n Q_n(x) = 1 - \frac{Q_\infty(x)}{\sum_{n=0}^{\infty} \pi_n^d}. \tag{5.41}$$

Case 2: Entrance boundary $(C = \infty,\ D < \infty)$

In this case, $A = \infty$ and $B < \infty$ from (5.4). As in Case 1, $Q_n^d(x)$ converges to an entire function $Q_\infty^d(x)$, say, and

$$0 < \xi_i^d = \xi_i < \xi_{i+1}^d, \quad i = 1, 2, \cdots.$$

Hence, we have (5.40) again. Also, from van Doorn (1986),

$$\lim_{n\to\infty} \frac{Q_n(x)}{\sum_{k=0}^{n} \pi_k^d} = Q_\infty^d(x).$$

Since $\sum_{n=0}^{\infty} \pi_n^d = 1 + \mu_0 A = \infty$, we conclude that $Q_n(x)$ diverges as $n \to \infty$ for $x \ne \xi_i^d = \xi_i$, $i = 1, 2, \cdots$.

Case 3: Exit boundary $(C < \infty,\ D = \infty)$

In this case, $A < \infty$ and $B = \infty$. As in Case 1, $Q_n(x)$ converges to an entire function $Q_\infty(x)$, say, and

$$0 = \xi_1^d < \xi_i = \xi_{i+1}^d < \xi_{i+1}, \quad i = 1, 2, \cdots.$$

Hence, we have (5.41) again. Also, it can be shown that

$$\lim_{n\to\infty} \frac{Q_n^d(x)}{\sum_{k=0}^n \pi_k} = -xQ_\infty(x),$$

from which $Q_n^d(x)$ diverges as $n \to \infty$ for $x \neq \xi_i = \xi_{i+1}^d$, $i = 1, 2, \cdots$.

Case 4: Natural boundary ($C = D = \infty$)

Since we are interested in the case where $\xi_1 > 0$, we confine our attention to the following two subcases. Note that if $A = B = \infty$ then $\xi_1 = 0$ (see Chihara, 1978). This ξ_1 is called the *first limit* of the zeros of $Q_n(x)$.

Case 4.1: ($A = \infty$, $B < \infty$ and $\xi_1 > 0$)

In this case, we have

$$0 < \xi_i^d = \xi_i \leq \xi_{i+1}^d, \quad i = 1, 2, \cdots;$$

see Case 2.

Lemma 5.8 *Let $x > 0$ and assume that $B < \infty$ and $D = \infty$. If $Q_n^d(x)$ tends to a limit as $n \to \infty$, then $\lim_{n\to\infty} Q_n^d(x) = 0$.*

Proof. Suppose that $Q_n^d(x) \to a$ as $n \to \infty$ where $0 < a \leq \infty$, and let $0 < b < a$. Choose N such that $Q_n^d(x) > b$ for all $n > N$. Then, for k sufficiently large,

$$\begin{aligned}
&\sum_{n=0}^{k} \pi_n \sum_{i=0}^{n} \pi_i^d Q_i^d(x) \\
&= \sum_{n=0}^{N} \pi_n \sum_{i=0}^{n} \pi_i^d Q_i^d(x) + \sum_{n=N+1}^{k} \pi_n \sum_{i=0}^{N} \pi_i^d Q_i^d(x) \\
&\qquad + \sum_{n=N+1}^{k} \pi_n \sum_{i=N+1}^{n} \pi_i^d Q_i^d(x) \\
&> \sum_{n=0}^{N} \pi_n \sum_{i=0}^{n} \pi_i^d Q_i^d(x) + \sum_{n=N+1}^{k} \pi_n \sum_{i=0}^{N} \pi_i^d \{Q_i^d(x) - b\} \\
&\qquad + b \sum_{n=N+1}^{k} \pi_n \sum_{i=0}^{n} \pi_i^d.
\end{aligned}$$

Since $B < \infty$, so that

$$D = \sum_{n=0}^{\infty} \tau_n \sum_{i=0}^{n-1} \frac{1}{\lambda_i \pi_i} = \frac{1}{\mu_0} \sum_{n=0}^{\infty} \pi_n \sum_{i=0}^{n} \pi_i^d - B.$$

The assumptions imply that $\sum_{n=0}^k \pi_k \sum_{i=0}^n \pi_i^d Q_i^d(x)$ tends to infinity as $k \to \infty$, which, together with (5.38), implies that $Q_n^d(x) \to -\infty$ as $n \to \infty$, yielding a contradiction. Similarly, the supposition $Q_n^d(x) \to a$ with $-\infty \leq a < 0$ leads to a contradiction. □

Theorem 5.11 *Let* $0 < x \leq \xi_1 = \xi_1^d$ *and assume that* $B < \infty$ *and* $D = \infty$. *Then*

$$\lim_{n\to\infty} Q_n^d(x) = 0 \quad \textit{and} \quad \lim_{n\to\infty} Q_n(x) = \infty.$$

Proof. First note that both $Q_n^d(x)$ and $Q_n(x)$ are positive for all $n \geq 0$ in $0 < x \leq \xi_1^d$. Hence, since $Q_n^d(x)$ is monotonically decreasing in n, from Theorem 5.3(i), $Q_n^d(x)$ tends to a limit so that the first result follows from Lemma 5.8. On the other hand, again from Theorem 5.3(i), $Q_n(x) \geq 1$ and it is monotonically increasing so that

$$\sum_{n=0}^{\infty} \pi_n \leq \sum_{n=0}^{\infty} \pi_n Q_n(x) = \frac{\mu_0}{x} < \infty,$$

where the equality follows from Theorem 5.3(i), together with the fact that $\lim_{n\to\infty} Q_n^d(x) = 0$. It follows that

$$Q_n^d(x) = \frac{x}{\mu_0} \sum_{k=n}^{\infty} \pi_k Q_k(x) \geq \frac{x}{\mu_0} \sum_{k=n}^{\infty} \pi_k > \frac{x}{\mu_0} \sum_{k=n+1}^{\infty} \pi_k.$$

Therefore, Theorem 5.3(i) implies that

$$\lim_{n\to\infty} Q_n(x) \geq x \sum_{n=0}^{\infty} \frac{1}{\lambda_n \pi_n} \sum_{k=n+1}^{\infty} \pi_k = xD,$$

which is infinity under the assumptions, completing the proof. □

As mentioned in the proof of Theorem 5.11, this case leads to

$$\sum_{n=0}^{\infty} \pi_n Q_n(x) = \frac{\mu_0}{x}, \quad 0 < x \leq \xi_1^d = \xi_1. \tag{5.42}$$

Note that, since $A = \infty$, we have $a_n = 1$ for all $n = 0, 1, \cdots$.

Case 4.2: ($A < \infty$, $B = \infty$ and $\xi_1 > 0$)

In this case, we have

$$0 = \xi_1^d < \xi_i = \xi_{i+1}^d \leq \xi_{i+1}, \quad i = 1, 2, \cdots;$$

see Case 3. The proof of the next lemma is similar to that of Lemma 5.8 and is omitted.

Lemma 5.9 *Let* $x > 0$ *and assume that* $A < \infty$ *and* $C = \infty$. *If* $Q_n(x)$ *tends to a limit as* $n \to \infty$, *then* $\lim_{n\to\infty} Q_n(x) = 0$.

Theorem 5.12 *Let* $0 < x \leq \xi_1 = \xi_2^d$ *and assume that* $A < \infty$ *and* $C = \infty$. *Then*

$$\lim_{n\to\infty} Q_n^d(x) = -\infty \quad \textit{and} \quad \lim_{n\to\infty} Q_n(x) = 0.$$

Proof. For $0 < x \le \xi_1$, we have $Q_n(x) > 0$ for all $n \ge 0$, while there is some N such that $Q_n^d(x) < 0$ for all $n \le N$. Hence, from Theorem 5.3(i), $Q_n(x)$ is decreasing in $n \ge N$ so that $\lim_{n\to\infty} Q_n(x)$ exists, which must be zero, from Lemma 5.9. On the other hand, $Q_n^d(x)$ is monotonically decreasing in n, so that there exist some M and $\ell > 0$ such that $Q_n^d(x) \le -\ell$ for all $n \ge M$. Note that, from Theorem 5.3(i), together with the fact that $\lim_{n\to\infty} Q_n(x) = 0$, we have

$$Q_n(x) = -\sum_{k=n+1}^{\infty} \pi_k^d Q_k^d(x), \quad n \ge 0.$$

It follows that

$$Q_n(x) \ge \ell \sum_{k=n+1}^{\infty} \pi_k^d = \ell \sum_{k=n}^{\infty} \frac{\mu_0}{\lambda_k \pi_k}, \quad n \ge M.$$

Therefore, Theorem 5.3(i) implies that

$$\lim_{n\to\infty} Q_n^d(x) \le 1 - \ell x \sum_{n=0}^{\infty} \pi_n \sum_{k=n}^{\infty} \frac{1}{\lambda_k \pi_k} = 1 - \ell x C,$$

proving the theorem. □

Theorem 5.12, in conjunction with Theorem 5.3(i), shows that

$$\sum_{n=0}^{\infty} \pi_n Q_n(x) = \infty, \quad 0 < x \le \xi_1, \tag{5.43}$$

while (5.37) and Theorems 5.9 and 5.12 together imply

$$\sum_{n=0}^{\infty} a_n \pi_n Q_n(x) = \frac{\mu_0}{x}, \quad 0 < x \le \xi_1, \tag{5.44}$$

where $a_n < 1$ for all $n = 0, 1, \cdots$. Equation (5.44) should be compared with (5.42), in which case we have $a_n = 1$ for all $n = 0, 1, \cdots$.

5.6 Quasi-stationary distributions

Consider the finite birth–death process $\{X_N(t)\}$ defined in Section 5 3 with the state space $\mathcal{N} = \{0, 1, \cdots, N-1\}$ and the infinitesimal generator $\mathbf{Q}_N$. If either $\mu_0 > 0$ or $\lambda_{N-1} > 0$, then $\mathbf{Q}_N$ is a lossy generator and it is of interest to study the conditional probabilities

$$q_{ij}(t) = P_i[X_N(t) = j | X_N(t) \in \mathcal{N}], \quad i,\ j \in \mathcal{N}; \quad t \ge 0.$$

Since the state space is finite, we know that $\{q_{ij}(t)\}$ converges as $t \to \infty$ to a nondefective distribution $\mathbf{q}_N = (q_i^N)$, called the *quasi-limiting distribution*, where the q_i^N are given by (5.27) with $r = \gamma_1$, the smallest positive eigenvalue of $-\mathbf{Q}_N$. The distribution $\mathbf{q}_N$ is the unique quasi-stationary distribution of $\{X_N(t)\}$ (see Theorem 4.27).

Suppose that $\mu_0 > 0$ and $\lambda_{N-1} = 0$. The case where $\mu_0 = 0$ and $\lambda_{N-1} > 0$ can be treated similarly. Let $\mathbf{r} = (\mu_0, 0, \cdots, 0)^\top$ and, as an analog of (3.48), we define

$$\mathbf{A}_i = \mathbf{Q}_N + \mathbf{r}\,\boldsymbol{\delta}_i^\top, \quad i = 0,\ N-1. \tag{5.45}$$

More precisely,

$$\mathbf{A}_0 = \begin{pmatrix} -\lambda_0 & \lambda_0 & 0 & \cdots & 0 \\ \mu_1 & -\lambda_1 - \mu_1 & \lambda_1 & \cdots & 0 \\ \vdots & \ddots & \ddots & \ddots & \vdots \\ 0 & \cdots & \mu_{N-2} & -\lambda_{N-2} - \mu_{N-2} & \lambda_{N-2} \\ 0 & \cdots & 0 & \mu_{N-1} & -\mu_{N-1} \end{pmatrix}$$

and

$$\mathbf{A}_{N-1} = \begin{pmatrix} -\lambda_0 - \mu_0 & \lambda_0 & 0 & \cdots & \mu_0 \\ \mu_1 & -\lambda_1 - \mu_1 & \lambda_1 & \cdots & 0 \\ \vdots & \ddots & \ddots & \ddots & \vdots \\ 0 & \cdots & \mu_{N-2} & -\lambda_{N-2} - \mu_{N-2} & \lambda_{N-2} \\ 0 & \cdots & 0 & \mu_{N-1} & -\mu_{N-1} \end{pmatrix}.$$

Each $\mathbf{A}_i$ is considered to be the generator of an ergodic continuous-time Markov chain $\{X^i(t)\}$ with the state space $\mathcal{N}$. The stationary distribution of $\{X^i(t)\}$ is denoted by $\boldsymbol{\pi}_i = (\pi_j^i)$. It is easy to see that the stationary distribution $\boldsymbol{\pi}_0$ is given by

$$\pi_j^0 = \frac{\pi_j}{\sum_{j=0}^{N-1} \pi_j}, \quad j \in \mathcal{N},$$

where the π_j are the potential coefficients.

Regarding the stationary distribution $\boldsymbol{\pi}_{N-1}$ of $\{X^{N-1}(t)\}$, we use the last-exit-decomposition formula

$$p_{i0}(t) = \mu_i \int_0^t p_{ii}(s)\, {}_i p_{i-1,0}(t-s)ds, \quad i = 1, \cdots, N-1, \tag{5.46}$$

where the $p_{ij}(t)$ are the transition probability functions of $\{X^{N-1}(t)\}$ and the ${}_k p_{ij}(t)$ denotes its transition probability functions with taboo state k; see (2.84) for the discrete-time case. By taking the Laplace transform of (5.46), we obtain

$$\pi_{i0}(s) = \mu_i\, \pi_{ii}(s)\, {}_i\pi_{i-1,0}(s), \quad i = 1, \cdots, N-1,$$

where the π denotes the Laplace transform of the p. From (5.22), we know that

$$\mu_i\, {}_i\pi_{i-1,0}(s) = \prod_{k=1}^{i} \frac{\mu_k}{s + x_{ik}}, \quad i = 1, \cdots, N-1,$$

where the x_{ik} are the zeros of the birth–death polynomial $Q_i(x)$. On the other hand, since the process $\{X^{N-1}(t)\}$ is ergodic, we have

$$\lim_{s\to 0} s\, \pi_{ij}(s) = \pi_j^{N-1}.$$

It follows that

$$\frac{\pi_0^{N-1}}{\pi_i^{N-1}} = \lim_{s\to 0} \frac{s\, \pi_{i0}(s)}{s\, \pi_{ii}(s)} = \prod_{k=1}^{i} \frac{\mu_k}{x_{ik}}, \quad i = 1, \cdots, N-1. \tag{5.47}$$

We are now ready to prove the next theorem. Recall from Sections 3.7 and 4.8 that

$$\boldsymbol{\pi}_{N-1} \geq_{\rm st} \mathbf{q}_N \geq_{\rm st} \boldsymbol{\pi}_0,$$

since any birth–death process is stochastically monotone. In the next two theorems, $\geq_{\rm slr}$ denotes the strict likelihood ratio ordering defined in Definition 5.3.

Theorem 5.13 *For a lossy birth–death process governed by* $\mathbf{Q}_N$ *with* $\mu_0 > 0$ *and* $\lambda_{N-1} = 0$*, we have*

$$\boldsymbol{\pi}_{N-1} \geq_{\rm slr} \mathbf{q}_N \geq_{\rm slr} \boldsymbol{\pi}_0.$$

Proof. We write $q_j = q_j^N$ for the sake of notational simplicity. For the first inequality, it suffices to prove

$$\frac{\pi_j^{N-1}}{\pi_{j+1}^{N-1}} < \frac{q_j}{q_{j+1}}, \quad j = 0, 1, \cdots, N-2. \tag{5.48}$$

From (5.47), one has

$$\frac{\pi_j^{N-1}}{\pi_{j+1}^{N-1}} = \frac{\mu_{j+1}}{x_{j+1,j+1}} \prod_{k=1}^{j} \frac{x_{jk}}{x_{j+1,k}},$$

while, from (5.27) with $\gamma_1 = x_{N1}^d$,

$$\frac{q_j}{q_{j+1}} = \frac{q_j(\gamma_1)}{q_{j+1}(\gamma_1)} = \frac{\pi_j\, Q_j(\gamma_1)}{\pi_{j+1} Q_{j+1}(\gamma_1)}.$$

Since the x_{jk} are the zeros of $Q_j(x)$, it can be shown (see Exercise 5.14) that

$$Q_j(x) = \prod_{k=1}^{j} \frac{x_{jk} - x}{\lambda_{k-1}}.$$

Using (5.2), it follows that

$$\frac{q_j}{q_{j+1}} = \frac{\mu_{j+1}}{x_{j+1,j+1} - \gamma_1} \prod_{k=1}^{j} \frac{x_{jk} - \gamma_1}{x_{j+1,k} - \gamma_1}.$$

Note that

$$\frac{a-c}{b-c} > \frac{a}{b} \quad \text{whenever} \quad a > b > c > 0.$$

Therefore, (5.48) follows from the interlacing property stated in Lemma 5.2 and Theorem 5.4(i).

To prove the second inequality, we claim that $Q_i(\gamma_1)$ is strictly increasing in $i = 0, 1, \cdots, N-1$. If so, we have

$$q_{j+1}\pi_j = \frac{\pi_{j+1}Q_{j+1}(\gamma_1)\pi_j}{K} > \frac{\pi_j Q_j(\gamma_1)\pi_{j+1}}{K} = q_j \pi_{j+1},$$

where K is the normalizing constant, and the result follows. To this end, we note that $Q_i^d(\gamma_1) > 0$ since $\gamma_1 = x_{N1}^d$. The claim now follows at once from Lemma 5.3(i). □

Another interesting problem is to compare $\mathbf{q}_N$ with $\mathbf{q}_{N+1}$. For this purpose, we assume that $\mathbf{q}_N = (q_i^N)$ is a vector of order $(N+1)$ with $q_N^N = 0$.

Theorem 5.14 *Suppose that $\lambda_{N-1} > 0$. Then, for any N, we have*

$$\mathbf{q}_{N+1} \geq_{\text{slr}} \mathbf{q}_N.$$

Proof. As in (5.28), define

$$\chi_i(r) = \lambda_{i-1}\frac{Q_i(r)}{Q_{i-1}(r)}, \quad i = 1, \cdots, N.$$

We saw in Section 5.3 that $\chi_i(r)$ is strictly decreasing in $r < x_{i1}$ and $\chi_N(x_{N1}) = 0$. Let $\gamma_1^k = x_{k1}$ for $k = N, N+1$. Then, from (5.27), we have

$$\frac{q_i^k}{q_{i-1}^k} = \frac{\pi_i Q_i(\gamma_1^k)}{\pi_{i-1}Q_{i-1}(\gamma_1^k)} = \frac{\chi_i(\gamma_1^k)}{\mu_i}.$$

The theorem follows since $\chi_i(r)$ is strictly decreasing in r and $\gamma_1^{N+1} < \gamma_1^N$ from the interlacing property. □

In what follows, we assume that $\mu_0 > 0$. From Theorem 5.14 above, the limit $\mathbf{q} = \lim_{N\to\infty} \mathbf{q}_N$ exists, possibly defective. The next result is a special case of Kijima (1993c) and provides the form of the limit $\mathbf{q}$ when absorption at -1 is certain.

Theorem 5.15 *Suppose $\xi_1 = \lim_{n\to\infty} x_{n1} > 0$. If $A = \infty$, i.e., absorption at -1 is certain, then $\mathbf{q}_N$ converges as $N \to \infty$ to $\mathbf{q} = (q_i)$, where*

$$q_j = \frac{\pi_j Q_j(\xi_1)}{\sum_{j=0}^{\infty} \pi_j Q_j(\xi_1)}, \quad j \in \mathcal{N}.$$

Proof. Note that, under the conditions, we have

$$\sum_{j=0}^{\infty} \pi_j Q_j(\xi_1) = \frac{\mu_0}{\xi_1} < \infty;$$

see Cases 2 and 4.1 in Section 5.5. Now, since $x_{n1} \to \xi_1$ monotonically from above as $n \to \infty$ and $Q_j(x)$ is continuous and monotonically decreasing in $x < x_{j1}$, one sees that $\lim_{N\to\infty} Q_i(x_{N1}) = Q_i(\xi_1)$ and

$$\lim_{N\to\infty} \sum_{j=0}^{N-1} \pi_j Q_j(x_{N1}) = \sum_{j=0}^{\infty} \pi_j Q_j(\xi_1),$$

by the dominated convergence theorem. Hence, from (5.27), one obtains

$$\lim_{N\to\infty} q_j^N = \lim_{N\to\infty} \frac{\pi_j Q_j(x_{N1})}{\sum_{j=0}^{N-1} \pi_j Q_j(x_{N1})} = \frac{\pi_j Q_j(\xi_1)}{\sum_{j=0}^{\infty} \pi_j Q_j(\xi_1)},$$

and the theorem follows. □

Let, in turn, $\mathcal{N} = \{0, 1, 2, \cdots\}$. As for the finite case (5.27), the preceding theorem suggests definition of

$$q_i(x) = \frac{x}{\mu_0} \pi_i Q_i(x), \quad i \in \mathcal{N}. \tag{5.49}$$

Write the infinite vector $\mathbf{q}(x) = (q_i(x))$. We shall shortly show that, under some circumstances, there is a one-parameter family of quasi-stationary distributions $\{\mathbf{q}(x)\}$. Recall that the probability vector $\mathbf{q} = (q_i)$ is a quasi-stationary distribution of the birth–death process $\{X(t)\}$ if

$$q_j = \frac{\sum_i q_i p_{ij}(t)}{\sum_j \sum_i q_i p_{ij}(t)}, \quad j \in \mathcal{N},$$

for all $t \geq 0$. That is, $\mathbf{q}$ is the probability vector with the property that, starting with $\mathbf{q}$, the conditional distribution at any time given that the birth–death process $\{X(t)\}$ is in $\mathcal{N}$ is equal to the initial distribution $\mathbf{q}$.

Let us denote the absorbing birth–death process by $\{X^a(t)\}$ when the absorbing state -1 is appended to $\{X(t)\}$. Then it is readily seen that $\mathbf{q}$ is a quasi-stationary distribution of $\{X(t)\}$ if and only if $q_i \geq 0$, $\sum_{i=0}^{\infty} q_i = 1$ and

$$\frac{\pi_j(t)}{1 - \pi_{-1}(t)} = q_j, \quad j \in \mathcal{N}; \quad t \geq 0, \tag{5.50}$$

where $\pi_j(t)$ denotes the state probabilities of the absorbing birth–death process $\{X^a(t)\}$ with state space $\mathcal{N} \cup \{-1\}$ and initial distribution $\mathbf{q}$, i.e.,

$$\pi_j(t) = P_q[X^a(t) = j] = \sum_{i=0}^{\infty} q_i p_{ij}(t), \quad j \in \mathcal{N} \cup \{-1\}.$$

The next result is due to van Doorn (1991) (see also Cavender, 1978).

Lemma 5.10 *Suppose $A = \infty$. Then $\mathbf{q} = (q_j)$ is a quasi-stationary distribution of $\{X(t)\}$ if and only if $q_j \geq 0$, $\sum_{j=0}^{\infty} q_j = 1$ and $\mathbf{q}$ is a solution of the system of equations*

$$-\mu_0 q_0 q_j = \lambda_{j-1} q_{j-1} - (\lambda_j + \mu_j) q_j + \mu_{j+1} q_{j+1}, \quad j \in \mathcal{N}, \tag{5.51}$$

where $q_{-1} = 0$.

Proof. Suppose $\mathbf{q} = (q_j)$ is a quasi-stationary distribution of $\{X(t)\}$. Then, from (5.50), we have $\pi_j(t) = q_j\{1 - \pi_{-1}(t)\}$ so that $\pi_j'(t) = -q_j\pi_{-1}'(t)$. By our earlier assumption, the $\pi_j(t)$ satisfy the Kolmogorov forward equation

$$\pi_j'(t) = \lambda_{j-1}\pi_{j-1}(t) - (\lambda_j + \mu_j)\pi_j(t) + \mu_{j+1}\pi_{j+1}(t) \tag{5.52}$$

for all $j \in \mathcal{N} \cup \{-1\}$, where $\lambda_{-2} = \lambda_{-1} = \mu_{-1} = 0$, from which we have $\pi_{-1}'(t) = \mu_0\pi_0(t)$. It follows that

$$\pi_j'(t) = -q_j\mu_0\pi_0(t) = -\mu_0 q_0 q_j\{1 - \pi_{-1}(t)\}, \quad j \in \mathcal{N}.$$

Substitution of these equations into (5.52) yields (5.51).

Conversely, let $\mathbf{q} = (q_j)$ be a solution of (5.51) satisfying $q_j \geq 0$ and $\sum_{j=0}^{\infty} q_j = 1$. Define $\pi_{-1}(t)$ such that $\pi_{-1}(0) = 0$ and

$$\pi_{-1}'(t) = \mu_0 q_0\{1 - \pi_{-1}(t)\}, \quad t > 0,$$

and let $\pi_j(t) = q_j\{1 - \pi_{-1}(t)\}$. Then it is easily shown that the $\pi_j(t)$ satisfy (5.52) and, since $\sum_{j=0}^{\infty} \pi_j(t) = 1 - \pi_{-1}(t)$, they must be the state probabilities of $\{X^a(t)\}$ with the initial distribution $\mathbf{q} = (q_j)$, where $q_{-1} = 0$. Since the $\pi_j(t)$ satisfy (5.50), $\mathbf{q}$ is a quasi-stationary distribution of $\{X(t)\}$. □

Theorem 5.16 *Suppose $\xi_1 > 0$ and $A = \infty$.*

(i) *If $D < \infty$ then there is precisely one quasi-stationary distribution $\mathbf{q}(\xi_1)$.*

(ii) *If $D = \infty$ then there is a one-parameter family of quasi-stationary distributions $\{\mathbf{q}(r),\ 0 < r \leq \xi_1\}$.*

Proof. From (5.51) and (5.8), any quasi-stationary distribution is given by $\mathbf{q}(x)$ for some $x > 0$. Here we must also have $x \leq \xi_1$, since, otherwise, $q_i(x)$ can be negative for some i. Suppose, first, that $D < \infty$. Then, consulting Case 2 in Section 5.5, we have from (5.40) that

$$\sum_{j=0}^{\infty} q_j(x) = \frac{x}{\mu_0}\sum_{j=0}^{\infty} \pi_j Q_j(x) = 1 - Q_{\infty}^d(x).$$

But, $Q_{\infty}^d(x) = 0$ if and only if $x = \xi_1 = \xi_1^d$, proving part (i). Part (ii) follows from (5.42). □

So far, we have assumed that $A = \infty$, i.e., absorption at -1 is certain. If this assumption is removed, we need to consider the conditional state probabilities of $\{X^a(t)\}$ with the initial distribution $\overline{\mathbf{q}}$, i.e.,

$$\overline{\pi}_j(t) = P_{\overline{q}}[X^a(t) = j | T < \infty], \quad j \in \mathcal{N} \cup \{-1\}, \tag{5.53}$$

where, as before, T denotes the time of absorption at -1. The distribution $\overline{\mathbf{q}} = (\overline{q}_j)$ may be called a *conditional quasi-stationary distribution* of $\{X(t)\}$

if $\overline{q}_j \geq 0$, $\sum_{j=0}^{\infty} \overline{q}_j = 1$ and

$$\frac{\overline{\pi}_j(t)}{1-\overline{\pi}_{-1}(t)} = \overline{q}_j, \quad j \in \mathcal{N}. \tag{5.54}$$

As in (5.49), we define

$$\overline{q}_j(x) = \frac{x}{\mu_0} a_j \pi_j Q_j(x), \quad j \in \mathcal{N}, \tag{5.55}$$

where a_j is the absorption probability at -1 with initial state j given by (5.34).

Note from (5.53) that

$$\overline{\pi}_j(t) = \frac{P_{\overline{q}}[X^a(t) = j,\, T < \infty]}{P_{\overline{q}}[T < \infty]} = \frac{a_j \pi_j(t)}{P_{\overline{q}}[T < \infty]}.$$

The proof of the next lemma can be obtained similarly to the proof of Lemma 5.10. This is left to the reader as an exercise (Exercise 5.15). The following two results are parallel to Lemma 5.10 and Theorem 5.16 respectively.

Lemma 5.11 *Suppose that $A < \infty$. Then, $\overline{\mathbf{q}} = (\overline{q}_j)$ is a conditional quasi-stationary distribution of $\{X(t)\}$ if and only if $\overline{q}_j \geq 0$, $\sum_{j=0}^{\infty} \overline{q}_j = 1$ and $\overline{\mathbf{q}}$ is a solution of the system of equations*

$$-\mu_0 \frac{\overline{q}_0}{a_0} \frac{\overline{q}_j}{a_j} = \lambda_{j-1} \frac{\overline{q}_{j-1}}{a_{j-1}} - (\lambda_j + \mu_j) \frac{\overline{q}_j}{a_j} + \mu_{j+1} \frac{\overline{q}_{j+1}}{a_{j+1}}, \quad j \in \mathcal{N},$$

where $\overline{q}_{-1} = 0$.

The next theorem can be proved using equations (5.41) and (5.44) instead of (5.40) and (5.42). The proof is omitted.

Theorem 5.17 *Suppose that $\xi_1 > 0$ and $A < \infty$.*

(i) *If $C < \infty$ then there is precisely one conditional quasi-stationary distribution $\overline{\mathbf{q}}(\xi_1)$.*

(ii) *If $C = \infty$ then there is a one-parameter family of conditional quasi-stationary distributions $\{\overline{\mathbf{q}}(r),\ 0 < r \leq \xi_1\}$.*

We next turn our attention to the quasi-limiting distribution of $\{X(t)\}$. If $\mu_0 > 0$, the birth–death process may escape from the state space $\mathcal{N} = \mathcal{Z}_+$ to the absorbing state -1 via state 0. Alternatively, the process may explode and subsequently escape from $\mathcal{N}$ in a finite time to the absorbing boundary at infinity. In any case, however, it is of interest to study the conditional probabilities

$$q_{ij}(t) = P_i[X(t) = j | X(t) \in \mathcal{N}], \quad i,\ j \in \mathcal{N}; \quad t \geq 0. \tag{5.56}$$

If $\{q_{ij}(t)\}$ converges as $t \to \infty$ to a nondefective distribution, we call it the *quasi-limiting distribution*.

There are also circumstances under which one may wish to study the following conditional probabilities rather than $\{q_{ij}(t)\}$:

$$\begin{aligned}\overline{q}_{ij}(t) &= P_i[X(t) = j | X(t) \in \mathcal{N}, \\ &\qquad X(t+s) = -1 \text{ for some } s > 0], \quad i,\ j \in \mathcal{N}; \quad t \geq 0.\end{aligned} \tag{5.57}$$

Obviously, $\overline{q}_{ij}(t) = q_{ij}(t)$ if absorption at -1 is certain, i.e., $A = \infty$. Hence, a separate study of $\overline{q}_{ij}(t)$ is of interest if $A < \infty$. Under this condition, one would like to know in particular whether $\{\overline{q}_{ij}(t)\}$ converges as $t \to \infty$ to a nondefective distribution, in which case this distribution is called the *conditional quasi-limiting distribution*.

Let $p_{ij}(t)$ be the transition probability functions of the birth–death process $\{X(t)\}$ and let T denote the time of absorption at -1. Obviously, the conditional probabilities $q_{ij}(t)$ in (5.56) satisfy

$$\begin{aligned}q_{ij}(t) &= P_i[X(t) = j | T > t] \\ &= \frac{P_i[X(t) = j]}{P_i[T > t]} \\ &= \frac{\pi_j}{\mu_0} \frac{\int_0^\infty e^{-xt} Q_i(x) Q_j(x) d\psi(x)}{\int_0^\infty e^{-xt} x^{-1} Q_i(x) d\psi(x)},\end{aligned} \tag{5.58}$$

provided that $A = \infty$, where we have used (5.10), (5.33) and the Markov property. Similarly, the conditional probabilities $\overline{q}_{ij}(t)$ in (5.57) satisfy

$$\begin{aligned}\overline{q}_{ij}(t) &= P_i[X(t) = j | t < T < \infty] \\ &= \frac{P_i[X(t) = j,\ t < T < \infty]}{P_i[t < T < \infty]} \\ &= \frac{a_j p_{ij}(t)}{P_i[t < T < \infty]} \\ &= \frac{a_j \pi_j}{\mu_0} \frac{\int_0^\infty e^{-xt} Q_i(x) Q_j(x) d\psi(x)}{\int_0^\infty e^{-xt} x^{-1} Q_i(x) d\psi(x)}.\end{aligned} \tag{5.59}$$

To proceed further, we need the following result. The proof is taken from Karlin and McGregor (1957b).

Lemma 5.12 *Suppose that $A + B = \infty$. Then, for any polynomial $f(x)$, we have*

$$\lim_{t \to \infty} \frac{\int_0^\infty e^{-xt} f(x) d\psi(x)}{\int_0^\infty e^{-xt} d\psi(x)} = f(\xi_1),$$

where $\xi_1 = \lim_{n \to \infty} x_{n1}$ is the first limit of the zeros of $Q_n(x)$.

Proof. First note that, under the condition, we have $\xi_1 = \gamma(\psi)$, the infimum of the support of ψ. Hence, writing $a = \xi_1$,

$$\frac{\int_0^\infty e^{-xt} f(x) d\psi(x)}{\int_0^\infty e^{-xt} d\psi(x)} = \frac{\int_a^\infty e^{-xt} f(x) d\psi(x)}{\int_a^\infty e^{-xt} d\psi(x)}.$$

Now consider

$$\frac{\int_a^\infty e^{-(x-a)t} f(x) d\psi(x)}{\int_a^\infty e^{-(x-a)t} d\psi(x)} = f(a) + \frac{\int_a^\infty e^{-(x-a)t}\{f(x)-f(a)\} d\psi(x)}{\int_a^\infty e^{-(x-a)t} d\psi(x)}.$$

Since $f(x)$ is continuous, given $\varepsilon > 0$ there is some $\delta > 0$ such that

$$|f(x) - f(a)| < \varepsilon, \quad a \le x \le a + \delta.$$

Hence

$$\begin{aligned}
&\left| \int_a^\infty e^{-(x-a)t}\{f(x)-f(a)\} d\psi(x) \right| \\
\le\ & \varepsilon \int_a^{a+\delta} e^{-(x-a)t} d\psi(x) + \int_{a+\delta}^\infty e^{-(x-a)t} |f(x)-f(a)| d\psi(x).
\end{aligned}$$

On the other hand,

$$\begin{aligned}
\int_a^\infty e^{-(x-a)t} d\psi(x) &\ge \int_a^{a+\delta} e^{-(x-a)t} d\psi(x) \\
&\ge \int_a^{a+\delta/2} e^{-(x-a)t} d\psi(x) \ge c\, e^{-\delta t/2},
\end{aligned}$$

where $c = \int_a^{a+\delta/2} d\psi(x) > 0$. It follows that

$$\begin{aligned}
&\left| \frac{\int_a^\infty e^{-(x-a)t}\{f(x)-f(a)\} d\psi(x)}{\int_a^\infty e^{-(x-a)t} d\psi(x)} \right| \\
\le\ & \varepsilon + \frac{e^{\delta t/2}}{c} \int_{a+\delta}^\infty e^{-(x-a)t} |f(x)-f(a)| d\psi(x),
\end{aligned}$$

the latter of which converges to zero as $t \to \infty$. Since $\varepsilon > 0$ is arbitrary, the lemma is proved. □

When absorption at -1 is certain, the limit of $q_{ij}(t)$ in (5.56) as $t \to \infty$ is given as follows. The result was obtained in Good (1968), Kijima and Seneta (1991) and van Doorn (1991).

Theorem 5.18 *Suppose that $A = \infty$ and $\xi_1 > 0$. Then the quasi-limiting distribution is given by*

$$\lim_{t\to\infty} q_{ij}(t) = \frac{\pi_j Q_j(\xi_1)}{\sum_{k=0}^\infty \pi_k Q_k(\xi_1)} > 0, \quad i,\ j \in \mathcal{N}.$$

Proof. From Lemma 5.12, we have

$$\lim_{t\to\infty} \frac{\int_0^\infty e^{-xt} Q_i(x) Q_j(x) d\psi(x)}{\int_0^\infty e^{-xt} d\psi(x)} = Q_i(\xi_1) Q_j(\xi_1),$$

while, if $\xi_1 > 0$, a similar argument leads to

$$\lim_{t\to\infty} \frac{\int_0^\infty e^{-xt} x^{-1} Q_i(x) d\psi(x)}{\int_0^\infty e^{-xt} d\psi(x)} = \frac{Q_i(\xi_1)}{\xi_1}.$$

It follows from (5.58) that

$$\lim_{t\to\infty} q_{ij}(t) = \frac{\xi_1 \pi_j Q_j(\xi_1)}{\mu_0} > 0.$$

The result follows from either (5.40) of Case 2 or (5.42) of Case 4.1 in Section 5.5; in either case we have $\xi_1 = \xi_1^d > 0$. □

For the conditional quasi-limiting distribution, we have the following. See Kijima, Nair, Pollett and van Doorn (1997, to appear) for more general results.

Theorem 5.19 *Suppose that $A < \infty$, $B = \infty$ and $\xi_1 > 0$. Then*

$$\lim_{t\to\infty} \overline{q}_{ij}(t) = \frac{a_j \pi_j Q_j(\xi_1)}{\sum_{k=0}^{\infty} a_k \pi_k Q_k(\xi_1)} > 0, \quad i,\ j \in \mathcal{N}.$$

Proof. Again referring to Lemma 5.12, we obtain

$$\lim_{t\to\infty} \overline{q}_{ij}(t) = \frac{\xi_1 a_j \pi_j Q_j(\xi_1)}{\mu_0} > 0,$$

provided that $\xi_1 > 0$. The result follows from either (5.41) of Case 3 or (5.44) of Case 4.2 in Section 5.5. □

Finally, we provide an illustrative example.

Example 5.7 Consider the M/M/1 queue given in Example 5.6. Suppose that the queue-size process $\{X(t)\}$ is observed up until time t and $X(\tau) \geq 1$ for all $\tau \leq t$. That is, the first busy period B_1 has not ended before t. Suppose, further, that we observe a statistical equilibrium in the system within this period. Then the conditional steady state probabilities can be approximated by either the quasi-limiting or the conditional quasi-limiting distributions of the lossy generator

$$\mathbf{Q} = \begin{pmatrix} -\lambda-\mu & \lambda & 0 & 0 & \cdots \\ \mu & -\lambda-\mu & \lambda & 0 & \cdots \\ 0 & \mu & -\lambda-\mu & \lambda & \cdots \\ \vdots & \vdots & \ddots & \ddots & \ddots \end{pmatrix}.$$

Suppose $\rho = \lambda/\mu < 1$. Then $A = \infty$ and we look into the quasi-limiting distribution

$$q_{j+1} = \frac{\xi_1}{\mu} \rho^j Q_j(\xi_1), \quad j = 0, 1, \cdots,$$

where $\xi_1 = (\sqrt{\lambda} - \sqrt{\mu})^2 > 0$, as we shall see later. The birth–death polynomials are given by $Q_{-1}(x) = 0,\ Q_0(x) = 1$ and

$$-x\, Q_n(x) = \mu\, Q_{n-1}(x) - (\lambda+\mu) Q_n(x) + \lambda\, Q_{n+1}(x), \quad n = 0, 1, \cdots;$$

see (5.8). It follows by an induction argument that

$$Q_n(\xi_1) = (n+1)\rho^{-n/2}, \quad n = 0, 1, \cdots,$$

so that

$$q_{j+1} = (1-\sqrt{\rho})^2 (j+1)\sqrt{\rho}^{\,j}, \quad j = 0, 1, \cdots.$$

See Seneta (1966) for another means of deriving the result.

Suppose, next, that $\rho > 1$. Then $A < \infty$ and $B = \infty$ so that we lock into the conditional quasi-limiting distribution. Since $a_j = \rho^{-(j+1)}$, from Example 5.6, we have

$$q_{j+1} = \left(1 - \frac{1}{\sqrt{\rho}}\right)^2 (j+1) \left(\frac{1}{\sqrt{\rho}}\right)^j, \quad j = 0, 1, \cdots.$$

It is interesting to note that these limiting distributions are symmetric with respect to λ and μ. See van Doorn (1991) for related results.

5.7 The decay parameter

Let $\{X(t)\}$ be a birth–death process with state space $\mathcal{N} = \{0, 1, 2, \cdots\}$, birth rates λ_n and death rates μ_n. Throughout this section, we assume that $A + B = \infty$. That is, the transition probability functions $p_{ij}(t)$ of $\{X(t)\}$ are uniquely determined by the birth and death rates. Let m_j be the limit of $p_{ij}(t)$ as $t \to \infty$. Then the *decay parameter* γ of $\{X(t)\}$ is defined by

$$\gamma = \sup\{\alpha \geq 0 : p_{ij}(t) - m_j = O(e^{-\alpha t}) \text{ as } t \to \infty \text{ for all } i,\ j \in \mathcal{N}\}.$$

The next result is due to van Doorn (1985); the reader should consult this paper for details.

Theorem 5.20 *The decay parameter of the birth–death process $\{X(t)\}$ is given by $\gamma = \xi_1$ if $\mu_0 > 0$ and $\gamma = \xi_1^d$ if $\mu_0 = 0$.*

When $\mu_0 = 0$, we call the process $\{X(t)\}$ *exponentially ergodic* if it is ergodic and $\gamma > 0$. The decay parameter in this case indicates the speed of convergence to ergodicity and is very important to know in practice. In this section, we consider the problem of finding the decay parameter γ or, equivalently, of finding the first limit point ξ_1 or ξ_1^d. For this purpose, we consider a general sign-symmetric, tridiagonal infinite matrix

$$\mathbf{T} = \begin{pmatrix} c_1 & b_2 & 0 & 0 & \cdots \\ a_2 & c_2 & b_3 & 0 & \cdots \\ 0 & a_3 & c_3 & b_4 & \cdots \\ \vdots & \vdots & \ddots & \ddots & \ddots \end{pmatrix},$$

where $\beta_i = a_i b_i > 0$ for all $i = 2, 3, \cdots$. Let us define a set of polynomials recursively by

$$P_n(x) = (x - c_n)P_{n-1}(x) - \beta_n P_{n-2}(x), \quad n = 1, 2, \cdots, \tag{5.60}$$

where $P_{-1}(x) = 0$ and $P_0(x) = 1$. The polynomial $P_n(x)$ has n real, distinct zeros $x_{n1} < x_{n2} < \cdots < x_{nn}$ and the zeros of $P_n(x)$ and $P_{n-1}(x)$

interlace (see Lemma 5.2). Hence, as before, the limit $\gamma = \lim_{n\to\infty} x_{n1}$ exists, possibly $\gamma = -\infty$.

The next representation result, which is parallel to Lemma 5.5, was obtained by van Doorn (1987b). The proof is omitted.

Lemma 5.13 *The first limit γ of $\mathbf{T}$ is given by*

$$\gamma = \max_{\mathcal{X}} \left\{ \inf_{i\geq 1} \left\{ c_i - \frac{\beta_i}{\chi_i} - \chi_{i+1} \right\} \right\},$$

where $\mathcal{X} = \{\chi_1, \chi_2, \cdots\}$ ranges over all sequences such that $\chi_1 = \infty$ and $\chi_i > 0$ for all $i = 2, 3, \cdots$.

As in Corollary 5.3, the next bounds are immediate from Lemma 5.13. See, for example, van Doorn (1987b) and Zeifman (1991) for other useful bounds.

Corollary 5.4 *Let $\beta_1 = 0$ and $\beta_i = a_i b_i > 0$ for $i = 2, 3, \cdots$. Then*

$$\inf_{i\geq 1}\{c_i - \sqrt{\beta_i} - \sqrt{\beta_{i+1}}\} \leq \gamma \leq \inf_{i\geq 1} c_i.$$

Example 5.8 Consider a birth–death process $\{X(t)\}$ with the infinitesimal generator $\mathbf{Q}$ given in (5.1). Suppose $\mu_0 = 0$. Then the decay parameter of $\{X(t)\}$ is given by the first limit ξ_1^d associated with $-\mathbf{Q}^d$, where $\mathbf{Q}^d$ is the dual generator given by (5.14). Hence, defining

$$c_i \equiv \lambda_{i-1} + \mu_i, \quad \beta_{i+1} \equiv \lambda_i \mu_i, \quad i = 1, 2, \cdots,$$

the lower bound in Corollary 5.4 is given by

$$\gamma \geq \inf_{i\geq 1} \left\{ \lambda_{i-1} + \mu_i - \sqrt{\lambda_{i-1}\mu_{i-1}} - \sqrt{\lambda_i \mu_i} \right\}.$$

In particular, when $\lambda_i = \lambda$, $i = 0, 1, \cdots$ and $\mu_i = \mu$, $i = 1, 2, \cdots$, we have

$$\gamma \geq \lambda + \mu - 2\sqrt{\lambda\mu} = \left(\sqrt{\lambda} - \sqrt{\mu}\right)^2.$$

As we shall see in Example 5.9 below, the decay parameter for this case is, in fact, given by $(\sqrt{\lambda} - \sqrt{\mu})^2$.

Let $\mathbf{T}_n$ be the $n \times n$ north-west corner truncation of $\mathbf{T}$, i.e., the matrix given in Lemma 5.5. Let $f_n(x) = \det(\mathbf{T}_n - x\,\mathbf{I})$, the determinant of the matrix $(\mathbf{T}_n - x\,\mathbf{I})$, with $f_0(x) = 1$, where $\mathbf{I}$ denotes the $n \times n$ identity matrix. It is readily seen that $f_n(x) = (-1)^n P_n(x)$ and that if $\gamma > -\infty$ then the maximum in Lemma 5.13 is attained by

$$\chi_i = \frac{f_{i-1}(\gamma)}{f_{i-2}(\gamma)}, \quad i = 2, 3, \cdots;$$

see (5.28). The source of the following results is Kijima (1992d).

Lemma 5.14 *There exists a sequence* $\{\chi_n\}$ *such that* $\chi_1 = \infty$, $\chi_n > 0$ *for* $n = 2, 3, \cdots$, *and*

$$y = c_i - \frac{\beta_i}{\chi_i} - \chi_{i+1}, \quad i = 1, 2, \cdots, \tag{5.61}$$

if and only if $y \leq \gamma$.

Proof. Suppose that $y \leq \gamma$. Then, from the interlacing property of the zeros x_{ni} and since $f_i(y) = (-1)^i P_i(y)$, one easily sees that $f_i(y) > 0$ for all i. Therefore, from (5.60), defining $\chi_i = f_{i-1}(y)/f_{i-2}(y) > 0$ for $i = 2, 3, \cdots$ with $\chi_1 = \infty$ yields (5.61). Conversely, suppose that the sequence $\{\chi_n\}$ satisfies the required conditions. Then, multiplying both sides of (5.61) by $\prod_{k=2}^{i} \chi_k$ and comparing the result with (5.60), it is readily seen that

$$\prod_{k=2}^{i+1} \chi_k = f_i(y) > 0, \quad i = 1, 2, \cdots.$$

But, from the same reasoning as given above, this happens only when $y < x_{i1}$ for all $i = 1, 2, \cdots$. Hence, since x_{n1} decreases monotonically in n and $\gamma = \lim_{n\to\infty} x_{n1}$, the lemma is proved. □

In what follows, we assume that there exists some $N \geq 1$ such that $\beta_n = \beta$ and $c_n = c$ for all $n \geq N$. In the birth–death context, such situations are often encountered in practice (see Examples 5.9 and 5.10 below). From Lemmas 5.13 and 5.14, γ is the maximum among the y satisfying

$$\begin{aligned} y &= c_1 - \chi_2 = c_2 - \frac{\beta_2}{\chi_2} - \chi_3 = \cdots = c_{N-1} - \frac{\beta_{N-1}}{\chi_{N-1}} - \chi_N \\ &= c - \frac{\beta}{\chi_N} - \chi_{N+1} = c - \frac{\beta}{\chi_{N+1}} - \chi_{N+2} = \cdots. \end{aligned} \tag{5.62}$$

Given y, consider the equation

$$x + y = c - \frac{\beta}{x}, \quad x > 0. \tag{5.63}$$

Let $a(y)$ and $b(y)$ be the positive solutions of (5.63), if they exist, where $a(y) \leq b(y)$. Given $\chi_N > 0$, the quantities χ_{N+i}, $i = 1, 2, \cdots$ must satisfy (5.62) so that

$$\chi_{N+i+1} + y = c - \frac{\beta}{\chi_{N+i}}, \quad i = 0, 1, \cdots. \tag{5.64}$$

If equation (5.63) has no positive solutions, one sees from (5.64) that there is some n such that $\chi_n < 0$ (see Figure 5.1(a)), which violates the required positivity condition. Thus, from (5.63) and Lemma 5.14, it must be the case that

$$\gamma \leq c - 2\sqrt{\beta}.$$

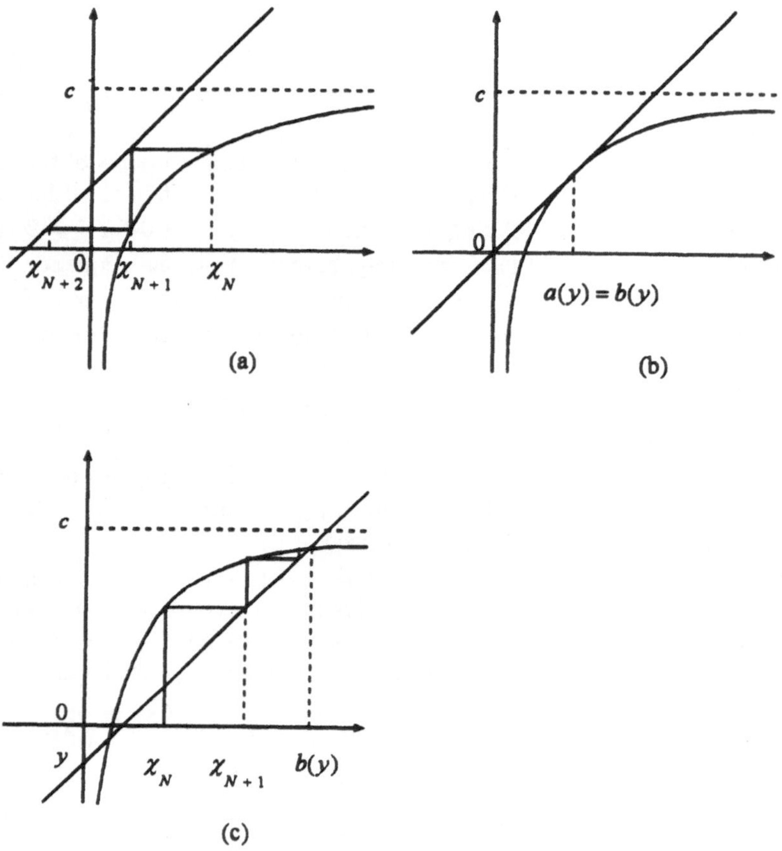

Figure 5.1 *Convergence of* χ_n.

If an equality holds, i.e., $\gamma = c - 2\sqrt{\beta}$, then we have $a(y) = b(y) = \sqrt{\beta}$ (see Figure 5.1(b)). Also, by the same reasoning as given above, the case where $\chi_N < a(y)$ is not allowed. Therefore, from (5.63) and (5.64), it follows that

$$\lim_{n\to\infty} \chi_n = \begin{cases} a(y), & \text{if } \chi_N = a(y), \\ b(y), & \text{if } \chi_N > a(y); \end{cases} \tag{5.65}$$

see Figure 5.1(c).

For real y, define $\chi_n(y)$ successively by

$$\chi_{n+1}(y) = c_n - \frac{\beta_n}{\chi_n(y)} - y, \quad n = 1, 2, \cdots, \tag{5.66}$$

where $\chi_1(y) = \infty$. Using the same argument as is given in the proof of Lemma 5.14, it can be shown that

$$\chi_n(y) = \frac{f_{n-1}(y)}{f_{n-2}(y)},$$

and $\chi_n(y) > 0$ for $n = 2, \cdots, N$ if and only if $y < x_{N-1,1}$. Also, as for the finite case (see Section 5.3), they are strictly decreasing in $y < x_{N-1,1}$. For $u > 0$, let $y = \xi(u)$ be the smallest solution of $\chi_N(y) = u$ or, equivalently, that of

$$f_{N-1}(y) = u\, f_{N-2}(y).$$

Then, $\xi(u) < x_{N-1,1}$ so that $\chi_n(\xi(u)) > 0$ for $n = 2, \cdots, N-1$. It should be noted that $\xi(u)$ is strictly decreasing in $u > 0$, since $\chi_N(y)$ is strictly decreasing in $y < x_{N-1,1}$. Fix y arbitrarily such that $y \leq c - 2\sqrt{\beta}$. Let $a(y)$ be determined as above. If there is an $n \leq N$ such that $\chi_n(y) \leq 0$ then $y \geq x_{N-1,1}$ so that this y cannot be the true value γ. Now suppose $\chi_n(y) > 0$ for $n = 2, \cdots, N$, and compare this y with $\xi(a(y))$. If $y > \xi(a(y))$, we have $\chi_N(y) < a(y)$ since $\chi_N(y)$ is strictly decreasing. But, then there is some n such that $\chi_n(y) < 0$, so that this y is too large to be equal to the true value γ. On the other hand, if $y < \xi(a(y))$, then $\chi_N(y) > a(y)$. From (5.65) and the discussion given above, the $\chi_n(y)$ are all positive and satisfy (5.62). But this y may not be the maximum. In fact, y can be increased until either $\chi_N(y) = a(y)$ or $y = c - 2\sqrt{\beta}$. Note that $a(y)$ is strictly increasing in $y \leq c - \sqrt{\beta}$ and $\chi_N(y)$ is strictly decreasing in $y \leq x_{N-1,1}$. Hence, the value y satisfying $\chi_N(y) = a(y)$ and $y < x_{N-1,1}$ is unique, if it exists. Also, $\gamma = c - 2\sqrt{\beta}$ if and only if $y^* \geq c - 2\sqrt{\beta}$, where y^* is the smallest solution of

$$f_{N-1}(y) = \sqrt{\beta}\, f_{N-2}(y). \tag{5.67}$$

This follows since $\gamma = c - 2\sqrt{\beta}$ if and only if

$$\begin{cases} \chi_n(c - 2\sqrt{\beta}) > 0, & n = 2, \cdots, N-1; \\ \chi_N(c - 2\sqrt{\beta}) \geq a(c - 2\sqrt{\beta}) = \sqrt{\beta}. \end{cases} \tag{5.68}$$

We summarize the above discussion in the next theorem.

Theorem 5.21 *Suppose that there exists some positive integer N such that $\beta_n = \beta$ and $c_n = c$ for all $n \geq N$. Let y^* be the smallest solution of (5.67). If $y^* \geq c - 2\sqrt{\beta}$ then the first limit is given by $\gamma = c - 2\sqrt{\beta}$. If $y^* < c - 2\sqrt{\beta}$, then the first limit is the smallest solution of*

$$f_{N-1}(y) = a(y) f_{N-2}(y).$$

To obtain the first limit γ numerically, we first check if the conditions in (5.68) hold, where $\chi_n(y)$ are recursively calculated by (5.66). If the conditions are satisfied, we have $\gamma = c - 2\sqrt{\beta}$. If not, the following bisection search is applied. The algorithm is similar to the algorithm given in Section 5.3.

Algorithm Let $\varepsilon > 0$ be a prespecified error.

Step 1 $L \leftarrow$ any lower bound and $R \leftarrow c - 2\sqrt{\beta}$.

Step 2 If $R - L < \varepsilon$ holds, then $y \leftarrow L$ and terminate. Otherwise, $y \leftarrow (L+R)/2$ and $i \leftarrow 1$.

Step 3 Calculate $\chi_{i+1}(y)$ by (5.66). If $\chi_{i+1}(y) \leq 0$ then $R \leftarrow y$ and go to Step 2.

Step 4 $i \leftarrow i+1$. If $i < N-1$ then go to Step 3.

Step 5 Calculate $\chi_N(y)$ by (5.66). If $\chi_N(y) > a(y)$ ($< a(y)$, respectively), then $L \leftarrow y$ ($R \leftarrow y$) and go to Step 2. If $\chi_N(y) = a(y)$, then $\gamma \leftarrow y$ and terminate.

Before concluding this section, we consider Markovian queueing systems.

Example 5.9 Consider an M/M/1 queue with arrival rate λ and service rate μ. This is a birth–death process with birth rates $\lambda_n = \lambda$, $n = 0, 1, \cdots$ and death rates $\mu_0 = 0$ and $\mu_n = \mu$, $n = 1, 2, \cdots$. In order to evaluate the decay parameter, we need to consider the dual generator (5.14). Then, taking $N = 2$ in Theorem 5.21, we must consider the smallest solution of $f_1(y) = \sqrt{\beta}\, f_0(y)$, where $\beta = \lambda\mu$, $f_0(y) = 1$ and $f_1(y) = \lambda + \mu - y$. It follows that

$$y^* = \lambda + \mu - \sqrt{\lambda\mu} \geq \lambda + \mu - 2\sqrt{\lambda\mu},$$

so that $\gamma = \lambda + \mu - 2\sqrt{\lambda\mu} = (\sqrt{\lambda} - \sqrt{\mu})^2$ is the decay parameter. Note that we do not assume the stability condition $\lambda < \mu$.

Example 5.10 Suppose, in turn, that the death rates are given by $\mu_0 = 0$, $\mu_1 = \mu$ and $\mu_n = 2\mu$, $n = 2, 3, \cdots$. In this M/M/2 case, taking $N = 3$ in Theorem 5.21, we need to consider the smallest solution of $f_2(y) = \sqrt{\beta}\, f_1(y)$, where $\beta = 2\lambda\mu$, $f_1(y) = \lambda + \mu - y$ and

$$f_2(y) = (\lambda + \mu - y)(\lambda + 2\mu - y) - \lambda\mu.$$

The smallest solution is given by

$$y^* = \frac{2\lambda + 3\mu - \sqrt{2\lambda\mu} - \sqrt{\mu^2 + 6\lambda\mu - 2\mu\sqrt{2\lambda\mu}}}{2}.$$

Solving the inequality

$$y^* \geq \lambda + 2\mu - 2\sqrt{2\lambda\mu},$$

we have $\rho \equiv \lambda/2\mu \geq 1/9$, in which case the decay parameter γ is given by $(\sqrt{\lambda} - \sqrt{2\mu})^2$. Such critical values for M/M/s queues are listed in van Doorn (table on page 61, 1981).

5.8 The M/M/1 queue

In this section, we consider a single-server queueing system with a Poisson arrival process and exponential service times as a special case of birth–death processes. For a discussion of many-server Markovian queues, the reader is referred to, e.g., Karlin and McGregor (1958), Parthasarathy and Sharafali (1989) and Kijima (1992b).

Let $X(t)$ represent the number of customers in the system at time t of an M/M/1 queue with arrival rate λ and service rate μ. The process $\{X(t)\}$ is a birth–death process with birth rates $\lambda_n = \lambda$ and death rates $\mu_n = \mu$, where $\mu_0 = 0$. The potential coefficients of $\{X(t)\}$ are given by $\pi_n = \rho^n$, $n = 0, 1, \cdots$, where $\rho = \lambda/\mu$, called the *traffic intensity*. Recall from Example 5.4 that $A = \infty$ if $\rho \le 1$ and $A < \infty$ if $\rho > 1$. Also, $B < \infty$ if $\rho < 1$ and $B = \infty$ if $\rho \ge 1$. Hence, the process $\{X(t)\}$ is recurrent if and only if $\rho \le 1$ and it is ergodic if $\rho < 1$. From Theorem 5.1, the boundary at infinity is natural for any $\rho > 0$.

In what follows, we assume that $\mu = 1$ so that the arrival rate equals the traffic intensity ρ. This does not cause any loss of generality because we can take the time scale in terms of the service rate μ. The birth–death polynomials are given by

$$-x\,Q_n(x) = Q_{n-1}(x) - (1+\rho)Q_n(x) + \rho\,Q_{n+1}(x), \quad n = 1, 2, \cdots,$$

where $Q_{-1}(x) = 0$ and $Q_0(x) = 1$. Let $p_{ij}(t)$ be the transition probability functions of $\{X(t)\}$. Since the boundary at infinity is natural for any $\rho > 0$, i.e., $C = D = \infty$, they are uniquely determined by the birth and death rates or, equivalently, by the traffic intensity, ρ in this case. The Karlin–McGregor representation of the transition probability function $p_{ij}(t)$ is given in Example 5.6.

Let T_{ij} denote the first passage time of $\{X(t)\}$ to state j with initial state i, and let $f_{ij}(t)$ be the probability density function of T_{ij}. In the queueing context, the random variable T_{10} is called the *busy period*. In this respect, a better notation for $f_{10}(t)$ may be $b(t)$. Note that, due to the spatial homogeneity, we have $b(t) = f_{i+1,i}(t)$ for all $i = 0, 1, \cdots$. Moreover, because of the skip-free property of birth–death processes, it follows that $f_{i+j,i}(t) = b^{(j)}(t)$, the j-fold convolution of $b(t)$ with itself. Namely, $b^{(1)}(t) = b(t)$ and

$$b^{(j+1)}(t) = \int_0^t b^{(j)}(s)\,b(t-s)\,ds, \quad j = 1, 2, \cdots.$$

Let $\beta(s)$ denote the Laplace transform of $b(t)$, i.e.,

$$\beta(s) = \int_0^\infty e^{-st} b(t)dt, \quad \mathrm{Re}\,(s) \ge 0,$$

and let $\phi_{ij}(s)$ be the Laplace transform of $f_{ij}(t)$. From Theorem B.11(viii), we then have $\phi_{i+j,i}(s) = \beta^j(s)$ for $j = 1, 2, \cdots$.

Lemma 5.15 *For each* $i = 1, 2, \cdots$,

(i) $p_{0i}(t) = \rho^i p_{i0}(t)$, $t \geq 0$,

(ii) $p_{i0}(t) = \int_0^t b^{(i)}(s)\, p_{00}(t-s) ds$, $t \geq 0$.

Proof. Assertion (i) follows since any birth–death process is reversible in time. To prove assertion (ii), we recall from Exercise 4.1 that, for $i \neq 0$,

$$p_{i0}(t) = \int_0^t f_{i0}(s)\, p_{00}(t-s) ds, \quad t \geq 0.$$

Since $f_{i0}(t) = b^{(i)}(t)$, the result follows. □

The density function of the busy period can be obtained as follows. Let Z be an exponentially distributed random variable with parameter $(1+\rho)$. Then T_{10} equals Z with probability $1/(1+\rho)$ or it equals $Z + T_{21} + \widehat{T}_{10}$ with probability $\rho/(1+\rho)$, where all the random variables involved are mutually independent and $\widehat{T}_{10}$ and T_{10} are equal in distribution. In the expression for the Laplace transform, this is equivalent to

$$\beta(s) = \frac{1}{1+\rho} \frac{1+\rho}{s+1+\rho} + \frac{\rho}{1+\rho} \frac{1+\rho}{s+1+\rho} \beta^2(s),$$

from which we obtain

$$\rho\, \beta^2(s) - (s+1+\rho)\beta(s) + 1 = 0, \quad \mathrm{Re}\,(s) \geq 0. \tag{5.69}$$

It follows that

$$\beta(s) = \frac{s+1+\rho - \sqrt{(s+1+\rho)^2 - 4\rho}}{2\rho}, \quad \mathrm{Re}\,(s) \geq 0, \tag{5.70}$$

since, otherwise, $\beta(0) > 1$ for the ergodic case $\rho < 1$. The Laplace transform $\beta(s)$ can be inverted (see, e.g., Abramowitz and Stegun, page 1024, 1965) and we have

$$b(t) = \rho^{-1/2} \frac{e^{-(1+\rho)t}}{t} I_1\left(2\sqrt{\rho} t\right), \quad t \geq 0,$$

where $I_n(y)$ denotes the modified Bessel function of order n which is given by

$$I_n(y) = \sum_{k=0}^{\infty} \frac{(y/2)^{n+2k}}{k!\,(n+k)!}, \quad n = 0, 1, \cdots.$$

Let $\pi_{ij}(s)$ be the Laplace transform of $p_{ij}(t)$. From the Karlin–McGregor representation (5.10), we have

$$\pi_{ij}(s) = \rho^j \int_0^\infty Q_i(x) Q_j(x) \frac{d\psi(x)}{s+x}, \quad \mathrm{Re}\,(s) > 0.$$

In particular,

$$\pi_{00}(s) = \int_0^\infty \frac{d\psi(x)}{s+x}, \quad \mathrm{Re}\,(s) > 0.$$

Abate and Whitt (1988) pointed out the following interesting relation between $p_{00}(t)$ and the busy period density $b(t)$. Note from (4.24) that

$$p_{00}(t) = e^{-\rho t} + \int_0^t \rho e^{-\rho s} p_{10}(t-s)ds, \quad t \geq 0.$$

It follows from Lemma 5.15(ii) with $i = 1$ that

$$\pi_{00}(s) = \frac{1}{s+\rho} + \frac{\rho}{s+\rho}\beta(s)\,\pi_{00}(s), \quad \mathrm{Re}\,(s) \geq 0,$$

whence

$$\pi_{00}(s) = \frac{1}{s+\rho-\rho\beta(s)}, \quad \mathrm{Re}\,(s) > 0. \tag{5.71}$$

From (5.69), we have

$$\frac{1}{s+\rho-\rho\beta(s)} = \frac{\beta(s)}{1-\beta(s)}, \quad \mathrm{Re}\,(s) > 0,$$

which, together with (5.71), leads to

$$\pi_{00}(s) = \frac{\beta(s)}{1-\beta(s)}, \quad \mathrm{Re}\,(s) > 0. \tag{5.72}$$

An interpretation of (5.72) is as follows. Let Y_i denote the random length of the ith busy period and consider a sequence $\{Y_i\}$. Let $m(t)$ be the renewal density associated with $\{Y_i\}$. It is well known (see, e.g., Ross, page 66, 1983) that the renewal density is given by $m(t) = \sum_{n=1}^{\infty} b^{(n)}(t)$ so that its Laplace transform is obtained as $\beta(s)/\{1-\beta(s)\}$. Thus, the transition probability function $p_{00}(t)$ of any M/M/1 queue can be characterized as the renewal density associated with a sequence of busy periods (see Kijima, 1992b, for more general results).

The next result is due to Abate and Whitt (1988).

Theorem 5.22 *For each $i = 0, 1, \cdots$,*

$$p_{i0}(t) = \int_0^t b^{(i)}(s)ds - \rho \int_0^t b^{(i+1)}(s)ds, \quad t \geq 0,$$

where $\int_0^t b^{(0)}(s)ds = 1$ for any $t \geq 0$.

Proof. From (5.69), we have

$$s\,\beta(s) = \{1-\beta(s)\}\{1-\rho\beta(s)\},$$

which, together with (5.72), yields

$$\pi_{00}(s) = \frac{\beta(s)}{1-\beta(s)} = \frac{1}{s} - \frac{\rho}{s}\beta(s), \quad \mathrm{Re}\,(s) > 0.$$

From Lemma 5.15(ii), the Laplace transform $\pi_{i0}(s)$ is then given by

$$\pi_{i0}(s) = \frac{1}{s}\beta^i(s) - \frac{\rho}{s}\beta^{i+1}(s), \quad \mathrm{Re}\,(s) > 0.$$

The theorem follows from the operational properties given in Theorem B.11 (v) and (viii). □

Another interesting result concerning the transition probability functions $p_{ij}(t)$ is the following. See Abate, Kijima and Whitt (1991) for further relations between $p_{ij}(t)$ and the busy period density $b(t)$. Such relations are very useful since they provide an insight into the transient behavior of M/M/1 queues.

Theorem 5.23 *For each* $i = 0, 1, \cdots,$

$$p_{i0}(t) - p_{i+1,0}(t) = b^{(i+1)}(t), \quad t \geq 0.$$

Proof. From (5.72), we have

$$\beta(s) = \pi_{00}(s) - \beta(s)\,\pi_{00}(s).$$

But, from Lemma 5.15(ii), $\pi_{i0}(s) = \beta^i(s)\,\pi_{00}(s)$, whence the result. □

5.9 Exercises

Exercise 5.1 In the machine repair problem given in Example 5.1, suppose that, initially, N machines are operating. Find the Laplace transform of the first time that there are two failed machines. In particular, find its distribution function for the case $\lambda = \mu$.

Exercise 5.2 In the machine repair problem, suppose, now, that there are s repairmen. Find the mean and variance of the number of failed machines in equilibrium.

Exercise 5.3 In Exercise 5.2, suppose, now, that at most $M \leq N$ can be operating at any time, i.e., the rest are 'spares'. Determine the stationary distribution of the number of failed machines.

Exercise 5.4 In the M/M/s queue given in Example 5.2, suppose that $s = 2$ and that the service times (though still exponential) have different means depending on the server. Let the service rate be given by μ_i, $i = 1, 2$, for the ith server and suppose that $\mu_1 \geq \mu_2$.

(i) Suppose that the second server works only when there are two or more customers in the system, so that if the first server becomes idle while the second is busy, then the customer switches to the first server immediately. Formulate this system by a birth–death process and find the mean number of customers in the system in equilibrium.

(ii) Suppose that the policy is the converse of case (i). That is, the first server works only when there are two or more customers in the system. Find the mean number of customers in equilibrium.

(iii) Suppose, now, that no policy is in operation and a customer who arrives when the servers are idle chooses server i with probability

α_i, where $\alpha_1 + \alpha_2 = 1$. Formulate this as a continuous-time Markov chain and obtain the mean number of customers in the system in equilibrium. Compare the means of the three cases.

Exercise 5.5 In an M/M/s queue, suppose that there are infinitely many servers, i.e., $s = \infty$, and no customers ever wait for service. Show that in equilibrium the number of customers in the system follows a Poisson distribution with parameter $\rho = \lambda/\mu$, where λ is the arrival rate and μ is the service rate.

Exercise 5.6 State the boundary classification and the state classification of an M/M/s queue (s may be infinity).

Exercise 5.7 A birth–death process is called a *linear growth* process if $\lambda_n = \lambda n + a$ and $\mu_n = \mu n$ with $\lambda,\ \mu > 0$ and $a \geq 0$. Such processes occur naturally in the study of biological reproduction and population growth. The factor λn represents a natural growth of the population depending on its current size, while the second factor a may be interpreted as the infinitesimal rate of increase of the population due to an external source such as immigration. The factor μn has an obvious interpretation.

(i) Suppose $\lambda < \mu$. Show that the linear growth birth–death process is positive recurrent and that the boundary at infinity is natural.

(ii) Suppose $\lambda = \mu$ and $a = 0$. Show that

$$u(t) = P_1[X(t) = 0]$$

satisfies the integral equation

$$u(t) = \int_0^t \lambda e^{-2\lambda\tau} d\tau + \int_0^t \lambda e^{-2\lambda\tau} u^2(t-\tau) d\tau.$$

Moreover, show that a solution of this integral equation is given by $u(t) = \lambda t/(1 + \lambda t)$ (Karlin and Taylor, 1975).

Exercise 5.8 Suppose that $\mu_0 > 0$. State the boundary classification at infinity of the dual birth–death process in terms of the quantities given in (5.3) of the original birth–death process.

Exercise 5.9 Let λ_i denote positive and distinct numbers, and suppose that the Laplace transform is given by

$$\phi(s) = \prod_{j=1}^{n} \frac{\lambda_j}{s + \lambda_j}, \qquad \operatorname{Re}(s) \geq 0.$$

Prove that the partial-fraction expansion of $\phi(s)$ is given by

$$\phi(s) = \sum_{j=1}^{n} \alpha_j \frac{\lambda_j}{s + \lambda_j}; \qquad \alpha_j = \prod_{k \neq j} \frac{\lambda_k}{\lambda_k - \lambda_j}.$$

Exercise 5.10 Let $\{X(t)\}$ be a finite birth–death process and suppose that $X(0) = 0$. Prove that $X(t) \geq_{\rm lr} X(s)$ for all $t > s \geq 0$, where $\geq_{\rm lr}$ denotes the likelihood ratio ordering.

Exercise 5.11 Let $p_{ij}(t)$ be the transition probability functions of a finite birth–death process. Prove that $p_{i0}(t)$ and $p_{0j}(t)$ are unimodal in t for all i and j.

Exercise 5.12 Prove Lemma 5.5 using the Gerschgorin circle theorem (see, e.g., Noble and Daniel, page 289, 1977): Each eigenvalue λ of an $n \times n$ matrix $\mathbf{A} = (a_{ij})$ satisfies at least one of the inequalities

$$|\lambda - a_{ii}| \leq r_i; \quad r_i \equiv \sum_{j \neq i} |a_{ij}|, \quad i = 1, \cdots, n,$$

that is, each eigenvalue lies in at least one of the discs with center a_{ii} and radius r_i in the complex plane. Moreover, if the union of k of the discs is disjoint from the remainder, then there are precisely k eigenvalues of $\mathbf{A}$ in the union of the k discs (van Doorn, 1987b).

Exercise 5.13 Suppose

$$p_{ij}(t) = \pi_j \int_0^\infty e^{-xt} Q_i(x) Q_j(x) d\psi(x).$$

Show that

$$p'_{ij}(t) = -\pi_j \int_0^\infty x e^{-xt} Q_i(x) Q_j(x) d\psi(x),$$

and that the $p_{ij}(t)$ satisfy the forward Kolmogorov equation

$$p'_{ij}(t) = \lambda_{j-1} p_{i,j-1}(t) - (\lambda_j + \mu_j) p_{ij}(t) + \mu_{j+1} p_{i,j+1}(t).$$

Exercise 5.14 Let $\mathbf{Q}_n$ denote the $n \times n$ north-west corner truncation of the generator $\mathbf{Q}$ given in (5.1). Let $\phi_n(x) = \det(x\mathbf{I} + \mathbf{Q}_n)$ be the characteristic polynomial of the matrix $-\mathbf{Q}_n$. By expanding $\phi_n(x)$ by the last row, show that

$$\phi_n(x) = (x - \lambda_{n-1} - \mu_{n-1})\phi_{n-1}(x) - \lambda_{n-2}\,\mu_{n-1}\phi_{n-2}(x).$$

Comparing this with (5.8), prove that $\phi_n(x) = (-1)^n Q_n(x) \prod_{i=0}^{n-1} \lambda_i$.

Exercise 5.15 Prove Lemma 5.11 and Theorem 5.17.

Exercise 5.16 In a birth–death process, we know that the transition probability functions $p_{ij}(t)$ satisfy the Karlin–McGregor representation (5.10). Prove the ratio limit theorem

$$\lim_{t \to \infty} \frac{p_{ij}(t)}{p_{k\ell}(t)} = \frac{\pi_j Q_i(\xi_1) Q_j(\xi_1)}{\pi_\ell Q_k(\xi_1) Q_\ell(\xi_1)},$$

where ξ_1 is the first limit of the zeros of $Q_n(x)$.

Exercise 5.17 In a birth–death process with birth rates $\lambda_n = \lambda(n+1)$ and death rates $\mu_n = \mu(n+1)$, suppose that $\mu > \lambda > 0$. It is known that the first limit is given by $\xi_1 = \mu - \lambda$. Show that the birth–death polynomials satisfy $Q_n(\xi_1) = n+1,\ n = 0, 1, \cdots$. Prove $A = \infty$ and obtain the quasi-limiting distribution (van Doorn, 1991).

Exercise 5.18 Apply the algorithm given in Section 5.7 to calculate the decay parameter of the M/M/3 queue.

Exercise 5.19 Let $p_{ij}(t)$ be the transition probability functions of the M/M/1 queue considered in Section 5.8.

(i) Prove that

$$\begin{aligned} p_{n+k,n}(t) - p_{n+k+1,n+1}(t) \\ = \rho^n p_{2n+k+1,0}(t) - \rho^{n+1} p_{2n+k+2,0}(t), \quad t \geq 0. \end{aligned}$$

(ii) Prove that if $\rho \leq 1$ then $p_{n+k,n}(t) > p_{n+k+1,n+1}(t)$ for all $t > 0$.

(iii) Show that $p_{i0}(t)$ is TP_2 in $i = 0, 1, \cdots$ and $t \geq 0$. Using this fact, prove that $p'_{i0}(t)/p_{i0}(t)$ is nondecreasing in i for each t and that

$$p_{i0}^2(t) - p_{i-1,0}(t)\, p_{i+1,0}(t) \geq \rho\{p_{i+1,0}^2(t) - p_{i0}(t)\, p_{i+2,0}(t)\} \geq 0$$

(Abate, Kijima and Whitt, 1991).

Exercise 5.20 In an M/M/1 queue-size process with arrival rate ρ and service rate 1, show that the Laplace transform of $p_{00}(t)$ is given by

$$\pi_{00}(s) = \frac{1 - \rho - s + \sqrt{(s+1+\rho)^2 - 4\rho}}{2s}, \quad \mathrm{Re}\,(s) > 0.$$

Invert this to obtain $p_{00}(t)$.

Exercise 5.21 In an M/M/1 queue with arrival rate λ and service rate μ, let X_n denote the number of customers just after the nth service completion. Determine the one-step transition matrix of the embedded Markov chain $\{X_n\}$ and, assuming $\mu > \lambda$, obtain its stationary distribution. Compare it with the stationary distribution of the continuous-time queue-size process.

A

Review of matrix theory

A.1 Nonnegative matrices

In this section we discuss nonnegative matrices $\mathbf{A} = (a_{ij})$, i.e., $a_{ij} \geq 0$ for all i and j, in which case we write $\mathbf{A} \geq \mathbf{O}$. If $a_{ij} > 0$ for all i and j, we write $\mathbf{A} > \mathbf{O}$. For two matrices $\mathbf{A}$ and $\mathbf{B}$, we write $\mathbf{A} \geq \mathbf{B}$ if and only if $\mathbf{A} - \mathbf{B} \geq \mathbf{O}$ and $\mathbf{A} > \mathbf{B}$ if and only if $\mathbf{A} - \mathbf{B} > \mathbf{O}$. Throughout this section, we assume that matrices are finite and square.

For a matrix $\mathbf{A}$, suppose that there exists a real or complex number λ such that

$$\lambda \mathbf{u}^\top = \mathbf{u}^\top \mathbf{A}, \quad \lambda \mathbf{v} = \mathbf{A}\mathbf{v}, \tag{A.1}$$

for some vectors $\mathbf{u}$ and $\mathbf{v}$. Then λ is called an *eigenvalue* of $\mathbf{A}$, and $\mathbf{u}$ ($\mathbf{v}$, respectively) is called a *left* (*right*) *eigenvector* associated with λ. In this section, we consider eigenvalues of nonnegative matrices that is largest in magnitude.

Definition A.1 A nonnegative matrix $\mathbf{A}$ is said to be *primitive* if there exists a positive integer k such that $\mathbf{A}^k > \mathbf{O}$.

In the context of discrete-time Markov chains, a primitive transition matrix is often called *regular*. For a primitive matrix, we have the following result, known as the *Perron–Frobenius theorem*. The source of the following results is Seneta (1981). The reader is referred to his excellent book for the proofs.

Theorem A.1 *Let* $\mathbf{A}$ *be a primitive matrix. Then there exists an eigenvalue* r *such that:*

(i) *r is positive;*

(ii) *the associated left and right eigenvectors can be chosen strictly positive componentwise;*

(iii) *the eigenvectors associated with r are unique up to constant multiples;*

(iv) *$r > |\lambda|$ for any other eigenvalue λ of $\mathbf{A}$;*

(v) *if $\mathbf{A} \geq \mathbf{B} \geq \mathbf{O}$ and β is an eigenvalue of $\mathbf{B}$, then $|\beta| \leq r$. Moreover, $|\beta| = r$ implies $\mathbf{B} = \mathbf{A}$;*

(vi) *r is a simple root of the characteristic function $\phi(x) = |x\mathbf{I} - \mathbf{A}|$ of $\mathbf{A}$, where $\mathbf{I}$ denotes the identity matrix.*

Definition A.2 The eigenvalue r in Theorem A.1 is called the *Perron–Frobenius eigenvalue* or, simply, the *PF eigenvalue*. The associated left and right eigenvectors $\mathbf{u}$ and $\mathbf{v}$, respectively, are called the *PF left* and *right eigenvectors* if $\mathbf{u}$ and $\mathbf{v}$ are positive componentwise and $\mathbf{u}^\top \mathbf{1} = \mathbf{u}^\top \mathbf{v} = 1$.

Assertion (iii) of Theorem A.1 states that the *geometric multiplicity* of the PF eigenvalue r is 1, whereas (vi) states that the *algebraic multiplicity* of the PF eigenvalue r is 1. Note that unit geometric multiplicity does not in general imply unit algebraic multiplicity.

To understand the PF eigenvalue r, consider for a nonnegative, nonzero vector $\mathbf{x} = (x_i)$

$$r(\mathbf{x}) \equiv \min_j \frac{\sum_i x_i a_{ij}}{x_j},$$

where $\mathbf{A} = (a_{ij})$ and the ratio is to be interpreted as ∞ if $x_j = 0$. Since

$$r(\mathbf{x})\, x_j \leq \sum_i x_i a_{ij}$$

for all j, it follows that

$$r(\mathbf{x}) \leq \frac{\sum_i x_i \sum_j a_{ij}}{\sum_i x_i} \leq \max_i \sum_j a_{ij}, \tag{A.2}$$

so $r(\mathbf{x})$ is uniformly bounded for all such $\mathbf{x}$. Define

$$r \equiv \sup_{\mathbf{x}\geq 0,\ \neq 0} r(\mathbf{x}) = \sup_{\mathbf{x}\geq 0,\ \mathbf{x}^\top\mathbf{x}=1} \min_j \frac{\sum_i x_i a_{ij}}{x_j}. \tag{A.3}$$

It can be shown that the supremum is actually attained by some $\mathbf{x}$ and that the r defined by (A.3) is the PF eigenvalue.

The next result is immediate from (A.2) and its dual.

Corollary A.1

$$\min_i \sum_j a_{ij} \leq r \leq \max_i \sum_j a_{ij},$$

with equality on either side implying equality throughout. The column version of the above inequalities also holds by taking the transpose.

Suppose that $\mathbf{A}$ is an $n \times n$ primitive matrix and has distinct eigenvalues $r, \lambda_2, \cdots, \lambda_t$, say, where $t \leq n$ and $r > |\lambda_2| \geq |\lambda_3| \geq \cdots \geq |\lambda_t|$. In the case where $|\lambda_2| = |\lambda_3|$, we assume that the algebraic multiplicity m_2 of λ_2 is not less than that of λ_3.

Theorem A.2 *For an $n \times n$ primitive matrix* $\mathbf{A}$, *let r be the PF eigenvalue and let* $\mathbf{u}$ *and* $\mathbf{v}$ *be the PF left and right eigenvectors respectively:*

(i) *If $\lambda_2 \neq 0$, then, as $k \to \infty$,*

$$\mathbf{A}^k = r^k \, \mathbf{v}\mathbf{u}^\top + O\left(k^{m_2-1}|\lambda_2|^k\right)$$

componentwise, where $O(f(k))$ means that $O(f(k))/f(k)$ converges to a matrix as $k \to \infty$.

(ii) *If $\lambda_2 = 0$, then, for $k \geq n-1$, $\mathbf{A}^k = r^k \, \mathbf{v}\mathbf{u}^\top$.*

We now turn to general nonnegative matrices. The irreducibility of a nonnegative (not necessarily stochastic) matrix is defined next.

Definition A.3 A finite, nonnegative matrix $\mathbf{A}$ is called *irreducible* if, for every pair of i and j, there exists some positive integer $k = k(i,j)$ such that $a_{ij}(k) > 0$, where $\mathbf{A}^k = (a_{ij}(k))$. An irreducible matrix is said to be *periodic* with period d if $d \geq 2$ is the greatest common divisor of all integers $k \geq 1$ for which $a_{jj}(k) > 0$ for any one of its indices. If $d = 1$ then an irreducible matrix is called *aperiodic*.

The Perron–Frobenius theorem for irreducible matrices is as follows.

Theorem A.3 *Suppose that* $\mathbf{A}$ *is an irreducible matrix. Then all of the assertions of Theorem A.1 hold, except that* (iv) *is replaced by the weaker statement:*

(iv$'$) *$r \geq |\lambda|$ for any eigenvalue λ of* $\mathbf{A}$.

For a periodic matrix, we have the following.

Theorem A.4 *For a periodic matrix* $\mathbf{A}$ *with period $d > 1$, there are precisely d distinct eigenvalues λ with $r = |\lambda|$. These eigenvalues are given by*

$$r \exp\left\{\frac{2\pi k i}{d}\right\}, \qquad k = 0, 1, \cdots, d-1,$$

where i denotes the imaginary number such that $i^2 = -1$.

From Theorems A.3 and A.4, there is only one real eigenvalue that is largest in magnitude for any irreducible matrix. Hence we call the real eigenvalue the Perron–Frobenius (PF) eigenvalue as in the primitive case.

Let $\mathbf{A}$ be an irreducible matrix with PF eigenvalue r. Sometimes we are interested in the existence of solutions to the system of equations

$$(s\mathbf{I} - \mathbf{A})\mathbf{x} = \mathbf{c} \tag{A.4}$$

for a nonnegative, nonzero vector $\mathbf{c}$. The next theorem answers the existence of such solutions.

Theorem A.5 *A nonnegative, nonzero solution* $\mathbf{x}$ *to* (A.4) *exists for any nonnegative, nonzero vector* $\mathbf{c}$ *if and only if $s > r$. In this case, there is only one solution, which is strictly positive componentwise and given by*

$$\mathbf{x} = (s\mathbf{I} - \mathbf{A})^{-1}\mathbf{c}.$$

In order to guarantee the existence of the inverse in Theorem A.5, we need the following general result.

Lemma A.1 *For a finite matrix* $\mathbf{A}$, *suppose* $\mathbf{A}^n \to \mathbf{O}$ *componentwise as* $n \to \infty$. *Then* $(\mathbf{I}-\mathbf{A})^{-1}$ *exists and*

$$(\mathbf{I}-\mathbf{A})^{-1} = \sum_{n=0}^{\infty} \mathbf{A}^n,$$

convergence being componentwise.

Proof. Note that

$$(\mathbf{I}-\mathbf{A})(\mathbf{I}+\mathbf{A}+\cdots+\mathbf{A}^{n-1}) = \mathbf{I}-\mathbf{A}^n.$$

Since $\mathbf{A}$ is a finite matrix and $\mathbf{A}^n \to \mathbf{O}$ as $n \to \infty$, $\mathbf{I}-\mathbf{A}^n$ is uniformly close to the identity matrix for sufficiently large n, so that it is nonsingular. Taking the determinants yields

$$\det(\mathbf{I}-\mathbf{A})\det(\mathbf{I}+\mathbf{A}+\cdots+\mathbf{A}^{n-1}) = \det(\mathbf{I}-\mathbf{A}^n) \neq 0,$$

whence $\det(\mathbf{I}-\mathbf{A}) \neq 0$ so that the inverse exists. Since $\mathbf{A}^n \to \mathbf{O}$ as $n \to \infty$, we have the desired result. □

For $s > r$, we have from Theorem A.2

$$\frac{1}{s^k}\mathbf{A}^k = \left(\frac{r}{s}\right)^k \mathbf{v}\mathbf{u}^\top + o(1),$$

where $o(1)$ means that $o(1) \to \mathbf{O}$ as $k \to \infty$. Hence the inverse $(\mathbf{I}-\mathbf{A}/s)^{-1}$ exists, which now confirms Theorem A.5.

A.2 ML-matrices

Recall that an infinitesimal generator of a continuous-time Markov chain has nonnegative off-diagonal elements and negative diagonal elements. In matrix theory, such a matrix is often called an *ML-matrix*, taking the names of mathematical economists Metzler and Leontief. A finite ML-matrix $\mathbf{B}$ is related to a nonnegative matrix $\mathbf{A}$ through the relation

$$\mathbf{A} = \mu\mathbf{I}+\mathbf{B},$$

where μ is sufficiently large to make $\mathbf{A}$ nonnegative.

Definition A.4 A finite ML-matrix $\mathbf{B}$ is called *irreducible* if an associated nonnegative matrix $\mathbf{A}$ is irreducible.

Note that, by taking μ sufficiently large, if $\mathbf{B}$ is irreducible then the corresponding $\mathbf{A}$ can be made aperiodic and thus primitive. The next result can be obtained from the Perron–Frobenius theorem for primitive matrices. As before, the eigenvalue r in Theorem A.6 is called the PF eigenvalue. The PF eigenvectors are defined in the same way.

Theorem A.6 *Let* $\mathbf{B}$ *be an irreducible ML-matrix. Then there exists an eigenvalue* r *such that:*

(i) r *is real;*

(ii) *the associated left and right eigenvectors can be chosen strictly positive componentwise,*

(iii) *the eigenvectors associated with* r *are unique up to constant multiples;*

(iv) $r > \mathrm{Re}(\lambda)$ *for any other eigenvalue* λ *of* $\mathbf{B}$*;*

(v) r *is a simple root of the characteristic function* $\phi(x) = |x\mathbf{I} - \mathbf{B}|$ *of* $\mathbf{B}$*;*

(vi) $r \leq 0$ *if and only if there exists a nonnegative, nonzero vector* $\mathbf{y}$ *such that* $\mathbf{By} \leq 0$*, in which case* $\mathbf{y} > 0$*;*

(vii) $r < 0$ *if and only if* $-\mathbf{B}^{-1} > \mathbf{O}$.

The proof of Theorem A.6 begins by writing $\mathbf{B} = \mathbf{A} - \mu\mathbf{I}$ for sufficiently large μ. If λ_i is an eigenvalue of $\mathbf{B}$ then $\mathbf{A}$ has the corresponding eigenvalue $\delta_i = \mu + \lambda_i$. To see assertion (iv) of Theorem A.6, for example, let $\lambda_j = x_j + iy_j$. Suppose $\lambda_j \neq r$ and $x_j \geq r$. If $x_j > r$ then, since $\delta_j = \mu + x_j + iy_j$, we have $|\delta_j| > \mu + x_j > \mu + r$, which is impossible. Similar arguments lead to a contradiction for the case where $x_j = r$ and $y_j \neq 0$. Hence assertion (iv) follows. Other parts of the theorem can be proved similarly using the results of Theorem A.1.

An ML-matrix $\mathbf{B}$ often appears in the form

$$\exp\{\mathbf{B}t\} \equiv \sum_{n=0}^{\infty} \frac{\mathbf{B}^n t^n}{n!}.$$

For the absolute convergence of the above infinite series, we need the following.

Lemma A.2 *For any* $n \times n$ *matrix* $\mathbf{B} = (b_{ij})$*, the series*

$$\sum_{k=0}^{\infty} \frac{\mathbf{B}^k t^k}{k!}$$

converges absolutely componentwise.

Proof. Let $\delta = \max_{i,j} |b_{ij}|$ and define $\mathbf{B}^k = (b_{ij}(k))$. We have $|b_{ij}(2)| \leq n\delta^2$ and, in general,

$$|b_{ij}(k)| \leq n^{k-1}\delta^k, \quad i,\ j = 1, \cdots, n,$$

and componentwise absolute convergence follows from the fact that

$$\sum_{k=1}^{\infty} t^k \frac{n^{k-1}\delta^k}{k!}$$

converges for any $t,\ \delta,\ n > 0$. □

Our final result is the asymptotic behavior of $\exp\{\mathbf{B}t\}$ for large $t > 0$.

Theorem A.7 *An ML-matrix* $\mathbf{B}$ *is irreducible if and only if* $\exp\{\mathbf{B}t\} > \mathbf{O}$ *for all* $t > 0$. *In this case, we have*

$$\exp\{\mathbf{B}t\} = e^{rt}\,\mathbf{v}\mathbf{u}^\top + O(e^{\tau t})$$

as $t \to \infty$, *where* r *is the PF eigenvalue,* $\mathbf{u}$ *and* $\mathbf{v}$ *are the associated left and right PF eigenvectors, respectively, and* $\tau < r$.

A.3 Infinite matrices

For infinite matrices, we need to take care in matrix multiplication. For two infinite matrices $\mathbf{A} = (a_{ij})$ and $\mathbf{B} = (b_{ij})$ whose indices run over $\mathcal{Z}_+$, the matrix product $\mathbf{C} = (c_{ij}) = \mathbf{AB}$ is *well defined* if

$$|c_{ij}| = \left|\sum_{k=0}^{\infty} a_{ik} b_{kj}\right| < \infty$$

for all $i,\ j \in \mathcal{Z}_+$. Let

$$\mathbf{U} = \begin{pmatrix} 1 & 0 & 0 & \cdots \\ 1 & 1 & 0 & \cdots \\ 1 & 1 & 1 & \cdots \\ \vdots & \vdots & \vdots & \ddots \end{pmatrix}$$

and

$$\mathbf{U}^{-1} = \begin{pmatrix} 1 & 0 & 0 & \cdots \\ -1 & 1 & 0 & \cdots \\ 0 & -1 & 1 & \cdots \\ \vdots & \vdots & \ddots & \ddots \end{pmatrix},$$

as defined in (3.5) and (3.6) respectively. It is easily seen that the products $\mathbf{U}\,\mathbf{U}^{-1}$ and $\mathbf{U}^{-1}\mathbf{U}$ are well defined and

$$\mathbf{U}\,\mathbf{U}^{-1} = \mathbf{U}^{-1}\mathbf{U} = \mathbf{I},$$

i.e. the identity matrix of infinite order. It is important to note that

$$1 = \mathbf{1}^\top\,(\mathbf{U}^{-1}\mathbf{1}) \neq (\mathbf{1}^\top\mathbf{U}^{-1})\,\mathbf{1} = 0,$$

although all the products involved are well defined. That is, the product of infinite matrices may *not* be associative.

The next results concern the associativity of matrix products. The first result may be found in Kemeny, Snell and Knapp (Section 1.1, 1976), while the second is due to van Doorn (1980).

Theorem A.8 (i) *The product of nonnegative matrices is associative.*

(ii) *The product of matrices is associative if the product of the absolute values is finite.*

Theorem A.9 *Let* $\mathbf{A} = (a_{ij})$, $\mathbf{B} = (b_{ij})$ *and* $\mathbf{C} = (c_{ij})$ *be infinite matrices defined on* $\mathcal{Z}_+$, *and suppose that* $\mathbf{AB}$ *and* $\mathbf{BC}$ *are well defined.*

(i) $\mathbf{A}(\mathbf{BC}) = (\mathbf{AB})\mathbf{C}$ *if and only if* $\mathbf{A}(\mathbf{BC})$ *is well defined and*

$$\lim_{m\to\infty} \sum_{k=0}^{\infty} a_{ik} \sum_{n=m}^{\infty} b_{kn} c_{nj} = 0$$

for all $i,\ j \in \mathcal{Z}_+$.

(ii) $\mathbf{A}(\mathbf{BC}) = (\mathbf{AB})\mathbf{C}$ *if and only if* $(\mathbf{AB})\mathbf{C}$ *is well defined and*

$$\lim_{m\to\infty} \sum_{n=0}^{\infty} c_{nj} \sum_{k=m}^{\infty} a_{ik} b_{kn} = 0$$

for all $i,\ j \in \mathcal{Z}_+$.

We note that, in general, if A, B and C are linear operators and if their range spaces and domains are suitably compatible, then associativity $(AB)C = A(BC)$ holds. The above theorems can be deduced from this general property.

B

Generating functions and Laplace transforms

All the probabilistic properties of a random variable X are contained in its distribution function $F(x)$. When working with the random variable X, therefore, we must start with the distribution function $F(x)$. However, it is often true that working with some transformation of $F(x)$ is much easier than working with $F(x)$ itself. In this appendix, we discuss two important transformations, generating functions and Laplace transforms, the former being particularly useful for discrete random variables and the latter for nonnegative, absolutely continuous random variables.

B.1 Generating functions

For a sequence of real numbers $\{a_n\}$, define

$$g(z) = \sum_{n=0}^{\infty} a_n z^n \tag{B.1}$$

for all complex z for which the series converges. The sum $\sum_{n=0}^{\infty} a_n z^n$ is called a *power series*. We begin with the following important theorem, called the Cauchy–Hadamard theorem. For the proof, see any standard textbook on analysis, e.g., Bartle (page 320, 1976).

Theorem B.1 *There exists a number ρ, $0 < \rho \leq \infty$, such that the power series $g(z)$ converges absolutely for $|z| < \rho$ and diverges for $|z| > \rho$. The convergence is uniform on every bounded, closed subset of the circle $|z| < \rho$.*

The quantity ρ in Theorem B.1 is called the *radius of convergence* and is given by

$$\rho = \frac{1}{\limsup_{n\to\infty} |a_n|^{1/n}}. \tag{B.2}$$

The *interval of convergence* is the open interval $(-\rho, \rho)$ of the real line R.

The following results are immediate from Theorem B.1.

Corollary B.1 (i) *A power series can be integrated term-by-term over any bounded, closed interval contained in* $(-\rho, \rho)$.

(ii) *A power series can be differentiated term-by-term within the interval of convergence, and*

$$g'(z) = \sum_{n=1}^{\infty} n a_n z^{n-1}$$

whose radius of convergence is equal to ρ.

By repeated application of Corollary B.1(ii), we conclude that if k is any positive integer, then the power series $\sum_{n=0}^{\infty} a_n z^n$ can be differentiated term-by-term k times and so we have

$$g^{(k)}(z) = \sum_{n=k}^{\infty} \frac{n!}{(n-k)!} a_n z^{n-k}, \tag{B.3}$$

which converges absolutely for $|z| < \rho$ and uniformly over any bounded, closed subset of the circle $|z| < \rho$. Here $g^{(k)}$ denotes the kth derivative of the power series g. Substituting $z = 0$ into (B.3), we have

$$g^{(k)}(0) = k!\, a_k.$$

Therefore we have the following.

Corollary B.2 *If both the power series* $\sum_{n=0}^{\infty} a_n z^n$ *and* $\sum_{n=0}^{\infty} b_n z^n$ *converge on some interval* $|z| < \rho$ *to the same function, then* $a_n = b_n$ *for all* $n = 0, 1, \cdots$. *That is, the power series* $g(z)$ *characterizes the sequence* $\{a_n\}$.

Note that the preceding theorem does not tell us anything about the convergence of the power series for $|z| = \rho$. This is indeed a difficult problem. The next theorem, called *Abel's theorem*, provides a partial result on this problem. In what follows, we assume without loss of generality that $\rho = 1$ by making the change of variable $z_1 = z/\rho$.

Theorem B.2 *Suppose that the power series* $g(z) = \sum_{n=0}^{\infty} a_n z^n$ *converges for* $|z| < 1$ *and that* $\sum_{n=0}^{\infty} a_n$ *converges. Then*

$$\lim_{z \to 1-} g(z) = \sum_{n=0}^{\infty} a_n.$$

where $\lim_{z\to 1-}$ *means that* z *approaches* 1 *from values less than* 1.

Abel's theorem states that if the power series converges at $z = 1$, the sum must equal the limit of $g(z)$ as $z \to 1-$. That is, the interchange of the infinite sum and the limit is permissible.

The converse of Abel's theorem is not true without further restrictions on the coefficients a_n. The limit $\lim_{z\to 1-} g(z)$ may exist, yet the series

$\sum_{n=0}^{\infty} a_n$ may diverge (see, e.g., Neuts, page 410, 1973). The next result, called the *Tauberian theorem*, is a partial converse to Abel's theorem. Recall that $f(x) = o(g(x))$ if

$$\lim_{x \to \infty} \frac{f(x)}{g(x)} = 0.$$

Theorem B.3 *Suppose that* $a_n = o(1/n)$. *If* $\lim_{z \to 1-} g(z)$ *exists, then* $\sum_{n=0}^{\infty} a_n$ *converges and*

$$\lim_{z \to 1-} g(z) = \sum_{n=0}^{\infty} a_n.$$

In probability theory, the sequence $\{a_n\}$ is often a probability distribution of a discrete random variable. The term 'generating function' is used for the power series in this context. Here is a formal definition of generating functions.

Definition B.1 Let X be a nonnegative, discrete random variable with probability distribution

$$p_n = P[X = n], \quad n = 0, 1, \cdots.$$

The *generating function* of X is given by

$$g_X(z) = E\left[z^X\right] = \sum_{n=0}^{\infty} p_n z^n$$

for all complex z for which the series converges.

From the preceding results, the generating function exists at least in the unit circle $|z| \leq 1$ since, for $|z| \leq 1$,

$$\sum_{n=0}^{\infty} p_n |z^n| \leq \sum_{n=0}^{\infty} p_n = 1.$$

Inside the unit circle, the generating function is absolutely and uniformly convergent and one can manipulate generating functions quite freely there. For example, from Corollary B.1(ii), we have

$$g'_X(z) = \sum_{n=1}^{\infty} n p_n z^{n-1}, \quad |z| < 1.$$

When $z = 1$, the right-hand side of the above equation is $E[X] = \sum_{n=1}^{\infty} n p_n$. Hence if $E[X] < \infty$ then, by Abel's theorem,

$$g'_X(1) = E[X].$$

Differentiating once more and applying (B.3) yields

$$V[X] = g''_X(1) + g'_X(1) - \{g'_X(1)\}^2,$$

where $V[X]$ denotes the variance of X, provided that $E[X^2] < \infty$.

The generating function $g_X(z)$ characterizes the probability distribution $\{p_n\}$ uniquely. That is, if two random variables X and Y taking values on $\mathcal{Z}_+$ have the same generating function $g_X(z) = g_Y(z)$ in a neighborhood of the origin, i.e., in $|z| < \varepsilon$ for some $\varepsilon > 0$, then they have the same distribution

$$P[X = n] = P[Y = n], \quad n = 0, 1, \cdots.$$

Finally, we provide some results concerning generating functions that are useful in stochastic modeling. For two nonnegative, discrete random variables X and Y with probability distributions $\{p_n\}$ and $\{q_n\}$, respectively, if X and Y are independent of each other then the distribution of $Z = X + Y$ is given by

$$r_n = \sum_{k=0}^{n} p_k q_{n-k}, \quad n = 0, 1, \cdots. \tag{B.4}$$

The operation in (B.4) is called the *discrete convolution* of $\{p_n\}$ and $\{q_n\}$.

Theorem B.4 *Suppose that nonnegative, discrete random variables X and Y are mutually independent. Then the generating function of $X+Y$ is given by*

$$g_{X+Y}(z) = g_X(z)\, g_Y(z)$$

for all complex z for which the generating functions exist.

Proof. By definition, we have

$$g_{X+Y}(z) = E\left[z^{X+Y}\right] = E\left[z^X\right] E\left[z^Y\right] = g_X(z)\, g_Y(z)$$

since X and Y are independent. In fact, since the distribution $\{r_n\}$ of $X+Y$ is given by (B.4), we have

$$\begin{aligned} g_{X+Y}(z) &= \sum_{n=0}^{\infty} z^n \sum_{k=0}^{n} p_k q_{n-k} \\ &= \sum_{k=0}^{\infty} p_k z^k \sum_{n=k}^{\infty} q_{n-k} z^{n-k} \\ &= g_X(z)\, g_Y(z), \end{aligned}$$

where we apply Theorem B.1 to ensure the interchange of the above two summations. □

Suppose that the generating function $g_X(z)$ converges beyond the unit circle, $|z| < \rho_X$ with $\rho_X > 1$, say. Suppose also that $g_Y(z)$ converges for $|z| < \rho_Y$ with $\rho_Y > 1$. Note that Theorem B.4 asserts that the radius of convergence of $g_{X+Y}(z)$ is *at least* $\min\{\rho_X, \rho_Y\}$. It can be shown, however, that if X and Y are nonnegative then, in fact, $\rho_{X+Y} = \min\{\rho_X, \rho_Y\}$.

Another useful property of generating functions is the following.

Theorem B.5 *Suppose that the limit* $\lim_{n\to\infty} a_n$ *exists. Then the generating function* $g(z)$ *of* a_n *exists for* $|z| < 1$ *and*

$$\lim_{z\to 1-} (1-z)\,g(z) = \lim_{n\to\infty} a_n .$$

B.2 Laplace transforms

Let X be a nonnegative random variable with distribution function F_X. For complex s, define

$$\phi_X(s) = \int_0^\infty e^{-sx}\, dF_X(x) \tag{B.5}$$

for which the integral exists. The function ϕ_X is often called the *Laplace–Stieltjes transform* of F_X. If F_X has the density function $f_X(x) = F'_X(x)$, then the function $\phi_X(s)$ is given by

$$\phi_X(s) = \int_0^\infty e^{-sx} f_X(x)\,dx. \tag{B.6}$$

In this case, the function ϕ_X is called the *Laplace transform* of f_X. In either case, we have

$$\phi_X(s) = E\left[e^{-sX}\right]$$

for all complex s for which the expectation exists, and we call ϕ_X the Laplace transform of X. Laplace transforms form a powerful tool in probability theory. In this section, we summarize their elementary properties and give some examples of them. The following results are taken from Feller (1971) and Widder (1946).

Theorem B.6 *The region of convergence of* (B.5) *or* (B.6) *is a half-plane.*

From Theorem B.6, there are three possibilities:

(a) the integral converges nowhere;

(b) the integral converges everywhere;

(c) the integral converges for s such that $\mathrm{Re}\,(s) > \sigma_c$ and diverges for $\mathrm{Re}\,(s) < \sigma_c$ for some σ_c, where $\mathrm{Re}\,(s)$ denotes the real part of s.

In case (c), there are again three possibilities concerning the existence of ϕ_X. That is, on the line $\mathrm{Re}\,(s) = \sigma_c$:

(c1) the integral converges nowhere;

(c2) the integral converges everywhere;

(c3) the integral converges for some values of s and diverges otherwise.

The real number σ_c is called the *abscissa of convergence* of ϕ_X. When F_X is a distribution function, it is clear that $\sigma_c \le 0$. If $\sigma_c < 0$, we have

$$1 - F_X(x) = o(e^{\sigma x}) \quad \text{or} \quad f_X(x) = o(e^{\sigma x})$$

as $x \to \infty$, where $\sigma > \sigma_c$.

The next result is similar to Corollary B.1(ii) in the preceding section.

Theorem B.7 *If the Laplace transform $\phi_X(s)$ exists for $\mathrm{Re}(s) > \sigma_c$, then $\phi_X(s)$ is analytic for $\mathrm{Re}(s) > \sigma_c$ and*

$$\phi_X^{(k)}(s) = \int_0^\infty e^{-sx}(-x)^k dF_X(x), \quad \mathrm{Re}(s) > \sigma_c.$$

The real point $s = \sigma_c$ is a singularity of $\phi_X(s)$.

It follows from Theorem B.7 that F_X possesses a finite kth moment if and only if a finite limit $\lim_{s\to 0+} \phi_X^{(k)}(s)$ exists, where $\lim_{s\to 0+}$ means that s approaches 0 from values larger than 0. In particular, we have

$$E[X] = -\phi'_X(0), \quad V[X] = \phi''_X(0) - \{\phi'_X(0)\}^2,$$

provided that the limit exists for $k = 2$.

As for the case of generating functions, the Laplace transform characterizes the distribution function uniquely.

Theorem B.8 *If $F_X \neq F_Y$ then $\phi_X \neq \phi_Y$. That is, distinct distribution functions have distinct Laplace transforms.*

In fact, the distribution function F_X is uniquely determined by the values of its Laplace transform $\phi_X(s)$ in some interval in $\sigma_c < s < \infty$.

The following basic result, called the *continuity theorem*, is a consequence of Theorem B.8. Before stating it, we need to provide a formal definition of weak convergence.

Definition B.2 Let $\{F_n\}$ be a sequence of distribution functions and let F be a (possibly defective) distribution function. We say that F_n *converges weakly* to F if

$$\lim_{n\to\infty} F_n(x) = F(x)$$

for every point of continuity of $F(x)$. In this case, we write $F_n \xrightarrow{w} F$.

Theorem B.9 *For each $n = 1, 2, \cdots$, let F_n be a distribution function with the Laplace transform ϕ_n. Let F be a (possibly defective) distribution function with the Laplace transform ϕ:*

(i) *If $F_n \xrightarrow{w} F$ then $\phi_n(s) \to \phi(s)$ for all s such that $\mathrm{Re}(s) > 0$.*

(ii) *If the sequence $\{\phi_n(s)\}$ converges for each s such that $\mathrm{Re}(s) > 0$ to a limit $\phi(s)$, then $\phi(s)$ is the Laplace transform of F and $F_n \xrightarrow{w} F$.*

(iii) *The limit F is not defective if and only if $\lim_{s\to 0+} \phi(s) = 1$.*

At this point, we provide some examples which appear quite often in this book. Other Laplace transform results may be found in, e.g., Abramowitz and Stegun (Chapter 29, 1965).

Example B.1 An exponential distribution with parameter $\lambda > 0$ is given by

$$F(x) = 1 - e^{-\lambda x}, \quad x \geq 0,$$

and the density function is $f(x) = \lambda e^{-\lambda x}$. Hence, the Laplace transform is given by

$$\phi(s) = \int_0^\infty e^{-sx} \lambda e^{-\lambda x} dx = \frac{\lambda}{s + \lambda},$$

which converges for $\mathrm{Re}(s) > -\lambda$. Hence, the abscissa of convergence is $\sigma_c = -\lambda$. Note that $s = -\lambda$ is a singularity of $\phi(s)$.

Example B.2 Let the density function f be of the form

$$f(x) = \sum_{k=1}^{n} p_k \lambda_k e^{-\lambda_k x}; \quad p_k > 0, \quad \sum_{k=1}^{n} p_k = 1, \tag{B.7}$$

where $\lambda_k > 0$, viz. f is a mixture of exponential densities. The corresponding Laplace transform is given by

$$\phi(s) = \sum_{k=1}^{n} p_k \frac{\lambda_k}{s + \lambda_k} \equiv \frac{Q(s)}{P(s)}, \tag{B.8}$$

where $P(s)$ is a polynomial of degree n with roots $-\lambda_k$, and $Q(s)$ is a polynomial of degree $n-1$. The abscissa of convergence is $\sigma_c = -\min_i \lambda_i$.

Conversely, for any polynomial $Q(s)$ of degree $n-1$, it is readily shown that the ratio $Q(s)/P(s)$ admits a partial-fraction expansion of the form (B.8) with

$$\lambda_k p_k = \frac{Q(-\lambda_k)}{P'(-\lambda_k)},$$

where $P'(s)$ is the derivative of $P(s)$. Suppose that $0 < \lambda_1 < \cdots < \lambda_n$. In order for (B.8) to correspond to a mixture (B.7) it is necessary and sufficient that $p_k > 0$ for all $k = 1, \cdots, n$ and that $Q(0)/P(0) = 1$. It is clear that $P'(-\lambda_k)$ and $P'(-\lambda_{k+1})$ are of opposite sign. Hence, for p_k to be positive, it is necessary that $Q(-\lambda_k)$ and $Q(-\lambda_{k+1})$ are of opposite signs. In other words, it is necessary that $Q(s)$ has a root $-\mu_k$ between $-\lambda_k$ and $-\lambda_{k+1}$. But, as $Q(s)$ cannot have more than $n-1$ roots, these must satisfy the *interlacing property*

$$0 < \lambda_1 < \mu_1 < \lambda_2 < \mu_2 < \cdots < \mu_{n-1} < \lambda_n,$$

which guarantees that all the p_k are of the same sign. It follows that, if $Q(0)/P(0) = 1$, then all the p_k are positive, and so $Q(s)/P(s)$ is the Laplace transform of a mixture of exponential densities (B.7).

Example B.3 For $n = 1, 2, \cdots$, let

$$f_n(x) = e^{-x} \frac{n}{x} I_n(x), \quad x > 0, \tag{B.9}$$

where $I_n(x)$ is the modified Bessel function of order n, i.e.,

$$I_n(x) = \sum_{k=0}^{\infty} \frac{(x/2)^{2k+n}}{k!\,\Gamma(k+n+1)}, \quad n > 0.$$

Here $\Gamma(\alpha)$, the gamma function, is defined by

$$\Gamma(\alpha) = \int_0^{\infty} x^{\alpha-1} e^{-x} dx, \quad \alpha > 0,$$

which satisfies the relation $\Gamma(\alpha+1) = \alpha\Gamma(\alpha)$ with $\Gamma(1) = 1$. The Laplace transform of $f_n(x)$ is given by

$$\phi_n(s) = \left(s + 1 - \sqrt{(s+1)^2 - 1}\right)^n, \quad \mathrm{Re}\,(s) > 0. \tag{B.10}$$

Since $\lim_{s\to 0+} \phi_n(s) = 1$, the function $f_n(x)$ is a probability density function, called a *Bessel distribution*. Note that $s = 0$ is a singularity of $\phi_n(s)$ since the complex function $\sqrt{z}$ is singular at $z = 0$. Hence the abscissa of convergence of $\phi_n(s)$ is $\sigma_c = 0$.

In what follows, we assume that the integral in (B.5) or (B.6) converges absolutely for $\mathrm{Re}\,(s) > \sigma_a$ for some $\sigma_a < \infty$. Then, it converges uniformly and absolutely in the half-plane. In general, $\sigma_a \geq \sigma_c$ and, indeed, σ_c and σ_a need not be coincident. When $\sigma_a < \infty$, we have the following *Laplace inversion formula.*

Theorem B.10 *Suppose* $\sigma_a < \infty$ *and* f_X *in* (B.6) *is continuous in* $(0, \infty)$. *Then, for* $x > \sigma_a$,

$$f_X(t) = \frac{1}{2\pi i} \int_{x-i\infty}^{x+i\infty} e^{st} \phi_X(s) ds, \quad t > 0.$$

Recall that the generating function of a sum of independent random variables is given by a product of generating functions of the random variables. The same is true for the Laplace transforms. Let X and Y be mutually independent, nonnegative random variables with distribution functions F_X and F_Y respectively. The distribution function of $X + Y$ is given by

$$F_{X+Y}(x) = \int_0^x F_X(x-y) dF_Y(y) \equiv F_X * F_Y(x), \quad x \geq 0. \tag{B.11}$$

The operation $F_X * F_Y$ is called the *convolution* of F_X and F_Y. When the density functions f_X and f_Y exist, the convolution reduces

$$f_{X+Y}(x) = \int_0^x f_X(x-y) f_Y(y) dy \equiv f_X * f_Y(x), \quad x \geq 0. \tag{B.12}$$

It is not difficult to show that the Laplace–Stieltjes transform of $F_X * F_Y$ or, equivalently, the Laplace transform of $f_X * f_Y$, if the densities exist, is given by $\phi_X(s)\,\phi_Y(s)$, $\mathrm{Re}\,(s) > 0$. In the next theorem, we summarize some properties of Laplace transforms that make these transforms a useful tool in probability theory. We write

$$\mathcal{L}[f(t)] = \int_0^\infty e^{-st} f(t)dt,$$

assuming that the integral exists.

Theorem B.11 *Let f_X and f_Y be bounded and of finite variation, and assume that the corresponding Laplace transforms ϕ_X and ϕ_Y, respectively, exist. Let a and b be real numbers:*

(i) $\mathcal{L}[af_X(t) + bf_Y(t)] = a\,\phi_X(s) + b\,\phi_Y(s)$.

(ii) $\mathcal{L}[f_X(at)] = \dfrac{1}{a}\,\phi_X(s/a)$ *for* $a > 0$.

(iii) $\mathcal{L}[f_X(t-a)] = e^{-as}\phi_X(s)$ *for* $a \geq 0$.

(iv) *Suppose that $f_X(t)$ is differentiable in $t > 0$ and that the Laplace transform of $f'_X(t)$ exists. Then* $\mathcal{L}[f'_X(t)] = s\,\phi_X(s) - f_X(0+)$.

(v) *Let F_X be an indefinite integral of f_X. Then the Laplace transform of F_X exists and* $\mathcal{L}[F_X(t)] = (\phi_X(s) + F_X(0+))/s$.

(vi) $\mathcal{L}[tf_X(t)] = -\phi'_X(s)$.

(vii) $\mathcal{L}[f_X(t)/t] = \int_s^\infty \phi_X(u)du$.

(viii) $\mathcal{L}[f_X * f_Y(t)] = \phi_X(s)\,\phi_Y(s)$.

If $\phi_X(s)$ exists for some real $s = s_0$, then the Laplace transform of the integral $\int_0^t f_X(y)dy$ exists for $s = s_0$ and

$$\lim_{t\to\infty} e^{-s_0 t} \int_0^t f_X(y)dy = 0.$$

Also, from (B.5), integration by parts leads to

$$\int_0^\infty e^{-sx}\{1 - F_X(x)\}dx = \frac{1-\phi_X(s)}{s}, \quad \mathrm{Re}\,(s) > 0. \tag{B.13}$$

When $F_X(t) = 0$ for $t < 0$, the transform (B.13) is more useful than the result given in Theorem B.11(v). When X is exponentially distributed with parameter λ, we have

$$\int_0^\infty e^{-sx}\{1 - F(x)\}dx = \frac{1}{s+\lambda},$$

which converges for $\mathrm{Re}\,(s) > -\lambda$.

Other useful properties of Laplace transforms are the following.

Theorem B.12 (i) *Suppose that the limit $\lim_{t\to 0+} f(t)$ exists and that $f(t)$ has the Laplace transform $\phi(s)$. Then*

$$\lim_{s\to\infty} s\,\phi(s) = \lim_{t\to 0+} f(t).$$

(ii) *Suppose that the limit* $\lim_{t\to\infty} f(t)$ *exists. Then the Laplace transform* $\phi(s)$ *of* $f(t)$ *exists and*

$$\lim_{s\to 0+} s\,\phi(s) = \lim_{t\to\infty} f(t).$$

Finally, we provide a characterization of Laplace transforms.

Definition B.3 A function ϕ on $(0,\infty)$ is said to be *completely monotone* if it possesses derivatives $\phi^{(n)}$ of all orders such that

$$(-1)^n \phi^{(n)}(t) \geq 0, \quad t > 0.$$

Theorem B.13 *A function* ϕ *on* $(0,\infty)$ *is completely monotone if and only if it is of the form*

$$\phi(t) = \int_0^\infty e^{-tx}\,dF(x), \quad t > 0,$$

for some nondecreasing function $F(x)$ *on* $[0,\infty)$.

When $F(x)$ is a step function, the function ϕ in Theorem B.13 can be written as in (B.7). Hence every mixture of exponential distributions is itself completely monotone, and so is its Laplace transform.

Based on the characterization given in Theorem B.13, the next useful result, called the *inversion formula*, can be proved.

Theorem B.14 *Let* ϕ *be the Laplace–Stieltjes transform of* $F(x)$. *Then, at all points of continuity of* F, *we have*

$$F(x) = \lim_{s\to\infty} \sum_{n\leq sx} \frac{(-s)^n}{n!}\phi^{(n)}(s).$$

C

Total positivity

In this appendix, we provide some information about total positivity. The theory of totally positive functions is very rich and the results provided here are indeed only 'the tip of the iceberg'. The reader interested in a complete discussion of the theory of total positivity should consult Karlin (1968). Throughout this appendix, X, Y and Z represent either intervals of the real line $R \equiv (-\infty, \infty)$ or a countable or finite set of discrete values along R.

C.1 TP$_r$ functions

For a real-valued function $K(x, y)$ of two variables ranging over X and Y, respectively, we write

$$K\begin{pmatrix} x_1, & x_2, \cdots, & x_m \\ y_1, & y_2, \cdots, & y_m \end{pmatrix}$$
$$\equiv \det \begin{pmatrix} K(x_1, y_1) & K(x_1, y_2) & \cdots & K(x_1, y_m) \\ K(x_2, y_1) & K(x_2, y_2) & \cdots & K(x_2, y_m) \\ \vdots & \vdots & \ddots & \vdots \\ K(x_m, y_1) & K(x_m, y_2) & \cdots & K(x_m, y_m) \end{pmatrix}$$

for $x_1 < x_2 < \cdots < x_m$; $y_1 < y_2 < \cdots < y_m$ and $x_i \in X$, $y_j \in Y$.

Definition C.1 A real-valued function $K(x, y)$ on $X \times Y$ is said to be *totally positive of order r*, or simply TP$_r$, denoted by $K \in \text{TP}_r$, if, for all $m = 1, 2, \cdots, r$, all $x_1 < x_2 < \cdots < x_m$, $x_i \in X$ and $y_1 < y_2 < \cdots < y_m$, $y_j \in Y$,

$$K\begin{pmatrix} x_1, & x_2, \cdots, & x_m \\ y_1, & y_2, \cdots, & y_m \end{pmatrix} \geq 0.$$

If strict inequality holds, then we say that $K(x, y)$ is *strictly totally positive of order r*, or simply STP$_r$, which we denote by $K \in \text{STP}_r$. In the case where $K \in \text{TP}_r$ ($K \in \text{STP}_r$, respectively) for all $r = 2, 3, \cdots$, $K(x, y)$ is said to be *totally positive* (*strictly totally positive*), or simply TP (STP), which we denote by $K \in \mathrm{TP}$ ($K \in \mathrm{STP}$).

Example C.1 The function $K(x,y) = e^{xy}$ where $X = Y = R$ is STP, while the function

$$K(x,y) = \begin{cases} 1, & x \le y, \\ 0, & x > y, \end{cases}$$

where $X = Y = [a,b]$ is TP.

Let K and L be measurable* functions of two variables and consider

$$M(x,y) = \int_Z K(x,z)L(z,y)d\sigma(z), \quad x \in X, \; y \in Y, \qquad \text{(C.1)}$$

where the integral is assumed to converge absolutely. Here σ is either a Lebesgue measure, in which case $d\sigma(z) = dz$ when Z is an interval of R, or a counting measure when Z consists of discrete values. In particular, when X, Y and Z consist of discrete values, (C.1) reduces to the usual matrix multiplication

$$\mathbf{M} = \mathbf{KL}, \qquad \text{(C.2)}$$

where $\mathbf{M} = (M(x,y))$ and so on.

The next result, called the *basic composition formula*, is an extension of the well-known Cauchy–Binet formula and plays a central role.

Theorem C.1

$$M\begin{pmatrix} x_1, & x_2, \cdots, & x_m \\ y_1, & y_2, \cdots, & y_m \end{pmatrix}$$

$$= \int \cdots \int_{z_1 < \cdots < z_m} K\begin{pmatrix} x_1, & x_2, \cdots, & x_m \\ z_1, & z_2, \cdots, & z_m \end{pmatrix} L\begin{pmatrix} z_1, & z_2, \cdots, & z_m \\ y_1, & y_2, \cdots, & y_m \end{pmatrix}$$

$$\times d\sigma(z_1) \cdots d\sigma(z_m).$$

An obvious consequence of the basic composition formula is the following.

Corollary C.1 *If* $K \in \mathrm{TP}_r$ *and* $L \in \mathrm{TP}_s$ *then* $M \in \mathrm{TP}_{\min\{r,s\}}$.

Any tridiagonal matrix with nonnegative off-diagonal elements has a surprising property, which we state in the next theorem. This property plays an important role in the study of birth–death processes. The proof is taken from Karlin and McGregor (1957a).

Theorem C.2 (i) *If* $\mathbf{A}$ *is a finite, tridiagonal matrix with positive off-diagonal elements, then* $\mathbf{P}(t) = \exp\{\mathbf{A}t\}$ *is* STP *for any* $t > 0$. *If the off-diagonal elements are nonnegative, then* $\mathbf{P}(t)$ *is* TP *for* $t \ge 0$.

(ii) *Conversely, if* $\mathbf{P}(t)$ *is* TP_2 *for all* $t > 0$, *then* $\mathbf{A}$ *must be a tridiagonal matrix with nonnegative off-diagonal elements.*

* Measurability is a purely technical requirement, designed to ensure that the integrals involved are well defined.

Proof. (i) Note that a tridiagonal matrix is TP if and only if all its elements and principal minors are nonnegative. Hence, given $t > 0$, the tridiagonal matrix

$$\mathbf{C} \equiv \mathbf{I} + \frac{t}{n}\mathbf{A}$$

is TP for sufficiently large n. The basic composition formula shows that a product of TP matrices is TP. Hence $\mathbf{C}^n$ is TP for all sufficiently large n and, therefore,

$$\exp\{\mathbf{A}t\} = \lim_{n\to\infty}\left(\mathbf{I} + \frac{t}{n}\mathbf{A}\right)^n \tag{C.3}$$

is TP. A formal proof of (C.3) will be found in Anderson (page 86, 1991). To prove STP, considerably more effort is required and the proof is omitted. The reader is referred to Karlin (1968).

(ii) Suppose $\mathbf{P}(t) \in \mathrm{TP}_2$ for all $t > 0$. For $\mathbf{A} = (a_{ij})$, recall that

$$a_{ij} = \lim_{t\to 0+} \frac{p_{ij}(t) - \delta_{ij}}{t}.$$

Since $p_{ij}(t) \geq 0$, we see that $a_{ij} \geq 0$ for $i \neq j$. For $j > i+1$, we have

$$\begin{aligned} 0 &\leq \det\begin{pmatrix} p_{i,i+1}(t) & p_{ij}(t) \\ p_{i+1,i+1}(t) & p_{i+1,j}(t) \end{pmatrix} \\ &= \det\begin{pmatrix} a_{i,i+1}t + o(t) & a_{ij}t + o(t) \\ 1 + a_{i+1,i+1}t + o(t) & a_{i+1,j}t + o(t) \end{pmatrix} \\ &= -a_{ij}t + o(t) \end{aligned}$$

as $t \to 0+$. Hence $a_{ij} = 0$ for $j > i+1$. The proof of the case where $j < i-1$ is similar. Therefore $\mathbf{A}$ must be tridiagonal. □

When $K(x,y) = f(x-y)$, we have the following important special case.

Definition C.2 A function $f(x)$ on X is said to be a *Pólya frequency of order r*, or simply PF_r, denoted by $f \in \mathrm{PF}_r$, if $K(x,y) = f(x-y)$ is TP_r on $X \times X$. If $f \in \mathrm{PF}_r$ holds for all $r = 2, 3, \cdots$, then $f(x)$ is said to be a *Pólya frequency function*, or simply PF, denoted by $f \in \mathrm{PF}$.

Example C.2 The function $f(x) = e^{-x^2}$ on R is PF since

$$e^{-(x-y)^2} = e^{-x^2}e^{-y^2}e^{2xy}$$

and e^{xy} on $R \times R$ is TP. Hence every normal density function is a Pólya frequency function. It is readily seen that if $f \in \mathrm{PF}_r$ then the function $g(x) = f(ax+b)$ is also PF_r, where a and b are constants.

Note that in this context (C.1) becomes

$$h(x-y) = \int_Z f(x-z)g(z-y)d\sigma(z). \tag{C.4}$$

If σ is a Lebesgue measure on $Z = R$, then

$$h(x) = \int_{-\infty}^{\infty} f(x-z)g(z)dz, \quad -\infty < x < \infty, \tag{C.5}$$

which we denote by $h = f * g$, the convolution of f and g. The basic composition formula applied to (C.4) yields the following.

Corollary C.2 *If* $f \in \mathrm{PF}_r$ *and* $g \in \mathrm{PF}_s$ *then* $f * g \in \mathrm{PF}_{\min\{r,s\}}$.

The class of PF_2 functions is particularly important and has applications to various fields. A key property that every PF_2 function possesses is the characterization that it has the form $f(x) = e^{-\phi(x)}$, where $\phi(x)$ is a convex function.

Theorem C.3 *Every positive* PF_2 *function on* R *is log-concave.*

Proof. If $f \in \mathrm{PF}_2$ then

$$f(x_1 - y_1)f(x_2 - y_2) \geq f(x_1 - y_2)f(x_2 - y_1)$$

for all $x_1 < x_2$ and $y_1 < y_2$. Hence, taking $y_2 = 0$ and $-y_1 = h > 0$, it follows that

$$\log f(x_1 + h) - \log f(x_1) \geq \log f(x_2 + h) - \log f(x_2), \quad x_1 < x_2,$$

for all $h > 0$, viz. $\log f(x)$ is concave. □

C.2 The variation-diminishing property

We next turn our attention to the *variation-diminishing property* (abbreviated to VDP) of totally positive functions. For a sequence $(x_1, \cdots, x_m)$, let $S(x_1, \cdots, x_m)$ count the number of sign changes of the sequence indicated, zero terms being discarded. For a function $f(t)$ defined on an ordered set I of R, we define

$$S(f) = \sup S(f(t_1), \cdots, f(t_m)), \tag{C.6}$$

where the supremum is taken over all sets $t_1 < \cdots < t_m$, $t_i \in I$, and m is arbitrary but finite. Figure C.1 shows some examples.

For a bounded, measurable function f and a measurable function K of two variables, consider the transformation

$$g(x) = (Kf)(x) \equiv \int_Y K(x,y)f(y)d\sigma(y), \quad x \in X, \tag{C.7}$$

assuming that the integral exists. When X and Y consist of discrete values, (C.7) reduces to the matrix transformation

$$\mathbf{g} = \mathbf{K}\mathbf{f}, \tag{C.8}$$

where $\mathbf{g} = (g(x))$, etc. The next theorem is the VDP in Karlin (Chapter 5, 1968) for details.

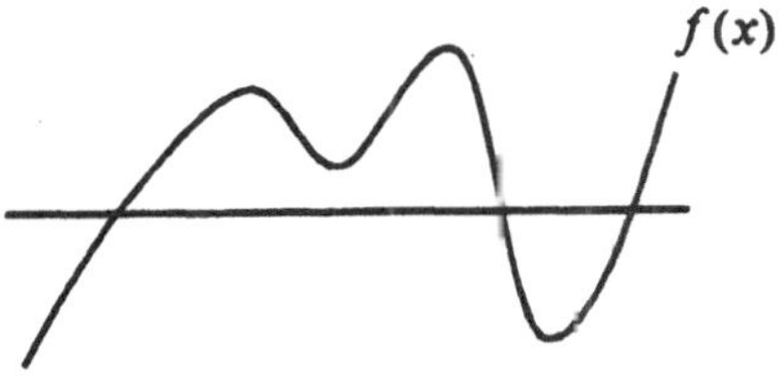

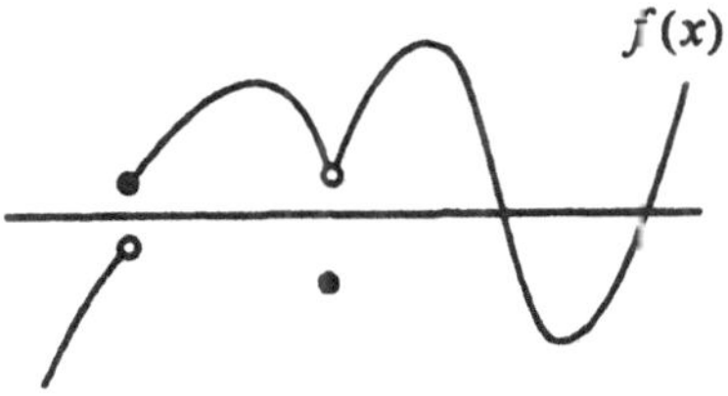

Figure C.1 *Examples of $S(f)$.*

Theorem C.4 *If $K \in \mathrm{TP}_r$, then $S(g) = S(Kf) \leq S(f)$ provided $S(f) \leq r-1$. If $f(x)$ is piecewise continuous and if $S(f) = S(g) = r-1$, then the values of the functions f and g exhibit the same sequence of signs when their respective arguments traverse the domain of definition from left to right.*

We provide two examples of applications of Theorem C.4. In the following examples, we assume that

$$\int_Y K(x,y)d\sigma(y) = 1, \quad x \in X, \tag{C.9}$$

or, in the matrix case, $\mathbf{K1} = \mathbf{1}$, i.e., $\mathbf{K}$ is stochastic.

Example C.3 Suppose $f(x)$ is monotonically increasing in x. Let α be any real number and consider the transformation

$$g(x) - \alpha = \int_Y K(x,y)\{f(y) - \alpha\}d\sigma(y), \quad x \in X.$$

For any α, $f(y) - \alpha$ changes sign at most once. Hence, if K is TP_2, then Theorem C.4 asserts that $g(x) - \alpha$ possesses the same property as $f(y) - \alpha$. Since α is arbitrary, this means that $g(x)$ is also monotonically increasing in x.

Example C.4 Suppose in addition to (C.9) that, for $a > 0$,

$$\int_Y y\,K(x,y)d\sigma(y) = ax + b.$$

Suppose that $f(x)$ is convex in x. Let α_1 and α_2 be any real numbers and consider the transformation

$$g(x) - \alpha_1 x - \alpha_2 = \int_Y K(x,y) \left\{ f(y) - \frac{\alpha_1 y}{a} - \alpha_2 + \frac{\alpha_1 b}{a} \right\} d\sigma(y),$$

where $x \in X$. Since $f(x)$ is convex,

$$f(y) - \frac{\alpha_1 y}{a} - \left(\alpha_2 - \frac{\alpha_1 b}{a} \right)$$

changes sign at most twice. Hence, if K is TP_3, then Theorem C.4 implies that every line crosses g at most twice and, if it does so twice, the two sign changes occur in the same order as f. This means that $g(x)$ is convex in x. If $f(x)$ is concave in x and $K \in \mathrm{TP}_3$ then $g(x)$ is concave, too.

Finally, we consider the convolution transformation (C.5), i.e.,

$$h(x) = \int_{-\infty}^{\infty} f(x-z) g(z) dz, \quad -\infty < x < \infty.$$

Suppose that $f(x)$ is a probability density function. The next result is a partial converse to Theorem C.4.

Theorem C.5 *Suppose $S(h) \leq S(g)$ for every bounded, measurable function $g(x)$ for which $S(g) \leq r-1$. Then $f \in \mathrm{PF}_r$.*

References

Abate, J., Kijima, M. and Whitt, W. (1991) Decompositions of the M/M/1 transition functions. *Queueing Systems*, **9**, 323–336.

Abate, J. and Whitt, W. (1988) Transient behavior of the M/M/1 queue via the Laplace transform. *Advances in Applied Probability*, **20**, 145–178.

Abate, J. and Whitt, W. (1989) Calculating time-dependent performance measures for the M/M/1 queue. *IEEE Transactions on Communications*, **37**, 1102–1104.

Abramowitz, M. and Stegun, I.A. (1965) *Handbook of Mathematical Functions*, Dover, New York.

Aho, A.V., Hopcroft, J.E. and Ullman, J.D. (1974) *The Design and Analysis of Computer Algorithms*, Addison-Wesley, Reading, Massachusetts.

Aldous, D.J. (1983) Random walk on finite groups and rapidly mixing Markov chains. *Séminaire de Probabilités XVII, Lecture Notes in Mathematics*, **986**, Springer, New York, 243–297.

Aldous, D.J. (1988) Finite-time implications of relaxation times for stochastically monotone processes. *Probability Theory and Related Fields*, **77**, 137–145.

Aldous, D.J. and Diaconis, P. (1987) Strong uniform times and finite random walks. *Advances in Applied Mathematics*, **8**, 69–97.

Aldous, D.J. and Thorisson, H. (1993) Shift-coupling. *Stochastic Processes and Their Applications*, **44**, 1–4.

Anderson, W.J. (1991) *Continuous-time Markov Chains: An applications-oriented approach*, Springer, New York.

Asmussen, S. (1987) *Applied Probability and Queues*, Wiley, New York.

Athreya, K.B. and Ney, P.E. (1972) *Branching Processes*, Springer, New York.

Barlow, R.E. and Proschan, F. (1975) *Statistical Theory of Reliability and Life Testing*, Holt, Rinehart and Winston, New York.

Bartle, R.G. (1976) *The Elements of Real Analysis*, Second Edition, Wiley, New York.

Boland, P.J., Proschan, F. and Tong, Y.L. (1992) A stochastic ordering of partial sums of independent random variables and of some random processes. *Journal of Applied Probability*, **29**, 645–654.

Box, G.E.P. and Jenkins, G.M. (1976) *Time Series Analysis: Forecasting and control*, Revised Edition, Prentice Hall, New Jersey.

Brown, M. (1980) Bounds, inequalities, and monotonicity properties for some specialized renewal processes. *Annals of Probability*, **8**, 227–240.

Brown, M. (1981) Further monotonicity properties for some specialized renewal processes. *Annals of Probability*, **9**, 891–895.

Brown, M. and Chaganty, N.R. (1983) On the first passage time distribution for a class of Markov chains. *Annals of Probability*, **11**, 1000–1008.

Callaert, H. and Keilson, J. (1973a) On exponential ergodicity and spectral structure for birth–death processes I. *Stochastic Processes and Their Applications*, **1**, 187–216.

Callaert, H. and Keilson, J. (1973b) On exponential ergodicity and spectral structure for birth–death processes II. *Stochastic Processes and Their Applications*, **1**, 217–235.

Cavender, J.A. (1978) Quasi-stationary distributions of birth–death processes. *Advances in Applied Probability*, **10**, 570–586.

Chihara, T.S. (1978) *An Introduction to Orthogonal Polynomials*, Gordon & Breach, New York.

Chung, K.L. (1967) *Markov Chains with Stationary Transition Probabilities*, Second Edition, Springer, New York.

Çinlar, E. (1969) Markov renewal theory. *Advances in Applied Probability*, **1**, 123–187.

Çinlar, E. (1975) *Introduction to Stochastic Processes*, Prentice-Hall, New Jersey.

Dambrine, S. and Moreau, M. (1981) Note on the stochastic theory of a self-catalytic chemical reaction, I. *Physica*, **106A**, 559–573.

Darroch, J.N. and Seneta, E. (1965) On quasi-stationary distributions in absorbing discrete-time finite Markov chains. *Journal of Applied Probability*, **2**, 88–100.

Darroch, J.N. and Seneta, E. (1967) On quasi-stationary distributions in absorbing continuous-time finite Markov chains. *Journal of Applied Probability*, **4**, 192–196.

Diaconis, P. (1988) *Group Representation in Probability and Statistics*, IMS Lecture Notes Series, **11**, IMS, Hayward.

Diaconis, P. and Fill, J.A. (1990) Strong stationary times via a new form of duality. *Annals of Probability*, **18**, 1483–1522.

Diaconis, P. and Stroock, D. (1991) Geometric bounds for eigenvalues of Markov chains. *Annals of Applied Probability*, **1**, 36–61.

Esary, J.D., Marshall, A.W. and Proschan, F. (1973) Shock models and wear processes. *Annals of Probability*, **1**, 627–649.

Feller, W. (1957) *An Introduction to Probability Theory and Its Applications, Volume I*, Second Edition, Wiley, New York.

Feller, W. (1959) The birth and death processes as diffusion processes. *Journal de Mathematiques Pures et Appliquees*, **38**, 301–345.

Feller, W. (1971) *An Introduction to Probability Theory and Its Applications, Volume II*, Second Edition, Wiley, New York.

Ferrari, P.A., Kesten, H., Martínez, S. and Picco, P. (1995) Existence of quasi-stationary distributions: A renewal dynamical approach. *Annals of Probability*, **23**, 501–521.

Fill, J.A. (1991) Eigenvalue bounds on convergence to stationarity for nonreversible Markov chains, with an application to the exclusion process. *Annals of Applied Probability*, 1, 62–87.

Foss, S.G. (1986) The method of renovating events and its applications in queueing theory. *Proceedings of the International Symposium on Semi-Markov Processes and Applications*, 337–350, Plenum Press.

Freedman, D. (1971) *Markov Chains*, Holden-Day, San Francisco.

Freedman, D. (1972) *Approximating Countable Markov Chains*, Holden-Day, San Francisco.

Friedland, S. and Gurvits, L. (1994) An upper bound for the real part of nonmaximal eigenvalues of nonnegative irreducible matrices. *SIAM Journal on Matrix Analysis and Applications*, **15**, 1015–1017.

Gibson, D. and Seneta, E. (1987a) Monotone infinite stochastic matrices and their augmented truncations. *Stochastic Processes and Their Applications*, **24**, 287–292.

Gibson, D. and Seneta, E. (1987b) Augmented truncations of infinite stochastic matrices. *Journal of Applied Probability*, **24**, 600–608.

Golub, G.H. and Van Loan, C.F. (1989) *Matrix Computations*, Second Edition, Johns Hopkins University Press, Baltimore.

Good, P. (1968) The limiting behavior of transient birth and death processes conditioned on survival. *Journal of Australian Mathematical Society*, **8**, 716–722.

Heyman, D.P. (1991) Approximating the stationary distribution of an infinite stochastic matrix. *Journal of Applied Probability*, **28**, 96–103.

Hillier, F.S. and Lieberman, G.J. (1990) *Introduction to Stochastic Models in Operations Research*, McGraw-Hill, New York.

Hirayama, T. and Kijima, M. (1992) Single machine scheduling problem when the machine capacity varies stochastically. *Operations Research*, **40**, 376–383.

Iosifescu, M. (1980) *Finite Markov Processes and Their Applications*, Wiley, New York.

Ito, H., Amari, S. and Kobayashi, K. (1992) Identifiability of hidden Markov information sources and their minimum degrees of freedom. *IEEE Transactions on Information Theory*, **38**, 324–333.

Kamae, T., Krengel, U. and O'Brien, G. (1977) Stochastic inequalities on partially ordered spaces. *Annals of Probability*, **5**, 899–912.

Karlin, S. (1964) Total positivity, absorption probabilities and applications. *Transactions of the American Mathematical Society*, **111**, 33–107.

Karlin, S. (1968) *Total Positivity*, Stanford University Press, Stanford, California.

Karlin, S. and McGregor, J.L. (1957a). The differential equations of birth-and-death processes, and the Stieltjes moment problem. *Transactions of the American Mathematical Society*, **85**, 489–546.

Karlin, S. and McGregor, J.L. (1957b) The classification of birth and death processes. *Transactions of the American Mathematical Society*, **86**, 366–400.

Karlin, S. and McGregor, J.L. (1958) Many server queueing processes with Poisson input and exponential service times. *Pacific Journal of Mathematics*, **8**, 87–118.

Karlin, S. and McGregor, J.L. (1959) Characterizations of birth and death processes. *Proceedings of the National Academy of Sciences of the United States of America*, **45**, 375–379.

Karlin, S. and Proschan, F. (1960) Pólya type distributions of convolutions. *Annals of Mathematical Statistics*, **31**, 721–736.

Karlin, S. and Taylor, H.M. (1975) *A First Course in Stochastic Processes*, Second Edition, Academic Press, New York.

Karlin, S. and Taylor, H.M. (1981) *A Second Course in Stochastic Processes*, Academic Press, New York.

Karr, A.F. (1978) Markov chains and processes with a prescribed invariant measure. *Stochastic Processes and Their Applications*, **7**, 277–290.

Keilson, J. (1969) On the matrix renewal function for Markov renewal processes. *Annals of Mathematical Statistics*, **40**, 1901–1907.

Keilson, J. (1971) Log-concavity and log-convexity of passage time densities of diffusion and birth–death processes. *Journal of Applied Probability*, **8**, 391–398.

Keilson, J. (1979) *Markov Chain Models – Rarity and Exponentiality*, Springer, New York.

Keilson, J. (1981) On the unimodality of passage time densities in birth–death processes. *Statistica Neerlandica*, **25**, 49–55.

Keilson, J. and Gerber, H. (1971) Some results for discrete unimodality. *Journal of American Statistical Association*, **66**, 386–389.

Keilson, J. and Kester, A. (1977) Monotone matrices and monotone Markov processes. *Stochastic Processes and Their Applications*, **5**, 231–241.

Keilson, J. and Kester, A. (1978) Unimodality preservation in Markov chains. *Stochastic Processes and Their Applications*, **7**, 179–190.

Keilson, J. and Sumita, U. (1982) Uniform stochastic ordering and related inequalities. *Canadian Journal of Statistics*, **10**, 181–198.

Keilson, J. and Wishart, D.M.G. (1964) A central limit theorem for processes defined on a finite Markov chain. *Proceedings of the Cambridge Philosophical Society*, **60**, 547–567.

Kelly, F.P. (1979) *Reversibility and Stochastic Networks*, Wiley, New York.

Kemeny, J.G. and Snell, J.L. (1960) *Finite Markov Chains*, Van Nostrand Reinhold Company, New York.

Kemeny, J.G., Snell, J.L. and Knapp, A.W. (1976) *Denumerable Markov Chains*, Springer, New York.

Kendall, D.G. (1966) Contribution to discussion in 'Quasi-stationary distributions and time-reversion in genetics' by E. Seneta [with discussion]. *Journal of the Royal Statistical Society, Series B*, **28**, 253–277.

Kesten, H. (1995) A ratio limit theorem for (sub)Markov chains on $\{1, 2, \cdots\}$ with bounded jumps. *Advances in Applied Probability*, **27**, 652–691.

Kijima, M. (1987a) Some results for uniformizable semi-Markov processes. *Australian Journal of Statistics*, **29**, 193–207.

Kijima, M. (1987b) Spectral structure of the first-passage-time densities for classes of Markov chains. *Journal of Applied Probability*, **24**, 631–643.

Kijima, M. (1988) On passage and conditional passage times for Markov chains in continuous time. *Journal of Applied Probability*, **25**, 279–290.

Kijima, M. (1989a) Upper bounds of a measure of dependence and the relaxation time for finite state Markov chains. *Journal of the Operations Research Society of Japan*, **32**, 93–102.

Kijima, M. (1989b) Uniform monotonicity of Markov processes and its related properties. *Journal of the Operations Research Society of Japan*, **32**, 475–490.

Kijima, M. (1989c) Some results for repairable systems with general repair. *Journal of Applied Probability*, **26**, 89–102.

Kijima, M. (1990a) On the unimodality of transition probabilities in Markov chains. *Australian Journal of Statistics*, **32**, 1–10.

Kijima, M. (1990b) On the largest negative eigenvalue of the infinitesimal generator associated with M/M/n/n queues. *Operations Research Letters*, **9**, 59–64.

Kijima, M. (1992a) A note on external uniformization for finite Markov chains in continuous time. *Probability in the Engineering and Informational Sciences*, **6**, 127–131.

Kijima, M. (1992b) On the transient solution to a class of Markovian queues. *Computers and Mathematics with Applications*, **24**, 17–24.

Kijima, M. (1992c) Further monotonicity properties of renewal processes. *Advances in Applied Probability*, **24**, 575–588.

Kijima, M. (1992d) Evaluation of the decay parameter for some specialized birth–death processes. *Journal of Applied Probability*, **29**, 781–791.

Kijima, M. (1993a) Numerical calculation of ruin probabilities for skip-free Markov chains. *SIAM Review*, **35**, 621–624.

Kijima, M. (1993b) Quasi-stationary distributions of single server phase-type queues. *Mathematics of Operations Research*, **19**, 423–437.

Kijima, M. (1993c) Quasi-limiting distributions of Markov chains that are skip-free to the left in continuous-time. *Journal of Applied Probability*, **30**, 509–517.

Kijima, M. (1994) On separation for birth–death processes. *Probability in the Engineering and Informational Sciences*, **8**, 51–68.

Kijima, M. (1995) Bounds for the quasi-stationary distributions of some specialized Markov chains. *Mathematical and Computer Modelling*, **22**, 141–147.

Kijima, M. (1996) Hazard rate and reversed hazard rate monotonicities in continuous-time Markov chains. Preprint.

Kijima, M. and Makimoto, N. (1992) Computation of the quasi-stationary distributions in M(n)/GI/1/K and GI/M(n)/1/K queues. *Queueing Systems*, **11**, 255–272.

Kijima, M., Morimura, H. and Suzuki, Y. (1988) Periodical replacement problem without assuming minimal repair. *European Journal of Operational Research*, **37**, 194–203.

Kijima, M., Nair, M.G., Pollett, P.K. and van Doorn, E.A. (1997) Limiting conditional distributions for birth–death processes. *Advances in Applied Probability*, **29**, to appear.

Kijima, M. and Ohnishi, M. (1996) Portfolio selection problems via the bivariate characterization of stochastic dominance relations. *Mathematical Finance*, **6**, 237–277.

Kijima, M. and Seneta, E. (1991) Some results for quasi-stationary distributions of birth–death processes. *Journal of Applied Probability*, **28**, 503–511.

Kijima, M. and Sumita, U. (1986a) A useful generalization of renewal theory: Counting processes governed by nonnegative Markovian increments. *Journal of Applied Probability*, **23**, 71–88.

Kijima, M. and Sumita, U. (1986b) On time reversibility of stationary semi-

Markov processes. Working Paper Series QM8628, Graduate School of Management, University of Rochester.

Kijima, M. and van Doorn, E.A. (1995) Weighted sums of orthogonal polynomials with positive zeros. *Journal of Computational and Applied Mathematics*, **65**, 195–206.

Kingman, J.F.C. (1972) *Regenerative Phenomena*, Wiley, London.

Kleinrock, L. (1975) *Queueing Systems, Volume I: Theory*, Wiley, New York.

Kurtz, T.G. (1981) The central limit theorem for Markov chains. *Annals of Probability*, **9**, 557–560.

Lacey, M.T. and Philipp, W. (1990) A note on the almost sure central limit theorem. *Statistics and Probability Letters*, **9**, 201–205.

Ledermann, W. and Reuter, G.E.H. (1954) Spectral theory for the differential equations of simple birth and death processes. *Philosophical Transactions of the Royal Society of London, Series A*, **246**, 321–369.

Li, H. and Shaked, M. (1995) Aging first-passage times. *Encyclopedia of Statistical Sciences*, Wiley, forthcoming.

Lindvall, T. (1992) *Lectures on the Coupling Method*, Wiley, New York.

Liu, H. (1994) On Cowan and Mecke's Markov chain. *Journal of Applied Probability*, **31**, 554–560.

Lund, R. and Tweedie, R.L. (1996) Geometric convergence rates for stochastically ordered Markov chains. *Mathematics of Operations Research*, **21**, 182–194.

Marshall, A.W. and Shaked, M. (1983) New better than used processes. *Advances in Applied Probability*, **15**, 601–615.

Massy, W.F., Montgomery, D.B. and Morrison, D.G. (1970) *Stochastic Models of Buying Behavior*, The MIT Press, Massachusetts.

Masuda, Y. (1988) First passage times of birth–death processes and simple random walks. *Stochastic Processes and Their Applications*, **29**, 51–63.

Meyer, C.D. (1994) Sensitivity of the stationary distribution of a Markov chain. *SIAM Journal on Matrix Analysis and Applications*, **15**, 715–728.

Meyn, S.P. and Tweedie, R.L. (1993) *Markov Chains and Stochastic Stability*, Springer, London.

Nair, M.G. and Pollett, P.K. (1993) On the relationship between μ-invariant measures and quasistationary distributions for continuous-time Markov chains. *Advances in Applied Probability*, **25**, 82–102.

Neuts, M.F. (1973) *Probability*, Allyn and Bacon, Boston.

Neuts, M.F. (1981) *Matrix-Geometric Solutions in Stochastic Models — An Algorithmic Approach*, Johns Hopkins University Press, Baltimore.

Noble, B. and Daniel, J.W. (1977) *Applied Linear Algebra*, Second Edition, Prentice-Hall, New Jersey.

Nollau, V. (1980) *Semi-Markovsche Prozesse*, Akademie-Verlin, Verlag.

Norman, F.M. (1974) Markovian learning processes. *SIAM Review*, **16**, 143–162.

Nummelin, E. (1984) *General Irreducible Markov Chains and Non-negative Operators*, Cambridge University Press, Cambridge.

Orey, S. (1971) *Limit Theorems for Markov Chain Transition Probabilities*, Van Nostrand Reinhold, London.

Parsons, R.W. and Pollett, P.K. (1987) Quasistationary distributions for autocatalytic reactions. *Journal of Statistical Physics*, **46**, 249–254.

Parthasarathy, P.R. and Sharafali, M. (1989) Transient solution to the many-server Poisson queue: A simple approach. *Journal of Applied Probability*, **26**, 584–594.

Paz, A. (1971) *Introduction to Probabilistic Automata*, Academic Press, New York.

Pollett, P.K. (1988) Reversibility, invariance and μ-invariance. *Advances in Applied Probability*, **20**, 600–621.

Pollett, P.K. (1989) The generalized Kolmogorov criterion. *Stochastic Processes and Their Applications*, **33**, 29–44.

Prabhu, N. (1965) *Queues and Inventories*, Wiley, New York.

Puterman, M.L. (1994) *Markov Decision Processes: Discrete Stochastic Dynamic Programming*, Wiley, New York.

Resnick, S.I. (1992) *Adventures in Stochastic Processes*, Birkhäuser, Boston.

Revuz, D. (1984) *Markov Chains*, Second Edition, North-Holland, Amsterdam.

Rosenthal, J.S. (1995) Convergence rates for Markov chains. *SIAM Review*, **37**, 387–405.

Ross, S.M. (1983) *Stochastic Processes*, Wiley, New York.

Ross, S.M. (1987) Approximating transition probabilities and mean occupation times in continuous-time Markov chains. *Probability in the Engineering and Informational Sciences*, **1**, 251–264.

Ross, S.M. (1989) *Introduction to Probability Models*, 4th Edition, Academic Press, New York.

Seneta, E. (1966) Quasi-stationary behaviour in the random walk with continuous time. *Australian Journal of Statistics*, **8**, 92–98.

Seneta, E. (1981) *Non-Negative Matrices and Markov Chains*, Second Edition, Springer, New York.

Seneta, E. (1991) Applications of ergodicity coefficients to homogeneous Markov chains. *Proceedings of the Doeblin Conference*, Edited by H. Cohn, AMS Series: Contemporary Mathematics.

Seneta, E. (1993) Sensitivity of finite Markov chains under perturbation. *Statistics and Probability Letters*, **17**, 163–168.

Seneta, E. and Vere-Jones, D. (1966) On quasi-stationary distributions in discrete-time Markov chains with a denumerable infinity of states. *Journal of Applied Probability*, **3**, 403–434.

Shaked, M. and Shanthikumar, J.G. (1987) IFRA properties of some Markov processes with general state space. *Mathematics of Operations Research*, **12**, 562–568.

Shaked, M. and Shanthikumar, J.G. (1994) *Stochastic Orders and Their Applications*, Academic Press, San Diego, California.

Shanthikumar, J.G. (1985) Bilateral phase-type distributions. *Naval Research Logistic Quarterly*, **32**, 119–136.

Shanthikumar, J.G. (1988) DFR property of first-passage times and its preservation under geometric compounding. *Annals of Probability*, **16**, 397–406.

Shanthikumar, J.G., Yamazaki, G. and Sakasegawa, H. (1991) Characterization of optimal order of services in a tandem queue with blocking. *Operations Research Letters*, **10**, 17–22.

Shanthikumar, J.G. and Yao, D.D. (1991) Bivariate characterization of some stochastic order relations. *Advances in Applied Probability*, **23**, 642–659.

Spieksma, F.M. (1993) Spectral conditions and bounds for the rate of convergence of countable Markov chains. Technical Report, University of Leiden.

Stewart, W.J. (1994) *Introduction to the Numerical Solution of Markov Chains*, Princeton University Press, Princeton.

Stieltjes, T.J. (1894) Recherches sur les fraction continues. *Oeuvres Complètes, Volume II*, 406–566, Springer, Berlin.

Stoyan, D. (1983) *Comparison Methods for Queues and Other Stochastic Models*, Wiley, Chichester.

Sumita, U. (1981) Development of the Laguerre transform method for numerical exploration of applied probability models. Ph.D. Thesis, University of Rochester.

Sumita, U. and Masuda, Y. (1985) On first passage time structure of random walks. *Stochastic Processes and Their Applications*, **20**, 133–147.

Sumita, U. and Shanthikumar, J.G. (1986) A software reliability model with multiple-error introduction and removal. *IEEE Transactions on Reliability*, **35**, 459–62.

Szegö, S. (1959) *Orthogonal Polynomials*, American Mathematical Society Colloquium Publications, **23**, Revised Edition, AMS, New York.

Szekli, R. (1995) *Stochastic Ordering and Dependence in Applied Probability*, Lecture Notes in Statistics, **97**, Springer, New York.

Thorisson, H. (1981) The coupling of regenerative processes. Ph.D. Thesis, University of Göteborg.

Tweedie, R.L. (1974a) R-theory for Markov chains on a general state space I: Solidarity properties and R-recurrent chains. *Annals of Probability*, **2**, 840–864.

Tweedie, R.L. (1974b) R-theory for Markov chains on a general state space II: R-subinvariant measures for R-transient chains. *Annals of Probability*, **2**, 865–878.

van Doorn, E.A. (1980) Stochastic monotonicity of birth–death processes. *Advances in Applied Probability*, **12**, 59–80.

van Doorn, E.A. (1981) *Stochastic Monotonicity and Queueing Applications of Birth–Death Processes*, Springer, Berlin.

van Doorn, E.A. (1985) Conditions for exponential ergodicity and bounds for the decay parameter of a birth–death process. *Advances in Applied Probability*, **17**, 514–530.

van Doorn, E.A. (1986) On orthogonal polynomials with positive zeros and the associated kernel polynomials. *Journal of Mathematical Analysis and Applications*, **113**, 441–450.

van Doorn, E.A. (1987a) The indeterminate rate problem for birth–death processes. *Pacific Journal of Mathematics*, **130**, 379–393.

van Doorn, E.A. (1987b) Representations and bounds for zeros of orthogonal polynomials and eigenvalues of sign-symmetric tri-diagonal matrices. *Journal of Approximation Theory*, **51**, 254–266.

van Doorn, E.A. (1991) Quasi-stationary distributions and convergence to quasi-stationarity of birth–death processes. *Advances in Applied Probability*, **23**, 683–700.

Vere-Jones, D. (1962). Geometric ergodicity in denumerable Markov chains. *Quarterly Journal of Mathematics, Oxford*, **13**, 7–28.

Widder, D.V. (1946) *The Laplace Transform*, Princeton University Press, Princeton.

Williams, D. (1991) *Probability with Martingales*, Cambridge University Press, Cambridge.

Wolff, R.W. (1989) *Stochastic Modeling and the Theory of Queues*, Prentice-Hall, New Jersey.

Yoon, B.S. and Shanthikumar, J.G. (1989) Bounds and approximations for the transient behavior of continuous-time Markov chains. *Probability in the Engineering and Informational Sciences*, **3**, 175–198.

Zeifman, A.I. (1991) Some estimates of the rate of convergence for birth and death processes. *Journal of Applied Probability*, **28**, 268–277.

Ziedins, I. (1987) Quasi-stationary distributions and one-dimensional circuit-switched networks. *Journal of Applied Probability*, **24**, 965–977.

Symbols

$R = (-\infty, \infty)$	real line
$R_+ = [0, \infty)$	nonnegative real line
$\mathcal{Z}_+ = \{0, 1, 2, \cdots\}$	
$\mathcal{Z} = \{\cdots, -2, -1, 0, 1, 2 \cdots\}$	
$\mathcal{N}$	state space
$\mathcal{T}$	index set representing time
$\mathcal{S}$	class of stochastic matrices
$\mathcal{P}$	class of probability vectors
$P[A]$	probability of event A
$P_\alpha[A]$	probability of event A given $\boldsymbol{\alpha}$
E	expectation with respect to P
E_α	expectation with respect to P_α
$\{X_n\}$	discrete-time process
$\{X(t)\}$	continuous-time process
$\stackrel{\mathrm{d}}{=}$	equality in distribution
IID	independent and identically distributed
VDP	variation diminishing property
TP_r	totally positive of order r
$\mathbf{A}^{-1}$	inverse of matrix $\mathbf{A}$
$\mathbf{A}^{\mathsf{T}}$	transpose of matrix $\mathbf{A}$
$\det(\mathbf{A})$	determinant of matrix $\mathbf{A}$
$\exp\{\mathbf{A}\}$	matrix exponential
$\mathbf{P}_{\mathrm{R}}$	dual of transition matrix $\mathbf{P}$
$\boldsymbol{\pi}_{\mathrm{D}}$	diagonal matrix with diagonal elements $\boldsymbol{\pi} = (\pi_i)$
$\boldsymbol{\pi}_{\mathrm{D}}^{1/2}$	diagonal matrix with diagonal elements $\pi_i^{1/2}$
$\boldsymbol{\pi}_{\mathrm{D}}^{-1/2}$	diagonal matrix with diagonal elements $\pi_i^{-1/2}$
$\mathbf{I}$	identity matrix
$\mathbf{O}$	zero matrix
$\mathbf{1}$	vector of 1s

$\mathbf{0}$	zero vector
$\boldsymbol{\delta}_i$	ith unit vector
$[\mathbf{A}]_{ij}$	(i,j)th element of matrix $\mathbf{A}$
I_A	indicator function of event A
δ_{ij}	Kronecker's delta
$\delta(t)$	Dirac's delta function
$U(x)$	unit step function
$\pi(s)$	Laplace transform of $p(t)$
$f^*(z)$	generating function of $f(n)$

$\{x\}_+ = \max\{0, x\}$

$[x]$	integer part of positive x
$\lVert\mathbf{x}\rVert_\alpha$	ℓ_α-norm of vector $\mathbf{x}$

$A_i = \sum_{k\le i} a_k$ for vector $\mathbf{a} = (a_i)$

$\overline{A}_i = \sum_{k\ge i} a_k$ for vector $\mathbf{a} = (a_i)$

Author index

Abate, J. 265, 289–90, 293
Abramowitz, M. 288, 308
Aho, A.V. 64
Aldous, D.J. 69, 73–4, 96
Amari, S. 12
Anderson, W.J. 175, 183–5, 188, 194, 226, 315
Asmussen, S. 9
Athreya, K.B. 10

Barlow, R.E. 112, 117-18
Bartle, R.G. 27, 174, 303
Box, G.E.P. 12
Brown, M. 143, 145, 155, 158, 164

Callaert, H. 245–7
Cavender, J.A. 275
Chaganty, N.R. 143, 145, 164
Chihara, T.S. 252, 269
Chung, K.L. 17, 84, 169, 185
Çinlar, E. 7, 37, 170, 230, 241

Dambrine, S. 92
Daniel, J.W. 62, 73, 83, 204, 256, 292
Darroch, J.N. 99, 215
Diaconis, P. 62, 65, 74–5, 140–1

Esary, J.D. 229

Feller, W. 150, 245, 307
Ferrari, P.A. 215
Fill, J.A. 65, 75, 140–1
Freedman, D. 20, 170
Friedland, S. 205

Gerber, H. 116
Gibson, D. 160
Golub, G.H. 64
Good, P. 279
Gurvits, L. 205

Hillier, F.S. 21
Hirayama, T. 155
Hopcroft, J.E. 64

Iosifescu, M. 6, 12, 16, 51, 74, 77, 174, 203, 236, 240
Ito, H. 12

Jenkins, G.M. 12

Kamae, T. 131
Karlin, S. 21, 39, 96, 98, 134, 136, 151, 170, 227, 237, 240, 245, 250, 257, 264–6, 278, 287, 291, 313–16
Karr, A.F. 58
Keilson, J. 62, 86, 103, 106, 115, 122, 131, 135, 137, 139, 164, 198, 200, 203, 213, 233, 244–7, 256
Kelly, F.P. 60
Kemeny, J.G. 12, 74, 77, 300
Kesten, H. 92, 215

Kester, A. 103, 106, 122, 131, 135, 137, 139, 164
Kijima, M. 73, 78, 92, 95, 103, 117, 128, 137, 140, 147, 154–6, 161, 164–5, 207–8, 221, 227, 234–5, 264, 268, 274, 279–80, 282, 287, 289–90, 293
Kleinrock, L. 9, 102, 164
Knapp, A.W. 300
Kobayashi, K. 12
Krengel, U. 131
Kurtz, T.G. 48

Lacey, M.T. 48
Ledermann, W. 261
Li, H. 143
Lieberman, G.J. 21
Lindvall, T. 69, 96
Liu, H. 37, 45
Lund, R. 69

Makimoto, N. 95
Markov, A.A. 3
Marshall, A.W. 143, 229
Martínez, S. 215
Massy, W.F. 23
Masuda, Y. 84
McGregor, J.L. 227, 245, 250, 257, 264–6, 278, 287, 314
Meyer, C.D. 67
Meyn, S.P. 6, 36, 48, 69, 187
Montgomery, D.B. 23
Moreau, M. 92
Morimura, H. 156
Morrison, D.G. 23

Nair, M.G. 215, 264, 280
Neuts, M.F. 84, 121, 157, 213, 305
Ney, P.E. 10
Noble, B. 62, 73, 83, 204, 256, 292
Nollau, V. 230
Norman, F.M. 11
Nummelin, E. 6, 88, 187

O'Brien, G. 131
Ohnishi, M. 128, 165
Orey, S. 87

Parsons, R.W. 92
Parthasarathy, P.R. 287
Paz, A. 67
Philipp, W. 48
Picco, P. 215
Pollett, P.K. 92, 215, 264, 280
Proschan, F. 112, 117–18, 151, 229
Puterman, M.L. 99

Reuter, G.E.H. 261
Revuz, D. 6
Rosenthal, J.S. 69
Ross, S.M. 7, 22, 95, 100, 150, 170, 221–2, 238–9, 289

Sakasegawa, H. 113
Seneta, E. 16, 37, 65, 68, 88–9, 91, 97, 99, 160, 215, 279, 281, 295
Shaked, M. 121, 131, 143, 145
Shanthikumar, J.G. 108, 113, 121, 126, 131, 143–5, 161, 214, 219, 221
Sharafali, M. 287
Snell, J.L. 12, 74, 77, 300
Spieksma, F.M. 69
Stegun, I.A. 288, 308
Stieltjes, T.J. 266
Stoyan, D. 121, 124, 158, 160, 162, 227
Stroock, D. 62, 65
Sumita, U. 84, 122, 156, 219, 232, 234
Suzuki, Y. 156
Szegö, S. 250
Szekli, R. 121, 158

Taylor, H.M. 21, 39, 96, 98, 170, 237, 240, 291
Tweedie, R.L. 6, 36, 48, 69, 88, 187

Ullman, J.D. 64

van Doorn, E.A. 92, 246, 250, 257, 264, 268, 275, 279–82, 286, 292, 300
Van Loan, C.F. 64
Vere-Jones, D. 89, 91, 99

Whitt, W. 265, 289–90, 293
Widder, D.V. 307
Williams, D. 220, 261
Wishart, D.M.G. 233
Wolff, R.W. 9

Yamazaki, G. 113
Yao, D.D. 126
Yoon, B.S. 221

Zeifman, A.I. 282
Ziedins, I. 94

Subject index

Page numbers appearing in *italic* refer to figures or tables.

Abel's theorem 304
Abscissa of convergence 307
Absorbing
  birth–death process 275
  Markov chain 75, 141–2, 208, 240
  state 32, 34, 53, 178, 231, 244
Absorption probability 75, 78–9, 210, 263–4
Accessible 32
Age 6, *153*
  process 6, *7*, 54, 97, 132, 153
  replacement 54
Aperiodic 32, 35, 297
  Markov chain 35
Arrival time 171, 236
Associative 300
Atom 262
Autoregressive (AR) process 12

Balance equation
  detailed 59, 200
  full 59
Basic
  composition formula 107, 314
  decomposition formula 85
  limit theorem 40
Bessel
  distribution 310
  function 288, 310
Binomial distribution 120, 163, 236
  negative 120, 163
Biorthonormal 63
Birth–death polynomial 248, 251, 266, 280
Birth–death process 193, 227
  absorbing 275
  dual 250–1, 291
  finite 178, 182, 201, 213, 259
  linear growth 291
  lossy 259
  minimal 261–2
Birth rate 178
Bisection search 79, 259, 286
Bivariate characterization 126
Boundary
  at infinity 245
  classification 245–7, *246*, 291
  entrance 245, *246*, 247, 268
  exit 245, *246*, 247, 268
  natural 245–7, *246*, 269, 291
  entrance 245, 247, 268
Branching process 10, *10*
Brown's conjecture 158
Busy period 144, 287
  number served during 144–5

Cesàro limit 41–2
Chain 4, *4*
Chapman–Kolmogorov equation 14, 17, 86, 168, 245
Classification
  boundary 245–7, *246*, 291
  state 30, *246*
Closed 32

Closure property 15, 116, 118
Coefficient
correlation 69–71, 73, 205–6
of ergodicity 65–8, 205
potential 244
Communicate 32, 231
Completely monotone (CM) 117–18, 163, 202, 213, 249, 312
Composition law 108, 110–11
Conditional
quasi-limiting distribution 278, 280–1
quasi-stationary distribution 276–7
Conditionally independent 4, 19
Conservative 185
Constant hazard rate (CHR) 117
Continuity theorem 308
Convergence
abscissa of 307
dominated 39
interval of 303
monotone 40
radius of 88, 303–4
rate of 56, 64, 69
weak 308
Convolution 115, 306, 310, 316
property 124, 163
Correlation coefficient 69–71, 73, 205–6
Coupling 68
time 68

Death rate 178
Decay parameter 56, 65, 71, 204, 257, 281, 286
Decomposition
first entrance 85
last exit 85, 272
spectral 63, 202, 254
Decreasing 112
hazard rate (DHR) 113, 118, 143–5, 154–5, 160, 163–4
hazard rate average (DHRA) 118
likelihood ratio (DLR) 114–16, 118–21, 152, 165
path 145
reversed hazard rate (DRHR) 114–18, 151, 153
Defective 25
Detailed balance equation 59, 200
Diagonalizable 217
Dirac's delta function 133
Distribution
Bessel 310
binomial 120, 163, 236
class 118, *119*
doubly limiting conditional 93–4, 99, 149, 216
Erlang 133, 188, 214, 221
exponential 8–9, 173, 179–81, 309
first-passage-time 25–6, 82–4, 211–13
geometric 82, 117, 120, 160, 163
holding-time 229–30
initial 15
invariant 44
limiting 41, 192–3, 234
negative binomial 120, 163
phase-type 84, 143, 151, 214, 240
Poisson 21, 121, 163, 173
quasi-limiting 92, 216, 271, 277, 279–80, 293
quasi-stationary 91, 147, 164, 215, 240, 259, 272, 276
residual 101, 162
return-time 25
state 15, 202
stationary 44, 192–3
taboo state 86
uniform 120
Dominated convergence theorem 39
Doubly limiting conditional distribution 93–4, 99, 149, 216
Doubly stochastic 62, 96
Dual 47, 73, 140–1
birth–death process 250–1, 291
generator 200, 250
polynomial 251

Ehrenfest urn model 53, 96
Eigenvalue 295
Perron–Frobenius (PF) 52, 87, 296–8

Eigenvector 295
  Perron–Frobenius (PF) 296, 298
Embedded
  Markov chain 8, 180, 188, 210, 231
  process 8
Entire function 268
Entrance boundary 245, *246*, 247, 268
Entropy 240
Ergodic 44
  exponentially 204, 281
  geometrically 69
  Markov chain 44, 47–8, 57, 192
Erlang distribution 133, 188, 214, 221
  generalized 144
Exit boundary 245, *246*, 247, 268
Explosive 246
Exponential distribution 8–9, 173, 179–81, 309
  mixture of 213, 309
External uniformization 221, *222*

First entrance decomposition 85
First limit 269, 285, 293
First passage time 19, 25, 80, 211, 235, 239, 249, 257
  distribution 25–6, 82–4, 211–13
  mean 247–8
First step analysis 26
Fubini's theorem 27–8
Full balance equation 59
Fundamental matrix 74–7, 81, 97, 203, 209–10, 239–40
Future 2

Gambler's ruin problem 34, 77, 97
Gamma function 310
Generating function 26, *46*, 305
Generator 175, 184
  dual 200, 250
  lossy 208
Geometric distribution 82, 117, 120, 160, 163
  mixture of 82, 117
Gerschgorin circle theorem 292
GI/M/1 queue 8–9, 18, 45, 55, 92–3
Graph *61*, 237
  random walk on 61–2

Hazard rate
  constant (CHR) 117
  decreasing (DHR) 113, 118, 143–5, 154–5, 160, 163–4
  function 55, 97, 112, 162–3
  increasing (IHR) 113, 115–16, 118, 133, 143–5, 151, 163–4
Hazard rate ordering ($\geq_{\text{hr}}$) 121–3, 125–6, 129, 131, 134, 153–4, 161, 163, 165
  monotone in the sense of ($\mathcal{M}_{\text{hr}}$) 129, 131–2, 134, 143
Hessenberg matrix 18
Hidden Markov process 12
Holding time 178, 229, 234
  distribution 229–30
Homogeneous 16, 168, 230
  spatially 31, 54, 131
  temporally 16
Hessenberg 18

Identifiability problem 12
Increment 171
Independent 2
  and identically distributed (IID) 6
  conditionally 4, 19
Increasing 112
  hazard rate (IHR) 113, 115–16, 118, 133, 143–5, 151, 163–4
  hazard rate average (IHRA) 117–18, 146
  likelihood ratio (ILR) 114, 118, 121
  path 145, 164
  reversed hazard rate (IRHR) 114, 163
Index set 1, 167
Indicator function 19
Infinitesimal generator 175, 184
Initial distribution 15
Instantaneous 178
Intensity 172, 175
  traffic 287
Interlacing property 250, 309
Inversion formula 312
  Laplace 310
Interval of convergence 303
Invariant 42–3, 50

distribution 44
measure 88–91, 94
vector 89–91, 94
Inverse problem 58
Irreducible 32, 297–8
Markov chain 32, 35–6

Karlin–McGregor representation 249, 262, 265
Kolmogorov's criterion 60, 201
Kolmogorov equation
backward 176, 186, 235, 245
forward 176, 189, 235, 245, 292
Kronecker's delta 26

ℓ_1-norm 65, 67
ℓ_2-norm 70, 72
ℓ_∞-norm 67
Laguerre transform 232
Laplace
inversion formula 310
transform 307, 309–10
Laplace–Stieltjes transform 307
Last exit decomposition 85, 272
Leaning model 10
linear 11, 23
Markovian 11
Lévy's theorem 170
Likelihood ratio
decreasing (DLR) 114–16, 118–21, 152, 165
function 114, 163
increasing (ILR) 114, 118, 121
Likelihood ratio ordering ($\geq_{\mathrm{lr}}$) 121–3, 125–6, 129, 131, 149, 152, 164
monotone in the sense of ($\mathcal{M}_{\mathrm{lr}}$) 129, 131, 134, 227, 257
strict 260, 273–4
Limit theorem
basic 40
ratio 87, 90, 292
Limiting distribution 41, 192–3, 234
Log-concave 316
Logistic process 244
Lossy
birth–death process 259
generator 208
Markov chain 86, 92, 214
Lumpable 12

Machine repair problem 243, 290
Markov chain 5
absorbing 75, 141–2, 208, 240
aperiodic 35
embedded 8, 180, 188, 210, 231
ergodic 44, 47–8, 57, 192
irreducible 32, 35–6
lossy 86, 92, 214
periodic 35, 38
recurrent 35, 50
subordinated 173, 198
transient 35–7
truncated 97
two-state 57, 61, 64, 66, 177, 238
uniformized 198
Markov decision process 99–100
Markov process 3, *4*, 5
hidden 12
Markov property 3, 167
second order 11–12
strong 19–20
Markov renewal process 230
Matrix
fundamental 74–7, 81, 97, 203, 209–10, 239–40
Hessenberg 18
nonnegative 295
normal 73
primitive 52, 75, 87, 295
product 103, 300
stochastic 14
substochastic 14
transition 13, 230
zero 15
Memoryless property 180–1
Measure
invariant 88–91, 94
spectral 249
subinvariant 88
M/G/1 queue 9, 18, 56, 95, 132–3, 144–5, 158–9, 162
M/M/1 queue 247, 265, 280–1, 286–7, 293
M/M/1/N queue 181–2, 203–4

M/M/s queue 243, 286, 290–1
Minimal
  birth–death process 261–2
  solution 194
  transition probability function 224
Mixture 118
  of exponential distributions 213, 309
  of geometric distributions 82, 117
ML-matrix 298
Mode 111
Moment generating function 21, 237
Monotone 112, 136
  completely (CM) 117–18, 163, 202, 213, 249, 312
  convergence theorem 40
  externally 101
  hazard rate ordering ($\mathcal{M}_{\mathrm{hr}}$) 129, 131–2, 134, 143
  internally 101
  likelihood ratio ordering ($\mathcal{M}_{\mathrm{lr}}$) 129, 131, 134, 227, 257
  path 145
  reversed hazard rate ordering ($\mathcal{M}_{\mathrm{rh}}$) 129–31, 134
  stochastically ($\mathcal{M}_{\mathrm{st}}$) 129–31, 134–6, 139–40, 146, 148, 158, 164, 224–6

Natural boundary 245–7, *246*, 269, 291
Negative binomial distribution 120, 163
New better than used (NBU) 118, 143
  in expectation (NBUE) 162
New worse than used (NWU) 118, 143
  in expectation (NWUE) 162
Nonexplosive 246
Nonhomogeneous 16, 168, 235–6
Nonnegative matrix 295
Normal matrix 73
Null recurrent 30–2, 35, 96, 245, *246*
Number of returns 30
Number of visits 28, 30, 76
  average 49

One-parameter family 276–7
Order statistics 236
Orthogonal polynomial 249
Orthonormal 63

Partial-sum process 6, 150
Past 2
Path 145
Periodic 32, 35, 297
  Markov chain 35, 38
Perron–Frobenius (PF)
  eigenvalue 52, 87, 296–8
  eigenvector 296, 298
  theorem 52, 295, 297, 299
Phase-type distribution
  (continuous) 214, 240
  (discrete) 84, 143, 151, 214
  generalized 214
Phase-type renewal process 157
Poisson
  distribution 21, 121, 163, 173
  shock model 229
Poisson process 171–3, 190, 194–5, 236
  compound 236
  nonhomogeneous 237
Policy 99
Pólya frequency (PF) 315
  of order r (PF_r) 106, 315
  of order 2 (PF_2) 115, 316
Pólya urn model 11, 13, 21
Polynomial
  birth–death 248, 251, 266, 280
  dual 251
  orthogonal 249
Portfolio selection problem 165
Positive recurrent 30, 35–6, 42, 245, *246*
Potential coefficient 244
Power series 303
Present 2
Primitive 52, 75, 87, 295
Probability
  absorption 75, 78–9, 210, 263–4
  renewal 155, 158, 165
  transition 13, 26
  vector 15
Process 1
  age 6, 7, 54, 97, 132, 153
  autoregressive (AR) 12

birth–death 178, 182, 193, 201, 213, 227, 250–1, 259, 275, 291
branching 10, *10*
embedded 8
logistic 244
Markov 3, *4*, 5
Markov decision 99–100
Markov renewal 230
partial-sum 6, 150
Poisson 171–3, 190, 194–5, 236
pure birth 187
pure death 237
renewal 150, *151*
residual lifetime 153
semi-Markov 180, 230
Yule 187
Pure
birth process 187
death process 237

Q-matrix 175
(q, Q)-policy 21
Quasi-limiting distribution 92, 216, 271, 277, 279–80, 293
conditional 278, 280–1
Quasi-stationary distribution 91, 147, 164, 215, 240, 259, 272, 276
conditional 276–7
Quasi-stationary equation 91, 93, 215
Queue
GI/M/1 8–9, 18, 45, 55, 92–3
M/G/1 9, 18, 56, 95, 132–3, 144–5, 158–9, 162
M/M/1 247, 265, 280–1, 286–7, 293
M/M/1/N 181–2, 203–4
M/M/s 243, 286, 290–1

R-null 88
R-positive 88, 90–1, 94, 99
R-recurrent 88–9
R-transient 88–9
Radius of convergence 88, 303–4
Random walk 18, 31, 44–5, 61, 66, 85–6, 95
finite 52–3, 82–4, 131–2, 149
on graph 61–2
Rate of convergence 56, 64, 69
Ratio limit theorem 87, 90, 292
Realization 2
Recurrent 30–1
Markov chain 35, 50
null 30–2, 35, 96, 245, *246*
positive 30, 35–6, 42, 245, *246*
Regular 52, 87, 187
boundary 245, 247, 268
Relaxation time 56–7, 71, 135, 204
Renewal
degenerate 154
density 289
equation 39
function 155
probability 155, 158, 165
Renewal process 150, *150*
alternating 177
generalized (g-) 156
Markov 230
phase-type 157
Residual *153*
distribution 101, 162
lifetime process 153
Return time
distribution 25
mean 40, 95, 187
Reversed hazard rate
decreasing (DRHR) 114–18, 151, 153
function 113
increasing (IRHR) 114, 163
Reversed hazard rate ordering ($\geq_{\rm rh}$) 121–3, 125–6, 130–1, 134, 153, 156, 163, 165
monotone in the sense of ($\mathcal{M}_{\rm rh}$) 129–31, 134
Reversible in time 59, 72, 139, 200–1, 206, 212, 288

Sample path 2, *2*, *7–8*, *10*, *150*, 178, *180*
Sensitivity 67
Separation 74, 139–42
Semi-Markov process 180, 230
Sign sequence 112
Singular value 72
Skip-free 78, 85, 287

Solidarity property 35
Spatially homogeneous 31, 54, 131
Spectral
  decomposition 63, 202, 254
  measure 249
(s, S)-policy 5
Stable 178, 185
Standard 169
State
  absorbing 32, 34, 53, 178, 231, 244
  classification 30, *246*
  distribution 15, 202
  space 1
  taboo 84
Stationary 44
  distribution 44, 192–3
  equation 44, 192
  waiting time 102, 162
Stirling's formula 31, 95
Stochastic 14
  doubly 62, 96
  matrix 14
  process 1
Stochastic ordering ($\geq_{st}$) 121–4, 126, 130–1, 134, 148, 151, 155–7, 163, 165, 187–8, 241
Stochastically
  dominate 158, 227, 241
  monotone ($\mathcal{M}_{st}$) 129–31, 134–6, 139–40, 146, 148, 158, 164, 224–6
Stopping time 19, 29, 241
Strong law of large numbers 48, 221
Strong Markov property 19–20
Subinvariant 42
  measure 88
  vector 89
Subordinated Markov chain 173, 198
Substochastic 14
  strictly 14

Taboo 84
  state distribution 86
  transition probability 84, 146
Tauberian theorem 305
Temporally homogeneous 16, 168, 230
Time 1
  reversal 47
  reversible 59, 72, 139, 200–1, 206, 212, 288
Totally positive (TP) 314
  of order r (TP_r) 106, 313
  of order 2 (TP_2) 106–8, 110–11, 122, 134, 136, 141, 317
  strictly (STP, STP_r) 257, 314
Trace 218
Traffic intensity 287
Transform
  Laguerre 232
  Laplace 307, 309–10
  Laplace–Stieltjes 307
Transient 30, 32, 245, *246*
  behavior 64
  Markov chain 35–7
Transition
  diagram 33, *33*
  rate 58
Transition matrix 13, 230
  function 168, 231
Transition probability 13, 26
  function 167, 231
  minimal 224
  taboo 84, 146
Truncated Markov chain 97
Truncation
  of infinite sum 219
  of state space 159
Two-state Markov chain 57, 61, 64, 66, 177, 238

Unidirectionality 2
Uniform distribution 120
Uniformizable 188, 195, 198–9
Uniformization 219, *220*
  external 221, *222*
Uniformized Markov chain 198
Unimodal 111, 115, 137, 292
  strictly 112, 115
  strongly (SU) 115
Unit
  step function 6
  vector 16
Upper set 150
Urn model
  Ehrenfest 53, 96

Pólya 11, 13, 21

Variation
diminishing property (VDP) 116, 228, 316–18
distance 74
Vector
invariant 89–91, 94
probability 15
subinvariant 89
unit 16
zero 15
Visit *29*
number of 28, 30, 76

Wald's identity 23
Waiting time 164
stationary 102, 162
Weak convergence 308

Yule process 187

Zero
matrix 15
vector 15